Foundations of Oscillator Circuit Design

For a listing of recent titles in the *Artech House Microwave Library*, turn to the back of this book.

Foundations of Oscillator Circuit Design

Guillermo Gonzalez

ARTECH HOUSE
BOSTON | LONDON
artechhouse.com

Library of Congress Cataloging-in-Publication Data
A catalog record for this book is available from the U. S. Library of Congress.

British Library Cataloguing in Publication Data
A catalogue record for this book is available from the British Library.

ISBN 10: 1-5963-162-0
ISBN 13: 978-1-59693-162-6

Cover design by Yekaterina Ratner

685 Canton Street
Norwood, MA 02062

10 9 8 7 6 5 4 3 2 1

Contents

Preface

My interest in oscillators started many years ago when I was an undergraduate student and one of the laboratory experiments was the design of a Colpitts oscillator. It was amazing to see how a sinusoidal signal appeared when the power supply was turned on. What an interesting way of controlling the motion of electrons in the circuit! My fascination with oscillators has remained to this date and, hopefully, this book will be a reflection of it.

Electronic oscillator theory and design is a topic that, in general, is barely covered in undergraduate electronic courses. However, since oscillators are one of the main components in many electronic circuits, engineers are usually required to design them. Sinusoidal carrier signals are needed in transmitters and receivers, and timing signals (square-wave signals) are needed in digital circuits.

The purpose of this book is to cover the foundations of oscillator circuit design in a comprehensive manner. The book covers the theory and design of oscillators in the frequency range that extends from the audio range to the microwave range at about 30 GHz. In this large range of frequencies the active element is usually a semiconductor, such as a BJT or FET, or an op amp. The techniques involved in the design of oscillators at the lower frequencies are different from those used at the higher frequencies. An important feature of this book is the wide and rather complete coverage of oscillators, from the low-frequency oscillator to the more complex oscillator found at radio frequencies (RF) and microwave (MW) frequencies. This book emphasizes the use of simulation techniques (i.e., CAD techniques) in the design of oscillators. In many cases the performance observed in the simulation is very similar to that obtained in the laboratory. This is mostly true for oscillators working at the lower frequencies and up to a few megahertz. As the frequency increases, the practical implementation is highly affected by the layout and by the parasitics associated with the components used. In such cases the simulation should provide a starting point to the associated practical implementation.

The advances in CAD techniques since the 1980s have certainly changed the approach to the design of many oscillators. Before the advent of advanced CAD techniques, oscillator design involved a significant amount of theoretical work, especially for those oscillators operating in the RF and MW-frequency regions. While a solid theoretical foundation is still needed, the modern CAD programs can perform a lot of nonlinear simulations that were once only a dream in oscillator analysis and design. In my experience the best oscillator designers are those who have a good understanding of the fundamental principles involved, experience with an appropriate CAD program, and a good practical sense.

In undergraduate courses I have used the transient simulator available in SPICE to analyze and design oscillators. Transient simulators work well, but in many cases it takes a lot of simulation time to get to the steady-state oscillatory waveform. As one matures in the field of oscillators, an advanced CAD program with harmonic balance capabilities is a must. The main program used in this book is the Advanced Design System (ADS) from Agilent. One of the many uses of this very powerful and state-of-the art program is for oscillator analysis and design since it contains a transient simulator, a harmonic balance simulator, a statistical design simulator, and an envelope simulator. The ADS program and associated licenses were donated by Agilent to the Department of Electrical and Computer Engineering at the University of Miami for teaching and research purposes.

One objective of this book is to cover the fundamentals of oscillator design using semiconductor devices as the active devices. A second objective, in spite of the fact that the material in electronic oscillators is volumetric, is to present the foundations of modern oscillators' design techniques. In this book the reader is first exposed to the theory of oscillators. Then, a variety of techniques that are used in the design of oscillators are discussed.

The Table of Contents clearly indicates the choice of material and the order of presentation. In short, Chapter 1 provides a general introduction to the theory of oscillators and discusses in detail several low-frequency oscillators. Chapter 2 discusses the oscillator characteristics such as frequency stability, quality factors, phase noise, and statistical considerations. Chapter 3 presents the design of tuned oscillators using BJTs, FETs, and op amps. Chapter 4 treats the design of oscillators using crystals, ceramic resonators, surface acoustic wave resonators, and dielectric resonators. The theory and design methods using the negative-resistance approach are presented in Chapter 5. Relaxation oscillators and other nonsinusoidal oscillators are discussed in Chapter 6.

This book can be used in a senior graduate-level course in oscillators. It is also intended to be used in industrial and professional short courses in oscillators. It should also provide for a comprehensive reference of electronic oscillators using semiconductors for electrical engineers.

Two large-signal simulators that are used to analyze and design oscillators are the harmonic balance simulator and the transient simulator.

The harmonic balance simulator in ADS performs a nonlinear steady-state analysis of the circuit. It is a very powerful frequency-domain analysis technique for nonlinear circuits. The simulator allows the analysis of circuits excited by large-signal sources. Also, ADS provides the function "ts" which calculates the time-domain signal from its frequency spectrum.

Transient-analysis simulation is performed entirely in the time domain. It also allows the analysis of nonlinear circuits and large-signal sources. The data displayed from the transient simulation shows the time-domain waveform. From the time-domain waveform, the oscillation build-up and the steady-state results can be viewed. The transient simulator requires an initial condition for the oscillator to begin. The initial condition can be an initial voltage across a capacitor, a voltage step for the power-supply component, or the use of a noise source. ADS provides the function "fs," which calculates the frequency spectrum from the time-domain signal.

I wish to thank all of my former students for their valuable input and helpful comments related to this book. Special thanks go to Mr. Jorge Vasiliadis who contributed to the section on DROs; to Mr. Hicham Kehdy for his contribution to the design of the GB oscillator in Chapter 5; to Mr. Orlando Sosa, Dr. Mahes M. Ekanayake, and Dr. Chulanta Kulasekere for reviewing several parts of the book; to Dr. Kamal Premaratne who provided input to the material in Chapter 1; and to Dr. Branko Avanic who did a lot of work with me on crystal oscillators. Also, I will always be grateful to Dr. Les Besser for his friendship and for the clarity that he has provided in the field of microwave electronics.

Thanks also go to the staff at Artech House, in particular for the help and guidance provided by Audrey Anderson (production editor) and Mark Walsh (acquisitions editor).

Finally, my love goes to the people that truly make my life busy and worthwhile, namely my wife Pat, my children Donna and Alex, my daughter-in-law Samantha, my son-in-law Larry, and my grandkids Tyler, Analise, and Mia. They were always supportive and put up with me during this long writing journey.

CHAPTER 1

Theory of Oscillators

1.1 Introduction

There are many types of oscillators, and many different circuit configurations that produce oscillations. Some oscillators produce sinusoidal signals, others produce nonsinusoidal signals. Nonsinusoidal oscillators, such as pulse and ramp (or sawtooth) oscillators, find use in timing and control applications. Pulse oscillators are commonly found in digital-systems clocks, and ramp oscillators are found in the horizontal sweep circuit of oscilloscopes and television sets. Sinusoidal oscillators are used in many applications, for example, in consumer electronic equipment (such as radios, TVs, and VCRs), in test equipment (such as network analyzers and signal generators), and in wireless systems.

In this chapter the feedback approach to oscillator design is discussed. The oscillator examples selected in this chapter, as well as the mix of theory and design information presented, help to clearly illustrate the feedback approach.

The basic components in a feedback oscillator are the amplifier, an amplitude-limiting component, a frequency-determining network, and a (positive) feedback network. Usually the amplifier also acts as the amplitude-limiting component, and the frequency-determining network usually performs the feedback function. The feedback circuit is required to return some of the output signal back to the input. Positive feedback occurs when the feedback signal is in phase with the input signal and, under the proper conditions, oscillation is possible.

One also finds in the literature the term negative-resistance oscillators. A negative-resistance oscillator design refers to a specific design approach that is different from the one normally used in feedback oscillators. Since feedback oscillators present an impedance that has a negative resistance at some point in the circuit, such oscillators can also be designed using a negative-resistance approach. For a good understanding of the negative resistance method, a certain familiarity with oscillators is needed. That is why the negative resistance method is discussed in Chapter 5.

1.2 Oscillation Conditions

A basic feedback oscillator is shown in Figure 1.1. The amplifier's voltage gain is $A_v(j\omega)$, and the voltage feedback network is described by the transfer function $\beta(j\omega)$. The amplifier gain $A_v(j\omega)$ is also called the open-loop gain since it is the

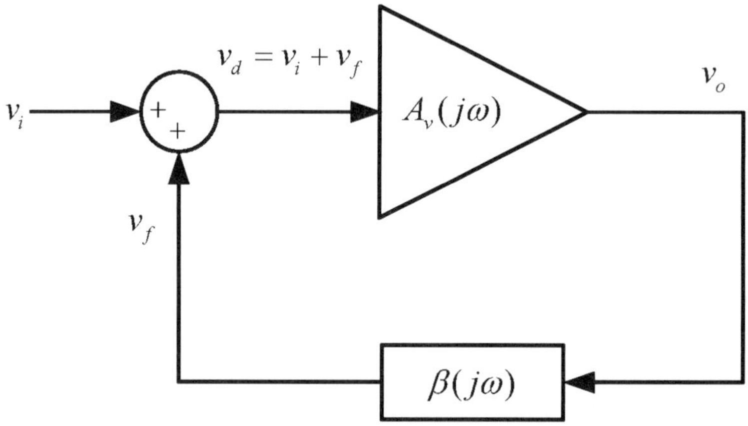

Figure 1.1 The basic feedback circuit.

gain between v_o and v_i when $v_f = 0$ (i.e., when the path through $\beta(j\omega)$ is properly disconnected).

The amplifier gain is, in general, a complex quantity. However, in many oscillators, at the frequency of oscillation, the amplifier is operating in its midband region where $A_v(j\omega)$ is a real constant. When $A_v(j\omega)$ is constant, it is denoted by A_{vo}.

Negative feedback occurs when the feedback signal subtracts from the input signal. On the other hand, if v_f adds to v_i, the feedback is positive. The summing network in Figure 1.1 shows the feedback signal added to v_i to suggest that the feedback is positive. Of course, the phase of v_f determines if v_f adds or subtracts to v_i. The phase of v_f is determined by the closed-loop circuit in Figure 1.1. If $A_v(j\omega) = A_{vo}$ and A_{vo} is a positive number, the phase shift through the amplifier is 0°, and for positive feedback the phase through $\beta(j\omega)$ should be 0° (or a multiple of 360°). If A_{vo} is a negative number, the phase shift through the amplifier is ±180° and the phase through $\beta(j\omega)$ for positive feedback should be $\pm 180° \pm n360°$. In other words, for positive feedback the total phase shift associated with the closed loop must be 0° or a multiple n of 360°.

From Figure 1.1 we can write

$$v_o = A_v(j\omega)v_d \tag{1.1}$$

$$v_f = \beta(j\omega)v_o \tag{1.2}$$

and

$$v_d = v_i + v_f \tag{1.3}$$

Thus, from (1.1) to (1.3), the closed-loop voltage gain $A_{vf}(j\omega)$ is given by

$$A_{vf}(j\omega) = \frac{v_o}{v_i} = \frac{A_v(j\omega)}{1 - \beta(j\omega)A_v(j\omega)} \tag{1.4}$$

The quantity $\beta(j\omega)A_v(j\omega)$ is known as the *loop gain*.

For oscillations to occur, an output signal must exist with no input signal applied. With $v_i = 0$ in (1.4) it follows that a finite v_o is possible only when the denominator is zero. That is, when

$$1 - \beta(j\omega)A_v(j\omega) = 0$$

or

$$\beta(j\omega)A_v(j\omega) = 1 \tag{1.5}$$

Equation (1.5) expresses the fact that for oscillations to occur the loop gain must be unity. This relation is known as the Barkhausen criterion.

With $A_v(j\omega) = A_{vo}$ and letting

$$\beta(j\omega) = \beta_r(\omega) + j\beta_i(\omega)$$

where $\beta_r(\omega)$ and $\beta_i(\omega)$ are the real and imaginary parts of $\beta(j\omega)$, we can express (1.5) in the form

$$\beta_r(\omega)A_{vo} + j\beta_i(\omega)A_{vo} = 1$$

Equating the real and imaginary parts on both sides of the equation gives

$$\beta_r(\omega)A_{vo} = 1 \Rightarrow A_{vo} = \frac{1}{\beta_r(\omega)} \tag{1.6}$$

and

$$\beta_i(\omega)A_{vo} = 0 \Rightarrow \beta_i(\omega) = 0 \tag{1.7}$$

since $A_{vo} \neq 0$. The conditions in (1.6) and (1.7) are known as the Barkhausen criteria in rectangular form for $A_v(j\omega) = A_{vo}$.

The condition (1.6) is known as the gain condition, and (1.7) as the frequency of oscillation condition. The frequency of oscillation condition predicts the frequency at which the phase shift around the closed loop is 0° or a multiple of 360°.

The relation (1.5) can also be expressed in polar form as

$$\beta(j\omega)A_v(j\omega) = |\beta(j\omega)A_v(j\omega)|\underline{/\beta(j\omega)A_v(j\omega)} = 1$$

Hence, it follows that

$$|\beta(j\omega)A_v(j\omega)| = 1 \tag{1.8}$$

and

$$\underline{/\beta(j\omega)A_v(j\omega)} = \pm n360^\circ \tag{1.9}$$

where $n = 0, 1, 2, \ldots$. Equation (1.9) expresses the fact that the signal must travel through the closed loop with a phase shift of 0° or a multiple of 360°. For $A_v(j\omega) = A_{vo}$, then $\underline{|\beta(j\omega)A_{vo}}$ is the angle of $\beta(j\omega)$, and the condition (1.9) is equivalent to saying that $\beta_i(j\omega) = 0$, in agreement with (1.7). Also, for $A_v(j\omega) = A_{vo}$ and with $\beta_i(j\omega) = 0$, (1.8) reduces to (1.6). The conditions in (1.8) and (1.9) are known as the Barkhausen criteria in polar form.

When the amplifier is a current amplifier, the basic feedback network can be represented as shown in Figure 1.2. In this case, $A_i(j\omega)$ is the current gain of the amplifier, and the current feedback factor $\alpha(j\omega)$ is

$$\alpha(j\omega) = \frac{i_f}{i_o}$$

For this network, the condition for oscillation is given by

$$\alpha(j\omega)A_i(j\omega) = 1 \tag{1.10}$$

which expresses the fact that loop gain in Figure 1.2 must be unity.

The loop gain can be evaluated in different ways. One method that can be used in some oscillator configurations is to determine $A_v(j\omega)$ and $\beta(j\omega)$ and to form the loop gain $A_v(j\omega)\beta(j\omega)$. In many cases it is not easy to isolate $A_v(j\omega)$ and $\beta(j\omega)$ since they are interrelated. In such cases a method that can usually be implemented is to represent the oscillator circuit as a continuous and repetitive circuit. Hence, the loop gain is calculated as the gain from one part to the same part in the following circuit. An alternate analysis method is to replace the amplifier and feedback network in Figure 1.1 by their ac models and write the appropriate loop equations. The loop equations form a system of linear equations that can be solved for the closed-loop voltage gain, which can be expressed in the general form

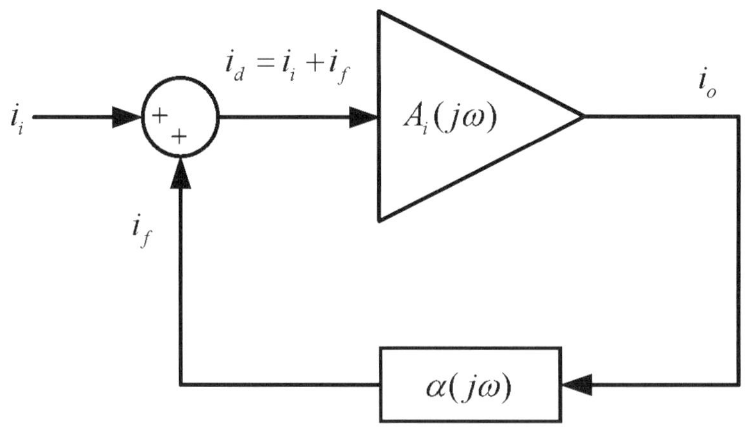

Figure 1.2 The current form of the basic feedback network.

$$A_{vf}(j\omega) = \frac{v_o}{v_i} = \frac{N(j\omega)}{D(j\omega)} \tag{1.11}$$

where $N(j\omega)$ represents the numerator polynomial and $D(j\omega)$ is the system determinant of the linear equations. In terms of (1.11) the conditions for oscillations are obtained by setting the system determinant equal to zero (i.e., $D(j\omega) = 0$). Setting $D(j\omega) = 0$ results in two equations: one for the real part of $D(j\omega)$ (which gives the gain condition), and one for the imaginary part of $D(j\omega)$ (which gives the frequency of oscillation).

From circuit theory we know that oscillation occurs when a network has a pair of complex conjugate poles on the imaginary axis. However, in electronic oscillators the poles are not exactly on the imaginary axis because of the nonlinear nature of the loop gain. There are different nonlinear effects that control the pole location in an oscillator. One nonlinear mechanism is due to the saturation characteristics of the amplifier. A saturation-limited sinusoidal oscillator works as follows. To start the oscillation, the closed-loop gain in (1.4) must have a pair of complex-conjugate poles in the right-half plane. Then, due to the noise voltage generated by thermal vibrations in the network (which can be represented by a superposition of input noise signals v_n) or by the transient generated when the dc power supply is turned on, a growing sinusoidal output voltage appears. The characteristics of the growing sinusoidal signal are determined by the complex-conjugate poles in the right-half plane. As the amplitude of the induced oscillation increases, the amplitude-limiting capabilities of the amplifier (i.e., a reduction in gain) produce a change in the location of the poles. The changes are such that the complex-conjugate poles move towards the imaginary axis. However, the amplitude of the oscillation was increasing and this makes the complex poles to continue the movement toward the left-half plane. Once the poles move to the left-half plane the amplitude of the oscillation begins to decrease, moving the poles toward the right-half plane. The process of the poles moving between the left-half plane and the right-half plane repeats, and some steady-state oscillation occurs with a fundamental frequency, as well as harmonics. This is a nonlinear process where the fundamental frequency of oscillation and the harmonics are determined by the location of the poles. Although the poles are not on the imaginary axis, the Barkhausen criterion in (1.5) predicts fairly well the fundamental frequency of oscillation. It can be considered as providing the fundamental frequency of the oscillator based on some sort of average location for the poles.

The movement of the complex conjugate poles between the right-half plane and the left-half plane is easily seen in an oscillator designed with an amplitude limiting circuit that controls the gain of the amplifier and, therefore, the motion of the poles. An example to illustrate this effect is given in Example 1.6.

The previous discussion shows that for oscillations to start the circuit must be unstable (i.e., the circuit must have a pair of complex-conjugate poles in the right-half plane). The condition (1.5) does not predict if the circuit is unstable. However, if the circuit begins to oscillate, the Barkhausen criterion in (1.5) can be used to predict the approximate fundamental frequency of oscillation and the gain condition. The stability of the oscillator closed-loop gain can be determined using the Nyquist stability test.

1.3 Nyquist Stability Test

There are several methods for testing the stability of a feedback amplifier. In general, (1.4) can be expressed in the form

$$A_{vf}(s) = \frac{v_o}{v_i} = \frac{A_v(s)}{1 - \beta(s)A_v(s)} \tag{1.12}$$

The stability $A_{vf}(s)$ is determined by the zeroes of $1 - \beta(s)A_v(s)$ provided there is no cancellation of right-half plane poles and zeroes when forming the product $\beta(s)A_v(s)$. In practical oscillators the previous pole-zero cancellation problems are unlikely to occur. If there are no pole-zero cancellation problems, the poles of $A_v(s)$ are common to those of $\beta(s)A_v(s)$ and of $1 - \beta(s)A_v(s)$. Therefore, the feedback amplifier is stable if the zeroes of $1 - \beta(s)A_v(s)$ lie in the left-half plane. In what follows we assume that there are no pole-zero cancellation problems.

The Nyquist stability test (or criterion) can be used to determine the right-half plane zeroes of $1 - \beta(s)A_v(s)$. A Nyquist plot is a polar plot of the loop gain $\beta(s)A_v(s)$ for $s = j\omega$ as the frequency ω varies from $-\infty < \omega < \infty$. Two typical Nyquist plots are shown in Figure 1.3. The Nyquist test states that the number of times that the loop-gain contour encircles the point $1 + j0$ in a clockwise direction is equal to the difference between the number of zeroes and the number of poles of $1 - \beta(s)A_v(s)$ with positive real parts (i.e., in the right-half plane). The point $1 + j0$ is called the *critical point.* To be specific, let N be the number of clockwise encirclements of the critical point by the Nyquist plot, let P be the number of right-half plane poles of $\beta(s)A_v(s)$ (which are the same as those of $1 - \beta(s)A_v(s)$), and let Z be the number of right-half plane zeroes of $1 - \beta(s)A_v(s)$. The Nyquist stability test states that $N = Z - P$ (or $Z = N + P$). If $Z > 0$ (or $N + P > 0$) the feedback amplifier is unstable and will oscillate under proper conditions. (Note: In the case that there is a right-half plane pole-zero cancellation, the Nyquist test is not sufficient to determine stability.)

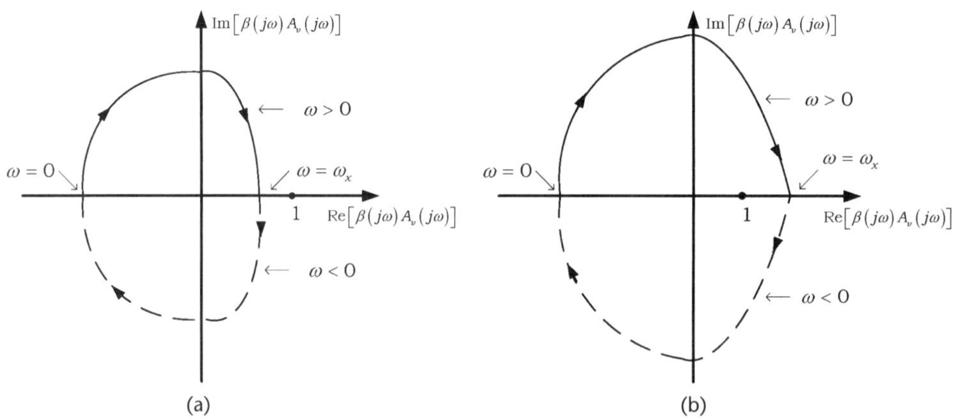

Figure 1.3 (a) A Nyquist plot of a stable feedback amplifier and (b) a Nyquist plot of an unstable feedback amplifier.

If $\beta(s)A_v(s)$ has no poles in the right-half plane, then it follows that $1 - \beta(s)A_v(s)$ has no poles in the right-half plane (i.e., $P = 0$). Thus, in this case $A_{vf}(s)$ is unstable (i.e., has right-half plane poles) only if $1 - \beta(s)A_v(s)$ has right-half plane zeroes (i.e., if $N > 0$). In other words, for $P = 0$ the feedback amplifier is unstable when $N > 0$ (since $N = Z$ when $P = 0$). When $\beta(s)A_v(s)$ is stable, the Nyquist test simply requires that the plot of $\beta(s)A_v(s)$ as a function of ω does not encircle the critical point for the feedback amplifier to be stable. An alternative way of stating the Nyquist test when $\beta(s)A_v(s)$ is stable is: "If $\beta(s)A_v(s)$ is stable, the feedback amplifier is stable if $|\beta(j\omega)A_v(j\omega)| < 1$ when the phase of $\beta(j\omega)A_v(j\omega)$ is 0° or a multiple of 360°." This condition ensures that the critical point is not enclosed.

In the case that $\beta(s)A_v(s)$ has a pole in the $j\omega$ axis, the contour in the s plane must be modified to avoid the pole. For example, if the pole is at $s = 0$, the path moves from $s = -j\infty$ to $s = j0$, then from $s = j0^-$ to $s = j0^+$ around a semicircle of radius ϵ (where ϵ approaches zero), and then from $s = j0^+$ to $s = j\infty$. From $s = j\infty$ the contour follows a semicircle with infinite radius and moves back to $s = -j\infty$. Hence, the contour encloses all poles and zeroes that $\beta(s)A_v(s)$ has in the right-half plane.

Two typical Nyquist plots for a feedback amplifier with a stable loop gain are shown in Figure 1.3. The solid curve corresponds to $\omega \geq 0$, and the dashed curve to $\omega \leq 0$. Since $\beta(j\omega)A_v(j\omega) = [\beta(j\omega)A_v(j\omega)]^*$ it follows that the dashed curve is simply the mirror image of the solid curve. In Figure 1.3(a) the Nyquist plot does not enclose the critical point. It is seen that at the frequency ω_x the phase of $\beta(j\omega)A_v(j\omega)$ is 0° and its magnitude is less than one. Hence, the amplifier associated with this Nyquist plot is stable. A typical Nyquist plot for an unstable feedback amplifier (with a stable $\beta(s)A_v(s)$) is shown in Figure 1.3(b). For this plot $N = Z = 1$, and the closed-loop response has one pole in the right-half plane.

Example 1.1

(a) Let $\beta(s) = \beta_o$ be a real number and

$$A_v(s) = \frac{K}{s(s+1)(s+2)}$$

Hence,

$$\beta(s)A_v(s) = \frac{\beta_o K}{s(s+1)(s+2)}$$

and it follows that the number of poles of the loop gain in the right-half plane is zero (i.e., $P = 0$). Therefore, the system is stable if the Nyquist plot of $\beta(s)A_v(s)$ does not encircle the point $1 + j0$ (i.e., if $N = Z = 0$).

The Nyquist plot of $\beta(s)A_v(s)$ for $\beta_o K = 3$ is shown in Figure 1.4(a). This plot shows that the system is stable since there are no encirclements of the $1 + j0$ point.

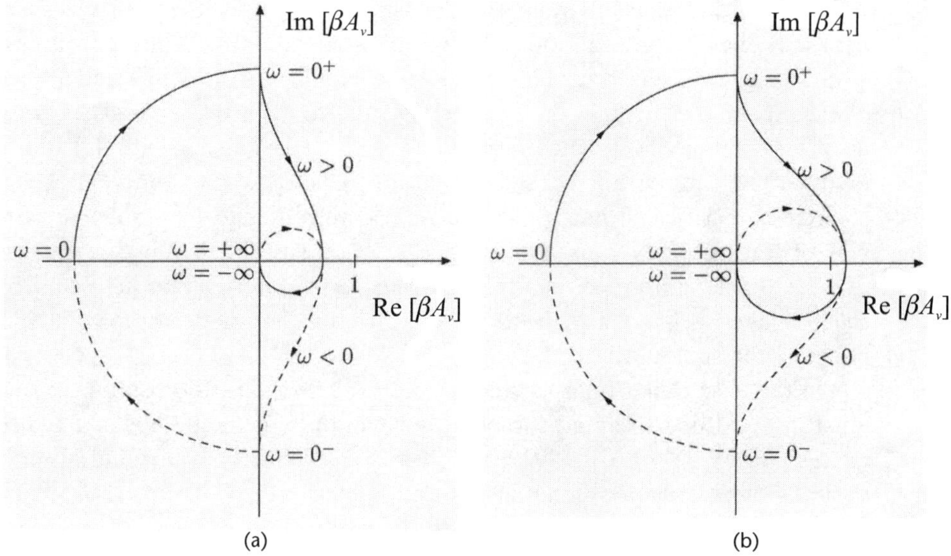

Figure 1.4 Nyquist plots for Example 1.1(a) with (a) $\beta_0 K = 3$ and (b) $\beta_0 K = 9$.

The resulting Nyquist plot for $\beta_0 K = 9$ is shown in Figure 1.4(b). In this case, the plot of $\beta(s)A_v(s)$ encircles the $1 + j0$ point twice in the clockwise direction. Hence, $N = Z = 2$, and the closed loop system is unstable because of two poles in the right-half plane.

In this part of the example the stability depended on the value of $\beta_0 K$.

(b) Let $\beta(s) = \beta_0$ be a real number and

$$A_v(s) = \frac{K}{s(s+1)(s-1)}$$

Hence,

$$\beta(s)A_v(s) = \frac{\beta_0 K}{s(s+1)(s-1)}$$

and it follows that $P = 1$, since there is a pole at $s = 1$. The Nyquist plots of the loop gain for $\beta_0 K > 0$ and $\beta_0 K < 0$ are shown in Figure 1.5. The solid curve in the plot corresponds to the mapping for $\omega > 0$, and the dashed curve for $\omega < 0$. Figure 1.5(a) shows that $N = 0$ when $\beta_0 K > 0$, and Figure 1.5(b) shows that $N = 1$ when $\beta_0 K < 0$; hence, the information in Table 1.1.

That is, the function $1 - \beta(s)A_v(s)$ for $\beta_0 K > 0$ has a zero in the right-half plane, and for $\beta_0 K < 0$ it has two zeroes in the right-half plane. Obviously, this feedback system is unstable for any real value of $\beta_0 K$.

The information displayed in the polar Nyquist diagram can also be shown using Bode plots. Thus, the stability of an amplifier can also be determined from the Bode magnitude and phase plots of the loop gain. In terms of the magnitude and phase Bode plots of a stable $\beta(j\omega)A_v(j\omega)$, it follows that the closed-loop gain

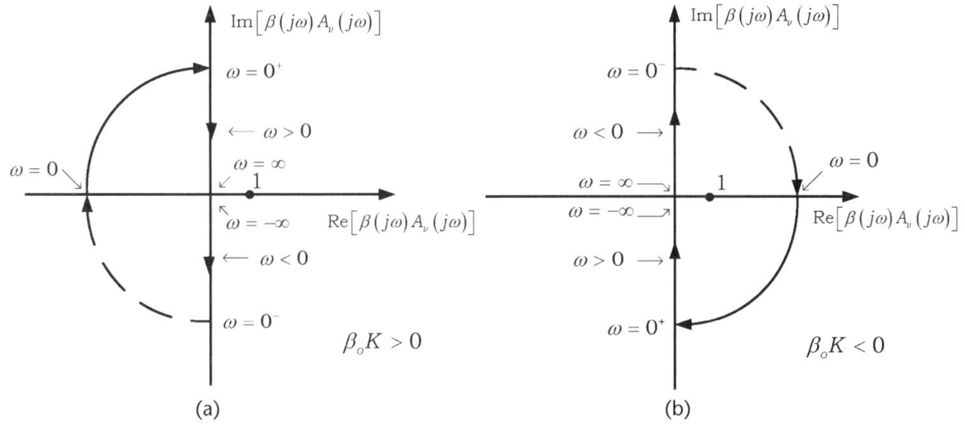

Figure 1.5 Nyquist plots for Example 1.1(b) when (a) $\beta_0 K > 0$ and (b) $\beta_0 K < 0$.

Table 1.1 Values of Z for Example 1.1(b)

	P	N	$Z = N + P$
$\beta_0 K > 0$	1	0	1
$\beta_0 K < 0$	1	1	2

is stable if $|\beta(j\omega)A_v(j\omega)|$ in dBs is smaller than 0 dB when the phase shift is 0° (or a multiple of 360°). In other words, the plot of $|\beta(j\omega)A_v(j\omega)|$ in dBs crosses the 0-dB axis at a frequency lower than the frequency at which the phase reaches 0° (or $\pm n360°$). Typical Bode plots of the magnitude and phase of a stable feedback amplifier are shown in Figure 1.6.

Two important quantities in the determination of stability are the gain margin and the phase margin (shown in Figure 1.6). The gain margin is the number of decibels that $|\beta(j\omega)A_v(j\omega)|$ is below 0 dB at the frequency where the phase is 0°. The phase margin is the number of degrees that the phase is above 0° at the frequency where $|\beta(j\omega)A_v(j\omega)|$ is 0 dB. A positive gain margin shows that the amplifier is potentially unstable. Similarly, a positive phase margin is associated with a stable amplifier. Of course, the gain margin and phase margin can also be shown in a Nyquist diagram.

Typical Bode plots of $\beta(j\omega)A_v(j\omega)$ for feedback amplifiers having one, two, and three poles with $\beta(0)A_v(0) =- K < 0$ are shown in Figure 1.7. The single-pole loop-gain function shown in Figure 1.7(a) has a minimum phase shift of 90°. Therefore, this amplifier is always stable. Figure 1.7(b) shows a loop gain having two poles. Again this amplifier is always stable because the phase shift is positive and approaches 0° only at $\omega = \infty$. Figure 1.7(c) shows a three-pole loop gain that is stable since $|\beta(j\omega)A_v(j\omega)|$ is below 0 dB at the frequency where the phase is 0° (i.e., the gain margin is negative). Figure 1.7(d) shows a three-pole loop gain that is unstable, since the phase is less than 0° at the frequency where $|\beta(j\omega)A_v(j\omega)|$ is 0 dB (i.e., the phase margin is negative).

It is of interest to see how the Nyquist and Bode plots portray the stability information and their relation to the closed loop and transient responses of the

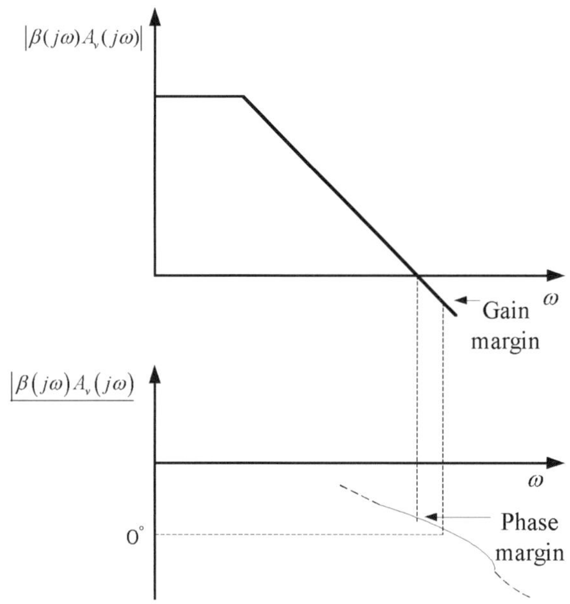

Figure 1.6 A typical Bode plot of the magnitude and phase of a stable feedback amplifier.

feedback amplifier. This is illustrated in Figure 1.8. In the Nyquist plots only the positive frequencies are shown. In the Bode plot the solid curve is for the magnitude of the closed loop response, and the dashed curve is for the phase. Figure 1.8(a) illustrates a stable feedback amplifier with a large positive phase margin. Observe the Bode plots, $|A_{vf}(j\omega)|$, and the transient response. In Figure 1.8(a), as well as in the other figures, the frequency at which $|\beta(j\omega)A_v(j\omega)| = 1$ is f_1, and the frequency at which $\angle\beta(j\omega)A_v(j\omega) = 0°$ is f_2. The phase margin in Figure 1.8(a) is positive. Figure 1.8(b) illustrates a stable feedback amplifier with a smaller positive phase margin. Observe the larger peak in the associated $|A_{vf}(j\omega)|$ response and in the transient response.

Figure 1.8(c) illustrates an ideal oscillator. The oscillation conditions are satisfied, since $|\beta(j\omega)A_v(j\omega)| = 1$ and $\angle\beta(j\omega)A_v(j\omega) = 0°$ at $f = f_1 = f_2$, which results in an ideal stable sinusoidal oscillation (see the plot of $v_o(t)$). Figure 1.8(d) illustrates an unstable oscillation. Observe that $|\beta(j\omega)A_v(j\omega)| > 1$ when $\angle\beta(j\omega)A_v(j\omega) = 0°$; hence, positive feedback occurs and $v_o(t)$ shows the associated growing sinusoidal response. Basically, Figure 1.8(c) shows what happens when the complex poles move to the imaginary axis, and Figure 1.8(d) shows what happens when the complex conjugate poles remain in the right-half plane. As we will see, there are ways to determine if the oscillation will be stable or not.

1.4 Root Locus

A root-locus plot is a convenient method to analyze the motion of the closed-loop gain poles in the complex s plane as a function of the amplifier gain, or as a function of the feedback factor. In order to use this method, the denominator of

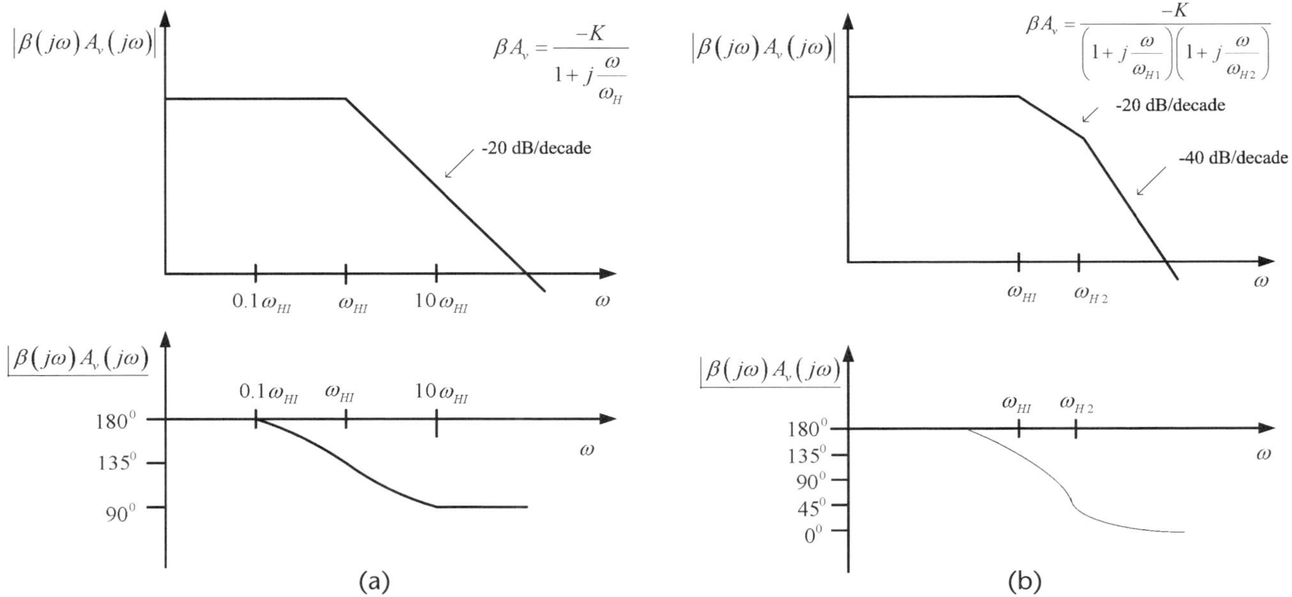

Figure 1.7 Bode plots for a loop gain having (a) one pole, (b) two poles, (c) three poles (stable case), and (d) three poles (unstable case).

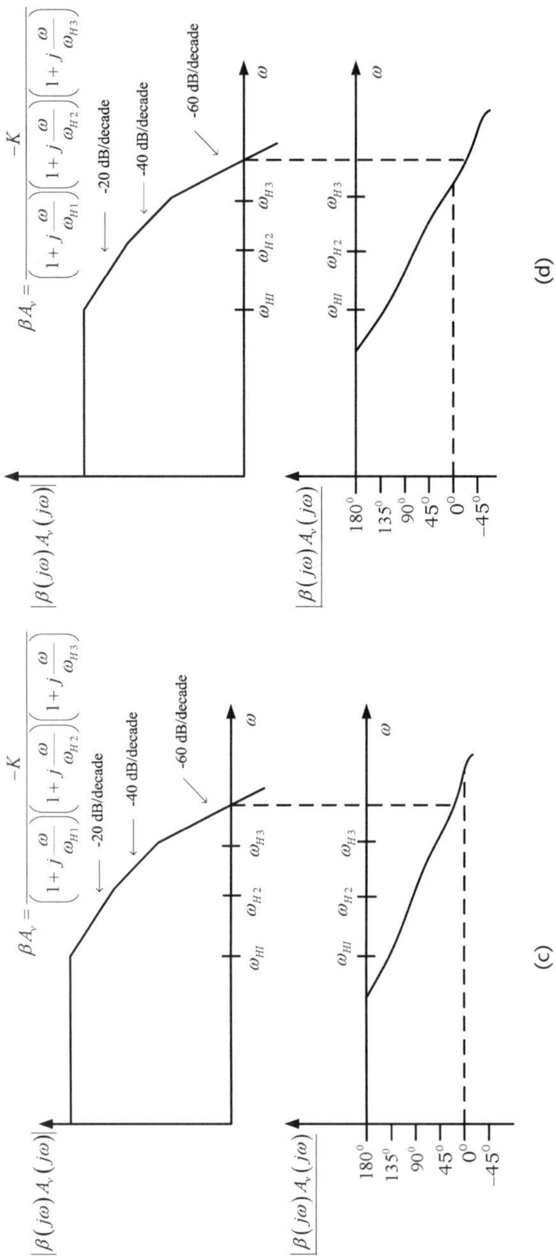

Figure 1.7 (Continued).

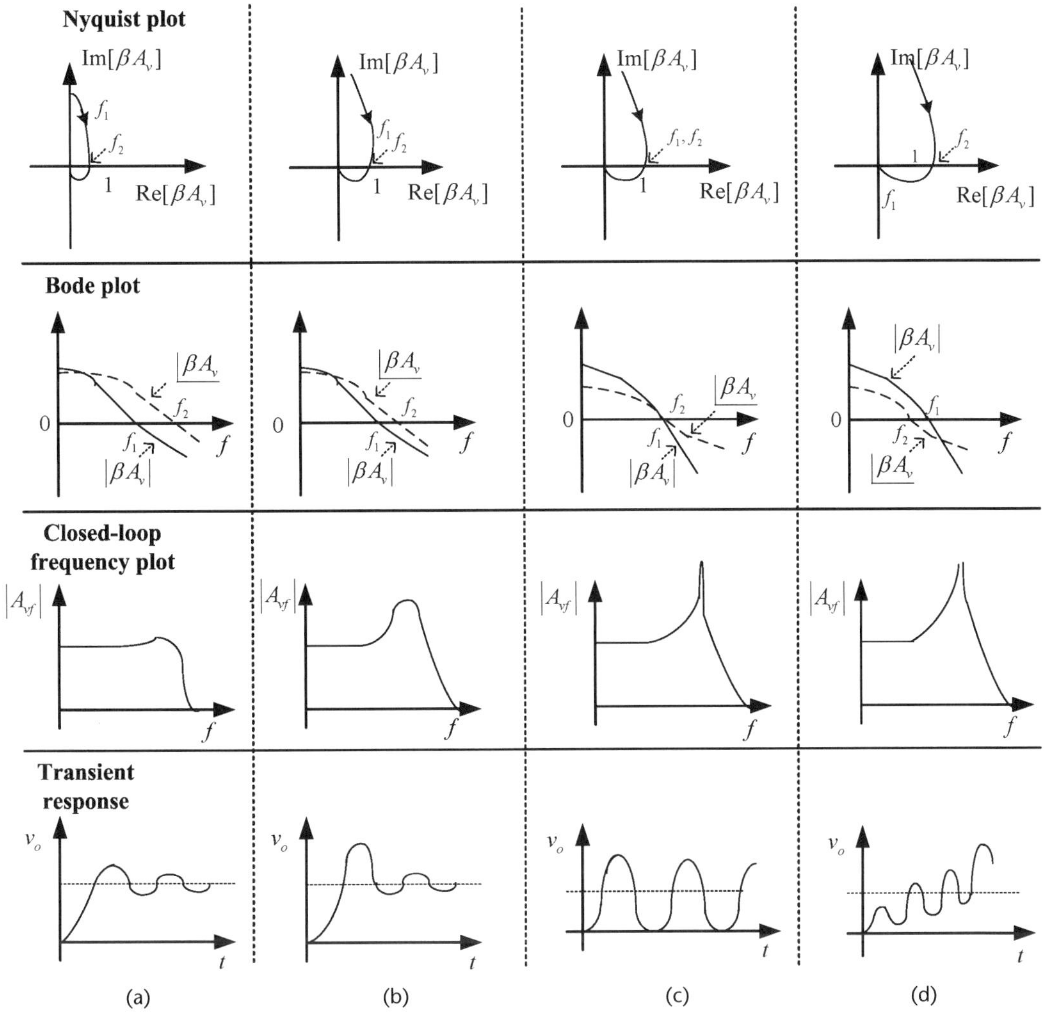

Figure 1.8 Nyquist, Bode, $|A_{vf}(j\omega)|$, and transient response plots of (a) a stable feedback amplifier with a large phase margin, (b) a stable feedback amplifier with a smaller phase margin, (c) a stable oscillator, and (d) an unstable oscillator.

$A_{vf}(s)$ is expressed in polynomial form. The stability of the feedback amplifier is analyzed by observing how the poles of $A_{vf}(s)$ move in the s plane. A typical analysis consists in studying the motion of the roots of $A_{vf}(s)$ as a function of the amplifier open-loop gain, and determining the value of gain that move the roots to the imaginary axis at $s = \pm j\omega_o$. The value of gain and the frequency ω_o are identical to the values predicted by the Barkhausen criterion (i.e., the gain condition and the frequency of oscillation condition).

Consider the root-locus analysis of a feedback amplifier with a two-pole $A_v(s)$ given by

$$A_v(s) = \frac{A_O}{\left(1 + \dfrac{s}{\omega_1}\right)\left(1 + \dfrac{s}{\omega_2}\right)}$$

and a constant feedback β_o, where $\beta_o < 0$, and A_o is the midband value of $A_v(s)$. Hence, $\beta_o A_o < 0$ and

$$\beta(s)A_v(s) = \frac{\beta_o A_o}{\left(1+\frac{s}{\omega_1}\right)\left(1+\frac{s}{\omega_2}\right)}$$

The closed-loop gain of the feedback amplifier is given by

$$A_{vf}(s) = \frac{A_v(s)}{1-\beta_o A_v(s)} = \frac{A_o\omega_1\omega_2}{s^2 + s(\omega_1+\omega_2) + \omega_1\omega_2(1-\beta_o A_o)} \quad (1.13)$$

The root locus of the poles in (1.13) follows from the analysis of

$$s^2 + s(\omega_1+\omega_2) + \omega_1\omega_2(1-\beta_o A_o) = 0$$

as A_o varies. The root locus is shown in Figure 1.9. This plot shows that for A_o approaching zero the roots are located at $s_1 =- \omega_1$ and $s_2 =- \omega_2$. As A_o increases, the roots move along the negative real axis as shown in Figure 1.9. At a specific value of A_o, denoted by the value of $A_o = A'_o$ (see Figure 1.9), the roots are identical, and for $A_o > A'_o$ the roots become complex but remain in the left-half plane. Therefore, this feedback amplifier is stable. The value of A'_o is given by

$$A'_o = \frac{1}{\beta_o}\left[1 - \frac{0.25(\omega_1+\omega_2)^2}{\omega_1\omega_2}\right]$$

Of course, if $\beta_o A_o > 0$ the feedback amplifier is unstable.

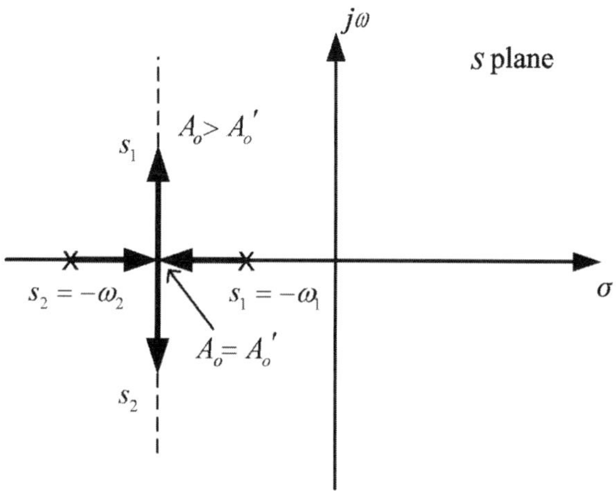

Figure 1.9 Root locus of a two-pole function $A_v(s)$.

Next, consider a three-pole $A_v(s)$ given by

$$A_v(s) = \frac{A_o}{\left(1 + \frac{s}{\omega_1}\right)\left(1 + \frac{s}{\omega_2}\right)\left(1 + \frac{s}{\omega_3}\right)}$$

with a constant feedback β_o, where $\beta_o < 0$ (i.e., $\beta_o A_o < 0$). In this case $A_{vf}(s)$ is given by

$$A_{vf}(s) = \frac{A_o}{a_3 s^3 + a_2 s^2 + a_1 s + (1 - \beta_o A_o)}$$

where

$$a_3 = \frac{1}{\omega_1 \omega_2 \omega_3}$$

$$a_2 = \frac{1}{\omega_1 \omega_2} + \frac{1}{\omega_1 \omega_3} + \frac{1}{\omega_2 \omega_3}$$

and

$$a_1 = \frac{1}{\omega_1} + \frac{1}{\omega_2} + \frac{1}{\omega_3}$$

The root locus of

$$a_3 s^3 + a_2 s^2 + a_1 s + (1 - \beta_o A_o) = 0$$

is shown in Figure 1.10. For A_o approaching 0, the poles are located at $s_1 =- \omega_1$, $s_2 =- \omega_2$, and $s_3 =- \omega_3$. As A_o increases, the pole s_3 moves along the negative real axis towards $-\infty$, and the poles s_1 and s_2 become complex conjugate poles. At a certain value of A_o the poles are located on the imaginary axis at $s_{1,2} =\pm j\omega_o$ and oscillation occurs. Figure 1.10 also shows that certain values of A_o move the poles into the right-half plane.

Another open-loop gain function that can lead to oscillations is

$$A_v(s) = \frac{A_o\left(1 - \frac{s}{\omega_3}\right)}{\left(1 + \frac{s}{\omega_1}\right)\left(1 + \frac{s}{\omega_2}\right)}$$

This function has two poles and a right-half plane zero at $s = \omega_3$. This type of transfer function occurs in the high-frequency analysis of several amplifier configurations.

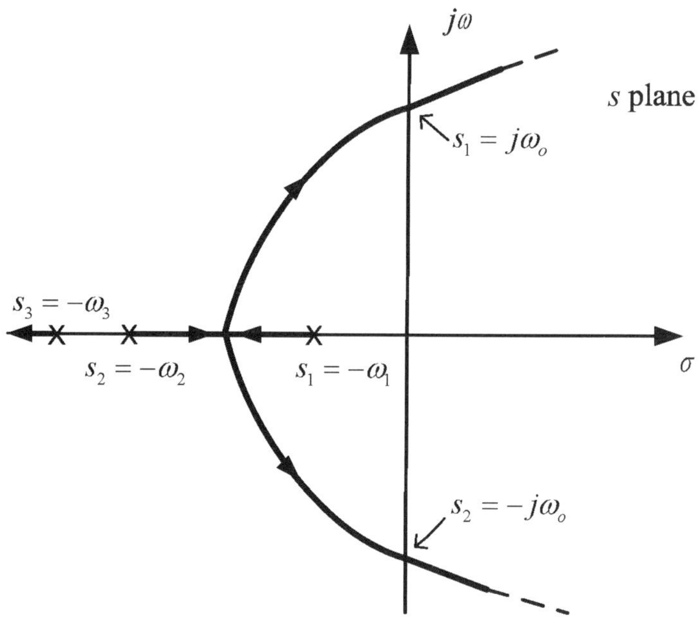

Figure 1.10 Root locus of a three-pole $A_v(s)$.

Example 1.2

Consider the open-loop gain of a CE amplifier given by

$$A_v(s) = \frac{A_o\left(1 - \frac{s}{\omega_3}\right)}{\left(1 + \frac{s}{\omega_1}\right)\left(1 + \frac{s}{\omega_2}\right)}$$

where $A_o = 2{,}200$, $\omega_1 = 10^6$ rad/s, $\omega_2 = 10^8$ rad/s, and $\omega_3 = 10^9$ rad/s. Assume that the feedback factor is constant and given by $\beta(j\omega) = \beta_o = -0.1$. The loop gain is

$$\beta(j\omega)A_v(j\omega) = \frac{-0.1(2{,}200)\left(1 - \frac{s}{10^9}\right)}{\left(1 + \frac{s}{10^6}\right)\left(1 + \frac{s}{10^8}\right)} \tag{1.14}$$

The Nyquist plot of (1.14) is shown in Figure 1.11(a). The behavior of the function in (1.14) around the critical point is difficult to see in Figure 1.11(a). A graph showing the behavior around the critical point is shown in Figure 1.11(b). From (1.14) it follows that $P = 0$, and from the Nyquist plots in Figure 1.11(a, b) we obtain $N = 0$, since the critical point is not enclosed. Therefore, the feedback amplifier with $A_o = 2{,}200$ is stable, since $Z = N + P = 0$.

Next, assume that A_o is given by $A_o = 22{,}000$. The Nyquist plot for this case is shown in Figure 1.11(c) and the behavior around the critical point in Figure

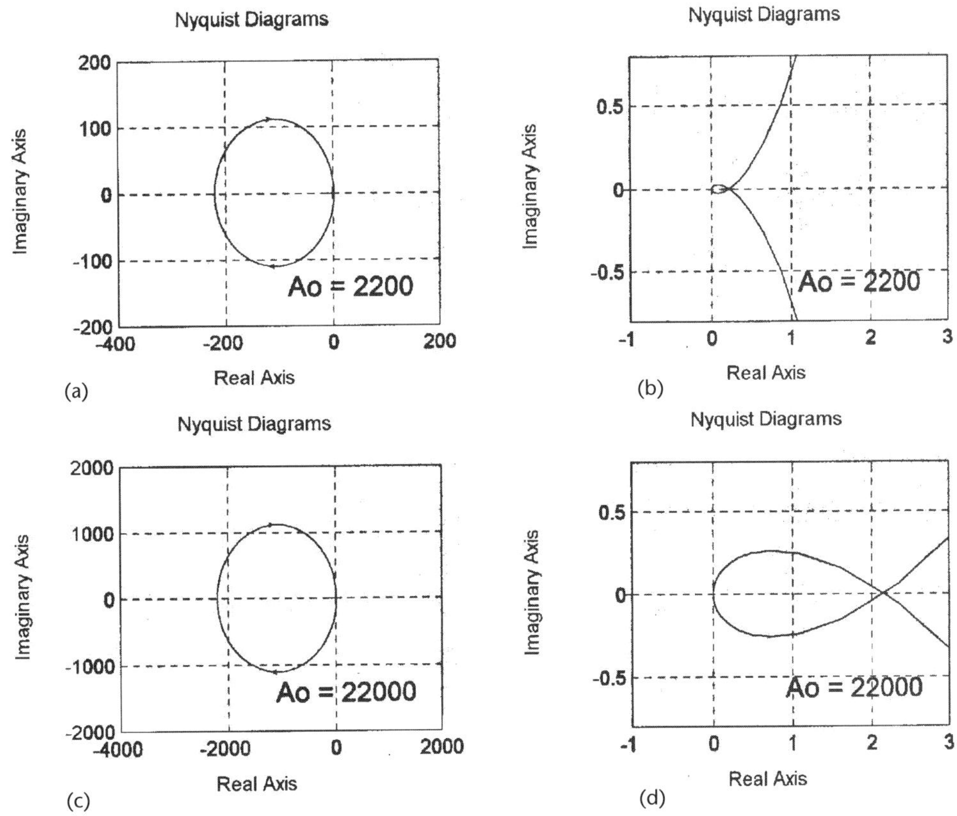

Figure 1.11 (a) Nyquist plot for Example 1.2 with $A_o = 2{,}220$, (b) the behavior of the loop gain (with $A_o = 2{,}220$) around the critical point, (c) Nyquist plot for Example 1.2 with $A_o = 22{,}000$, and (d) the behavior of the loop gain (with $A_o = 22{,}000$) around the critical point.

1.11(d). Figure 1.11(d) shows that the critical point is enclosed twice, or $N = 2$. Hence, there are two roots in the right-half plane and the amplifier is unstable with $A_o = 22{,}000$.

The root locus of $A_{vf}(s)$ for this example follows from the analysis of

$$s^2 + s\left(\omega_2 + \omega_1 + \frac{\omega_1 \omega_2 \beta_o A_o}{\omega_3}\right) + \omega_1 \omega_2 (1 - \beta_o A_o) = 0$$

or

$$s^2 + s(101 \times 10^6 - 10^4 A_o) + 10^{14}(1 + 0.1 A_o) = 0 \tag{1.15}$$

The resulting root-locus plot is shown in Figure 1.12. For A_o approaching 0 the roots are located at $s_1 =- 10^6$ and $s_2 =- 10^8$. At the value of $A_o = 10{,}100$ the complex poles are on the imaginary axis at $s_1 = j318 \times 10^6$ and $s_2 =- j318 \times 10^6$, and oscillations occur.

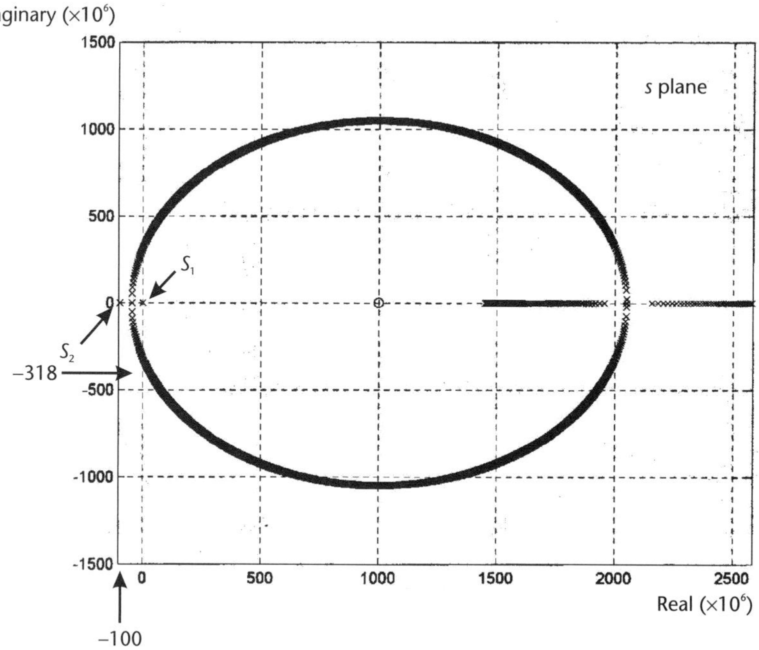

Figure 1.12 Root locus for Example 1.2.

1.5 Routh-Hurwitz Method

Another method that can be used to determine if a polynomial has right-half plane roots is the Routh-Hurwitz method. In the Routh-Hurwitz method the denominator polynomial of $A_{vf}(s)$ is written in the form

$$D(s) = a_n s^n + a_{n-1} s^{n-1} + a_{n-2} s^{n-2} + \ldots + a_1 s + a_o$$

The coefficients are then arranged in the following array form (known as the Routh-Hurwitz array):

$$\begin{array}{c|ccccc}
s^n & a_n & a_{n-2} & a_{n-4} & \cdot & \cdot \\
s^{n-1} & a_{n-1} & a_{n-3} & a_{n-5} & \cdot & \cdot \\
s^{n-2} & b_{n-1} & b_{n-3} & b_{n-5} & \cdot & \cdot \\
s^{n-3} & c_{n-1} & c_{n-3} & c_{n-5} & \cdot & \cdot \\
\cdot & \cdot & \cdot & \cdot & \cdot & \cdot \\
s^1 & \cdot & & & & \\
s^0 & \cdot & & & &
\end{array}$$

which shows that a polynomial of degree n has $n + 1$ rows in the Routh-Hurwitz array. The terms b_{n-1}, b_{n-3}, b_{n-5}, etc., are the first, second, third, etc., entries in the $(n - 2)$ row. The terms c_{n-1}, c_{n-3}, c_{n-5}, etc., are the first, second, third, etc., entries in the $(n - 3)$ row. The entries are defined by

$$b_{n-1} = \frac{a_{n-1}a_{n-2} - a_n a_{n-3}}{a_{n-1}}$$

$$b_{n-3} = \frac{a_{n-1}a_{n-4} - a_n a_{n-5}}{a_{n-1}}$$

$$\vdots$$

$$c_{n-1} = \frac{b_{n-1}a_{n-3} - a_{n-1}b_{n-3}}{b_{n-1}}$$

$$\vdots$$

The entries in the first column (i.e., a_n, a_{n-1}, b_{n-1}, c_{n-1}, etc.) are called the leading entries in the array. If none of the leading entries vanish, the number of roots of $D(s)$ in the right-half plane is equal to the number of sign changes in the leading entries.

Example 1.3

Determine if the following denominator polynomials of $A_{vf}(s)$ have right-half plane roots:

(a) $$D(s) = s^4 + 10s^3 + 35s^2 + 50s + 24 \quad (1.16)$$

(b) $$D(s) = s^3 + s^2 + 2s + 24 \quad (1.17)$$

Solution

(a) The Routh-Hurwitz array for (1.16) is shown in Figure 1.13(a). The leading entries (i.e., 1, 10, 30, 42, and 24) are positive. Hence, $D(s)$ in (1.16) has no right-half plane roots, or $A_{vf}(s)$ has no right-half plane poles.

The polynomial in (1.16) can be shown to be equal to

$$D(s) = (s+1)(s+2)(s+3)(s+4)$$

which obviously has no right-half plane roots.

(b) The Routh-Hurwitz array for the polynomial in (1.17) is shown in Figure 1.13(b). In this case, the leading entries have two sign changes. Hence, the polynomial in (1.17) has two right-half plane roots, and therefore, $A_{vf}(s)$ is unstable. In fact, the polynomial in (1.17) can be shown to be equal to

s^4	1	35	24
s^3	10	50	
s^2	30	24	
s^1	42		
s^0	24		

(a)

s^3	1	2
s^2	1	24
s^1	−22	
s^0	24	

(b)

Figure 1.13 (a) Routh-Hurwitz array for the polynomial in (1.16) and (b) Routh-Hurwitz array for the polynomial in (1.17).

$$D(s) = (s + 3)(s - 1 + j2.6458)(s - 1 - j2.6458)$$

which has two roots in the right-half plane.

The Routh-Hurwitz method can also be used to determine at what value of the open-loop gain or feedback factor value is the closed-loop gain unstable. Example 1.4 illustrates this point.

Example 1.4

In an oscillator, the denominator polynomial of $A_{vf}(s)$ is

$$D(s) = s^2 + (3 - A_o)s + \omega_o^2 \tag{1.18}$$

where A_o is the amplifier gain and ω_o is the frequency of oscillation. Determine for what values of A_o there are right-half plane roots.

Solution

The Routh-Hurwitz array for (1.18) is shown in Figure 1.14.

The leading entries are 1, $3 - A_o$, and 2. Hence, the feedback system is stable if $3 - A_o > 0$ or $A_o < 3$. The feedback system is unstable when $3 - A_o < 0$ or $A_o > 3$. Oscillations can occur when $A_o = 3$. This oscillator is analyzed in detail in the next section.

There are some degenerate cases that can occur in the Routh-Hurwitz array. One case occurs when a leading entry vanishes, and at least one entry in the corresponding row is nonzero. The other case occurs when a complete row vanishes. For these degenerate cases the reader is referred to an appropriate textbook in control systems.

Next, the loop-gain associated with an oscillator is analyzed. The oscillator selected is the Wien bridge, which provides an example where the feedback factor and the open-loop voltage gain are evaluated separately. In many oscillators the feedback network is loaded by the amplifier, and this effect must be taken into consideration in the analysis of the loop gain.

1.6 The Wien-Bridge Oscillator

The Wien-bridge oscillator is shown in Figure 1.15. The four arms of the bridge are R_1, R_2, Z_a, and Z_b. The op amp maintains the voltage across two of the arms equal, since $v_- = v_+$. This oscillator can be used to analyze in closed form the conditions for oscillation, its stability, and the location of the complex poles.

s^2	1	2
s^1	$3 - A_o$	
s^0	2	

Figure 1.14 Routh-Hurwitz array for Example 1.4.

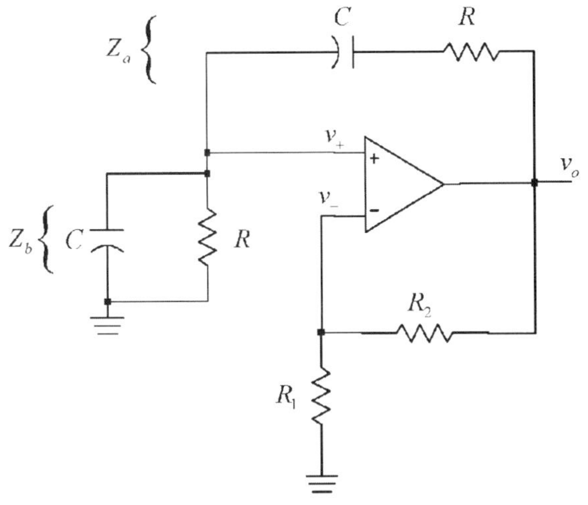

Figure 1.15 The Wien-bridge oscillator.

From Figure 1.15 it is seen that there are two feedback paths. Positive feedback occurs between v_o and v_+ through the voltage divider formed by Z_a and Z_b, and negative feedback occurs between v_o and v_- through R_1 and R_2.

The Wien-bridge oscillator in Figure 1.15 uses an op amp in an inverting configuration (i.e., the negative feedback path) to provide the open-loop gain. The open-loop gain is constant and given by

$$A_v(j\omega) = A_{vo} = \frac{v_o}{v_+} = 1 + \frac{R_2}{R_1} \tag{1.19}$$

The voltage v_+ is given by

$$v_+ = v_o \frac{Z_b}{Z_b + Z_a} = v_o \frac{\left(\dfrac{R}{1 + j\omega RC}\right)}{R + \dfrac{1}{j\omega C} + \left(\dfrac{R}{1 + j\omega RC}\right)}$$

$$= v_o \frac{1}{3 + j\left(\omega RC - \dfrac{1}{\omega RC}\right)}$$

Therefore, the voltage-feedback transfer function $\beta(j\omega)$ can be written in the form

$$\beta(j\omega) = \frac{v_+}{v_o} = \frac{1}{3 + j\left(\omega RC - \dfrac{1}{\omega RC}\right)} \tag{1.20}$$

From (1.19) and (1.20) it follows that the loop gain is

$$\beta(j\omega)A_{vo} = \frac{1}{3 + j\left(\omega RC - \dfrac{1}{\omega RC}\right)}\left(1 + \frac{R_2}{R_1}\right) \tag{1.21}$$

For oscillation the loop gain must be unity. The frequency of oscillation and the gain condition are given by (1.7) and (1.6), respectively.

The condition (1.7) requires that the imaginary part of $\beta(j\omega)$ be zero. From (1.20) it is seen that $\beta_i(j\omega) = 0$ when

$$\omega RC - \frac{1}{\omega RC} = 0 \Rightarrow \omega = \omega_o = \frac{1}{RC}$$

or

$$f_o = \frac{1}{2\pi RC} \tag{1.22}$$

The frequency f_o is the frequency of oscillation.

At $\omega = \omega_o$ the real part of $\beta(j\omega)$ is

$$\beta_r(\omega_o) = \frac{1}{3}$$

Therefore, from (1.6) the gain condition is

$$A_{vo} = \frac{1}{\beta_r(\omega_o)} = 3$$

and from (1.19) it follows that $R_2 = 2R_1$. In practice, the gain should be greater than 3 to start the oscillation. Values of 3.1 to 3.3 are reasonable.

It is interesting to observe that the op amp uses negative feedback through R_1 and R_2 to provide the required gain A_{vo}, and it uses positive feedback through the RC networks to obtain the required closed-loop phase shift. The attenuation in the positive feedback loop must be equal to the gain A_{vo}, making the loop gain equal to one. Since the phase shift through the op amp is zero at ω_o, the phase shift through the RC networks must also be zero so the feedback signal v_+ is in phase with v_o [see (1.20) at $\omega = \omega_o$].

The frequency of oscillation can be varied by simultaneously changing the capacitance values using a ganged capacitor arrangement. In addition, different frequency ranges can be selected by simultaneously switching different values of the resistors R.

A final observation is that op amps in this book are assumed to operate from a dual power supply. Of course, op amps with a single power supply can be used if proper single-supply techniques are used.

Further insight into the Wien-bridge oscillator is obtained by analyzing the oscillator in terms of the circuit poles. The poles of the closed-loop gain $A_{vf}(s)$ are the roots of $1 - \beta(s)A_{vo} = 0$. Using (1.21) with $s = j\omega$ the roots of $A_{vf}(s)$ are given by

$$1 - \frac{A_{vo}}{3 + \frac{s}{\omega_o} + \frac{\omega_o}{s}} = 0$$

or

$$s^2 + (3 - A_{vo})\,\omega_o s + \omega_o^2 = 0 \tag{1.23}$$

The two roots of (1.23) are shown in Figure 1.16 in a root-locus plot as a function of A_{vo}. As A_{vo} varies from 0 to 1, the poles move along the negative real axis. For $A_{vo} = 1$, the poles meet at $-\omega_o$. As A_{vo} varies from 1 to 3, the poles are complex and move in a semicircular path towards the imaginary axis. For $A_{vo} = 3$, the complex poles are located at $j\omega_o$ and $-j\omega_o$, respectively. As A_{vo} increases above 3, the complex poles move into the right-half plane; and at $A_{vo} = 5$ the poles meet on the positive real axis at ω_o. For $A_{vo} > 5$, they move along the positive real axis.

The conditions that produce complex poles are now analyzed. Observe that $1 < A_{vo} < 3$ corresponds to a loop gain of $\beta(j\omega_o)A_{vo} < 1$, and $3 < A_{vo} < 5$ corresponds to a loop gain of $\beta(j\omega_o)A_{vo} > 1$. For $1 < A_{vo} < 3$ the poles produce an output voltage with an exponentially damped sinusoidal response, and for $3 < A_{vo} < 5$ an exponentially growing sinusoidal response results. In order to start the oscillation, a value of gain slightly greater than 3 is used. Thus, at the start of oscillation the complex poles are in the right-half plane and an exponentially growing sinusoidal oscillation is produced. As the amplitude of the oscillation increases, the op amp saturates and its gain decreases. When the gain is 3, the poles are on the imaginary axis (i.e., to $s = \pm\, j\omega_o$); and when the gain goes below 3, the poles move into the left-half plane. In the Wien-bridge oscillator in Figure 1.15, the amplitude of the sinusoidal oscillation is limited by the saturation of the op amp. Hence, the oscillation amplitude varies between approximately $V^+ - 1$ and $V^- + 1$.

Wien-bridge oscillators work very well for frequencies up to about 1 MHz. With an amplitude-limiting circuit the harmonic distortion can be less than 5%.

The Wien bridge can also be constructed with different values of the branch values of R and C. If

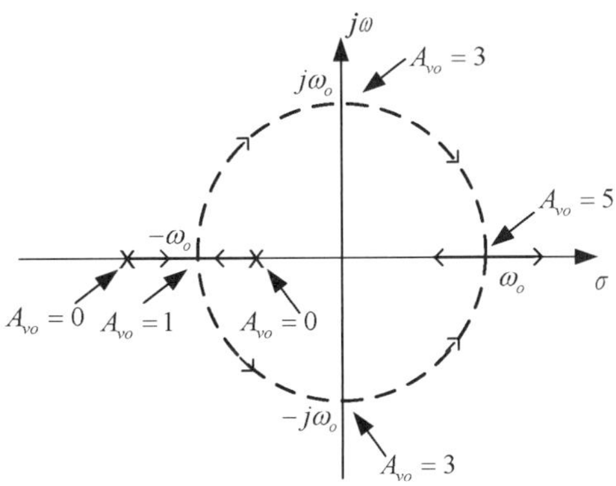

Figure 1.16 Poles of the closed-loop gain as a function of A_{vo}.

$$Z_a = R_a + \frac{1}{j\omega C_a}$$

and

$$Z_b = \frac{R_b}{1 + j\omega C_b}$$

it follows that for oscillation: $R_a C_a = R_b C_b$, and the frequency of oscillation is given by

$$f_o = \frac{1}{2\pi R_a C_a} = \frac{1}{2\pi R_b C_b}$$

The required gain of the amplifier is

$$A_{vo} = \frac{1}{\beta(j\omega_o)} = 1 + \frac{2R_1}{R_2}$$

Example 1.5

Design the Wien-bridge oscillator shown in Figure 1.15 to oscillate at 5 kHz.

Solution

A practical value of 0.01 μF for the capacitors can be selected. Then, using (1.22), the value of R is

$$R = \frac{1}{2\pi f_o C} = \frac{1}{2\pi(5 \times 10^3)(0.01 \times 10^{-6})} = 3.18\ \text{k}\Omega$$

A practical value of 3 kΩ can be used in series with a trimming potentiometer to set the frequency of oscillation at 5 kHz.

To start the oscillation, a value of $A_{vo} = 3.2$ is used. From (1.19), a gain of 3.2 is obtained with $R_2 = 22$ kΩ and $R_1 = 10$ kΩ. The supply voltages of the op amp can be selected as 12V and –12V. The transient simulation of the oscillator, using a 741 op amp is shown in Figure 1.17. Observe that the output voltage reaches saturation producing a clipping in the output waveform and, therefore, a significant amount of distortion. This occurs because the starting condition requires $A_{vo} > 3$, and the gain of the amplifier changes when its output reaches saturation. The fundamental frequency of oscillation (i.e., f_o = freq[1] = 5.0002 kHz) is close to the predicted value using $\beta(j\omega_o)A_{vo} = 1$.

Example 1.5 shows that some sort of amplitude-limiting mechanism is needed to reduce the harmonic distortion. There are several ways of accomplishing amplitude limiting. The amplitude of oscillation is determined by the loop gain, which is made to be greater than 1 in order to start the oscillation, and by the nonlinearities

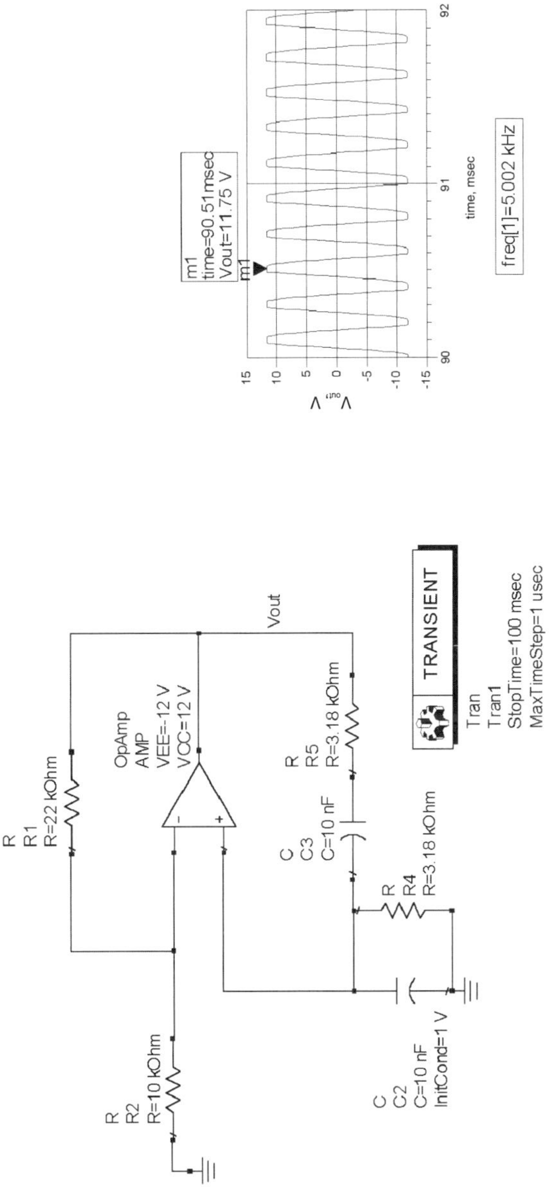

Figure 1.17 Simulation of the Wien-bridge oscillator in Example 1.5.

of the amplifier and feedback network. If the feedback network is a passive network, the nonlinearities of the amplifier determine the amplitude of oscillation. This procedure, as seen in Figure 1.17, generates unwanted harmonics since the amplitude of the oscillation is limited by the saturation voltage of the op amp. In order to remove these harmonics, a bandpass filter that passes only the oscillation frequency can be used after the amplifier. Of course, the nonlinearities of the amplifier can be avoided by limiting somehow the amplitude of the oscillation before the amplitude reaches the amplifier's saturation value. There are many circuits that can be used to limit the amplitude of the oscillator. Some of these circuits are shown in Figure 1.18.

Figure 1.18(a) shows a limiting circuit using back-to-back Zener diodes (usually $V_{Z1} = V_{Z2}$). The output is limited to $-(V_{Z2} + 0.7) < v_o < (V_{Z1} + 0.7)$. When the Zeners are not conducting, the op-amp gain is $-R_2/R_1$. The transfer function of this circuit is illustrated in Figure 1.18(a). Another limiting circuit is shown in Figure 1.18(b) with its transfer function. In this circuit the gain between v_o and v_{IN} changes from A_{v1} (when the Zeners conduct) to A_{v2} (when the Zeners are not conducting).

An amplitude-limiting mechanism is basically an automatic gain control (AGC) circuit that forces the amplifier gain to decrease when the amplitude of the

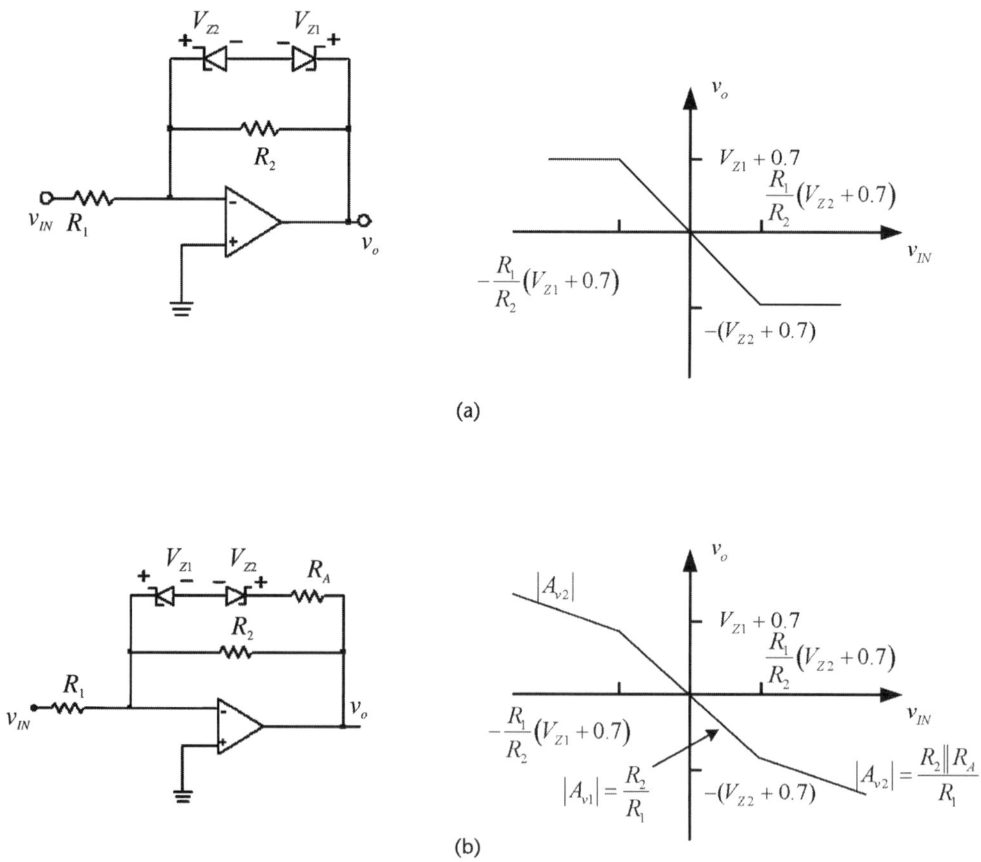

Figure 1.18 (a) A limiting circuit and the associated transfer functions and (b) another limiting circuit.

oscillation increases. Figure 1.19 shows three Wien-bridge oscillators with amplitude-limiting mechanisms. In Figure 1.19(a), when the diodes are off, the gain is $1 + R_2||R_1$; and when a diode is on, the gain is reduced to $1 + (R_2||R_3)/R_1$. The start up condition requires a gain slightly greater than 3 or

$$\frac{R_2}{R_1} > 2 \tag{1.24}$$

and when a diode is on, the gain should be slightly less than 3, or

$$\frac{R_2||R_3}{R_1} < 2 \tag{1.25}$$

The inequality in (1.24) can be satisfied by making it equal to a value between 2.1 to 2.2 and in (1.25) using a value between 1.8 and 1.9.

When a diode is conducting, the amplitude of the output voltage is limited. Since $v_+ = v_- = v_o/3$, a nodal equation gives

$$\frac{v_o}{3R_1} = \frac{v_o - \dfrac{v_o}{3}}{R_2} + \frac{v_o - \dfrac{v_o}{3} - V_D}{R_3}$$

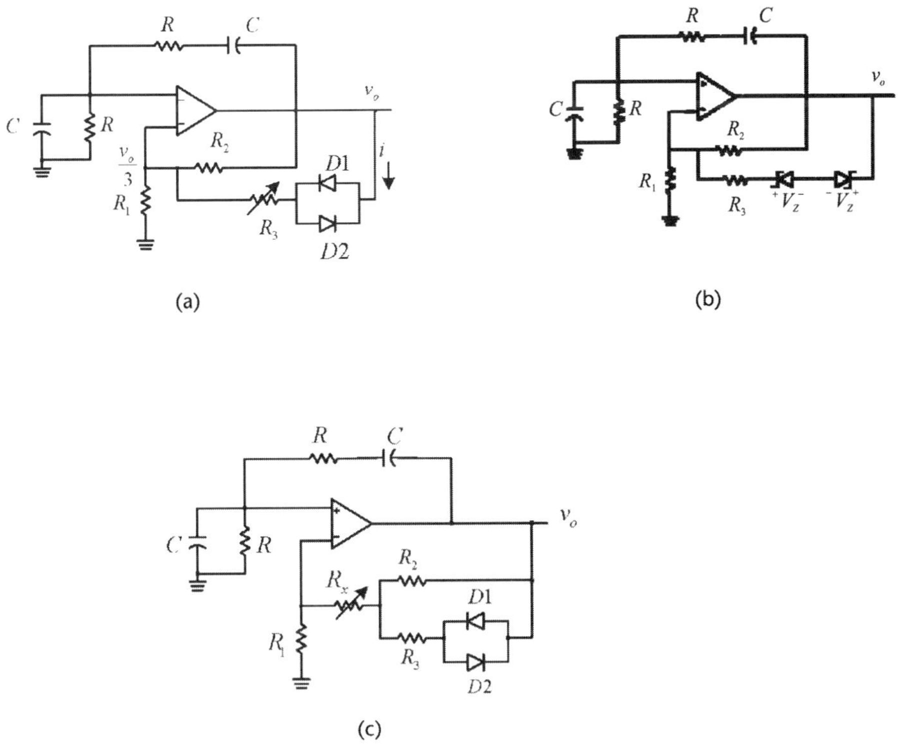

Figure 1.19 Three Wien-bridge oscillators with (a, b, c) amplitude-limiting mechanism.

or

$$v_o = \frac{3V_D}{2\left(1 + \frac{R_3}{R_2}\right) - \frac{R_3}{R_1}} \tag{1.26}$$

where $V_D \approx 0.5$V for an actual diode.

Equation (1.26) shows that the amplitude of v_o depends on R_3. This equation can be used to obtain the value of R_3 for a given amplitude of v_o. Since in this oscillator the gain is controlled by the circuit, it provides an opportunity to verify that the location of the poles changes between the right-half plane and the left-half plane, and how the pole movement is related to the Barkhausen criterion in the determination of the frequency of oscillation (see Example 1.6).

In Figure 1.19(b) the amplitude-limiting mechanism is implemented with Zener diodes. The analysis of this oscillator is similar to the previous analysis. The output voltage is given by (1.26) with V_D replaced by $V_Z + V_D$. A simple way of designing this oscillator is to let $R_2 || R_1 = 2.15$ and $(R_2 || R_3)/R_1 = 1.8$, which can be satisfied with $R_2 = 2R_a$, $R_1 = 0.93R_a$, and $R_3 = 10R_a$ (R_a is a scaling factor). As an example, with $V_Z = 4.2$V, and selecting $R_a = 5$ kΩ, it follows that $R_1 = 4.65$ kΩ, $R_2 = 10$ kΩ, and $R_3 = 50$ kΩ. The amplitude of the oscillation is limited to $v_o = 9.7$V.

In Figure 1.19(c) a modification of the amplitude control circuit that produces smaller values of the control resistors is shown. For this circuit the design equations are

$$\frac{R_2 + R_x}{R_1} > 2$$

$$\frac{(R_2 || R_3) + R_x}{R_1} < 2$$

and

$$v_o = \frac{3V_D}{\left(2 - \frac{R_x}{R_1}\right)\left(1 + \frac{R_3}{R_2}\right) - \frac{R_3}{R_1}} \tag{1.27}$$

Obviously, as $R_x \to 0$, (1.27) reduces to (1.26).

Usually, in these oscillators a buffer stage (i.e., a unity gain amplifier) is used to couple the oscillation signal.

The following example helps to explain why, although not exact, the Barkhausen criterion is useful in predicting the expected frequency of oscillation. As Lindberg [1] mentioned, the placement of the poles on the imaginary axis is an impossible act of balance.

Example 1.6

(a) Design the Wien-bridge oscillator shown in Figure 1.20(a) to oscillate at ω_o = 10 krad/s with an amplitude of 2V.
(b) Repeat part (a) for an amplitude of 10V.

Solution

(a) From (1.22) with $C = 0.01\ \mu$F and $R = 10$ kΩ, the desired $\omega_o = 10$ krad/s is obtained.

To start the oscillation, a value of $A_{vo} = 3.2$ (or $R_2 \| R_1 = 2.2$) is used. This can be obtained with $R_2 = 11$ kΩ and $R_1 = 5$ kΩ. From (1.26), if $V_D = 0.45$V, an amplitude of 2V is obtained with $R_3 = 73$ kΩ; and if $V_D = 0.6$V, the value is R_3 = 60.5 kΩ.

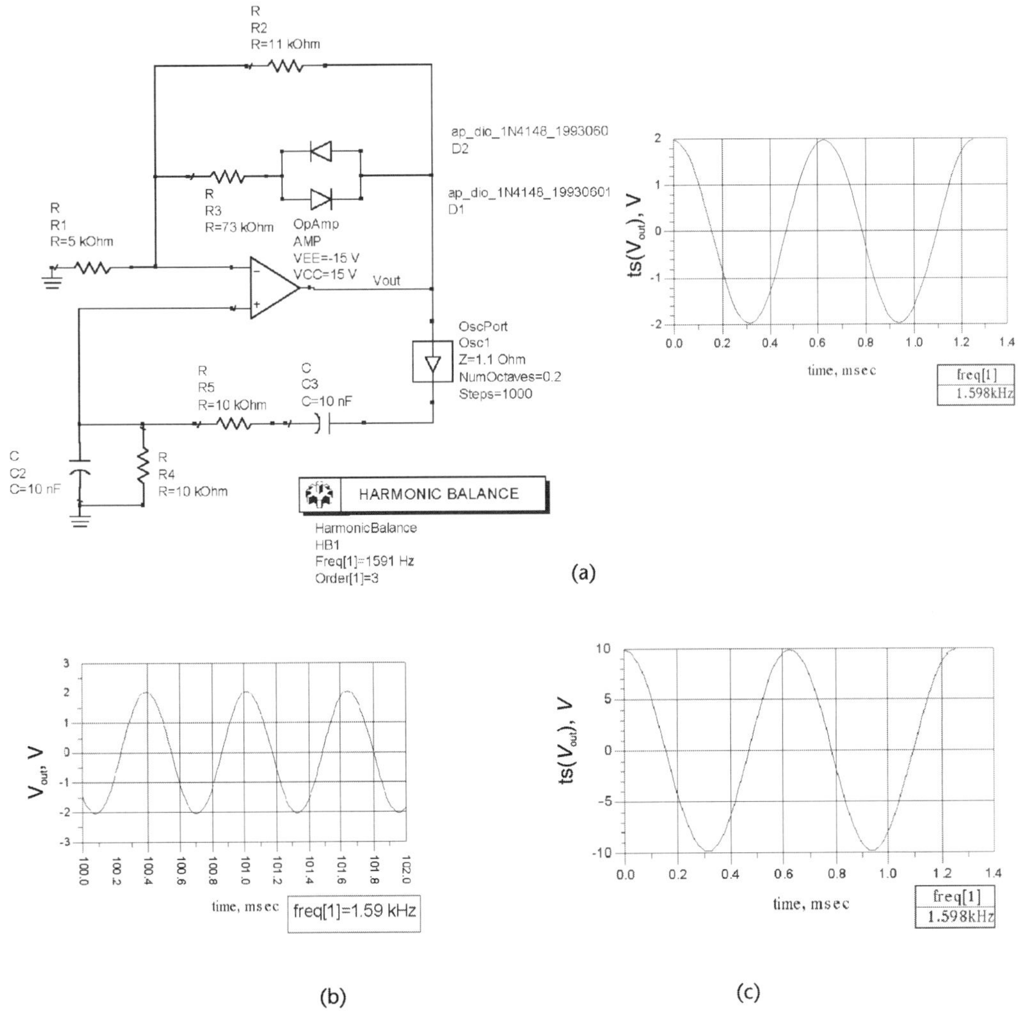

Figure 1.20 (a) ADS simulation for Example 1.6 for v_o = 2V, (b) transient simulation results, and (c) simulation result for v_o = 10V.

The harmonic-balance simulation of this oscillator is shown in Figure 1.20(a) with $R_3 = 73$ kΩ, and the transient simulation data is shown in Figure 1.20(b). The fundamental frequency of oscillation is freq[1] = 1.598 kHz (or $\omega_o = 10.04$ krad/s). The current waveform in the diodes shows the on/off states of the diodes.

For this design, when the diodes are off the gain is $A_{vo} = 3.2$. From (1.23), the poles of the closed-loop gain are located at

$$s^2 - 2{,}000s + 10^8 = 0$$

or

$$s_{1,2} = 1{,}000 \pm j9{,}949.9 \tag{1.28}$$

Therefore, the poles are in the right-half plane.

When a diode conducts, the gain is reduced to

$$A_{vo} = 1 + \frac{73 \times 10^3 || 11 \times 10^3}{5 \times 10^3} = 2.912$$

and the new location of the poles, using (1.23), is

$$s^2 + 880s + 10^8 = 0$$

or

$$s_{1,2} = -440 \pm j9{,}990.3 \tag{1.29}$$

From (1.28) and (1.29) it is seen that the poles move between the right-half plane and the left-half plane. The frequency associated with the poles in the right-half plane is $\omega_{o,1} = 9{,}949.9$ rad/s, and the frequency associated with the poles in the left-half plane is $\omega_{o,2} = 9{,}990.3$ rad/s. The frequency of oscillation predicted by the Barkhausen criterion is $\omega_o = 10$ krad/s, which occurs if the poles are on the $j\omega$ axis. However, the poles of this oscillator are moving between $\omega_{o,1}$ and $\omega_{o,2}$. This variation in frequency produces a fundamental frequency of oscillation and the associated harmonics.

(b) An amplitude of $v_o = 10$V can be obtained with $R_1 = 5$ kΩ, $R_2 = 11$ kΩ, and $R_3 = 101$ kΩ. When the diodes are not conducting, the poles are in the right-half plane and are given by (1.28). When a diode conducts the poles move to the left-half plane and $A_{vo} = 2.9821$. Their location is calculated using (1.23), namely,

$$s^2 + 179s + 10^8 = 0$$

or

$$s_{1,2} = -89.5 \pm j9{,}979.6$$

Hence, the poles of this oscillator are moving between $\omega_{o,1} = 9{,}949.9$ rad/s and $\omega_{o,2} = 9{,}979.6$ rad/s.

The simulation results are shown in Figure 1.20(c). The fundamental frequency of oscillation is 1.598 kHz.

Another Wien-bridge oscillator with an amplitude-limiting circuit is shown in Figure 1.21. In this oscillator the amplitude-limiting circuit consists of the diodes $D1$ and $D2$, and the resistors R_3, R_4, R_5, and R_6. To understand the operation of the amplitude-limiting circuit, observe that as v_o increases, the voltage at node v_y will exceed the voltage v_1, forcing $D2$ to conduct. When $D2$ conducts, the value of v_y is $v_y = v_1 + 0.7$, and v_o is clamped at the value $v_{o(\max)}$, given by

$$v_y = v_1 + 0.7 = \frac{v_{o(\max)} R_6}{R_5 + R_6} + \frac{V^- R_5}{R_5 + R_6} \tag{1.30}$$

Since v_1 is approximately $v_o/3$, it follows from (1.24) that

$$\frac{v_{o(\max)}}{3} + 0.7 = \frac{v_{o(\max)} R_6}{R_5 + R_6} + \frac{V^- R_5}{R_5 + R_6} \tag{1.31}$$

Similarly, as v_o decreases, the voltage v_x will drop below v_1, forcing $D1$ to conduct. When $D1$ conducts, the voltage v_x is $v_x = v_1 - 0.7$, and v_o is clamped at the value $v_{o(\min)}$, given by

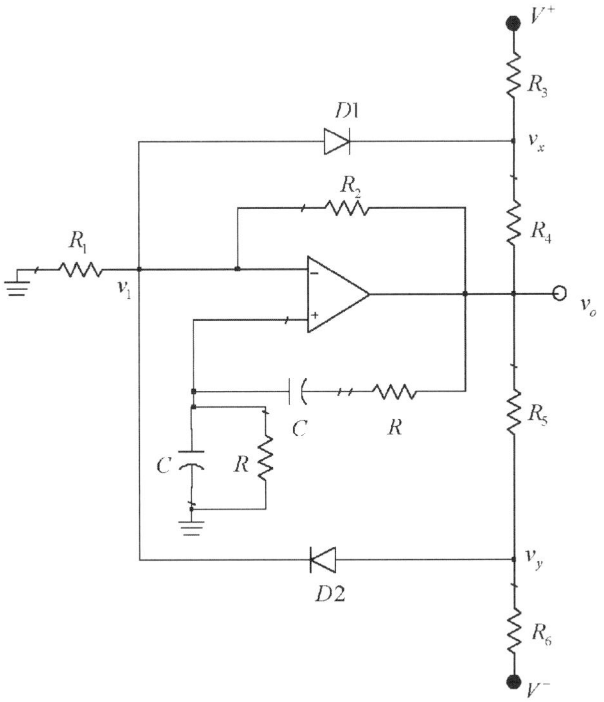

Figure 1.21 Wien-bridge oscillator with an amplitude-limiting circuit.

$$v_x = \frac{v_{o(\min)}}{3} - 0.7 = \frac{v_{o(\min)}R_3}{R_3 + R_4} + \frac{V^- R_4}{R_3 + R_4} \tag{1.32}$$

The simultaneous solution of (1.31) and (1.32) gives the value of the resistors that limit the output voltage to $v_{o(\min)} < v_o < v_{o(\max)}$. In order to obtain a symmetrical sinusoidal voltage, the selection $R_3 = R_6$ and $R_4 = R_5$ is usually made.

Example 1.7

Design an amplitude-limiting circuit for the 5-kHz Wien-bridge oscillator in Example 1.5. The amplitude of the sinusoidal output voltage is to be limited to $|v_o| =$ 5V.

Solution

With $v_{o(\max)} = 5\text{V}$ and $v_{o(\min)} =- 5\text{V}$, it follows from (1.31) and (1.32) that $R_3 = R_6 = 10.9\ \text{k}\Omega$ and $R_4 = R_5 = 2\ \text{k}\Omega$.

To start the oscillation a voltage gain of 3.2 is obtained with $R_2 = 22\ \text{k}\Omega$ and $R_1 = 10\ \text{k}\Omega$. The simulation of the oscillator is shown in Figure 1.22. Obviously, the harmonic content of the oscillator in Figure 1.22 is much less than the one in Figure 1.17.

Another way of stabilizing the amplitude of oscillation is to use a thermistor in the circuit. Thermistors with positive- and negative-temperature coefficients are available. A negative-temperature coefficient thermistor has a resistance that decreases as the temperature increases. For example, in Figure 1.15 a negative-temperature coefficient thermistor can be used for R_2. Hence, as the amplitude of oscillation increases, the current in R_2 increases and its resistance decreases, resulting in a decrease in A_{vo} which makes $\beta(j\omega_o)A_{vo} = 1$.

A positive-temperature coefficient device that can also be used to stabilize the oscillation is a tungsten filament lamp. In low-wattage lamps, cold resistances of a few ohms to hundreds of ohms are available, with hot resistances values being about 5 to 10 times larger than the cold values. The schematic of a Wien-bridge oscillator using a filament lamp is shown in Figure 1.23. The nonlinearities associated with the lamp resistance provide a challenge in the control of v_o.

In oscillators that use op amps the maximum frequency of operation is limited by the frequency response of the op amp. The op amp frequency response is limited by either its gain-bandwidth product or by its slew rate. The gain-bandwidth product is a *small-signal* limitation, which limits the frequency of oscillation to

$$f_o < \frac{f_T}{|A_{vo}|} \tag{1.33}$$

where f_T is the gain-bandwidth product frequency, and A_{vo} is the midband gain of the amplifier. For an inverting amplifier $A_{vo} =- R_2/R_1$, and for a noninverting amplifier $A_{vo} = 1 + R_2/R_1$.

The slew rate is a large-signal limitation, which limits the frequency of oscillation to

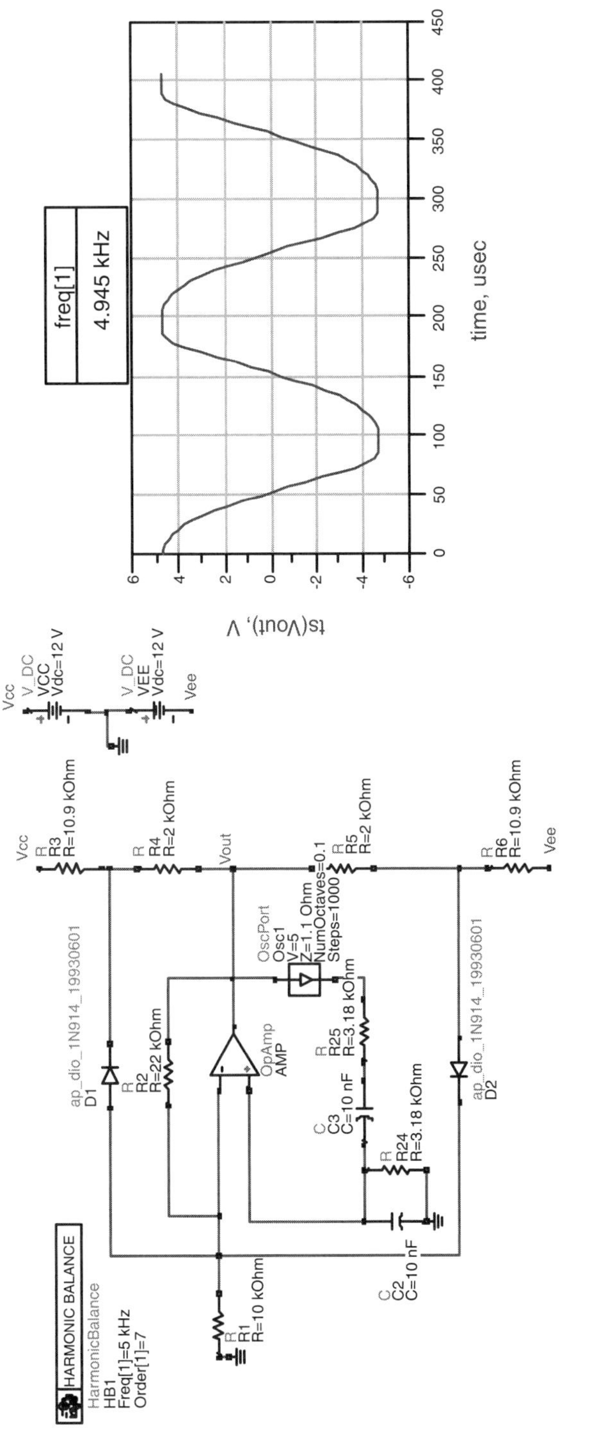

Figure 1.22 ADS simulation of the oscillator in Example 1.7.

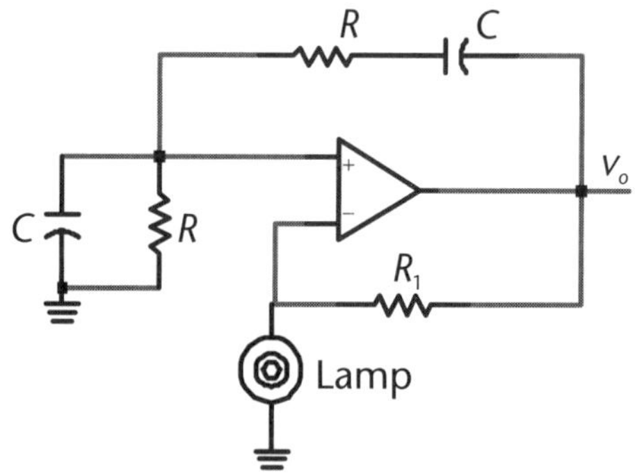

Figure 1.23 A Wien-bridge oscillator using an incandescent lamp for stability.

$$f_o < \frac{SR}{2\pi|v_o|} \tag{1.34}$$

where SR is the op amp slew rate, and $|v_o|$ is the magnitude of the output voltage.

In the absence of an amplitude-limiting circuit, $|v_o|$ is limited by the op amp saturation. A limiting circuit keeps the operation of the op amp in its linear region (i.e., away from saturation). For a designed value of $|v_o|$, the smaller of the two f_o values in (1.33) or (1.34) limits the maximum oscillation frequency. It also follows that larger values of f_o are obtained, if the amplitude of the oscillation is limited using an amplitude limiting circuit.

1.7 The Phase-Shift Oscillator

Phase-shift oscillators usually use RC networks in the feedback path. A phase-shift oscillator is shown in Figure 1.24(a). The op amp is used in an inverting configuration with a gain of $-R_2/R$. Thus, the signal experiences a phase shift of −180° through the amplifier, and the phase shift from each RC section is 60° at the frequency of oscillation, for a total phase shift in the feedback path of 180°. The total phase shift around the closed loop is 0° (i.e., −180° + 180°). If the gain condition is satisfied, the circuit will oscillate at the frequency where the total phase shift is zero.

The phase-shift network is composed of three RC sections. The following analysis provides insight into the behavior of the phase-shift network. The transfer function of the single RC section shown in Figure 1.25 is

$$\frac{v_2}{v_1} = \frac{j\omega RC}{1 + j\omega RC} = \frac{j\dfrac{\omega}{\omega_L}}{1 + j\dfrac{\omega}{\omega_L}}$$

where

$$\omega_L = \frac{1}{RC}$$

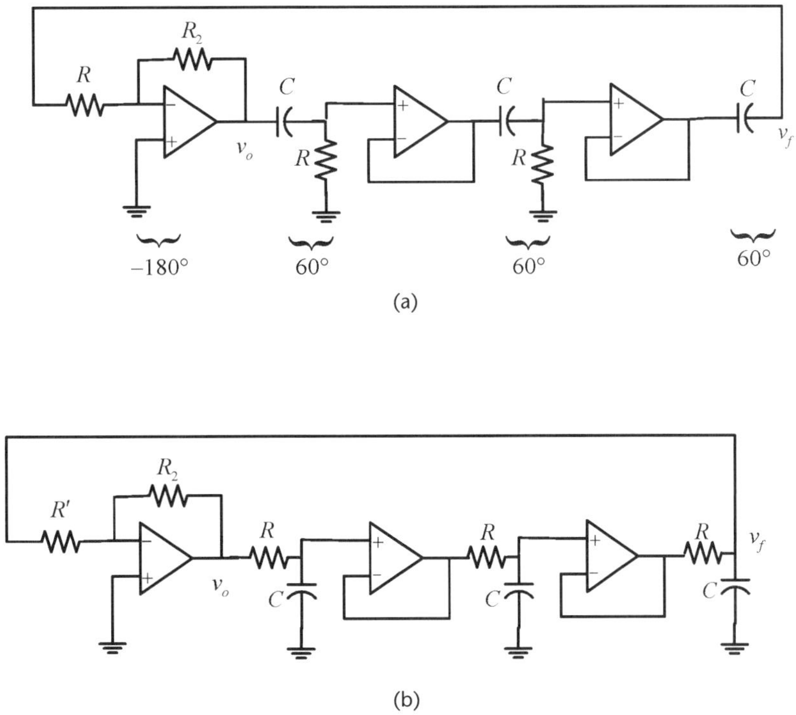

Figure 1.24 (a) A phase-shift oscillator and (b) with RC interchanged.

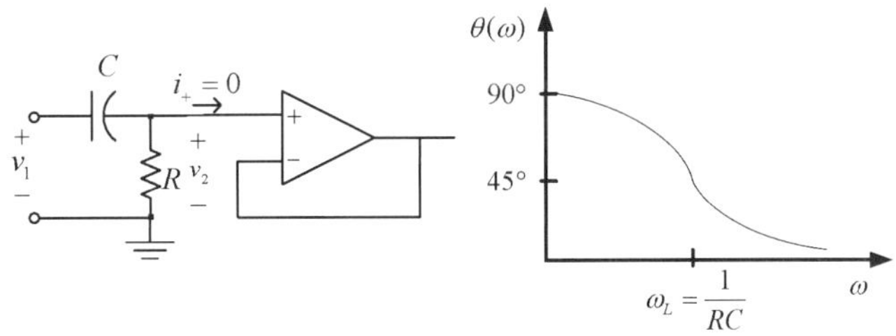

Figure 1.25 A single *RC* section.

Hence, the phase shift is

$$\theta(\omega) = 90° - \tan^{-1} \frac{\omega}{\omega_L} \tag{1.35}$$

Equation (1.35) shows that the largest phase shift that can be obtained from a single RC section is 90°. Two RC sections will have a phase shift smaller than 180°. Therefore, a minimum of three RC sections is needed to obtain a phase shift of 180° at the frequency of oscillation.

The phase-shift network in Figure 1.24(a) is described by the feedback factor, namely,

$$\beta(j\omega) = \frac{v_f}{v_o} = \left(\frac{j\dfrac{\omega}{\omega_L}}{1 + j\dfrac{\omega}{\omega_L}} \right)^3 = \frac{-\left(\dfrac{\omega}{\omega_L}\right)^3}{\dfrac{\omega}{\omega_L}\left[3 - \left(\dfrac{\omega}{\omega_L}\right)^2\right] - j\left[1 - 3\left(\dfrac{\omega}{\omega_L}\right)^2\right]} \quad (1.36)$$

The imaginary part of $\beta(j\omega)$ will vanish at the frequency

$$\omega_o = \frac{1}{\sqrt{3}RC}$$

which is the frequency of oscillation. At the frequency of oscillation it follows from (1.35) that each RC section produces a phase shift of 60°.

From (1.36) the real part of $\beta(j\omega)$ at ω_o is $\beta_r = -1/8$. Therefore, from (1.6), the gain condition is

$$A_{vo} = \frac{1}{\beta_r} = -8$$

Hence, it follows that

$$\frac{R_2}{R} = 8$$

The phase-shift oscillator in Figure 1.24(a), implemented with R and C interchanged in each RC section, is shown in Figure 1.24(b). The resistor R should be large so that it does not load the third RC section. Of course, a buffer amplifier (i.e., a unity gain amplifier) can be used between the third section and the resistor R to avoid the loading. In the configuration shown in Figure 1.24(b) the loop gain is

$$\beta(j\omega)A_{vo} = \frac{A_{vo}}{\left(1 + j\dfrac{\omega}{\omega_L}\right)^3}$$

Hence, it follows that the frequency of oscillation is given by

$$\omega_o = \frac{\sqrt{3}}{RC}$$

and the gain condition is satisfied with $R_2/R' = 8$.

Another phase-shift oscillator is shown in Figure 1.26. In this oscillator the RC sections are connected without isolation and, therefore, there is loading. In the last stage the resistors R and R_1 appear in parallel. The loading of R_1 can be neglected if $R_1 || R \approx R$, or in some cases by removing R in the third stage and letting $R_1 = R$.

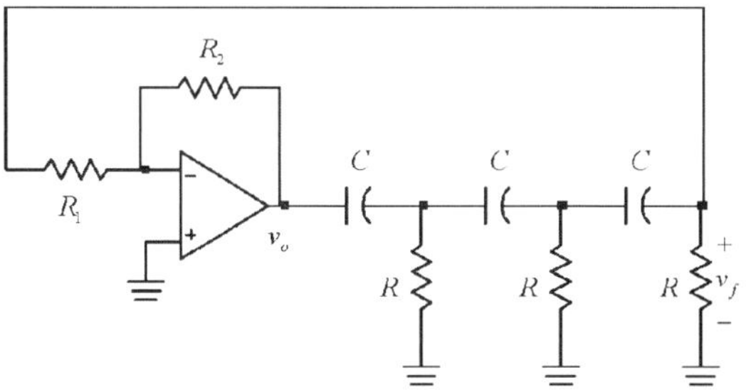

Figure 1.26 A phase-shift oscillator using a single op amp.

The analysis of the phase-shift network in Figure 1.26 with $R_1||R \approx R$ gives

$$\beta(j\omega) = \frac{v_f}{v_o} = \frac{1}{\left(\frac{1}{j\omega RC}\right)^3 + 5\left(\frac{1}{j\omega RC}\right)^2 + 6\left(\frac{1}{j\omega RC}\right) + 1} \tag{1.37}$$

The imaginary part of $\beta(j\omega)$ comes from the first and third term in the denominator, namely the odd-power terms. The imaginary part of $\beta(j\omega)$ will vanish when

$$\left(\frac{1}{j\omega RC}\right)^3 + 6\left(\frac{1}{j\omega RC}\right) = 0$$

or at

$$\omega = \omega_o = \frac{1}{\sqrt{6}RC} \tag{1.38}$$

At the frequency of oscillation the phase-shift network produces a phase shift of 180°.

From (1.37) the real part of $\beta(j\omega)$ at ω_o is

$$\beta_r(\omega_o) = \frac{1}{\frac{5}{(j\omega_o RC)^2} + 1} \tag{1.39}$$

Substituting (1.38) into (1.39) gives

$$\beta_r(\omega_o) = \frac{1}{5(-6) + 1} =- \frac{1}{29}$$

and from (1.6), the gain condition is

$$A_{vo} = \frac{1}{\beta_r(\omega_o)} =- 29 \tag{1.40}$$

In Figure 1.26 the resistors R_1 and R_2 provide an inverting gain of

$$A_{vo} = \frac{v_o}{v_f} = -\frac{R_2}{R_1}$$

Hence, the start of oscillator condition is satisfied if R_1 and R_2 are selected to provide $|A_{vo}| > 29$.

To summarize, the phase-shift oscillator in Figure 1.26 will oscillate at the frequency ω_o given by (1.38) if the gain is $|A_{vo}| > 29$. The loading of the op amp is minimized by making $R_1 > 10R$.

The phase-shift oscillator in Figure 1.26 could have been implemented by interchanging R and C in the phase-shift sections. It follows that for such an oscillator the frequency of oscillation is given by

$$\omega_o = \frac{\sqrt{6}}{RC}$$

and the gain condition is $|A_{vo}| > 29$.

Example 1.8

Design the RC oscillator shown in Figure 1.26 to oscillate at 1 kHz.

Solution

Selecting the capacitor values to be 0.1 μF, then from (1.38)

$$R = \frac{1}{\omega_o \sqrt{6} C} = \frac{1}{(2\pi \times 10^3)\sqrt{6}(0.1 \times 10^{-6})} = 650\Omega$$

The resistors R_1 and R_2 must provide the gain $|A_{vo}| > 29$ in order to prevent loading $R_1 || R \approx R$. Letting $R_1 = 15$ kΩ, then $R_2 = 29R_1 = 435$ kΩ. A 495-kΩ resistor was used to implement R_2. This will allow for some extra gain to satisfy the start of oscillation condition (i.e., $|A_{vo}| > 29$).

This oscillator is easy to construct. However, some adjustments in the RC network (due to component variations) are necessary in order to obtain the desired frequency of oscillation. The use of precision resistors and capacitors is recommended. The simulation of this oscillator is shown in Figure 1.27. The fundamental frequency of oscillation is 1.006 kHz. Observe that the output voltage reaches saturation (i.e., $|v_o| \approx 11$V), since the circuit uses the nonlinearities of the operational amplifier at saturation to determine the amplitude of oscillation. The clipping in the output waveform produces harmonic distortion.

The harmonic distortion can be significantly reduced with an amplitude-limiting circuit. One such oscillator circuit is shown in Figure 1.28. The amplitude-limiting circuit is designed using (1.30) and (1.32), with v_1 set equal to zero (i.e., $v_y = 0.7$V and $v_x = 0.7$V).

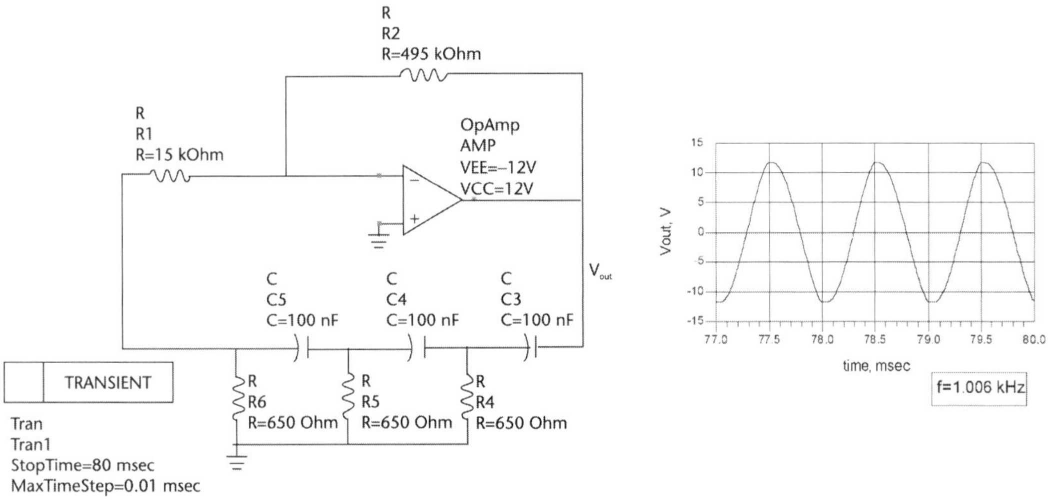

Figure 1.27 ADS simulation of the phase-shift oscillator in Example 1.8.

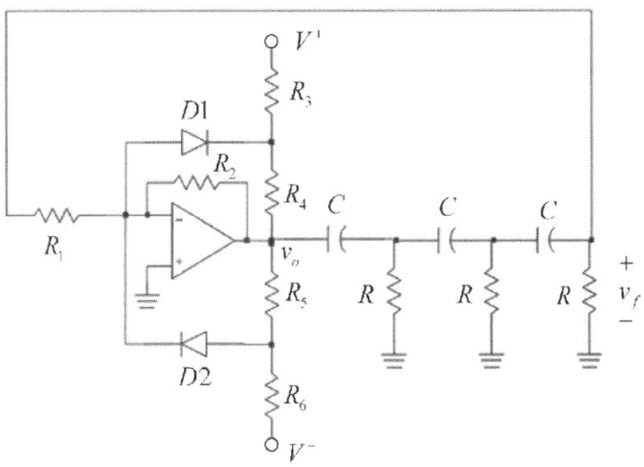

Figure 1.28 A phase-shift oscillator with an amplitude-limiting circuit.

Example 1.9

Design the phase-shift oscillator in Figure 1.28 to oscillate at 1 kHz. The amplitude of the output voltage is to be limited to ±5V.

Solution

The design of the amplitude-limiting circuit is similar to the one in Example 1.7. From (1.30), with $v_1 = 0$ and $v_{o(\max)} = 5$V, we obtain

$$0.7 = \frac{5R_6}{R_5 + R_6} - \frac{12R_5}{R_5 + R_6}$$

which can be satisfied with $R_5 = 2$ kΩ and $R_6 = 5.9$ kΩ. From (1.32), for symmetry, we obtain $R_3 = R_6 = 5.9$ kΩ and $R_4 = R_5 = 2$ kΩ.

The simulation is shown in Figure 1.29. The fundamental frequency of oscillation is at 1 kHz, and the total harmonic content of this oscillator is certainly less than that in Figure 1.27.

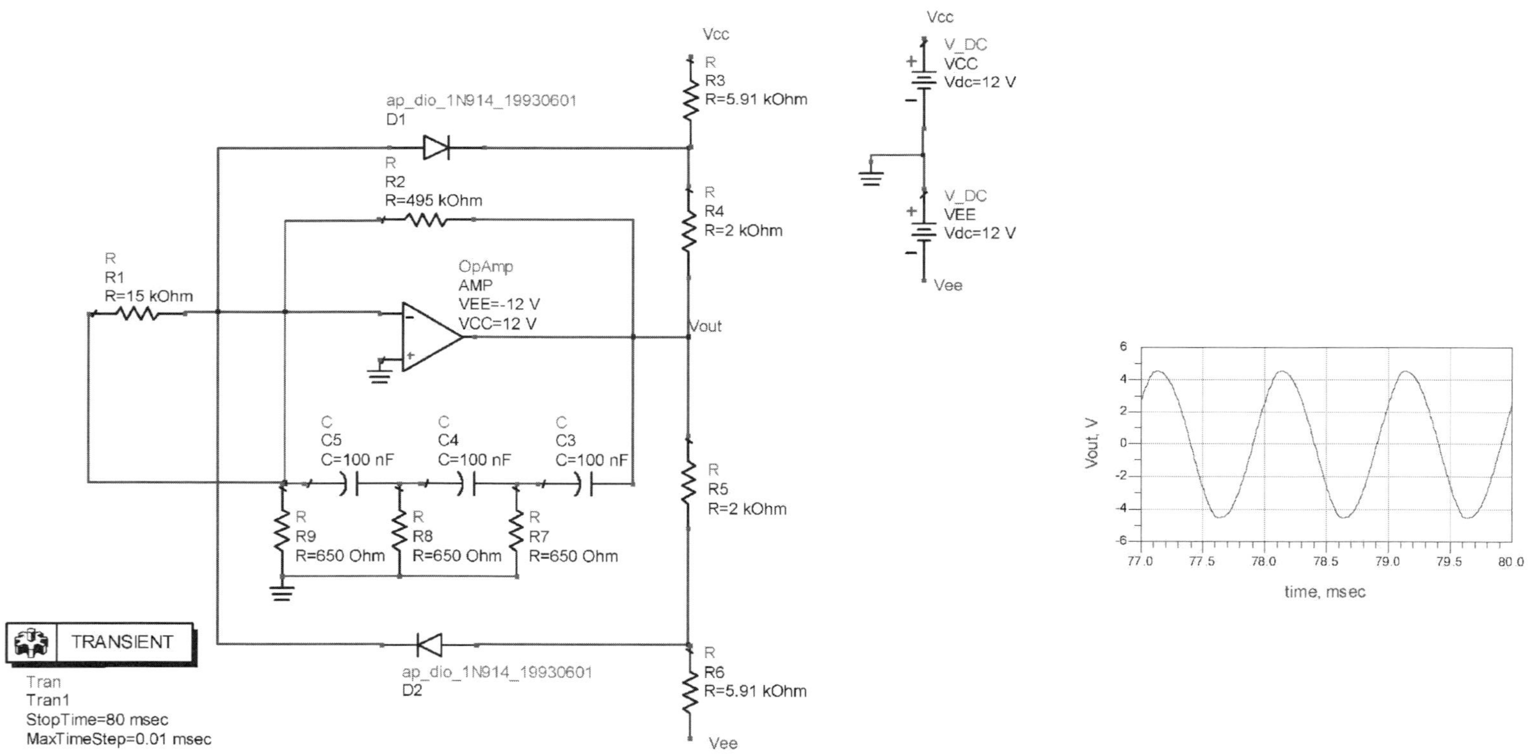

Figure 1.29 Phase-shift oscillator with an amplitude-limiting circuit.

An interesting phase-shift oscillator is shown in Figure 1.30. It is composed of three integrators with transfer function of

$$A_v(j\omega) = \frac{-\dfrac{1}{R_1 C}}{j\omega + \dfrac{1}{R_2 C}}$$

In this oscillator the gain and phase shift is distributed throughout the loop. The loop gain is

$$\beta A_v = \left(\frac{-\dfrac{1}{R_1 C}}{j\omega + \dfrac{1}{R_2 C}} \right)^3 = \frac{-\left(\dfrac{R_2}{R_1}\right)^3}{[1 - 3(\omega R_2 C)^2] + j\omega R_2 C[3 - (\omega R_2 C)^2]} \tag{1.41}$$

From (1.41) the imaginary part vanishes at the frequency

$$\omega_o = \frac{\sqrt{3}}{R_2 C}$$

and the gain condition follows from setting the loop gain at $\omega = \omega_o$ equal to unity, namely,

$$\beta A_v = \frac{1}{8}\left(\frac{R_2}{R_1}\right)^3 = 1$$

or $R_2 = 2R_1$.

In this oscillator the phase shift from each integrator is 120° at ω_o. Since at ω_o the gain of each stage is $1\underline{|120^\circ}$, the three output voltages are equal in magnitude but at 120° from each other.

A quadrature oscillator produces two signals having a phase shift of 90°. Some phase-shift oscillators can produce signals in quadrature. For example, an additional

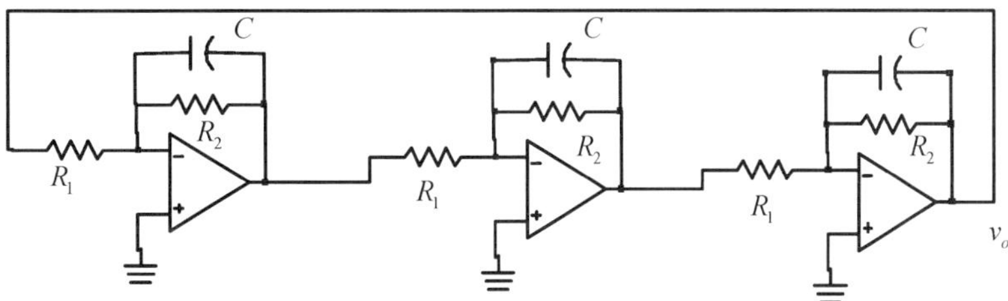

Figure 1.30 A phase-shift oscillator using op amp integrators.

RC section could have been added to the phase-shift oscillators in Figure 1.24(b), so that each section produces a phase shift of –45°, or –90° every two sections. The resulting quadrature oscillator that produces a sine and a cosine signal is shown in Figure 1.31. The amplifier gain is

$$A_{vo} = \frac{v_o}{v_f} = -\frac{R_2}{R'}$$

and the RC phase-shift network transfer function is

$$\beta(j\omega) = \frac{v_f}{v_o} = \frac{1}{\left(1 + j\frac{\omega}{\omega_L}\right)^4} \tag{1.42}$$

Hence, the loop gain is given by

$$\beta(j\omega)A_{vo} = \frac{-\frac{R_2}{R'}}{\left(1 + j\frac{\omega}{\omega_L}\right)^4} \tag{1.43}$$

It is observed that at

$$\omega_o = \omega_L = \frac{1}{RC}$$

each RC section produces a phase shift of –45°, for a total of –180°. Hence, the loop-gain phase shift is –360°. At ω_o, (1.42) gives

$$\beta(j\omega) = \frac{1}{(1 + j)^4} = \frac{1}{4}e^{-j\pi}$$

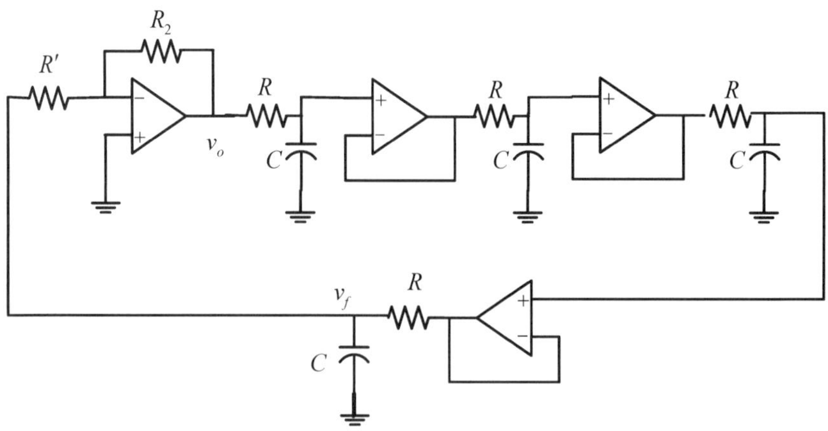

Figure 1.31 A quadrature oscillator.

Therefore, from (1.43) the required gain is

$$A_{vo} =-4 \Rightarrow \frac{R_2}{R'} = 4$$

A gain greater than 4 is required to start the oscillation (say, $A_{vo} = 4.2$).

Another quadrature oscillator is shown in Figure 1.32. The second op amp is connected as an inverting integrator; that is,

$$A_{v2} = \frac{v_{o2}}{v_{o1}} =- \frac{1}{j\omega R_2 C_2} \tag{1.44}$$

This inverter produces a phase shift of –270° (or 90°).

The first op amp is connected as a noninverting integrator; that is,

$$A_{v1}(j\omega) = \frac{v_{o1}}{v_f} = 1 + \frac{1}{j\omega R_1 C_1} = \frac{1 + j\omega R_1 C_1}{j\omega R_1 C_1} \tag{1.45}$$

and the feedback network transfer function is

$$\beta(j\omega) = \frac{v_f}{v_{o2}} = \frac{1}{1 + j\omega R_3 C_3} \tag{1.46}$$

Therefore, from (1.45) and (1.46), with $R_1 C_1 = R_3 C_3$, we obtain

$$\frac{v_{o1}}{v_{o2}} = \frac{1}{j\omega R_1 C_1} \tag{1.47}$$

which shows that the phase shift from v_{o2} to v_{o1} is –90°. From (1.44) and (1.47) the loop phase shift is 0° and oscillations occur with v_{o2} being a sine signal and v_{o1} a cosine signal.

Using (1.44), (1.45), and (1.46), the loop-gain condition is

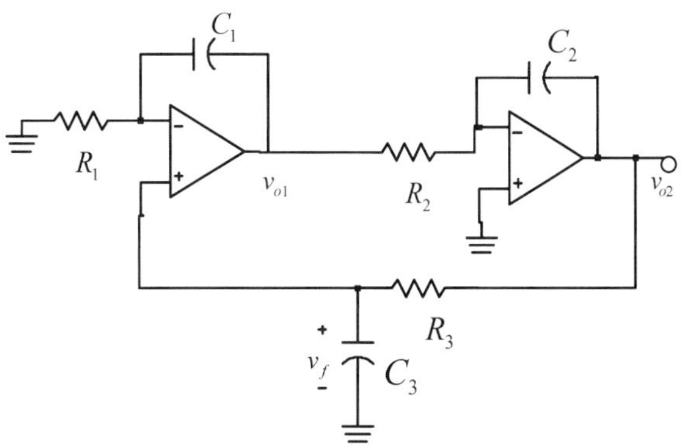

Figure 1.32 A quadrature oscillator.

$$\beta A_{v1}A_{v2} = -\frac{1 + j\omega R_1C_1}{j\omega R_1C_1}\left(\frac{1}{j\omega R_2C_2}\right)\frac{1}{1 + j\omega R_3C_3} = 1$$

If $R_1C_1 = R_2C_2 = R_3C_3$, the loop-gain condition reduces to

$$\beta A_{v1}A_{v2} = \frac{1}{(\omega R_1C_1)^2} = 1$$

and the frequency of oscillation is

$$\omega_o = \frac{1}{R_1C_1}$$

In a practical quadrature oscillator circuit, the component variations affect the required RC equality, and therefore, some tuning mechanism is needed, such as a variable potentiometer to implement the resistors. In addition, an amplitude-limiting circuit might be needed to reduce the distortion.

A phase-shift oscillator using a JFET amplifier is shown in Figure 1.33. The input resistance of the amplifier (i.e., R_G) is very large and does not load the output of the phase-shift network. However, the output resistance of the amplifier is $r_d||R_D \approx R_D$ and a certain amount of loading occurs. Good results are obtained by neglecting the loading and designing the oscillator based on (1.38) and (1.40), where $A_{vo} \approx -g_mR_D$.

A phase-shift oscillator using a BJT amplifier is shown in Figure 1.34(a). The BJT phase-shift oscillator works well at a frequency below 1 MHz. This is the current-controlled equivalent of the JFET phase-shift oscillator. Since the BJT is a current-controlled device, the oscillation frequency and required gain can be obtained using (1.10). The ac model for this circuit is shown in Figure 1.34(b), where it was assumed that $R_1||R_2 \gg h_{ie}$ and $1/h_{oe} = \infty$. While an approximate

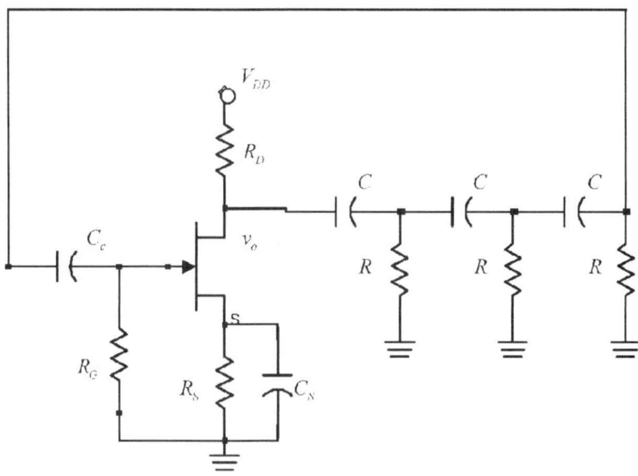

Figure 1.33 A phase-shift oscillator using a JFET amplifier.

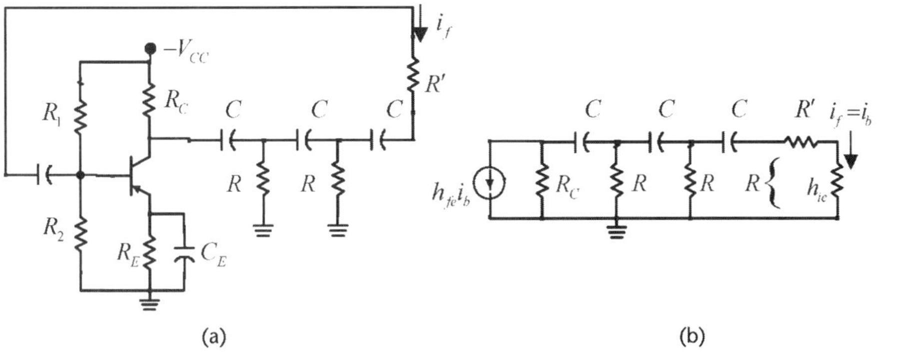

Figure 1.34 (a) A phase-shift oscillator using a BJT amplifier and (b) the phase-shift network.

design can be made by neglecting the effects of R_C and h_{ie} and the use of a CAD program to optimize the oscillator, it is of interest to show the complexities in the analysis due to their presence in the ac model.

In Figure 1.34(b) the resistor R' is selected so that $R = R' + h_{ie}$. The feedback current is the current into the base of the transistor ($i_f = i_b$), and the input current ($h_{fe} i_b$) is the collector current. From Figure 1.34(b) it can be shown that the loop gain is

$$\beta A_v = \frac{i_f}{i_b} = \frac{-h_{fe}}{3 - \dfrac{1}{\omega^2 R^2 C^2} + \dfrac{R}{R_c} - \dfrac{5}{\omega^2 R R_c C^2} - j\dfrac{4}{\omega RC} - j\dfrac{6}{\omega R_c C} + j\dfrac{1}{\omega^3 R^2 R_c C^3}} \tag{1.48}$$

Setting the imaginary part of (1.48) equal to zero gives the frequency of oscillation, namely,

$$\omega_o = \frac{1}{RC\sqrt{6 + \dfrac{4R_c}{R}}} \tag{1.49}$$

At ω_o, setting the loop gain in (1.48) equal to unity gives

$$\frac{-h_{fe}}{3 - \dfrac{1}{\omega_o^2 R^2 C^2} + \dfrac{R}{R_c} - \dfrac{5}{\omega_o^2 R R_c C^2}} = 1 \tag{1.50}$$

Substituting (1.49) into (1.50) results in the following gain condition:

$$h_{fe} = 23 + 29\frac{R}{R_c} + 4\frac{R_c}{R} \tag{1.51}$$

This equation can be solved for R/R_c as a function of h_{fe} to determine the minimum value of R/R_c for oscillation. Alternatively, this value can be determined by differentiating (1.51) with respect to R/R_c and setting the result equal to zero. That is,

$$29 - 4\left(\frac{R_c}{R}\right)^2 = 0 \Rightarrow \frac{R}{R_c} = \sqrt{\frac{4}{29}} = 0.372$$

With $R/R_c = 0.372$, (1.49) gives $h_{fe} = 44.5$. This is the minimum value of h_{fe} for oscillations. For $h_{fe} = 44.5$ the circuit will oscillate at the frequency given by (1.49). In practical circuits, the transistors' h_{fe} are much larger than 44.5, and the tolerances in the resistors and the capacitors usually makes them unequal (especially in the capacitors). Hence, a variable resistor can be used for R_c, and its value changed until the loop gain is slightly larger than unity and the circuit oscillates at the desired frequency with low distortion. The frequency of oscillation will still be closely predicted by (1.49). Another practical way of building this oscillator is to use a variable resistor for R'.

The RC phase-shift network is one specific case of the general phase-shift network shown in Figure 1.35. It can be shown that for this network

$$\beta(j\omega) = \frac{v_f}{v_o} = \frac{1}{\left(\frac{Z_1}{Z_2}\right)^3 + 5\left(\frac{Z_1}{Z_2}\right)^2 + 6\left(\frac{Z_1}{Z_2}\right) + 1} \tag{1.52}$$

Equation (1.52) shows that oscillations can be obtained with other combinations of Z_1 and Z_2, provided that $\beta_i(\omega) = 0$. For example, Z_1 can represent a resistor and Z_2 an inductor. Of course, the frequency of oscillation depends on the impedances used. Observe that Z_1 and Z_2 cannot both be reactive, because in such case $\beta(j\omega)$ will not have an imaginary part.

1.8 Active-Filter Oscillators

Many oscillators that use an active-filter feedback path have been proposed. Basically, in these oscillators the poles of an active filter are adjusted to lie on the $j\omega$-axis at the desired frequency of oscillation. To start the oscillation the poles must be slightly on the right-half plane. A summary of these types of oscillators can be found in [2].

In this section three active-filter oscillators are analyzed to illustrate their design procedure. A simple second-order low-pass filter with positive feedback is shown in Figure 1.36. For this oscillator

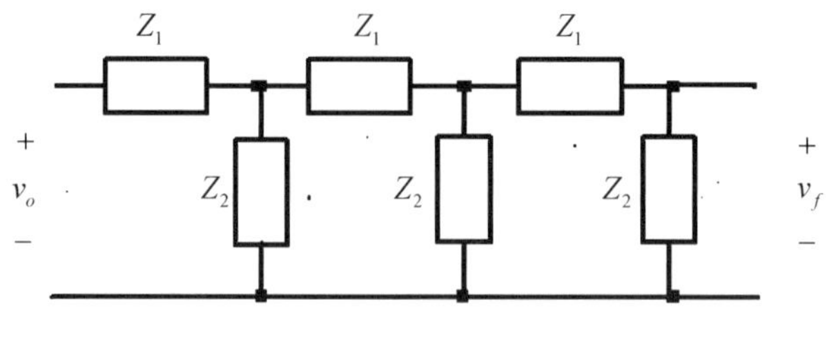

Figure 1.35 A general phase-shift network.

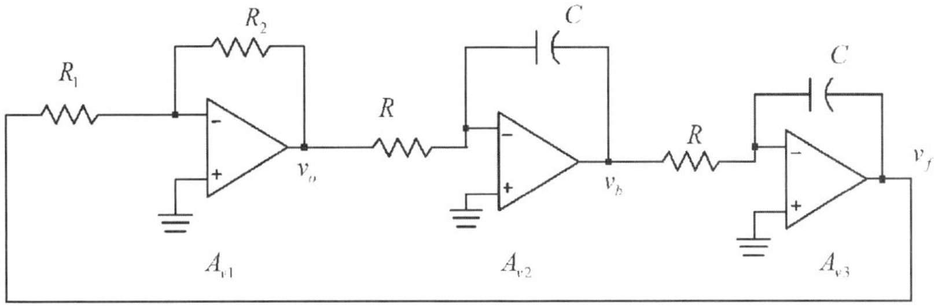

Figure 1.36 A simple active-filter oscillator.

$$A_{v1} = \frac{v_o}{v_f} =- \frac{R_2}{R_1}$$

$$A_{v2} = \frac{v_b}{v_o} =- \frac{1}{j\omega RC}$$

and

$$A_{v3} = \frac{v_f}{v_b} =- \frac{1}{j\omega RC}$$

Therefore, the loop gain is

$$\beta A_{v1} A_{v2} A_{v3} =- \left(\frac{R_2}{R_1}\right)\left(\frac{1}{\omega RC}\right)^2$$

and the condition for oscillation (i.e., loop gain equal to one) is satisfied at

$$\omega_o = \frac{1}{RC}\sqrt{\frac{R_2}{R_1}}$$

Next we consider the use of bandpass filters in the implementation of oscillators. Figure 1.37 illustrates an oscillator that uses a bandpass filter in the feedback loop. The voltage transfer function for this filter is

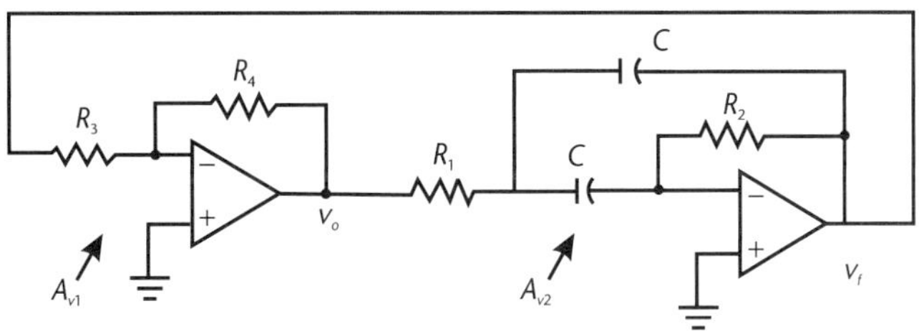

Figure 1.37 An oscillator using a bandpass filter.

$$A_{v2} = \frac{v_f}{v_o} = \frac{-\left(\frac{1}{R_1 C}\right)s}{s^2 + \frac{2}{R_2 C}s + \frac{1}{R_1 R_2 C^2}} = \frac{-2Q\omega_o s}{s^2 + \left(\frac{\omega_o}{Q}\right)s + \omega_o^2}$$

where

$$Q = \frac{1}{2}\sqrt{\frac{R_2}{R_1}}$$

and

$$\omega_o = \frac{1}{C\sqrt{R_1 R_2}} \tag{1.53}$$

The loop gain is

$$\beta A_{v1} A_{v2} = \left(\frac{R_4}{R_3}\right)\frac{2Q\omega_o \omega}{\omega\left(\frac{\omega_o}{Q}\right) + j\left(\omega^2 - \omega_o^2\right)}$$

which shows that the frequency of oscillation is given by (1.53), and the gain condition is

$$\frac{R_4}{R_3}2Q^2 = 1 \Rightarrow \frac{R_4}{R_3} = 2\frac{R_1}{R_2}$$

An oscillator configuration that uses a Twin-T filter in the feedback path is shown in Figure 1.38. The Twin-T filter consists of two Tee-shaped networks

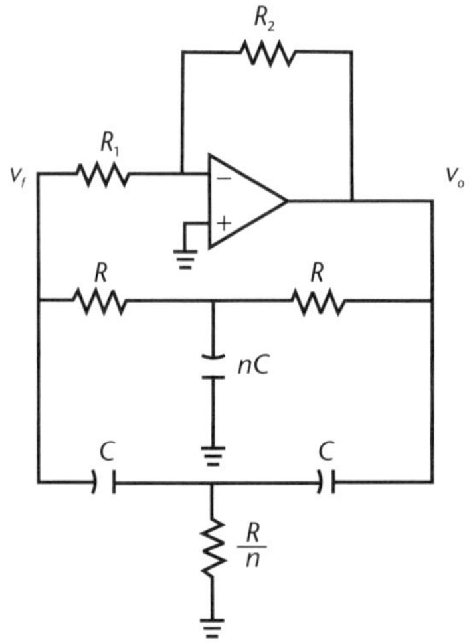

Figure 1.38 An oscillator using a Twin-T filter.

connected in parallel. These Twin-T filters are also known as band-reject filters or notch filters. In the Twin-T filter shown in Figure 1.38, the elements connected to ground have values of nC and R/n, respectively.

The transfer function of the Twin-T filter in Figure 1.38 is

$$\beta(s) = \frac{v_f}{v_o} = \frac{s^2 + \left(\frac{2}{n} - 1\right)\omega_o^2}{s^2 + \omega_o\left(n + \frac{2}{n} + 1\right)s + \omega_o^2} \tag{1.54}$$

where

$$\omega_o = \frac{1}{RC} \tag{1.55}$$

and n is a number such that $n \geq 2$.

Observe that if $n = 2$, (1.54) reduces to

$$\beta(s) = \frac{s^2 + \omega_o^2}{s^2 + 4\omega_o s + \omega_o^2} \tag{1.56}$$

which is the transfer function of a Twin-T filter commonly used as notch filter. Of course, the transfer function in (1.56) has a zero at ω_o, which is responsible for the "deep" notch of this filter.

The gain of the op amp is $A_{vo} =- R_2/R_1$. Setting the loop gain of the oscillator equal to 1 gives

$$-A_{vo}\frac{s^2 + \left(\frac{2}{n} - 1\right)\omega_o s + \omega_o^2}{s^2 + \omega_o\left(n + \frac{2}{n} + 1\right)s + \omega_o^2} = 1$$

or

$$(1 - A_{vo})s^2 + \left[n + \frac{2}{n} + 1 - A_{vo}\left(\frac{2}{n} - 1\right)\right]\omega_o s + (1 - A_{vo})\,\omega_o^2 = 0$$

Therefore, the required gain is

$$A_{vo} =- \frac{n + \frac{2}{n} + 1}{1 - \frac{2}{n}} \tag{1.57}$$

and the frequency of oscillation is given by (1.55). Equation (1.57) shows that an oscillator designed with $n = 2$ in the Twin-T filter requires an infinite gain. Hence, for oscillator purposes $n \geq 2.5$ is more practical.

Example 1.10

Design the Twin-T filter in Figure 1.38 to oscillate at $\omega_o = 1$ krad/s (or $f_o = 159$ Hz) with $n = 4$.

Solution

Letting $R = 10\Omega$, it follows from (1.55) that $C = 0.1\ \mu$F. From (1.57), the value of A_{vo} with $n = 4$ is $A_{vo} = -11$. To start the oscillation we let $R_2/R_1 > 11$, or $R_1 = 50$ kΩ and $R_2 = 660$ kΩ. The op amp supply voltages are ±15V.

The simulation of the oscillator is shown in Figure 1.39. The resulting oscillation is at 161.9 Hz. The notch associated with the Twin-T filter (high Q) makes the resulting oscillation fairly clean of harmonics.

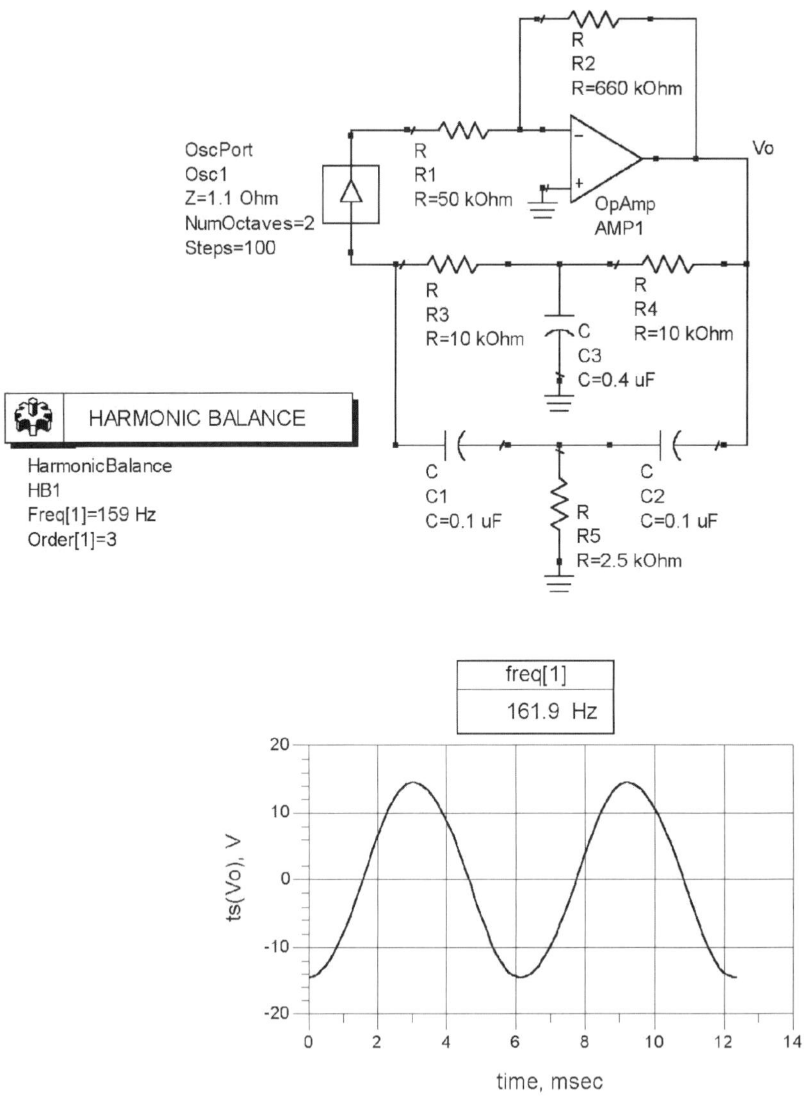

Figure 1.39 ADS simulation for the Twin-T oscillator in Example 1.10.

References

[1] Lindberg, E., "Oscillators—An Approach for a Better Understanding," *Proceedings of the 2003 European Conference on Circuit Theory and Design,* Krakow, Poland, 2003.

[2] Lindquist, C. S., *Active Network Design with Signal Filtering Applications,* Santa Cruz, CA: Steward & Sons, 1977.

CHAPTER 2

Oscillator Characteristics

2.1 Introduction

This chapter discusses several oscillators' characteristics that apply to all types of oscillators. Oscillators can have frequency variations, which can be classified as long term and short term. Long-term frequency stability problems can be due to component variation or aging, while short-term frequency variation can be caused by a variety of factors, such as mechanical vibrations, parasitic reactances, and random noise.

The spectral characteristics of an oscillator usually show that, in addition to the frequency of oscillation, a certain amount of harmonics is also present due to the nonlinear characteristics of the oscillator. The harmonic contents can range from severe to minimal

There are several sources that contribute to the oscillator noise, namely thermal noise, shot noise, and flicker noise. The theory of oscillator noise is presented, and the single sideband noise of an oscillator is discussed. Techniques for measuring the phase noise of an oscillator are described.

2.2 Frequency Stability

Frequency stability or phase stability refers to the ability of the oscillator to maintain a constant frequency of oscillation. Several terms are used when referring to the frequency stability, such as long-term and short-term stability. Long-term frequency stability is the ability of the oscillator to maintain a frequency of oscillation that does not depend on changes in temperature, aging of components, and so on. Those frequency variations that occur over a long period of time result from changes in the components that directly affect the frequency of oscillation ω_O. For example, in an oscillator where

$$\omega_O = \frac{1}{\sqrt{LC}}$$

the variation in ω_O due to changes in L and C (which can be produced by temperature variations or aging) is

$$\Delta\omega_O = \frac{\partial \omega_O}{\partial L}\Delta L + \frac{\partial \omega_O}{\partial C}\Delta C$$

or

$$\Delta\omega_o = -\frac{1}{2}\omega_o\left(\frac{\Delta L}{L} + \frac{\Delta C}{C}\right)$$

This equation shows how variations in L (i.e., ΔL) and in C (i.e., ΔC) produce variations in the frequency of oscillation $\Delta\omega_o$. For example, a 1% increase in C produces a –0.5% variation (i.e., a 0.5% decrease) in ω_o. The long-term stability of an oscillator is predictable and is commonly expressed as a frequency drift. That is, as a change in frequency per unit time.

Short-term stability describes the oscillator's frequency variation over a short period of time. Short-term frequency variations are usually produced by noise, mechanical vibrations, and temperature-dependent parasitic reactances. For example, the reactances of the active device are functions of temperature and bias variations, and they can introduce or affect the location of the poles and zeroes of the loop gain. These parasitic poles and zeroes introduce an additional phase shift (i.e., $\Delta\phi$), and the frequency of oscillation must shift an amount $\Delta\omega$ so that the loop gain counteracts with a phase shift of $-\Delta\phi$. Obviously, the larger the phase variation of the loop gain, the greater the frequency variation (usually seconds).

Consider Figure 2.1 where the phase response of two oscillators' loop gains are shown, denoted by ϕ_1 and ϕ_2, respectively. At ω_o the phase shift is –360° (or 0°) for both loop gains. If parasitic reactances affect the loop gain by suddenly introducing a phase shift $\Delta\phi$, then the loop gain adjusts (i.e., the frequency changes) to supply a phase shift of $\Delta\phi$, so that the overall phase shift stabilizes at –360° (or 0°). This is accomplished by a decrease in frequency $\Delta\omega_1$ in $\beta_1(j\omega)A_{v1}(j\omega)$ and by $\Delta\omega_2$ in $\beta_2(j\omega)A_{v2}(j\omega)$, as shown in Figure 2.1. For example, if the amplifier in an oscillator introduces an instantaneous phase shift of 1°, the operating frequency will change until the phase shift in the feedback circuit changes by –1°.

From Figure 2.1 it is seen that the phase response ϕ_2 has better frequency stability than ϕ_1, since $\Delta\omega_1 > \Delta\omega_2$. That is, the frequency variation associated with the loop gain $\beta_1(j\omega)A_{v1}(j\omega)$ is greater than the frequency variation associated with $\beta_2(j\omega)A_{v2}(j\omega)$. The frequency stability can be measured in terms of the phase variation $d\phi/d\omega$. A frequency stability factor S_F is defined as the change in phase divided by the normalized change in frequency from ω_o (i.e., $\Delta\omega/\omega_o$). Hence,

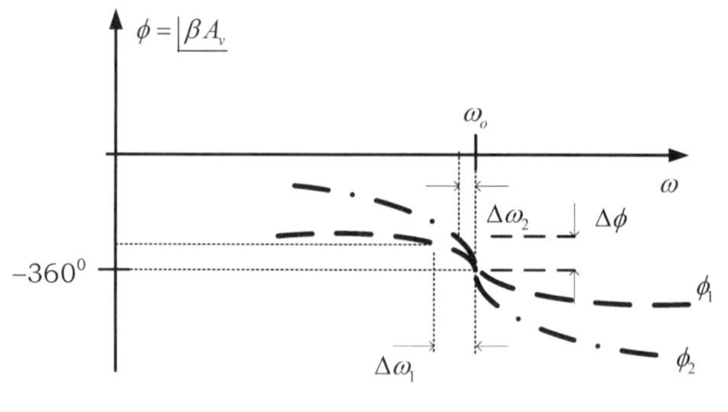

Figure 2.1 Phase response of two loop gain functions.

$$S_F = \frac{\Delta\phi}{\left(\dfrac{\Delta\omega}{\omega_O}\right)} = \omega_O \left.\frac{d\phi}{d\omega}\right|_{\omega=\omega_o} \tag{2.1}$$

The factor S_F provides a qualitative way for comparing the stability of oscillators. The larger the value of S_F, the smaller its frequency variation, since

$$\Delta\omega = \frac{\omega_O \Delta\phi}{S_F}$$

The expression (2.1) can be evaluated for the phase responses commonly found in oscillator circuits. To this end we consider next the frequency responses of a parallel resonant circuit and a series resonant circuit.

A parallel resonant circuit is shown in Figure 2.2, where g_m represents the transconductance of the active device. The output voltage is given by

$$v_o = -(g_m v) Z_R$$

where Z_R is the input impedance of the resonator, namely,

$$Z_R = \frac{1}{\dfrac{1}{sL} + sC + \dfrac{1}{R}} = \frac{\dfrac{s}{C}}{s^2 + \dfrac{1}{RC}s + \dfrac{1}{LC}} \tag{2.2}$$

The voltage gain is

$$A_v = \frac{v_o}{v} = -g_m Z_R \tag{2.3}$$

Equations (2.2) and (2.3) show that the characteristics of the parallel resonant circuit are determined by the frequency response of its input impedance.

Defining

$$\omega_O = \frac{1}{\sqrt{LC}} \tag{2.4}$$

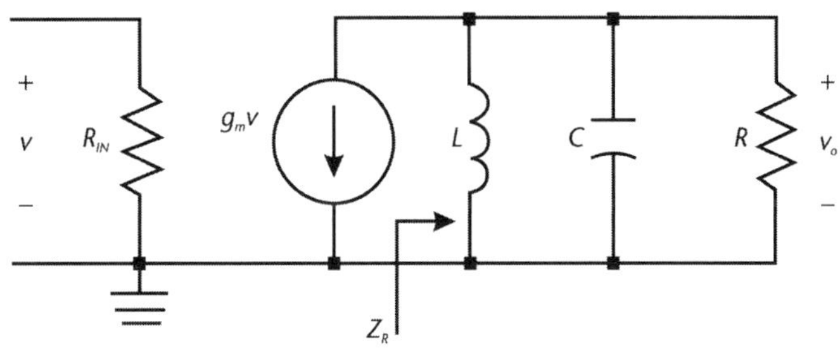

Figure 2.2 A parallel resonant circuit.

and

$$2\alpha = \frac{1}{RC} \tag{2.5}$$

(2.2) can be written in the form

$$Z_R = \frac{\frac{s}{C}}{s^2 + 2\alpha s + \omega_o^2} \tag{2.6}$$

Letting $s = j\omega$, (2.6) can be expressed in the form

$$Z_R = \frac{R}{1 + j\frac{\omega_o}{2\alpha}\left(\frac{\omega}{\omega_o} - \frac{\omega_o}{\omega}\right)} \tag{2.7}$$

From (2.7), at the frequency ω_o the input impedance is R, which shows that at ω_o the RLC circuit appears as a pure resistor because parallel resonance occurs (i.e., the impedance of the parallel combination of L and C is infinite). Obviously, the frequency ω_o given by (2.4) is the resonant frequency of the circuit.

The impedance of the resonator in (2.7) can be expressed in the form

$$Z_R = |Z_R(j\omega)| \underline{/Z_R(j\omega)}$$

where the magnitude of the impedance is

$$|Z_R(j\omega)| = \frac{R}{\sqrt{1 + \left[\frac{\omega_o}{2\alpha}\left(\frac{\omega}{\omega_o} - \frac{\omega_o}{\omega}\right)\right]^2}} \tag{2.8}$$

and the phase is

$$\phi(j\omega) = -\tan^{-1}\left[\frac{\omega_o}{2\alpha}\left(\frac{\omega}{\omega_o} - \frac{\omega_o}{\omega}\right)\right] \tag{2.9}$$

Typical plots of (2.8) versus ω for two values of $\omega_o/2\alpha$ are shown in Figure 2.3(a). It is observed that as the ratio of ω_o to 2α becomes larger, the sharpness of the resonant response increases. At the frequency ω_o the magnitude Z_R is R.

Figure 2.3(b) illustrates the half-power frequencies ω_1 and ω_2 for a typical tuned-amplifier response. The half-power frequencies can be determined by solving (2.8) for the values of ω that result in a magnitude of Z_R equal to $R/\sqrt{2}$ (or 3 dB down from its maximum value). Thus, from (2.8) setting

$$\left[\frac{\omega_o}{2\alpha}\left(\frac{\omega}{\omega_o} - \frac{\omega_o}{\omega}\right)\right]^2 = 1$$

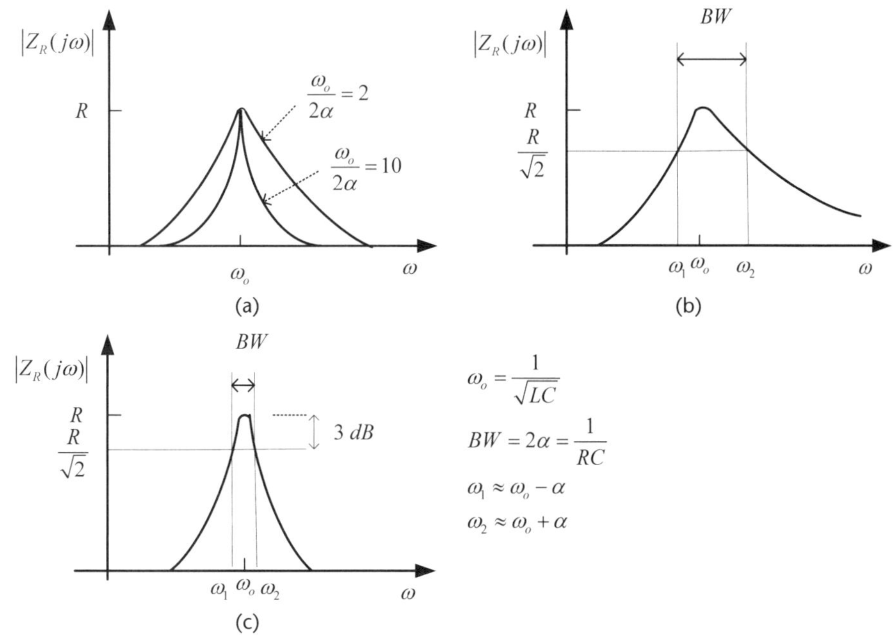

Figure 2.3 (a) Frequency response of $|Z_R(j\omega)|$, (b) the half-power frequencies ω_1 and ω_2, and (c) details of the characteristics for a high Q_L.

it follows that the half-power frequencies are given by

$$\omega_1 = -\alpha + \omega_o\sqrt{1 + \left(\frac{\alpha}{\omega_o}\right)^2} \tag{2.10}$$

and

$$\omega_2 = \alpha + \omega_o\sqrt{1 + \left(\frac{\alpha}{\omega_o}\right)^2} \tag{2.11}$$

Also, the following relationship is obtained:

$$\omega_o = \sqrt{\omega_1 \omega_2}$$

The frequencies ω_1 and ω_2 are not symmetrically located about ω_o. In fact, ω_o is at the geometric mean of ω_1 and ω_2.

Using (2.10) and (2.11), the bandwidth of the amplifier is

$$BW = \omega_2 - \omega_1 = 2\alpha = \frac{1}{RC}$$

which shows that the parameter 2α is simply the bandwidth of the tuned circuit [see Figure 2.3(b)].

The ratio of the resonant frequency to the bandwidth for the circuit in Figure 2.2 is known as its quality factor Q, namely,

$$Q = \frac{\omega_o}{BW} = \frac{\omega_o}{2\alpha}$$

If the resonant frequency and bandwidth are given in hertz, then Q is expressed in the form

$$Q = \frac{f_o}{BW}$$

In a high-Q circuit (i.e., $Q > 10$) we can approximate (2.10) and (2.11) as

$$\omega_1 \approx \omega_o - \alpha$$

and

$$\omega_2 \approx \omega_o + \alpha$$

The frequency response for a high-Q resonant circuit is shown in Figure 2.3(c). In this case, ω_o approaches the arithmetic mean of ω_1 and ω_2.

When the resistor R in Figure 2.2 represents the loss associated with the LC portion of the resonant circuit (i.e., the equivalent parallel loss resistance of the inductor), the quality factor is referred to as the *unloaded* Q (i.e., Q_U). The nomenclature *loaded* Q (i.e., Q_L) is used when the resistor R represents the parallel combination of the loss resistance and an external load resistor.

Using (2.2) and (2.3), the Q (either Q_U or Q_L) of the parallel RLC circuit in Figure 2.2 can be expressed in a variety of forms, namely,

$$Q = \frac{\omega_o}{BW} = \omega_o CR = \frac{R}{\omega_o L} \tag{2.12}$$

This relation shows that Q represents the ratio of the resistance to the magnitude of the reactance of C or L at ω_o.

The tuned circuit in Figure 2.4 helps to understand the different Qs. The resistor R_p represents the loss of the inductor (in parallel), and R_L is an external load resistor. The unloaded Q is

$$Q_U = \omega_o CR_p = \frac{R_p}{\omega_o L}$$

and the loaded Q is

$$Q_L = \omega_o CR = \frac{R}{\omega_o L}$$

where $R = R_p || R_L$.

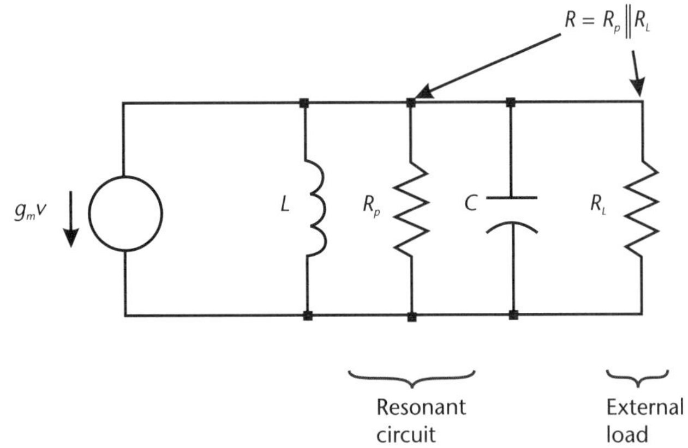

Figure 2.4 Tuned circuit.

An external Q can be defined in terms of R_L; namely,

$$Q_E = \omega_o C R_L = \frac{R_L}{\omega_o L}$$

and it follows that

$$\frac{1}{Q_L} = \frac{1}{Q_U} + \frac{1}{Q_E}$$

This expression shows how Q_U and Q_E affect Q_L. Of course, $Q_L < Q_U$.

In terms of Q, the impedance expression in (2.6) can be written as

$$Z_R = R \frac{\frac{\omega_o}{Q} s}{s^2 + \frac{\omega_o}{Q} s + \omega_o^2} \tag{2.13}$$

Figure 2.5 shows a plot of the magnitude of (2.13) versus frequency for several values of Q. The plot is normalized to a value of 1 at ω_o (i.e., $R = 1$).

It is evident that a resonant circuit is a narrowband filter. As the Q of the circuit increases, the bandwidth becomes smaller, and the selectivity of the circuit improves. In other words, only a selective range of frequencies is passed by the resonator.

The phase variation given by (2.9) can be expressed in the form

$$\phi(j\omega) =- \tan^{-1}\left[Q\left(\frac{\omega}{\omega_o} - \frac{\omega_o}{\omega}\right)\right] \tag{2.14}$$

At resonance (i.e., at $\omega = \omega_o$) the phase is 0°. At ω_1, it follows that $\phi(j\omega_1) = 45°$, and at ω_2 the phase is $\phi(j\omega_2) =- 45°$. The phase plot of (2.14) is shown in Figure 2.6.

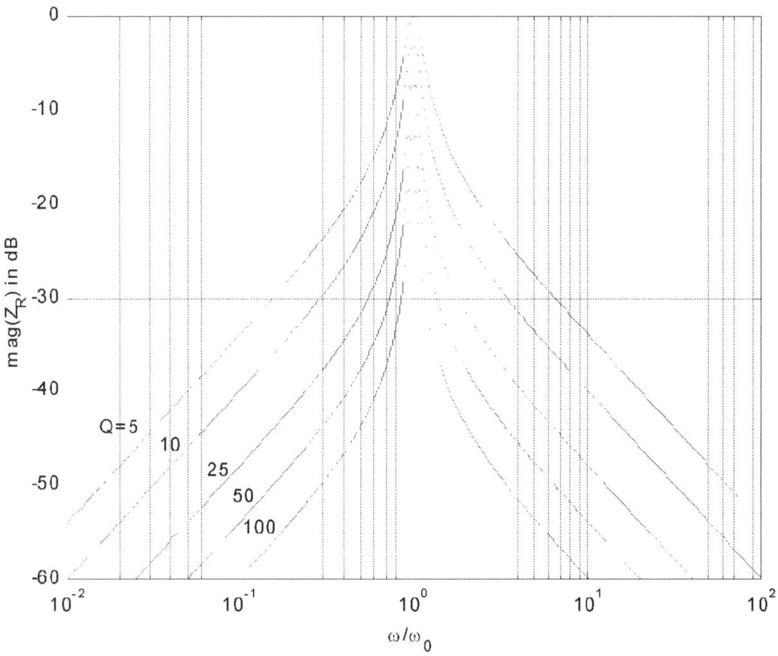

Figure 2.5 A plot of $|Z_R(j\omega)|$ in decibels for several values of Q.

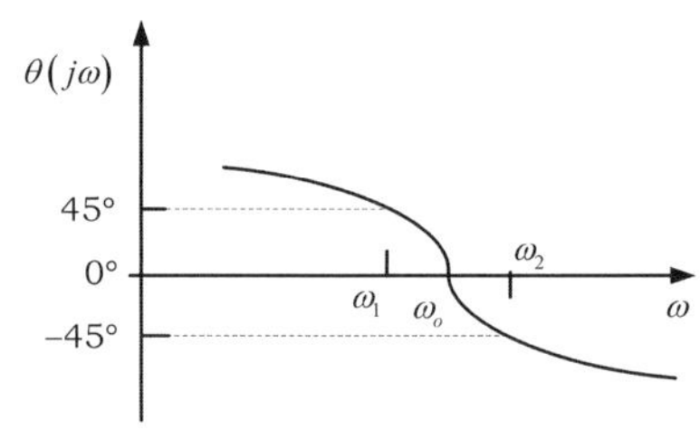

Figure 2.6 The phase plot of the tuned amplifier in Figure 2.2.

The frequency stability of an oscillator that uses the tuned circuit in Figure 2.2 depends on the derivative of $\phi(j\omega)$ with respect to ω; that is,

$$\frac{d\phi}{d\omega} = \frac{-\dfrac{1}{Q}}{\dfrac{1}{Q^2} + \left[\dfrac{\left(\omega_o^2 - \omega^2\right)}{\omega_o \omega}\right]^2} \left(\frac{\omega^2 + \omega_o^2}{\omega_o \omega^2}\right)$$

Therefore,

$$\left.\frac{d\phi}{d\omega}\right|_{\omega = \omega_o} = \frac{-2Q}{\omega_o} \tag{2.15}$$

and, from (2.1), the frequency stability factor is

$$S_F = -2Q \tag{2.16}$$

Equation (2.16) shows that the frequency stability depends directly on the resonator's Q. That is, the higher the value of Q, the smaller the changes in frequency due to phase changes. The negative sign indicates that $\Delta\phi < 0$ for $\Delta\omega > 0$.

A summary for a series resonant circuit is now given. For the series resonant circuit shown in Figure 2.7, its input admittance Y_R has a form similar to (2.2), namely,

$$Y_R = \frac{\dfrac{s}{L}}{s^2 + \dfrac{R}{L}s + \dfrac{1}{LC}} = \frac{1}{R\left[1 + jQ\left(\dfrac{\omega}{\omega_o} - \dfrac{\omega_o}{\omega}\right)\right]} \tag{2.17}$$

or

$$Z_R = \frac{1}{Y_R} = R\left[1 + jQ\left(\frac{\omega}{\omega_o} - \frac{\omega_o}{\omega}\right)\right]$$

where

$$R = R_s + R_L$$

$$Q = \frac{1}{\omega_o CR} = \frac{\omega_o L}{R}$$

$$\omega_o = \frac{1}{\sqrt{LC}}$$

and

$$BW = \frac{\omega_o}{Q} = \frac{R}{L}$$

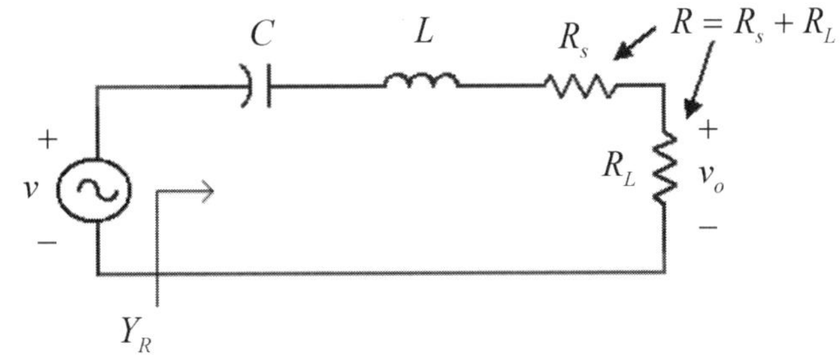

Figure 2.7 Series resonant circuit.

The voltage transfer function is $v_o = iR_L = vR_LY_R$. If R in (2.17) represents the inductor losses (i.e., $R = R_s$), the Q is called the unloaded Q (i.e., Q_U). If R represents the series combination of the inductor losses and the external load resistor, the Q is called the loaded Q (i.e., Q_L). Of course, $Q_L < Q_U$. An external Q can also be defined in terms of R_L.

2.3 Expressions for the Quality Factor

Actual inductors have a loss associated with them that can be represented by a resistance in series with the inductor, as shown in Figure 2.8(a). These so-called *coil losses* are taken into account by obtaining an equivalent parallel representation, as shown in Figure 2.8(b).

The circuit in Figure 2.8(a) consists of the coil resistance, denoted by R_s, in series with the inductance impedance (i.e., $j\omega L_s$). The circuit in Figure 2.8(b) consists of the equivalent parallel coil resistance (denoted by R_p) in parallel with the equivalent parallel inductance impedance (denoted by $j\omega L_p$). These circuits behave similarly at a given frequency if

$$R_s + j\omega L_s = \frac{R_p(j\omega L_p)}{R_p + j\omega L_p} = \frac{R_p(\omega L_p)^2}{R_p^2 + (\omega L_p)^2} + j\frac{\omega L_p R_p^2}{R_p^2 + (\omega L_p)^2}$$

Equating the real and imaginary parts gives

$$R_s = \frac{R_p(\omega L_p)^2}{R_p^2 + (\omega L_p)^2} = \frac{R_p}{1 + Q_U^2} \tag{2.18}$$

and

$$L_s = \frac{L_p R_p^2}{R_p^2 + (\omega L_p)^2} = \frac{L_p}{1 + \dfrac{1}{Q_U^2}} \tag{2.19}$$

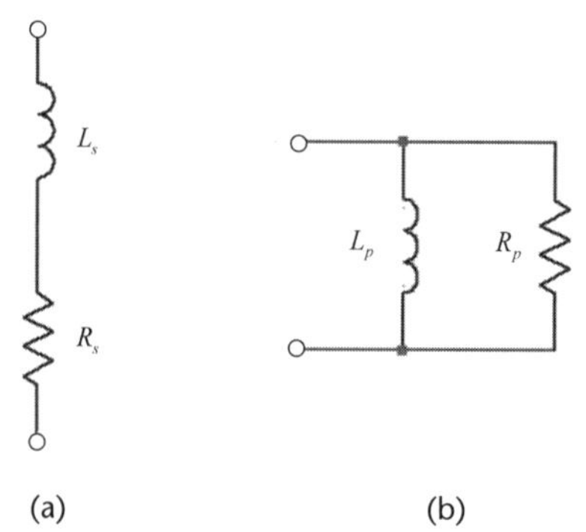

Figure 2.8 (a) Series representation of a lossy inductor and (b) the equivalent parallel representation.

where the unloaded Q is

$$Q_U = \frac{R_p}{\omega L_p} = \frac{\omega L_s}{R_s} \tag{2.20}$$

If $Q_U > 10$, (2.18) and (2.19) are approximated by

$$R_p \approx R_s Q_U^2 \tag{2.21}$$

and

$$L_p \approx L_s \tag{2.22}$$

which are the desired relations for oscillator design where high values of Q are used. Equations (2.21) and (2.22) show that the coil losses can be represented by a resistor R_p in parallel with a lossless inductor $L_p = L_s$.

A parallel resonant circuit with coil losses is drawn in Figure 2.9(a). Using (2.21) and (2.22), the equivalent circuit for $Q_U > 10$ is shown in Figure 2.9(b). The circuit shows the losses in parallel. For frequencies around the resonant frequency of the resonator, it follows that

$$\omega = \frac{1}{\sqrt{LC}}$$

and

$$BW = \frac{1}{R_p C}$$

The transformations in (2.21) and (2.22) make the tuned circuits in Figure 2.9 equivalent in the passband.

Some expressions for the Q of a resonator are now derived. A parallel resonant circuit is drawn in Figure 2.10(a) with an external load resistor R_L. From Figure 2.10(a) it follows that

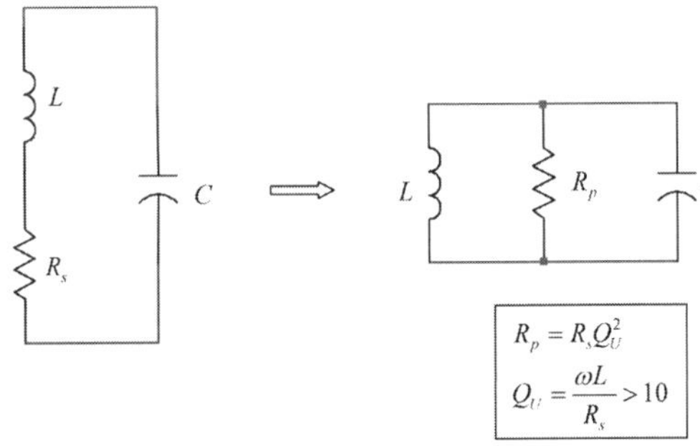

Figure 2.9 Passband equivalent circuits.

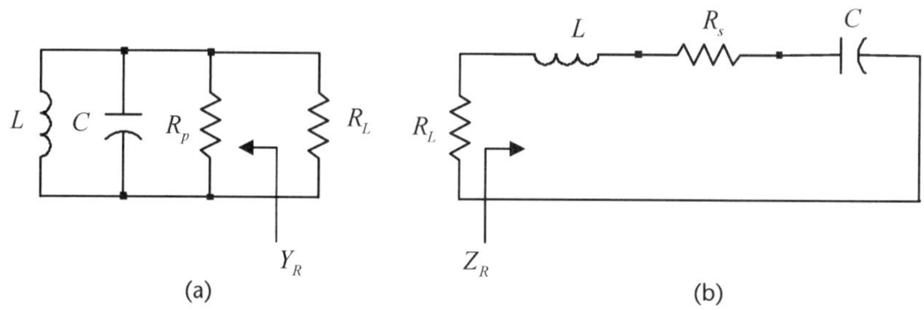

Figure 2.10 (a) Parallel resonant circuit with an external load resistor R_L and (b) a series resonant circuit with an external load resistor R_L.

$$Y_R(j\omega) = \frac{1}{R_p} + j\left(\omega C - \frac{1}{\omega L}\right)$$

which can be expressed in the form

$$Y_R(j\omega) = G_p + jQ_U G_p\left(\frac{\omega}{\omega_o} - \frac{\omega_o}{\omega}\right) \tag{2.23}$$

where

$$G_p = \frac{1}{R_p}$$

and

$$Q_U = \omega_o C R_p = \frac{R_p}{\omega_o L}$$

Since

$$\frac{dB}{d\omega} = \frac{Q_U G_p}{\omega_o}\left[1 + \left(\frac{\omega_o}{\omega}\right)^2\right]$$

where B is the susceptance of Y_R; then, at $\omega = \omega_o$, we obtain

$$Q_U = \frac{\omega_o}{2G_p}\frac{dB}{d\omega}\bigg|_{\omega=\omega_o} \tag{2.24}$$

With an external load resistor R_L, the resonator is loaded and its Q is expressed in the form

$$Q_L = \frac{\omega_o}{2G_T}\frac{dB}{d\omega}\bigg|_{\omega=\omega_o} \tag{2.25}$$

where

$$G_T = G_p + G_L$$

and

$$G_L = \frac{1}{R_L}$$

In (2.25) the quantity

$$\frac{dB}{G_T} \approx \frac{\Delta B}{G_T}$$

evaluated around ω_o is simply the phase of the admittance. Calling this quantity $\tan \Delta\theta$, it follows that $\tan \Delta\theta \approx \Delta\theta$ for small variation of ΔB around ω_o, and (2.25) is written as

$$Q_L \approx \frac{\omega_o}{2} \frac{\Delta\theta}{\Delta\omega}\bigg|_{\omega=\omega_o} = \frac{\omega_o}{2} \frac{d\theta}{d\omega}\bigg|_{\omega=\omega_o} \tag{2.26}$$

This form is convenient for use with CAD programs.

Similarly, for the series resonator in Figure 2.10(b) the input impedance is

$$\begin{aligned} Z_R(j\omega) &= R_s + j\left(\omega L - \frac{1}{\omega C}\right) \\ &= R_s + jQ_U R_s\left(\frac{\omega}{\omega_o} - \frac{\omega_o}{\omega}\right) \end{aligned}$$

where

$$Q_U = \frac{\omega_o L}{R_s}$$

Then, at $\omega = \omega_o$, we obtain

$$Q_U = \frac{\omega_o}{2R_s} \frac{dX}{d\omega}\bigg|_{\omega=\omega_o} \tag{2.27}$$

where X is the reactance of Z_R.

For the series loaded resonator with R_L, the loaded Q can be expressed as

$$Q_L = \frac{\omega_o}{2R_T} \frac{dX}{d\omega}\bigg|_{\omega=\omega_o} \tag{2.28}$$

where

$$R_T = R_s + R_L$$

Equation (2.28) shows how a small variation in ΔX affects the frequency since

$$\Delta\omega = \omega_o \frac{\Delta X}{2R_T Q_L}$$

This relation shows that the higher the Q_L, the smaller the change in frequency around resonance. Equation (2.28) can also be expressed in the form (2.26) where θ in this case is the phase of the reactance.

The relation (2.24) shows that in a parallel resonator high values of Q_U are associated with low losses (G_p) and a high value of $dB/d\omega$. Similar considerations apply to the series resonator where Q_U is described by (2.27). Also, loading a resonator with R_L decreases the quality factor from Q_U to Q_L.

Example 2.1

The tuned circuit of a certain oscillator is shown in Figure 2.11(a) where $g_m = 2$ mS, $r_o = 50$ kΩ, $C = 100$ pF, $L = 500$ μH, $R_s = 50$Ω, and the load resistor is $R_L = 50$ kΩ. The resistor R_s represents the coil losses, and r_o is the output resistance of the active device. Find the equivalent passband resonant circuit, evaluate ω_o, the gain at resonance, and plot the frequency response.

Solution

The series to parallel transformations can be applied to the LC portion of the circuit in Figure 2.11(a), resulting in the model shown in Figure 2.11(b), where $L = L_s = L_p = 500$ μH and $C = 100$ pF.

The resonant frequency is

$$\omega_o = \frac{1}{\sqrt{LC}} = \frac{1}{\sqrt{500 \times 10^{-6}(100 \times 10^{-12})}} = 4.472 \text{ Mrad/s}$$

or $f_o = 711.76$ kHz.

Using (2.20), the unloaded Q of the inductor is

$$Q_U = \frac{\omega_o L_s}{R_s} = \frac{4.472 \times 10^6 (500 \times 10^{-6})}{50} = 44.72$$

Using the transformations in (2.27) we find that

$$R_p = Q_U^2 R_s = (44.72)^2 50 = 100 \text{ k}\Omega$$

The equivalent parallel resonant circuit is shown in Figure 2.11(b). The total load resistance of the parallel RLC circuit is

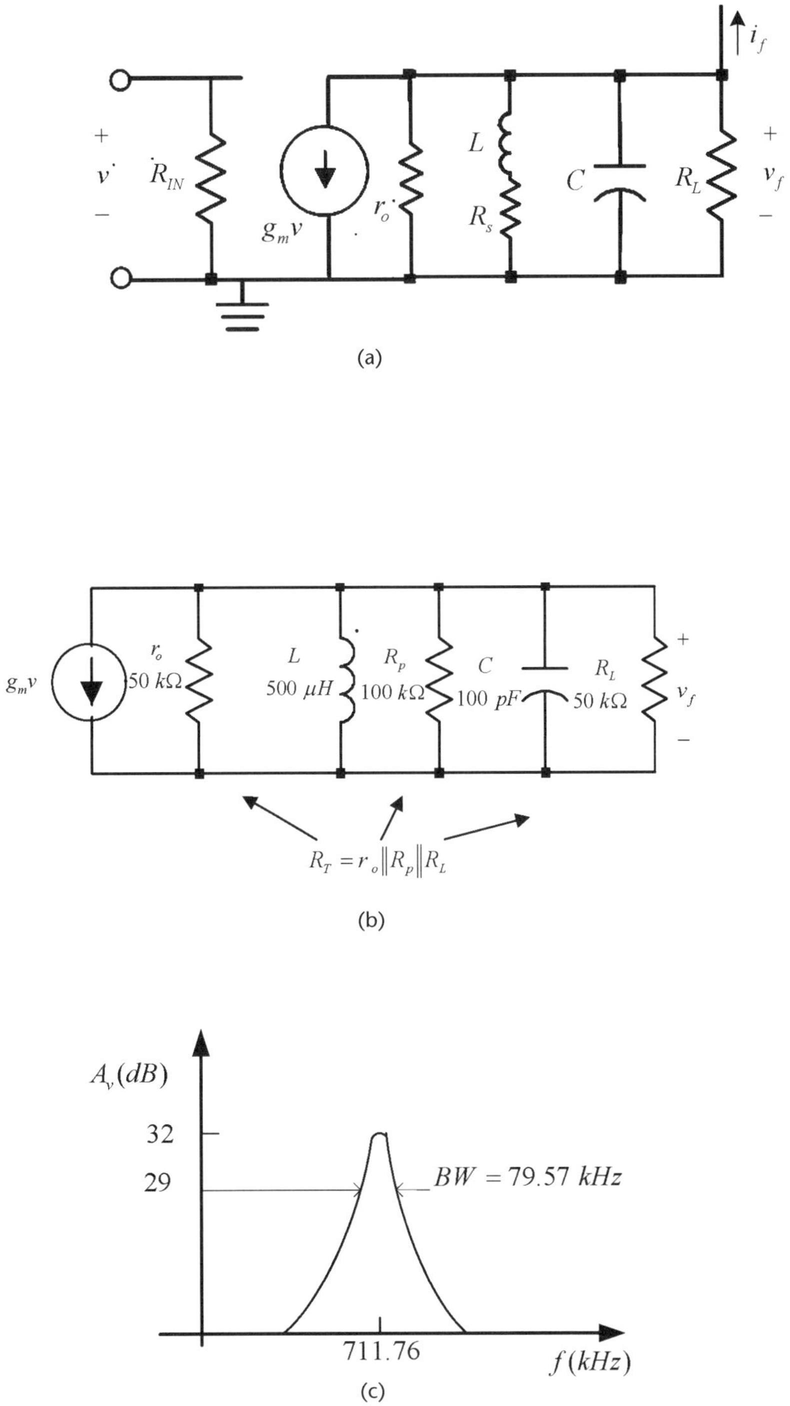

Figure 2.11 (a) Tuned circuit for Example 2.1, (b) the equivalent parallel resonant circuit, and (c) the frequency response.

$$R_T = r_o \parallel R_p \parallel R_L = 50 \times 10^3 \parallel 100 \times 10^3 \parallel 50 \times 10^3 = 20\ \text{k}\Omega$$

The bandwidth of the parallel-resonant circuit is

$$BW = \frac{1}{R_T C} = \frac{1}{20 \times 10^3 (100 \times 10^{-12})} = 500\ \text{krad/s}$$

or $BW = 79.57$ kHz, and its loaded Q is

$$Q_L = \frac{\omega_o}{BW} = \frac{4.472 \times 10^6}{500 \times 10^3} = 8.9$$

The frequency response is shown in Figure 2.11(c) where the gain at resonance is

$$|A_v(j\omega_o)| = g_m R_T = 2 \times 10^{-3}(20 \times 10^3) = 40 \text{ (or 32 dB)}$$

It is interesting to observe in this example how the coil series resistance ($R_s = 50\Omega$) appears in the parallel equivalent circuit as a resistor of value $R_p = 100$ kΩ. This resistor loads the circuit further since the total load resistance becomes $r_o \parallel R_p \parallel R_L$, resulting in $Q_L = 8.9$, while $Q_U = 44.7$.

2.4 Noise in Oscillators

An ideal oscillator will produce an output voltage given by

$$v_o(t) = A \cos(\omega_o t)$$

where $\omega_o = 2\pi f_o$ is the frequency of oscillation. A nonideal oscillator output voltage is

$$v_o(t) = A(t) \cos[\omega_o t + \phi(t)] \tag{2.29}$$

where $A(t)$ represents the amplitude fluctuations as a function of time (commonly known as AM noise), and $\phi(t)$ represents the random phase variations (commonly known as phase noise). Since the instantaneous frequency is

$$\omega(t) = \frac{d}{dt}[\omega_o t + \phi(t)] = \omega_o + \frac{d\phi(t)}{dt}$$

or

$$f(t) = f_o + \frac{1}{2\pi}\frac{d\phi(t)}{dt}$$

it follows that the random phase fluctuation [i.e., $\phi(t)$] produces a random frequency fluctuation [i.e., $d\phi(t)/dt$]. Hence, these terms are related, and both are used to describe the noise in oscillators.

In oscillators the phase noise is the specification of choice, since amplitude fluctuations are easily controlled with a limiter. The amplitude-limiting mechanism of the oscillator, which can be a self-limiting mechanism or an automatic gain control (AGC) mechanism, also limits the amplitude fluctuations. Hence, the oscillator noise is mainly due to the random phase fluctuations $\phi(t)$. The phase noise (or FM noise) is produced by thermal noise, shot noise, and flicker noise. Therefore, it is a short-term frequency stability problem. Thermal noise is a function of the temperature, bandwidth, and noise resistance. Shot noise is a function of the dc bias current, and flicker noise is a function of the characteristics of the active device. Figure 2.12 shows a signal with a small random phase fluctuation, denoted by $\Delta\phi(t)$. This is a typical time domain display of the signal.

The short-term random phase fluctuations are best viewed in the frequency domain. A spectrum analyzer provides the amplitude versus frequency display of the signal, from which the power spectrum follows. A typical spectral distribution for an oscillator is shown in Figure 2.13(a). The signal power at f_o is denoted by P_S, and the power at the single sideband (SSB) frequency f_m in a bandwidth of 1 Hz is denoted by P_{SSB} (P_{SSB} is the single-sideband phase-noise power).

A common characterization of phase noise is to measure the ratio of a single-sideband power in a bandwidth of 1 Hz at a frequency f_m away from f_o to the signal power at f_o. This quantity is usually denoted by $\mathcal{L}(f_m)$, namely

$$\mathcal{L}(f_m) = \frac{\text{single-sideband phase-noise power}}{\text{signal power}} = \frac{P_{SSB}}{P_S} \tag{2.30}$$

The quantity $\mathcal{L}(f_m)$ is known as the single-sideband phase-noise power-to-carrier ratio (or simply referred to as the SSB phase noise). $\mathcal{L}(f_m)$ provides a measure of the noise energy, and it is easily measured from the spectrum analyzer display of the signal, as shown in Figure 2.13(a). The quantity $\mathcal{L}(f_m)$ is commonly found

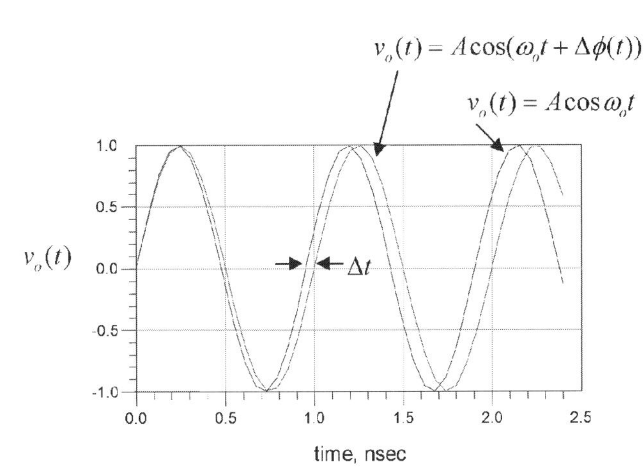

Figure 2.12 Time domain display of a signal with a random phase fluctuation $\Delta\phi(t) = \omega_0\Delta t$.

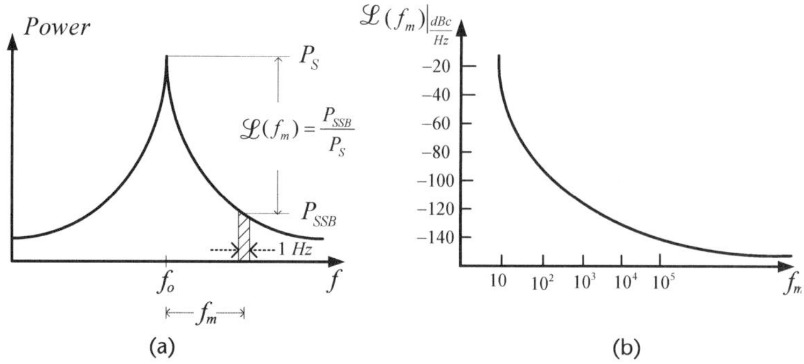

Figure 2.13 (a) Power spectrum display of a signal with random-phase fluctuations and (b) a typical $\mathcal{L}(f_m)$ in dBc/Hz.

expressed in decibels relative to the signal power (or carrier power) per hertz (i.e., in dBc/Hz), as shown in Figure 2.13(b). That is,

$$\mathcal{L}(f_m) = 10 \log \frac{P_{SSB}}{P_S} \quad \frac{\text{dBc}}{\text{Hz}}$$

If the oscillator signal f_o is modulated by a single sine wave with frequency f_m, we can express the phase modulated (PM) signal [i.e., (2.29)] in the form

$$v(t) = A \cos(\omega_o t + m_p \sin \omega_m t) \tag{2.31}$$

where

$$\phi(t) = m_p \sin \omega_m t$$

The factor m_p (in radians) is known as the phase modulation index. The phase modulation index gives the maximum phase deviation that the phase can attain from its nominal value of $\omega_o t$. The factor m_p is also denoted by $\Delta\phi$ in the literature, since the notation $\Delta\phi$ suggests a small phase deviation.

The phase variation in (2.31) produces an instantaneous frequency variation (denoted by $\omega_{PM}(t)$), which is given by

$$\omega_{PM}(t) = \frac{d(\omega_o t + m_p \sin \omega_m t)}{dt} = \omega_o + m_p \omega_m \cos \omega_m t$$

where $m_p \omega_m$ measures the maximum frequency deviation of the PM signal. This term is usually denoted by $\Delta\omega$ (i.e., $\Delta\omega = m_p \omega_m$). The instantaneous frequency can also be expressed as

$$f_{PM}(t) = f_o + \Delta f \cos 2\pi f_m t$$

where

$$\Delta f = m_p f_m$$

From (2.31) we can write

$$v(t) = A \cos \omega_o t \cos(m_p \sin \omega_m t) - A \sin \omega_o t \sin(m_p \sin \omega_m t) \quad (2.32)$$

For a narrowband phase modulating signal, the value of m_p is small ($m_p \ll 1$ radian), and (2.32) can be approximated by

$$\begin{aligned} v(t) &= A \cos \omega_o t - A \sin \omega_o t (m_p \sin \omega_m t) \\ &= A \cos \omega_o t - \frac{A m_p}{2} [\cos(\omega_o + \omega_m)t - \cos(\omega_o - \omega_m)t] \end{aligned} \quad (2.33)$$

Equation (2.33) shows that for small m_p values, the phase modulation results in frequency components at $\omega_o \pm \omega_m$, as shown in Figure 2.14. This type of signal is referred to as *narrowband FM*. For more than one modulating signal, the oscillator spectrum can be considered to be a superposition of a large number of modulating signals.

From (2.33) it is seen that the ratio of the peak sideband voltage (i.e., $Am_p/2$) to the peak carrier voltage (i.e., A) is $m_p/2$. Hence, the power ratio is $(m_p/2)^2$ and from (2.30) it follows that the SSB phase noise $\mathcal{L}(f_m)$ is

$$\mathcal{L}(f_m) = \left(\frac{m_p}{2}\right)^2 = \frac{m_{p,\text{rms}}^2}{2} \quad (2.34)$$

where

$$m_{p,rms} = \frac{m_p}{\sqrt{2}}$$

For completeness, the spectrum of a PM signal for an arbitrary value of m_p is now presented. From (2.31), we have

$$v(t) = \text{Re}\left[A e^{j(\omega_o t + m_p \sin \omega_m t)}\right] \quad (2.35)$$

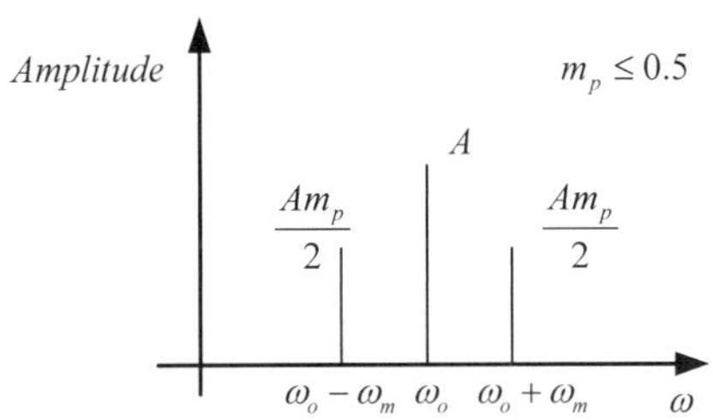

Figure 2.14 Magnitude spectrum of a narrowband FM signal.

The term $e^{jm_p \sin \omega_m t}$ can be expanded in terms of a set of functions, called Bessel functions. The appropriate expansion is

$$e^{jm_p \sin \omega_m t} = \sum_{n=-\infty}^{n=\infty} J_n(m_p) e^{jn\omega_m t} \tag{2.36}$$

where $J_0(m_p)$ is the Bessel function of order zero and argument m_p, $J_1(m_p)$ is the Bessel function of order one and argument m_p, and so on.

Substituting (2.36) into (2.35) gives

$$v(t) = \text{Re}\left[A \sum_{n=-\infty}^{n=\infty} J_n(m_p) e^{j(\omega_o + n\omega_m)t} \right] \tag{2.37}$$

$$= A \sum_{n=-\infty}^{n=\infty} J_n(m_p) \cos(\omega_o + n\omega_m)t$$

Equation (2.37) shows that the PM signal contains an infinite number of frequencies, namely: ω_o, $\omega_o \pm \omega_m$, $\omega_o \pm 2\omega_m$, $\omega_o \pm 3\omega_m$, etc. In other words, the number of sidebands is infinite. Using the identity $J_{-n}(m_p) = (-1)^n J_n(m_p)$, we can plot the amplitude spectrum in Figure 2.15 with $A = 1$. The amplitude component at ω_o is $J_0(m_p)$, at $\omega_o \pm \omega$ is $J_1(m_p)$, and so on.

The spectrum of the PM signal depends on the values of the Bessel functions. Table 2.1 lists some specific values of the functions.

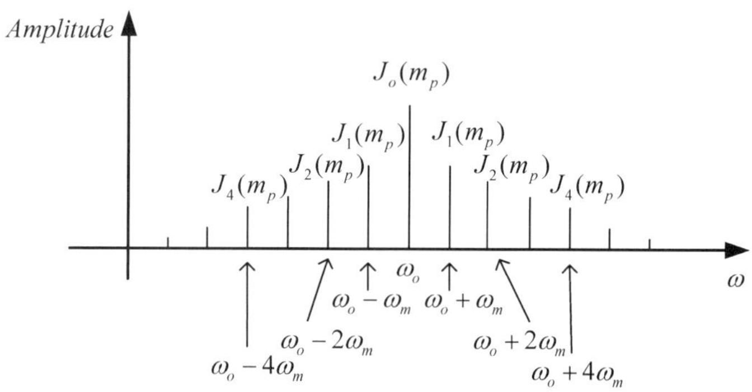

Figure 2.15 Amplitude spectrum of an angle-modulated signal.

Table 2.1 Specific Values of Some Bessel Functions

	$m_p = 0.1$	$m_p = 0.5$	$m_p = 1$	$m_p = 2$	$m_p = 5$	$m_p = 10$
$J_0(m_p)$	0.998	0.938	0.765	0.224	–0.178	–0.246
$J_1(m_p)$	0.049	0.242	0.440	0.577	–0.328	0.043
$J_2(m_p)$	0.0012	0.031	0.115	0.353	0.047	0.255
$J_3(m_p)$	—	0.003	0.020	0.129	0.365	0.058
$J_4(m_p)$	—	—	0.002	0.034	0.391	–0.220
$J_5(m_p)$	—	—	—	0.007	0.261	–0.234

From Table 2.1 it is observed that for $m_p < 0.5$, only the $J_0(m_p)$ and $J_1(m_p)$ functions have significant values. Thus, for $m_p < 0.5$, (2.37) reduces to

$$v(t) \approx A[J_0(m_p)\cos\omega_o t + J_1(m_p)\cos(\omega_o + \omega_m)t - J_1(m_p)\cos(\omega_o - \omega_m)t] \tag{2.38}$$

The spectrum of (2.38) is similar to the one shown in Figure 2.14, except that the amplitude at ω_o is $J_0(m_p)$, and at the sidebands is $J_1(m_p)$. The bandwidth of this signal is $2\omega_m$. Since for small m_p, $J_0(m_p) \approx 1$ and $J_1(m_p) \approx m_p/2$, we can express (2.38) as

$$v(t) \approx A\left[\cos\omega_o t + \frac{m_p}{2}\cos(\omega_o + \omega_m)t - \frac{m_p}{2}\cos(\omega_o - \omega_m)t\right] \tag{2.39}$$

which is the narrowband FM signal in (2.33) whose spectrum is shown in Figure 2.14.

The derivation of the phase noise in (2.34) was based on a single modulating signal. For the general case, a basic quantity that describes the distribution of phase-noise power as a function of frequency is the spectral density of the phase fluctuation $S_\phi(f_m)$. That is,

$$S_\phi(f_m) = \frac{\phi^2_{\text{rms}}(f_m)}{BW \text{ used to measure } \phi_{\text{rms}}(f_m)} \ \frac{\text{rad}^2}{\text{Hz}} \tag{2.40}$$

Observe that for a single-tone phase-modulation signal $m_{p,\text{rms}} = \Delta\phi_{\text{rms}}$, and the quantity $S_\phi(f_m)$ in (2.40) is 3 dB higher than $\mathcal{L}(f_m)$ in (2.34). This occurs because in (2.40) the power in the two sidebands is added.

A typical plot of $S_\phi(f_m)$ is shown in Figure 2.16. The phase-noise power in a bandwidth $f_2 - f_1$, denoted by $P_\phi(f_1, f_2)$, can be calculated using

$$P_\phi(f_1, f_2) = \int_{f_1}^{f_2} S_\phi(f_m)\, df_m \ \text{rad}^2 \tag{2.41}$$

Equation (2.41) shows that the total phase noise is given by

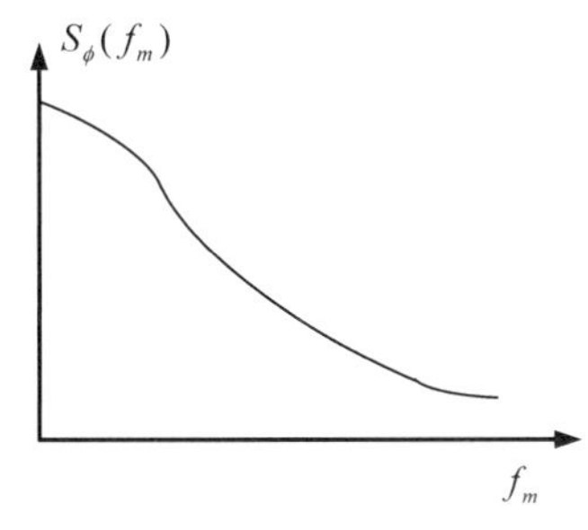

Figure 2.16 Plot of $S_\phi(f_m)$.

$$P_\phi(0, \infty) = \int_0^\infty S_\phi(f_m)\, df_m \text{ rad}^2 \tag{2.42}$$

In logarithmic form $S_\phi(f_m)$ is expressed in dBr/Hz (decibels above one radian per hertz bandwidth), namely,

$$S_\phi(f_m) = 10 \log \Delta\phi_{\text{rms}}^2 = 20 \log \Delta\phi_{\text{rms}} \frac{\text{dBr}}{\text{Hz}}$$

The spectral density of the phase fluctuation for the case illustrated in Figure 2.14 (i.e., small values of m_p) is simply the total phase-noise power in both sidebands. That is,

$$S_\phi(f_m) = 2\mathcal{L}(f_m) = m_{p,\text{rms}}^2 = \Delta\phi_{\text{rms}} \frac{\text{rad}^2}{\text{Hz}} \tag{2.43}$$

or in logarithmic form

$$S_\phi(f_m) = 10 \log m_{p,\text{rms}}^2 \frac{\text{dBr}}{\text{Hz}}$$

To summarize, $\mathcal{L}(f_m)$ and $S_\phi(f_m)$ are defined in (2.30) and (2.40), respectively. If the modulation sidebands are such that m_p is a small value, then $\mathcal{L}(f_m)$ and $S_\phi(f_m)$ are related by (2.43). Observe that the relation (2.43) is valid only when the noise sidebands at $f_o \pm f_m$ are correlated (i.e., produced by the same modulation) and the small-angle approximation applies.

The spectral density of the frequency fluctuation Δf is defined as

$$S_f(f_m) = \frac{\Delta f_{\text{rms}}^2(f_m)}{BW \text{ used to measure } \Delta f_{\text{rms}}(f_m)} \frac{\text{Hz}^2}{\text{Hz}} \text{ (or Hz)}$$

In log form the units are in decibels above one hertz per hertz bandwidth (dBHz/Hz). That is,

$$S_f(f_m) = 10 \log \Delta f_{\text{rms}}^2 = 20 \log \Delta f_{\text{rms}} \frac{\text{dBHz}}{\text{Hz}}$$

The spectral density of the frequency fluctuations is related to the spectral density of the phase fluctuations. Since

$$\Delta\omega(t) = 2\pi\Delta f(t) = \frac{d\Delta\phi(t)}{dt} \tag{2.44}$$

it follows that (2.44) in the frequency domain reads

$$\Delta f(f_m) = f_m \Delta\phi(f_m)$$

Hence,

$$S_f(f_m) = \Delta f_{\text{rms}}^2(f_m) = f_m^2 \Delta\phi_{\text{rms}}^2(f_m) = f_m^2 S_\phi(f_m)$$

For the case of a narrowband FM signal (i.e., for $\Delta\phi = m_p$ with $m_p < 0.5$) it follows that

$$S_f(f_m) = f_m^2 m_{p,\text{rms}}^2 = 2f_m^2 \mathscr{L}(f_m)$$

There are other ways in which the frequency stability of oscillators is specified. One of them is the residual FM. The residual FM [denoted by $(\Delta f)_{\text{residual}}$] is the total rms frequency deviation in a specified bandwidth. That is,

$$(\Delta f)_{\text{residual}} = \sqrt{2}\sqrt{\int_{f_1}^{f_2} \mathscr{L}(f_m) f_m^2 \, df}$$

where $BW = f_2 - f_1$.

A better understanding of the relation between the random phase fluctuations and the associated $S_\phi(f_m)$ requires a knowledge of the theory of random variables. The random fluctuations of the phase can be characterized by a probability distribution function $p(\phi)$. That is, the phase $\phi(t)$ is described by the distribution function $p(\phi)$, which gives the probability that $\phi(t)$ attains a certain value of ϕ. The function $p(\phi)$ is Gaussian; therefore, it can be expressed in the form

$$p(\phi) = \frac{1}{\sqrt{2\pi}\sigma_\phi} e^{-\frac{\phi^2}{2\sigma_\phi^2}}$$

where σ_ϕ is the standard deviation of $p(\phi)$. That is,

$$\sigma_\phi = \sqrt{\overline{\phi^2(t)}} \text{ rad}_{\text{rms}}$$

In other words, assuming that the noise is ergodic, the standard deviation can be calculated using a statistical average or time average. The standard deviation σ_ϕ is also referred as the rms value of the phase fluctuation.

A typical phase fluctuation about the oscillator signal phase $\phi_o = \omega_o t$ is shown in Figure 2.17, as well as the function $p(\phi)$ and its standard deviation σ_ϕ. A Gaussian function has the property that about 99.7% of the phase fluctuation values are in the interval $\pm 3\sigma_\phi$. Hence, a good approximation for the peak-to-peak value of the phase fluctuations (ϕ_{pp}) is

$$\phi_{pp} = 6\sigma_\phi$$

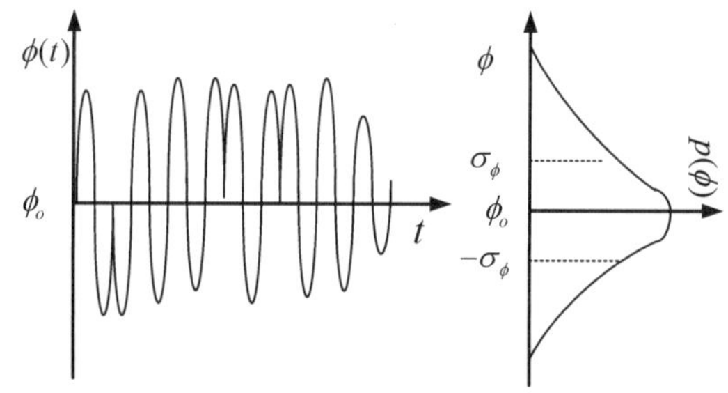

Figure 2.17 A random phase $\phi(t)$ and its probability distribution function $p(\phi)$.

In the time domain the random phase fluctuations are characterized by the standard deviation σ_ϕ. In the frequency domain $S_\phi(f_m)$ is defined as the Fourier transform of the expected value of $\phi(t)\phi(t+\tau)$. Hence, for $\tau = 0$ the inverse Fourier transform gives

$$\phi^2(t) = \sigma_\phi^2 = \int_0^\infty S_\phi(f_m)\, df_m$$

From (2.42), the right-hand side is recognized as the total phase noise power. Hence, we obtain

$$P_\phi(0, \infty) = \sigma_\phi^2 \text{ rad}^2$$

For completeness the time-domain characterization method is also mentioned. In some cases, such as that of a highly stable crystal oscillator, the characterization of phase noise is best done in the time domain. For example, the frequency fluctuations can be measured using a frequency counter. These measurements produce a set of random data for the frequency instability (say, $\Delta f_n(t)$, $n = 1, 2, \ldots$). The statistical analysis of the frequency-fluctuation samples $\Delta f_n(t)$ provides the characterization of the frequency instability in the time domain.

2.5 Oscillator Phase Noise

In the previous section the phase-noise characteristics of an oscillator were discussed. The following example illustrates why phase noise is so important. Consider the down-converter system shown in Figure 2.18(a), where the radio frequency (RF) signals are mixed with the local oscillator (LO) signal. The amplitude of the RF signal at f_1 is larger than the amplitude of the signal at f_2. The intermediate-frequency (IF) amplifier filters and amplifies the signals at $f_0 - f_1$ and $f_0 - f_2$. If the phase noise of the oscillator is low, or in other words the oscillation signal is fairly pure, the IF signal shows that the signal corresponding to $f_0 - f_1$ and $f_0 - f_2$ are distinct, and the system can process both signals. On the other hand, consider the situation illustrated in Figure 2.18(b) where the LO phase noise is

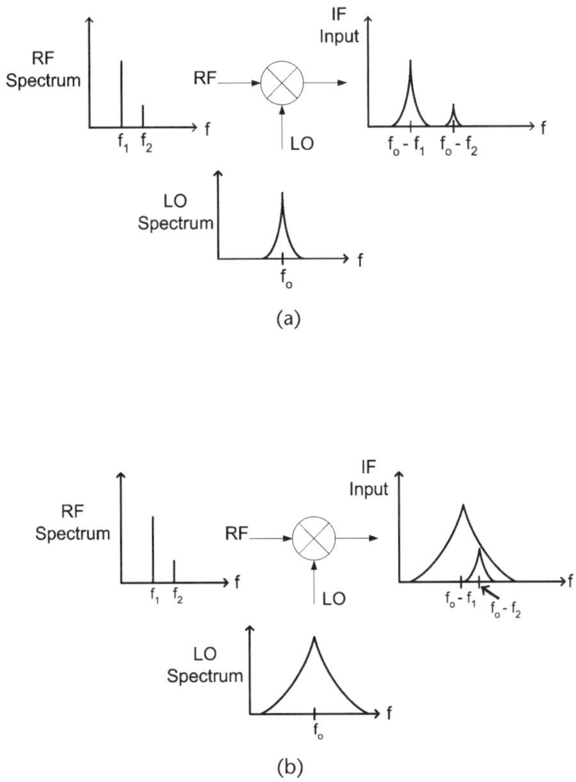

Figure 2.18 A down-converter system with (a) a low phase noise LO and (b) a high phase noise LO.

high and, therefore, has a wide spectrum. In this case, the IF signal at $f_0 - f_1$ has a large noise component at $f_0 - f_2$. Therefore, the IF signal at $f_0 - f_2$ is said to be drowned in the phase noise of $f_0 - f_1$. While the IF amplifier might be able to amplify and recover the IF signal at $f_0 - f_1$, the IF signal at $f_0 - f_2$ cannot be recovered by the system. This example illustrates that when an oscillator is used in a mixer system, the phase noise of the oscillator determines the ability of the mixer system to down convert and separate the desired signal. Hence, the phase noise of the oscillator can degrade the system selectivity, as well as its sensitivity.

Other systems affected by phase noise include a Doppler radar system and a digital communication system. In a Doppler radar the phase noise of the oscillator sets the minimum detectable target signal. In a communication system the phase noise affects the system bit-error rate.

The effect of the oscillator's components on its phase noise, based on a linear time-invariant model, was described by Leeson [1]. Leeson's model, although simple to follow, has severe limitations. An improvement over Leeson's model is the Lee and Hajimiri model [2, 3]. The Lee and Hajimiri model is based on the linear time-varying properties of the current flow in the active device of the oscillator.

First, the Leeson's model is considered. The feedback model used by Leeson for the oscillator is shown in Figure 2.19. It consists of an amplifier with noise figure F, and a resonant circuit in the feedback loop. The amplifier noise power

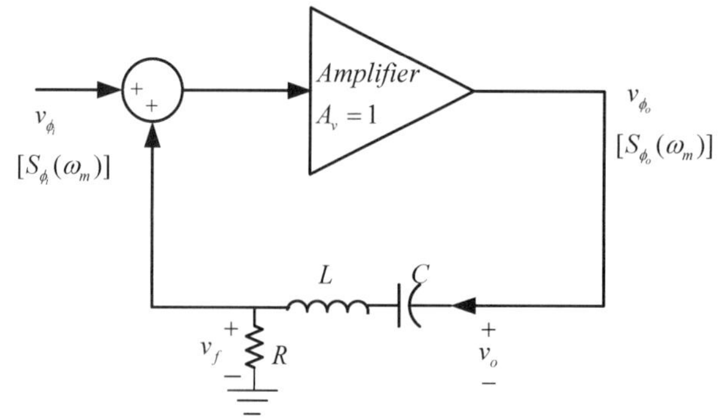

Figure 2.19 Feedback model of an oscillator.

spectral density referred to the input of the amplifier is denoted by $S_{\phi_i}(\omega)$. It is produced by random-phase fluctuations. The associated noise voltage at the input is denoted by v_{ϕ_i}, and the noise voltage at the output is v_{ϕ_o}. The function $H_R(j\omega)$ represents the transfer function of the resonator.

A series resonator was used for convenience. The voltage transfer function of the series resonator in Figure 2.19 is

$$H_R(j\omega) = \frac{1}{1 + jQ_U\left(\frac{\omega}{\omega_o} - \frac{\omega_o}{\omega}\right)} \tag{2.45}$$

where the resonant frequency is

$$\omega_o = \frac{1}{\sqrt{LC}}$$

and its unloaded Q is

$$Q_U = \frac{\omega_o L}{R} = \frac{1}{\omega_o RC}$$

In (2.45), Q_U is used for the resonator because when the resonator is loaded by the amplifier, and the circuit oscillates, the total resistance cancels and $Q_L \to \infty$.

At $\omega = \omega_o$, $|H_R(j\omega)| = 1$. Therefore, for the circuit to oscillate, the gain of the amplifier at ω_o must also be set equal to unity.

For $\omega \approx \omega_o + \omega_m$, where ω_m is the phase-noise modulation frequency and $\omega_o \gg \omega_m$, it follows that

$$\frac{\omega}{\omega_o} - \frac{\omega_o}{\omega} = \frac{\omega^2 - \omega_o^2}{\omega_o \omega} \approx \frac{2\omega_m}{\omega_o}$$

Therefore, (2.45) reduces to the form

$$H_R(j\omega_m) = \frac{1}{1 + jQ_U \dfrac{2\omega_m}{\omega_o}} = \frac{1}{1 + j\dfrac{\omega_m}{\omega_L}} \tag{2.46}$$

where

$$\omega_L = \frac{\omega_o}{2Q_U} \tag{2.47}$$

The frequency ω_L represents the half-power bandwidth of the resonator.

From (2.45) and (2.46) it can be stated that the transfer function of a modulated signal passing through a bandpass filter is the same as the transfer function of the modulating signal passing through an equivalent low-pass filter.

In Figure 2.19 the closed-loop voltage transfer function that relates the input and output phase-noise voltage is

$$H(j\omega_m) = \frac{v_{\phi_o}}{v_{\phi_i}} = \frac{A_v}{1 - A_v H_R(j\omega_m)} = \frac{1}{1 - H_R(j\omega_m)} \tag{2.48}$$

where $A_v = 1$. Substituting (2.46) into (2.48), we obtain

$$\left|H(j\omega)\right|^2 = 1 + \left(\frac{\omega_L}{\omega_m}\right)^2 = 1 + \left(\frac{\omega_o^2}{2Q_U\omega_m}\right) \tag{2.49}$$

The output spectral density $S_{\phi_o}(\omega_m)$ (measured at the output of the amplifier) is given by

$$S_{\phi_o}(\omega_m) = S_{\phi_i}(\omega_m)\left|H_T(j\omega_m)\right|^2$$

and using (2.49) we obtain

$$S_{\phi_o}(f_m) = \left[1 + \left(\frac{f_L}{f_m}\right)^2\right] S_{\phi_i}(f_m) = \left[1 + \frac{1}{f_m^2}\left(\frac{f_o}{2Q_U}\right)^2\right] S_{\phi_i}(f_m) \tag{2.50}$$

This relation shows how the input phase noise (i.e., the amplifier phase noise) is affected by the resonator. Obviously, the phase noise that falls within the half-power bandwidth of the resonator, given by (2.47), is significantly affected by the resonator.

The expression for $S_{\phi_i}(\omega_m)$ in (2.50) requires a bit of analysis. To this end, let us examine how phase noise is added to a signal passing through an amplifier with noise figure F, as shown in Figure 2.20(a). The oscillator's available signal power at the input of the amplifier is denoted by P_s and the thermal noise power referred to the input of the amplifier by $N_{o(in)}$. The signal voltage $v_{s,\text{rms}}$ can be expressed in terms of the oscillator's available signal power and the amplifier input resistance R as

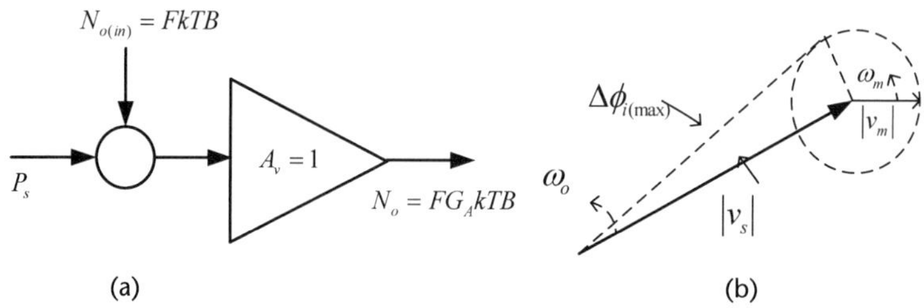

Figure 2.20 (a) Model for the phase perturbation analysis of the amplifier and (b) vector representation of the signal plus noise and the resulting phase perturbation at $\omega_o + \omega_m$.

$$v_{s,\text{rms}} = \sqrt{P_s R} \tag{2.51}$$

The available input thermal-noise power is given by $N_i = kTB$, where $k = 1.374 \times 10^{-23}$ J/K is Boltzmann's constant, T is the resistor noise temperature in degrees kelvin, and B is the noise bandwidth. The amplifier noise figure is the ratio of the input signal-to-noise ratio (i.e., S_i/N_i) to the output signal-to-noise ratio (i.e., S_o/N_o); namely,

$$F = \frac{\left(\dfrac{S_i}{N_i}\right)}{\left(\dfrac{S_o}{N_o}\right)}$$

or

$$F = \frac{N_o}{G_A N_i} = \frac{N_o}{G_A kTB}$$

where $G_A = S_o/S_i$ is the available gain of the amplifier. Therefore, the amplifier output noise power is given by

$$N_o = FG_A kTB$$

The noise power N_o referenced to the input of the amplifier is

$$N_{o(\text{in})} = FkTB$$

This thermal-noise power can be expressed in terms of the phase-noise voltage $v_{n,\text{rms}}$ as

$$N_{o(\text{in})} = \frac{v_{n,\text{rms}}^2}{R}$$

or

$$v_{n,\text{rms}} = \sqrt{FkTBR} \tag{2.52}$$

Equation (2.52) shows that for each 1-Hz band (i.e., $B = 1$ Hz) the noise can be represented by an equivalent signal $v_{n,\text{rms}} = \sqrt{FkTR}$. This noise signal adds to the oscillator signal producing a phase perturbation in each 1-Hz band, which is best described by the spectral density of the phase noise.

Consider Figure 2.20(b) where a vector representation of the additive signal and noise at the frequency $\omega_o + \omega_m$ is given. The phasor associated with the signal v_s rotates at the frequency ω_o, and the phasor associated with the sideband noise signal v_n rotates at ω_m. The phasor diagram illustrates the phase perturbation $\Delta\phi_i$ produced by the thermal noise v_n. The phase perturbation is a maximum (i.e., $\phi_{i(\max)}$) when the phasor v_n is perpendicular to v_s. Denoting the maximum phase fluctuation produced by v_n, at the frequency $\omega_o + \omega_m$ by $\Delta\phi_{i(\max),1}$ we obtain

$$\tan \Delta\phi_{i(\max),1} = \frac{v_{n,\text{rms}}}{v_{s,\text{rms}}} = \frac{|v_n|}{|v_s|} \approx \Delta\phi_{i(\max),1} \tag{2.53}$$

where it was assumed that $v_{n,\text{rms}} \ll v_{s,\text{rms}}$.

From (2.51), (2.52), and (2.53), with $B = 1$ Hz, it follows that

$$\Delta\phi_{i(\max),1} = \sqrt{\frac{FkT}{P_s}}$$

and the rms value is

$$\Delta\phi_{i(\text{rms}),1} = \frac{1}{\sqrt{2}}\sqrt{\frac{FkT}{P_s}} \tag{2.54}$$

The noise signal at $\omega_o - \omega_m$ also produces a phase fluctuation term $\Delta\phi_{i(\max),2}$, given by (2.54). Using rms addition, the total phase fluctuation is

$$\Delta\phi_{i(\text{rms})} = \sqrt{\Delta\phi^2_{i(\text{rms}),1} + \Delta\phi^2_{i(\text{rms}),2}} = \sqrt{\frac{FkT}{P_s}}$$

Therefore, the input phase-noise spectral density in a 1-Hz bandwidth is

$$S_{\phi_i}(f_m) = (\Delta\phi_{i(\text{rms})})^2 = \frac{FkT}{P_s} \tag{2.55}$$

and the SSB phase noise $\mathcal{L}(f_m)$ is

$$\mathcal{L}(f_m) = \frac{FkT}{2P_s} \tag{2.56}$$

Equations (2.55) and (2.56) determine the noise floor of $S_{\phi_i}(f_m)$ and $\mathcal{L}(f_m)$. For example, a 1-mW signal (i.e., if $P_s = 0$ dBm) passing through an amplifier with

$F = 4$ dB at a room temperature of $T = 290°K$ has an SSB phase-noise floor value of

$$\mathcal{L}(f_m)\big|_{floor} = 10 \log \frac{FkT}{2P_s}$$
$$= 10 \log(kT) + 10 \log F - 10 \log P_s - 10 \log 2$$
$$=-174 \text{ dBm} + 4 \text{ dB} - 0 \text{ dBm} - 3 \text{ dB}$$
$$=-173 \text{ dBc/Hz}$$

The noise floor of $S_{\phi_i}(f_m)$ is 3 dB higher than that of $\mathcal{L}(f_m)$.

In an actual transistor amplifier the noise floor occurs after some offset frequency f_m. In fact, the actual phase-noise spectrum of a transistor amplifier has the typical shape shown in Figure 2.21. This spectrum shows that in addition to the flat thermal-noise spectral density, the phase-noise spectrum shows a flicker characteristic (i.e., $1/f_m$ noise). The frequency f_c is known as the flicker-noise corner frequency.

While the actual physics of flicker noise is complicated, the flicker noise of a transistor can be measured. Its noise spectra has the $1/f_m$ slope characteristics. A typical dependence of f_c on dc bias current for a BJT is shown in Figure 2.22. In BJT silicon technology typical f_c values are between 1 kHz and 10 kHz. For JFETs typical values of f_c are from 50 to 100 Hz, and GaAs FETs have f_c values above 20 MHz.

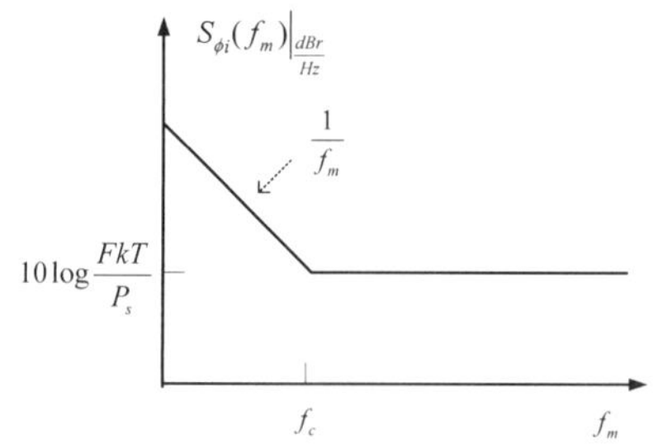

Figure 2.21 Noise-power spectrum of a transistor.

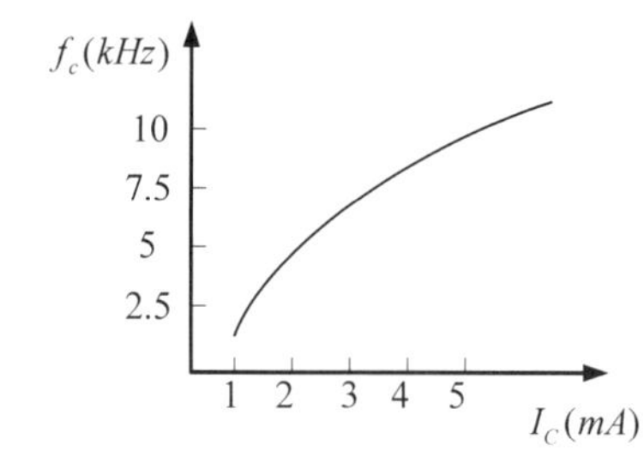

Figure 2.22 Typical dependence of f_c on I_C for a BJT.

The flicker noise of a BJT can be reduced by using negative feedback, such as the feedback produced by an emitter resistor. The reason for this is that flicker noise depends on the dc bias current, and the oscillation amplitude will produce a shift in the dc bias current due to the nonlinearity associated with the base-to-emitter junction. This noise contribution is basically an AM-to-PM effect that can be reduced using negative feedback to stabilize the bias point. A reduction in phase noise of up to 40 dB has been reported [4].

In addition to thermal and flicker noise, the other type of noise that can affect the phase noise of an oscillator is shot noise. The total noise in a circuit is due to the sum of all the noise sources. Shot noise has a flat spectrum (like thermal noise). Hence, the resulting noise spectral density has the shape shown in Figure 2.21.

The input phase-noise spectral density (see Figure 2.21) can be expressed in the form

$$S_{\phi_i}(f_m) = \frac{FkT}{P_s}\left(1 + \frac{f_c}{f_m}\right)\frac{\text{radian}^2}{\text{Hz}} \tag{2.57}$$

From (2.50) and (2.57), the output phase-noise spectral density is given by

$$S_{\phi_o}(f_m) = \frac{FkT}{P_s}\left(1 + \frac{f_c}{f_m}\right)\left[1 + \left(\frac{f_L}{f_m}\right)^2\right] \tag{2.58}$$

Hence, the SSB phase noise $\mathscr{L}(f_m)$ can be expressed in the form

$$\mathscr{L}(f_m) = \frac{FkT}{2P_s}\left(1 + \frac{f_c}{f_m}\right)\left[1 + \left(\frac{f_L}{f_m}\right)^2\right] \tag{2.59}$$

Equation (2.58) [or (2.59)] is known as Leeson's equation. Leeson's equation describes the phase-noise spectrum at the output of the amplifier. It shows how the input phase-noise spectrum [i.e., $S_{\phi_i}(f_m)$] is affected by the feedback circuit for frequencies up to f_L (where $f_L = f_o/2Q_U$). For $f_m \gg f_L$ the phase noise is not attenuated by the resonator.

Equation (2.59) can also be expressed in the form

$$\mathscr{L}(f_m) = \frac{FkT}{2P_s}\left[\frac{1}{f_m^3}\left(\frac{f_o^2 f_c}{4Q_U^2}\right) + \frac{1}{f_m^2}\left(\frac{f_o}{2Q_U}\right)^2 + \frac{1}{f_m}(f_c) + 1\right] \tag{2.60}$$

This relation shows that the phase noise of an oscillator, according to Leeson's model, is produced by four causes (i.e., the four terms in the brackets multiplied by the noise floor term $FkT/2P_s$). The first term represents the up-converted $1/f_m$ noise of the transistor (also called the flicker FM noise); the second term represents the up-converted thermal noise (also called thermal FM noise); the third term is due to flicker noise; and the fourth term is the thermal-noise floor [see (2.56)].

Certain approximations can be made in (2.59). In fact, the slope of the Bode plot of (2.59) depends on whether f_c is greater or smaller than $f_L = f_o/2Q_U$. If Q_U

is such that $f_L \gg f_c$ (usually low Q_U or small f_c value), the Bode plot of (2.59) is as shown in Figure 2.23(a). This plot shows a $1/f_m^3$ and a $1/f_m^2$ dependence. That is, the resulting SSB phase noise decreases at 30 dB/decade (or 9 dB per octave) until f_m is equal to f_c. As f_m increases beyond f_c, the noise power density decreases at 20 dB/decade (or 6 dB per octave) until the half-power bandwidth of the resonator, at which point the noise floor is reached. For example, in a 1-GHz oscillator with $Q_U = 1{,}000$ and $f_c = 5$ kHz, it follows that the plot is as shown in Figure 2.23(a), since $f_L = f_o/2Q_U = 0.5$ MHz (or $f_L \gg f_c$).

If Q_U is high, such that $f_c \gg f_L$ (usually high Q_U or large f_c value), the plot is as shown in Figure 2.23(b). In this case the plot has a $1/f_m^3$ and a $1/f$ dependence. That is, the resulting phase-noise spectral density decreases at 30 dB/decade (or 9 dB per octave) until f_m is equal to f_L. Then, the SSB phase noise decreases at 10 dB/decade (or 3 dB per octave) until f_m is equal to f_c, at which point the noise

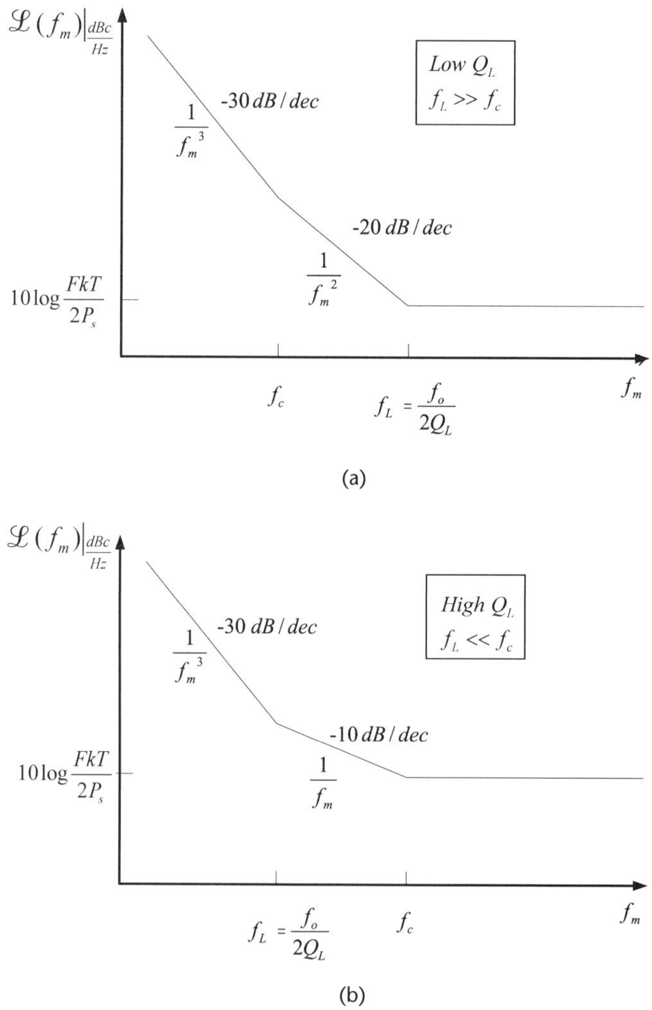

Figure 2.23 Plots of the SSB phase noise $\mathcal{L}(f_m)$ for (a) $f_L \gg f_c$ (i.e., low Q_L) and (b) $f_L \ll f_c$ (i.e., high Q_L).

floor is reached. For example, in a 1-GHz oscillator with $Q_U = 10{,}000$ and $f_c = 20$ MHz, it follows that $f_L = 50$ kHz, and the plot is as shown in Figure 2.23(b).

Equation (2.60) (or Figure 2.23) shows the various contributions to the phase noise. According to this relation the phase noise can be lowered by:

1. Selecting a transistor with low flicker noise;
2. Using a resonator configuration with a high Q_U;
3. Selecting a transistor with a low-noise figure value at the operating input impedance of the amplifier;
4. Designing the oscillator for a large P_s/FkT ratio and avoiding saturation of the transistor.

The limitations of the previous results have been observed. Lee and Hajimiri [2, 3] noted the fact that F cannot be predicted and is basically an empirically fitting parameter determined by phase-noise measurement. They also noted that the ω_{1/f^3} (i.e., the $1/f^3$ corner frequency) is equal to the $1/f$ noise corner frequency of the active device, which does not agree with phase-noise measurements. Hence, ω_{1/f^3} must also be an empirically fitted parameter. Another problem observed is that the frequency at which the noise floor is reached is not always equal to f_L. Since F and ω_{1/f^3} must be determined by measuring the oscillator spectrum, it follows that Leeson's equation cannot be used to quantitative predict the phase noise of an oscillator.

Next, the Lee and Hajimiri method is discussed. They looked at the oscillator noise as a linear time-varying system. They analyzed how a noise impulse affected the oscillator signal and found excellent agreement with the phase-noise spectrum of the oscillator.

According to Lee and Hajimiri's theory, minimal phase noise is obtained if the noise impulses coincide in time with the peaks of the oscillation signal, which shows that the system is periodic and time varying. Linearity does apply to the noise-to-phase transfer function, since doubling the injected noise will double the phase noise.

The analysis by Lee and Hajimiri is involved. They introduced an impulse sensitivity function (ISF), denoted by $\Gamma(\omega_o \tau)$. The ISF function has its maximum value near the zero crossing of the oscillation. The ISF is given by

$$\Gamma(\omega_o \tau) = \frac{c_o}{2} + \sum_{n=1}^{\infty} c_n \cos(n\omega_o \tau)$$

In terms of $\Gamma(\omega_o \tau)$, the SSB phase noise is given by

$$\mathcal{L}(f_m) = 10 \log \left(\frac{\dfrac{\overline{i_n^2}}{\Delta f} \Gamma_{\text{rms}}^2}{2 q_{\max}^2 \Delta\omega^2} \right)$$

in the $1/f^2$ region, and by

$$\mathcal{L}(f_m) = 10 \log \left(\frac{\frac{\overline{i_n^2}}{\Delta f} c_o^2}{8 q_{max}^2 \Delta\omega^2} \frac{\omega_{1/f}}{\Delta\omega} \right)$$

in the $1/f^3$ region, where $\overline{i_n^2}$ is the mean-square spectral density of the noise current, Δf is the noise bandwidth, $\omega_{1/f}$ is the $1/f$ noise corner frequency of the active device, q_{max} is the maximum charge on the resonator capacitors, and Γ_{rms} is the rms value of the ISF.

An important conclusion from the previous results is that the designer has control over how the noise is injected into the oscillator. The oscillator voltage produces a series of current pulses (say, in the collector current). The noise that depends on the device current will exist only during these current pulses. If these current pulses are wide, the associated noise pulses will contribute significantly to the phase noise. On the other hand, if the drive level is increased, so that the current pulses are narrower, the resulting phase noise will be less.

Lee and Hajimiri point out that the ISFs are an indication of the sensitive and insensitive moments to phase noise in an oscillation cycle. A low phase-noise oscillator results when the transistor remains off most of the time, conducting with a small pulse width of current at the signal peaks of each cycle. In addition, in common with Lesson's model, the signal power and the resonator Q_U should be maximized.

In Section 3.7 the relation between collector current pulses and the associated output voltage of the oscillator is discussed.

Designers of an oscillator with a certain phase noise requirement select a configuration (i.e., transistor, LC resonator, crystal resonator, dielectric resonator, and so on) that places their design in the required phase-noise region. The designer should not select an LC resonator oscillator when the phase-noise requirement can only be met with, say, a crystal oscillator. Once the design is done, the phase noise of the oscillator is measured and, if necessary, further improvements can be implemented following the previous recommendations for phase-noise reduction.

Manufacturers of oscillators specify the resulting SSB phase noise. Some typical specifications are: –110 dBc/Hz at a 1-kHz offset for a BJT oscillator operating at 100 MHz; –85 dBc/Hz at a 10-kHz offset for a BJT oscillator in the 1- to 3-GHz range; –90 dBc/Hz at a 10-kHz offset for a GaFET oscillator in the 3- to 7-GHz range; –110 to –120 dBc/Hz at a 10-Hz offset for a crystal oscillator in the 10- to 300-MHz range; –140 dBc/Hz at a 10-kHz offset for a dielectric resonator (DR) oscillator in the 3- to 7-GHz range; and –160 dBc/Hz at a 10-kHz offset for a surface acoustic wave (SAW) oscillator in the 500-MHz range.

Equation (2.59) can be used to approximately compare the phase noise between two oscillators. For example, a phase noise of –110 dBc/Hz at a 1-kHz offset at 100 MHz scales to –90 dBc/Hz at a 1-kHz offset at 1 GHz.

Many electronic systems also have SSB phase-noise specifications. For example, a wireless LAN system operating at 2.4 GHz might require an oscillator with at least –115 dBc/Hz at a 4-MHz offset. In the Personal Communication Service (PCS) a typical specification for a transmitter power of 30 dBm is –151 dBc/Hz in the

2-GHz range at a frequency offset of 20 MHz. A signal generator lists –134 dBc/Hz at 1 GHz at a frequency offset of 20 kHz.

In many applications the oscillator phase noise far from f_o can affect an incoming signal as it gets through the IF. Specifications of the phase noise for these oscillators are usually given in the $1/f_m^2$ region.

When frequency multiplication is used, the phase noise is also increased. Let $f_o = mf_i$ where m is the multiplier that relates f_i to f_o. It follows that the frequency change Δf_o, the phase change $\Delta\phi_o$, the phase-noise spectral density of f_o, and the SSB phase noise of f_o are given by

$$\Delta f_o = m\Delta f_i$$

$$\Delta\phi_o = m\Delta\phi_i$$

$$S_{\phi_o}(f_m) = m^2 S_{\phi_i}(f_m)$$

$$\mathcal{L}_{\phi_o}(f_m) = m^2 \mathcal{L}_{\phi_i}(f_m)$$

These relations show that frequency multiplication increases the phase noise by m^2 (or 20 log m). This is why crystal oscillators used in the microwave region have higher phase noise due to the frequency multiplication effect.

Example 2.2

The SSB phase noise of a 500-MHz oscillator at a frequency offset of 10 kHz is –100 dBc/Hz. The oscillator output at 500 MHz is 0 dBm. What is the oscillator output at a frequency of 500.01 MHz if: (a) a filter with a 1-Hz bandwidth is used; (b) a filter with a 2-kHz bandwidth is used?

Solution

(a) Since the frequency 500.01 MHz is at a 10-kHz offset from 500 MHz, the output at 500.01 MHz is –100 dBm in a 1-Hz bandwidth.

(b) For a small bandwidth (i.e., 2 kHz) we can assume that the SSB phase noise is constant at the value of –100 dBc/Hz. From (2.30) the output power in the 2-kHz bandwidth is approximately given by

$$P_{\text{out}} = [\mathcal{L}_\phi(f_m)P_c]BW$$

or

$$P_{\text{out}}\ [\text{dBm}] = \mathcal{L}_\phi(f_m)\left[\frac{\text{dBc}}{\text{Hz}}\right] + P_c\ [\text{dBm}] + 10\log(BW)$$

$$=-\ 100 + 0 + 10\log(2\times 10^3)$$

$$=-\ 67\ \text{dBm}$$

Example 2.3

The SSB phase noise of a 10-MHz crystal oscillator is shown in Figure 2.24. Find $S_\phi(f_m)$, the phase-noise power from 1 Hz to 1 kHz and from 1 kHz to 100 kHz.

Solution

A crystal oscillator is a low-noise oscillator. Its phase noise satisfies the condition $m_p < 0.5$ radians. Therefore, (2.43) can be used, and it follows that

$$S_\phi(f_m)\big|_{\text{dBr/Hz}} = \mathcal{L}(f_m)\big|_{\text{dBc/Hz}} + 3\text{ dB}$$

The plot of $S_\phi(f_m)$ is identical in shape to that of $\mathcal{L}(f_m)$, except that the values are 3 dB higher. That is, $S_\phi(f_m)$ at $f_m = 1$ Hz is –60 dBr/Hz.

In this example the shape of the $\mathcal{L}(f_m)$ response is of the type shown in Figure 2.23(b), where the frequency band where $\mathcal{L}(f_m)$ has a $1/f_m$ dependence is very small. Hence, the phase-noise power from 1 Hz to 1 kHz is taken as having a $1/f_m^3$ slope, and we write

$$P_\phi(1, 10^3) = \int_1^{10^3} S_\phi(f_m)\, df_m = \int_1^{10^3} \frac{10^{-6}}{f_m^3}\, df_m = 0.5\ \mu\text{rad}^2$$

and from 1 kHz to 100 kHz the phase-noise power is

$$P_\phi(10^3, 10^5) = \int_{10^3}^{10^5} S_\phi(f_m)\, df_m = \int_{10^3}^{10^5} 10^{-15}\, df_m = 0.0001\ \mu\text{rad}^2$$

Observe that the total phase-noise power from 1 Hz to 100 kHz is 0.5 + 0.0001 ≈ 0.5 μrad^2. Hence, for the given $\mathcal{L}(f_m)$ in this example, the phase-noise power depends on the spectrum from 1 Hz to 1 kHz.

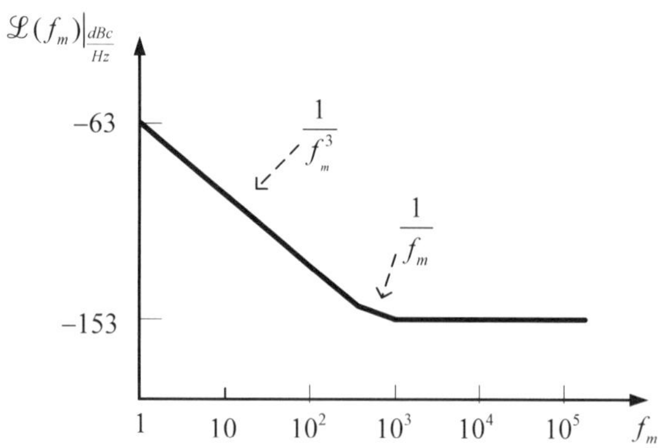

Figure 2.24 SSB phase noise $\mathcal{L}(f_m)$ for the crystal in Example 2.2.

The phase noise of an oscillator is calculated in ADS using a small-signal mixing of noise method. The SSB phase noise is accessible in the data. For example, the ADS variable

hb_noise_Vout.pnmx

displays the phase noise for the variable V_{out}, where *pnmx* means phase noise using the mixing method. The mixing analysis should be done with the option for *all-small signal frequencies* selected (i.e., set all=yes). For example, see the phase-noise plot in Figure 5.59.

2.6 Oscillator Noise Measurements

There are several methods for measuring the phase noise of an oscillator. A good reference is the Hewlett Packard Product Note 11729B-1 [5]. In this section we will discuss three methods for measuring the phase noise: the direct method, the phase-detector method, and the delay-line/frequency-discriminator method.

2.6.1 The Direct Method

This method is the simplest and most straightforward since it consists of a direct measurement. The oscillator under test is connected directly to a spectrum analyzer as shown in Figure 2.25. This method is limited by the spectrum analyzer characteristics, such as its LO phase noise, dynamic range, and resolution. Oscillators that have lower SSB phase noise than that of the spectrum analyzer can usually be measured using this method. With today's spectrum analyzers this method is suitable for many oscillators. In general, the phase noise of low-noise oscillators, such as a crystal oscillator, cannot be measured using this method.

2.6.2 The Phase-Detector Method

This is one of the most-used techniques for SSB phase noise measurements. The basic block diagram used in this method is shown in Figure 2.26.

Figure 2.26 shows two signals $v_{ref}(t)$ and $v_{osc}(t)$ applied to the mixer. The mixer (a double-balance mixer) is used as a phase detector. The low-pass filter selects the difference signal of $v_m(t)$, which is denoted by $v(t)$. If the two signals have identical frequencies and are in phase quadrature (i.e., 90° phase difference), the output voltage $v(t)$ is zero. If noise produces a small phase fluctuation $\Delta\Phi(t)$,

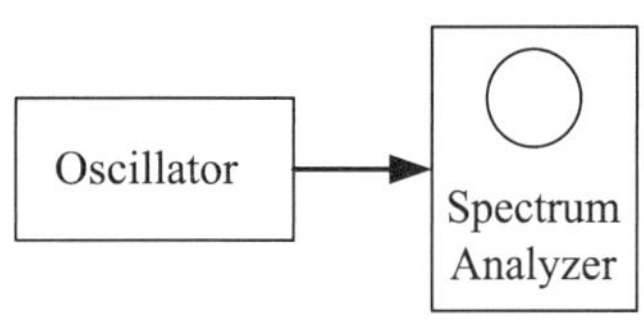

Figure 2.25 The direct method for measuring the SSB phase noise.

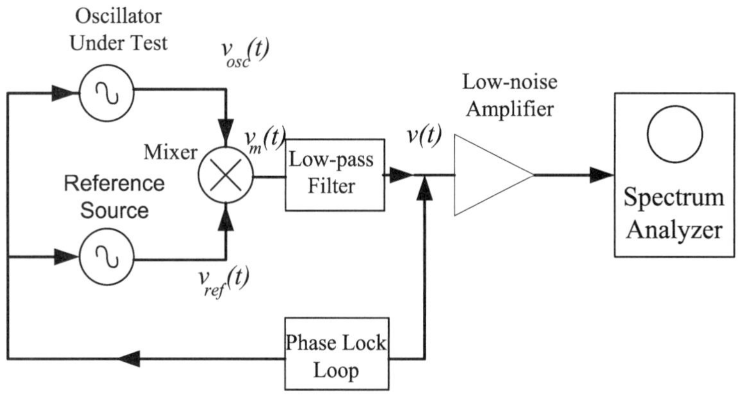

Figure 2.26 The phase-detector method for phase-noise measurements.

such that the phase difference between $v_{ref}(t)$ and $v_{osc}(t)$ is $\phi(t) = 90° + \Delta\phi(t)$, then a small fluctuating voltage appears at the output of the mixer. This voltage is proportional to the phase fluctuations $\Delta\phi(t)$. The phase-lock-loop locks the signals so they remain in quadrature.

The mixing process produces an output that contains power from both the lower and upper sidebands. For example, if the oscillator signal is described by (2.33), the output $v(t)$ contains the difference signal, which is a function of f_m. The difference signal is dc, but the lower sideband has been folded into the upper sideband around 0 Hz. These amplitudes add linearly at the output of the mixer producing four times (or 6 dB) the power of one sideband. Since the measured $\mathcal{L}(f_m)$ involves the noise power in one sideband, the measured noise power in the spectrum analyzer (i.e., in both sidebands) is 6 dB higher than that of the single sideband. Hence,

$$\begin{aligned}\mathcal{L}(f_m)\big|_{\text{dBc/Hz}} &= 10 \log P_{SSB} - 10 \log P_s \\ &= 10 \log P_m(f_m) - 6 \text{ dB} - 10 \log P_s\end{aligned} \tag{2.61}$$

where $P_m(f_m)$ represents the measured power at the frequency f_m.

Additional correction factors are usually needed in (2.61). For example, the spectrum analyzer measures noise in a bandwidth, denoted by B_{sa}. Since $\mathcal{L}(f_m)$ is defined in a 1-Hz bandwidth, (2.61) must be modified to read

$$\mathcal{L}(f_m)\big|_{\text{dBc/Hz}} = 10 \log P_m(f_m) - 6 \text{ dB} - 10 \log P_s - 10 \log B_{sa} \tag{2.62}$$

The noise bandwidth in many spectrum analyzers is approximately equal to 1.2 times the resolution bandwidth of the analyzer. Digital spectrum analyzers describe procedures to be used in noise measurements.

The analysis of the phase detector circuit in Figure 2.26 is as follows. Let

$$v_{\text{ref}} = V_r \cos \omega_r t$$

and

$$v_{\text{osc}} = V_o \cos[\omega_o t + \phi(t)]$$

Then, the output of the mixer [i.e., $v_m(t)$] is

$$v_m(t) = KA \cos[(\omega_o - \omega_r)t + \phi(t)] + KA \cos[(\omega_o + \omega_r)t + \phi(t)] + \ldots$$

where K is the gain constant of the mixer.

The low-pass filter passes the $\omega_o - \omega_r$ component (usually referred to as the *beat* signal). Hence, the input signal to the amplifier $v(t)$ can be expressed in the form

$$v(t) = KV_r V_o \cos[(\omega_o - \omega_r)t + \phi(t)]$$

The peak amplitude of $v(t)$ (i.e., $KV_r V_o$) is denoted by $V_{b,peak}$, and we can write

$$v(t) = V_{b,peak} \cos[(\omega_o - \omega_r)t + \phi(t)] \tag{2.63}$$

When the mixer is operated as a phase detector, the reference frequency must be equal to the frequency of the oscillator under test (i.e., $\omega_r = \omega_o$) and the phase shift between the two must be 90° (i.e., in quadrature). That is, let

$$\phi(t) = 90° + \Delta\phi(t)$$

where $\Delta\phi(t)$ represents the instantaneous phase fluctuation. Under these conditions, (2.63) can be expressed in the form

$$\begin{aligned} v(t) &= V_{b,peak} \cos[90° + \Delta\phi(t)] \\ &= V_{b,peak} \sin \Delta\phi(t) \end{aligned} \tag{2.64}$$

The phase fluctuation $\Delta\phi(t)$ is very small, usually much less than 1 radian. With $\Delta\phi(t) \ll 1$ rad it follows that $\sin \Delta\phi(t) \approx \Delta\phi(t)$, and (2.64) is conveniently written as

$$\Delta v(t) = V_{b,peak} \Delta\phi(t) \tag{2.65}$$

where $\Delta v(t)$ is the instantaneous voltage fluctuation (around 0V) due to the phase fluctuation. Equation (2.65) shows the linear relationship between the voltage fluctuations at the low-pass filter output and the phase noise fluctuations of the signal v_{osc}. Observe that the units of $V_{b,peak}$ are volts per radians. In the frequency domain (2.64) is

$$\Delta v(f_m) = V_{b,peak} \Delta\phi(f) \tag{2.66}$$

A typical detector characteristic is shown in Figure 2.27. $V_{b,peak}$ is known as the phase detector constant, it is simply the slope of the phase detector response

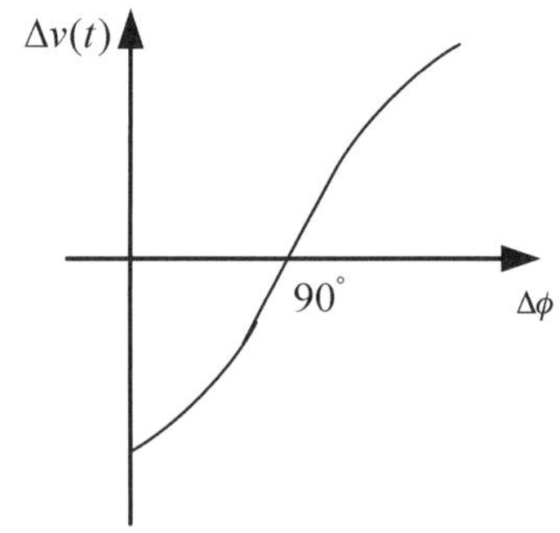

Figure 2.27 Typical phase detector characteristics of a double balance mixer.

at the zero crossing. The value of $V_{b,peak}$ can be determined by operating the mixer with the inputs at two different frequencies, such that $v(t)$ is given by (2.63). This signal can be measured with a spectrum analyzer to obtain $V_{b,peak}$. The spectrum analyzer measures the rms value of the signal (i.e., $V_{b,\text{rms}}$), where

$$V_{b,\text{rms}} = \frac{V_{b,peak}}{\sqrt{2}}$$

Once $V_{b,peak}$ is known and with the mixer operating in quadrature, the output voltage of the mixer as a function of frequency is directly proportional to the input phase deviation [see (2.66)]. Hence,

$$\Delta v(f_m) = \sqrt{2} V_{b,\text{rms}} \Delta\phi(f_m)$$

or

$$\Delta\phi_{\text{rms}}(f_m) = \frac{\Delta v_{\text{rms}}(f_m)}{\sqrt{2} V_{b,\text{rms}}}$$

The spectral density of the phase fluctuation is

$$S_\phi(f) = [\Delta\phi_{\text{rms}}(f_m)]^2 = \frac{1}{2}\left[\frac{\Delta v_{\text{rms}}(f_m)}{V_{b,\text{rms}}}\right]^2$$

and the SSB phase noise is given by

$$\mathcal{L}(f) = \frac{S_\phi(f)}{2} = \frac{1}{4}\left[\frac{\Delta v_{\text{rms}}(f)}{V_{b,\text{rms}}}\right]^2$$

In logarithmic form

$$\mathcal{L}(f)|_{\text{dBc/Hz}} = 10 \log \left[\frac{\Delta v_{\text{rms}}^2 (f_m)}{4 V_{b,\text{rms}}^2} \right]$$
$$= 20 \log \Delta v_{\text{rms}}(f_m) - 20 \log V_{b,\text{rms}} - 6 \text{ dBc/Hz}$$

The measured phase noise is equal to the combined phase noise of the two sources (i.e., the reference source and the oscillator under test). Hence, there is a measurement error introduced by the finite phase noise of the reference source. This error is minimal if the phase noise of the reference source is much lower than that of the oscillator under test. In general, the phase noise of the reference signal should be at least 10 dBc/Hz better than that of the oscillator under test. In some very-low-noise oscillators two equal oscillators with the same noise characteristics are used in the setup shown in Figure 2.26. In this case, the measured phase noise has the contribution from both sources, and a –3-dB factor should be used to account for the equal phase-noise contribution from each signal.

2.6.3 The Delay-Line/Frequency-Discriminator Method

A third method for measuring phase noise is the delay-line/frequency-discriminator method. In this method a second source is not needed. One way for implementing this measuring procedure is shown in Figure 2.28. The delay line converts the frequency fluctuations into phase fluctuations. Frequency changes in the delay line are associated with a phase shift $\Delta\phi$, as indicated in the figure. When the phase-shifted signal is mixed with the undelayed signal, the output of the doubly balanced mixer produces a voltage fluctuation ΔV which is proportional to $\Delta\phi$. Hence, the mixer acts as a frequency discriminator since ΔV is proportional to Δf. If the two input signals to the mixer are in quadrature the output voltage fluctuations are proportional to Δf. The voltage fluctuations can be measured in a spectrum analyzer and converted to phase noise.

A brief description of the system in Figure 2.28 follows. Let the signal from the oscillator under test be

$$v_s(t) = V_o \cos(\omega_o t + m_p \sin \omega_m t)$$

where $m_p = \Delta f / f_m$.

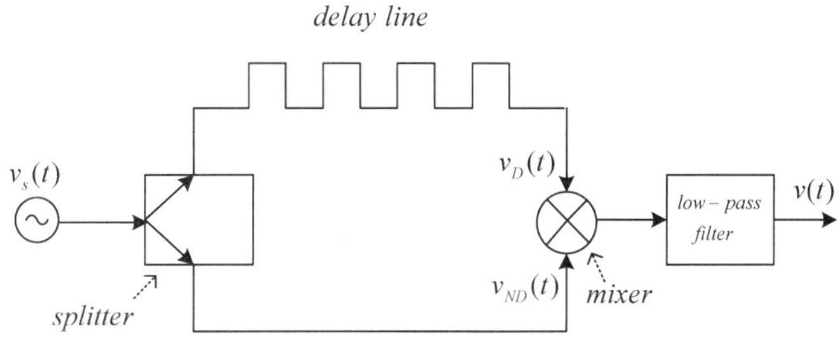

Figure 2.28 A delay-line/frequency-discriminator measuring system.

The signal after the delay line is

$$v_D(t) = KV_o \cos[\omega_o(t - \tau_d) + m_p \sin \omega_m(t - \tau_d)]$$

where τ_d is the delay time. The undelayed signal is

$$v_{ND}(t) = KV_o \cos(\omega_o t + m_p \sin \omega_m t)$$

After the mixer the low-pass filter passes the difference frequency, which is

$$\begin{aligned} v(t) &= K_\phi \cos[\omega_o(t - \tau_d) + m_p \sin \omega_m(t - \tau_d) - \omega_o t - m_p \sin \omega_m t] \\ &= K_\phi \cos\left[\omega_o \tau_d + 2m_p \sin \omega_m \frac{\tau_d}{2} \cos \omega_m \left(t - \frac{\tau_d}{2}\right)\right] \end{aligned}$$

where K_ϕ represents the amplitude of the signal after the filter.

Quadrature requires that $\omega_o \tau_d = \pi/2$ (or an odd multiple of 90°). Therefore, $v(t)$ can be written as

$$v(t) = K_\phi \sin\left[2m_p \sin \omega_m \frac{\tau_d}{2} \cos \omega_m \left(t - \frac{\tau_d}{2}\right)\right]$$

which for small m_p values, reduces to

$$v(t) = K_\phi 2m_p \sin \omega_m \frac{\tau_d}{2} \cos \omega_m \left(t - \frac{\tau_d}{2}\right) \tag{2.67}$$

The amplitude of $v(t)$ is ΔV. Hence, the transfer response is

$$\begin{aligned} \Delta V &= K_\phi 2m_p \sin \omega_m \frac{\tau_d}{2} \\ &= K_\phi 2m_p \pi f_m \tau_d \frac{\sin \pi f_m \tau_d}{\pi f_m \tau_d} \end{aligned}$$

For $f_m < 1/2\pi\tau_d$, (2.67) reduces to

$$\Delta V = K_\phi 2m_p \pi f_m \tau_d = K_\phi 2\pi\tau_d \Delta f$$

which is the desired relation.

2.7 Statistical Design Considerations

Statistical design considerations account for the random variations of the component values and how they affect the oscillator performance. Such considerations can be very useful in the design of oscillators. The effect of these random variations can be measured with an appropriate computer simulation. Finally, in order to meet the specifications, the oscillator design can be modified to account for the statistical variations.

A yield analysis simulation involves the changing of the components values and the calculation of the yield. The yield is basically the ratio of the number of designs that pass the specifications to the total number of designs in the trials, or simply the probability that a design sample meets the specifications. As the number of trials is increased, the accuracy of the yield probability increases.

Yield optimization, also known as design centering, is used to optimize the nominal values of the components so that the yield probability is improved.

The yield analysis simulation is based on the Monte-Carlo method. Basically, in a Monte-Carlo simulation the component values in a trial are randomly varied (within a specified range) and a simulation is performed. The results of the simulation are checked to see if the design passed or failed. The yield probability is calculated from the number of simulations that passed and failed in the set of trials.

The component variations can be specified in a variety of ways. Typically, a Gaussian, a uniform, or a discrete variation can be specified.

The data produced by the yield analysis can be used to generate histograms. Basically, a histogram is a bar graph that is used to display the number of times that a certain measurement or value occurred. This type of histogram is known as measurement histogram or yield histogram.

Another histogram that can be generated is the sensitivity histogram. In a sensitivity histogram the effect of a selected statistical value on a certain statistical response is observed.

In a yield optimization the nominal values of the components are adjusted in order that the yield is minimized.

Postproduction tuning can be used with yield analysis and yield optimization. This feature permits certain parameters to be tuned at the end of a yield analysis and optimization. For example, a variable capacitor could be tuned at the end of an optimization in order to meet the specifications.

The following example (using ADS) illustrates the use of statistical methods in the design of an oscillator. Statistical considerations are applied to some designs in the book to illustrate their usefulness.

Example 2.4

(a) Consider the Wien-bridge oscillator in Figure 1.20(a) where ω_O = 10 krad/s and $|v_O| = 2$V. Assume that the specification requires that the frequency of oscillation does not to vary by more than 1%. Perform a statistical analysis to calculate the yield if the components values vary by ±5%.

(b) Calculate the yield if it is also required that $1.8\text{V} \leq |v_O| \leq 2.2\text{V}$. Use components with 1% tolerance.

(c) Perform a yield optimization.

(d) Perform a postproduction tuning yield analysis.

Solution

(a) The yield analysis simulation is shown in Figure 2.29(a). The yield specification (YIELD SPEC) shows that the frequency range (i.e., MIN = 1,575; MAX = 1,606)

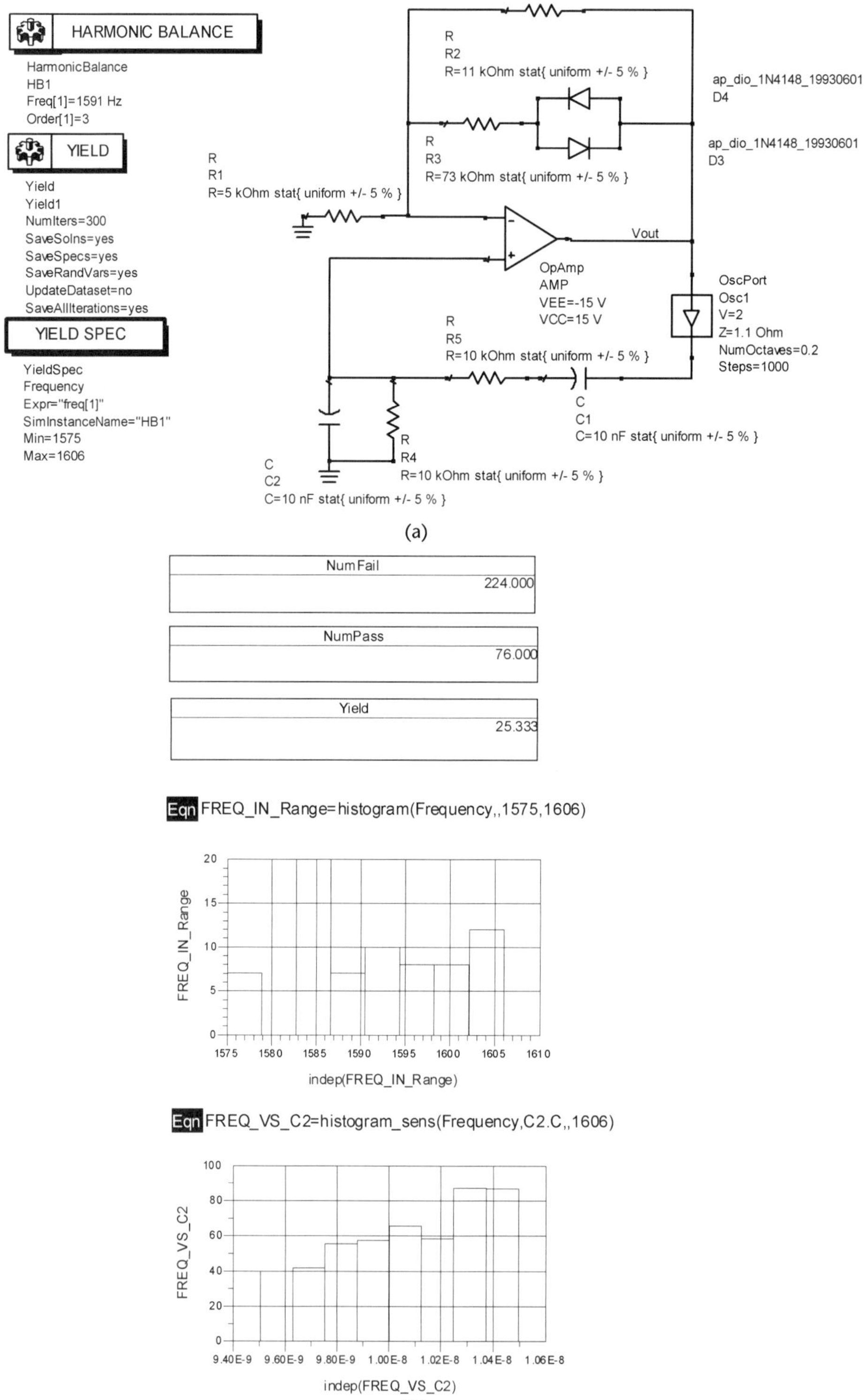

Figure 2.29 (a) The yield analysis simulation and (b) the simulation data.

should be within 1% of $f_o = 1{,}591$ Hz. The yield analysis (YIELD) shows that the number of trials used is 300.

The simulation data is shown in Figure 2.29(b). The number of trials that failed was 224 and the number that passed was 76. Hence, the yield is 25.3%. The number of frequencies that occurred in the range 1,575 to 1,606 Hz is displayed in the histogram named FREQ_IN_RANGE.

Many statistical analyses can be performed. The histogram labeled FREQ_VS_C2 shows the number of frequencies in the range 1,575 to 1,606 Hz as a function of the capacitance C2. This histogram shows that values of C2 around 10.4 nF improves the yield. This type of histograms or yield sensitivity is used to see the effect of a particular component on the yield.

The previous resulting yield of 25.3% is low for mass production of the oscillator. If the component's tolerances are changed from 5% to 1%, the yield improves to 73%.

(b) The simulation that also specifies a yield analysis for the amplitude of v_o is shown in Figure 2.30(a). The amplitude of v_o meets the amplitude yield specification if it is between 1.8V and 2.2V. The simulation data is shown in Figure 2.30(b). The number of amplitudes that occurred in the range 1.8V to 2.2V is shown in the histogram named AMPLITUDE_IN_RANGE. The number of trials that passed both specifications was 57, and those that failed were 243, for a yield of 19%. The resulting yield is poor and a yield optimization or a postproduction tuning simulation is appropriate.

(c) The yield optimization simulation is shown in Figure 2.31(a). Basically, the yield-analysis component is changed to a yield-optimization component. The yield optimization uses all the yield specifications in the simulation. The number of design iterations (i.e., shown as 20) is the number of yield improvements that the simulation will perform. The number of trials in each design iteration is set by the simulator. The simulation was performed assuming that the components can vary by ±10% from their nominal values, except R_3 which was allowed to vary by ±20%.

The resulting yield of 30%, with the component values shown in Figure 2.31(b), is low. A histogram sensitivity analysis suggests that a post-production tuning with the resistor R_3, which controls v_o, should improve the yield.

(d) The postproduction tuning simulation is shown in Figure 2.32(a). The resistor R_3 was selected for postproduction tuning by allowing a variation of ±20% from its nominal value. The data shown in Figure 2.32(b) shows that the pre-tune yield is 22.6%, and the postproduction yield is 79.6%. The use of postproduction tuning, in this case, produces a significant improvement of the yield. Further yield improvements can be obtained if postproduction tuning is also used with R_4 (which can adjust the frequency).

Associated with a statistical simulation is the number of trials required to obtain a certain yield. The number of times N is given by

$$N = \left(\frac{C_\sigma}{\epsilon}\right)^2 Y(1 - Y)$$

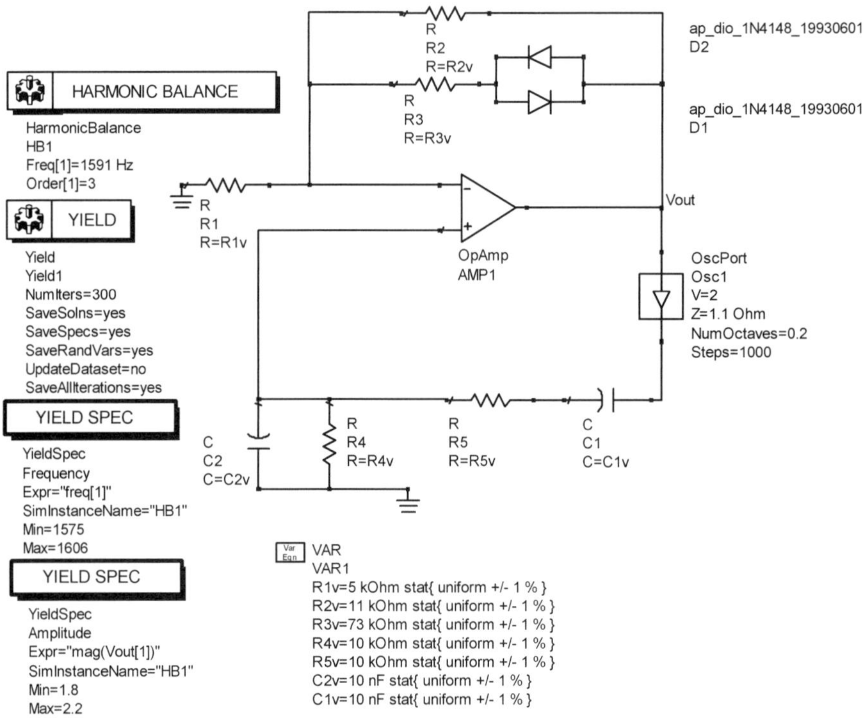

(a)

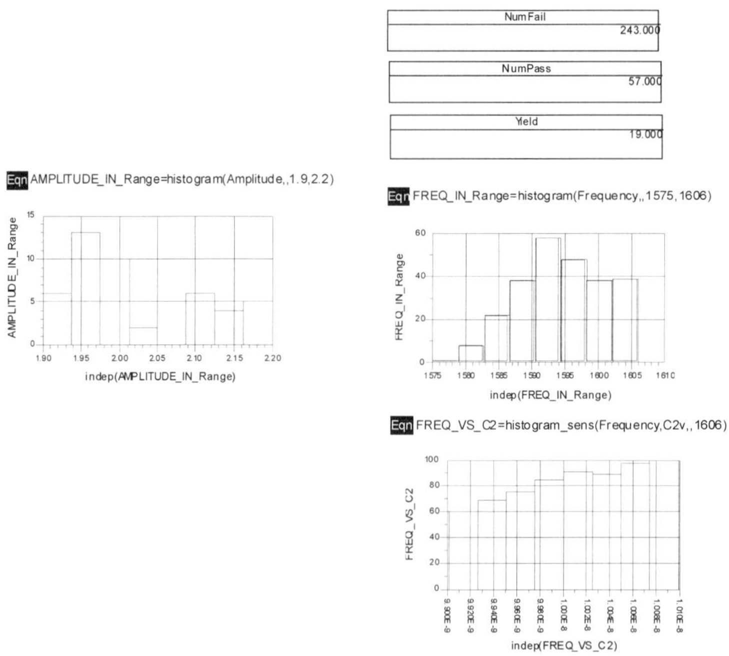

(b)

Figure 2.30 (a) Yield analysis simulation for frequency and amplitude and (b) the simulation data.

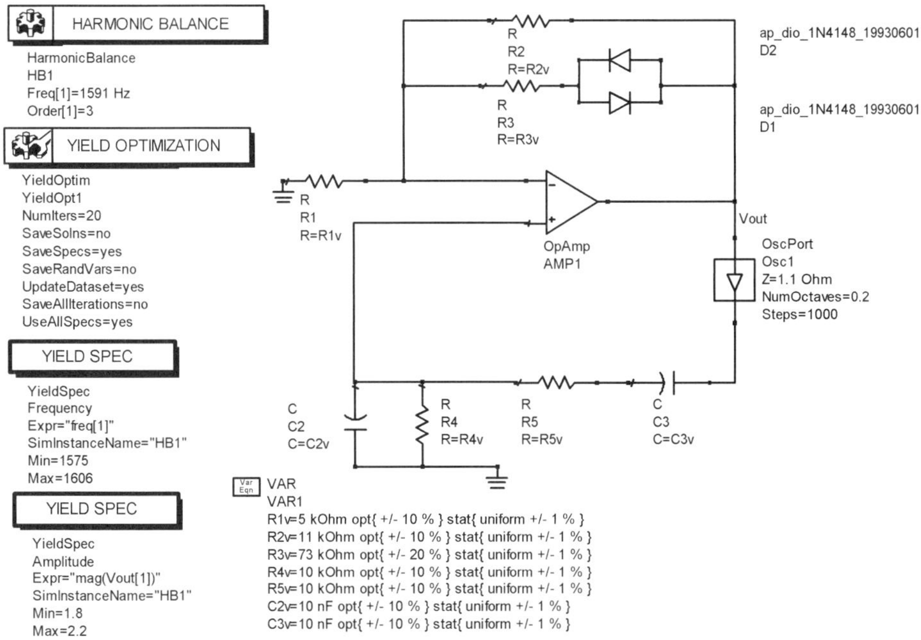

(a)

C2v	C3v	R1v	R2v	R3v	R4v	R5v
9.962E-9	1.003E-8	5021.307	11041.963	73168.533	10015.537	10069.935

InitialYield	FinalYield
18.000	30.000

(b)

Figure 2.31 (a) The yield optimization simulation and (b) the simulation data.

where Y is the actual yield, ϵ is the error, and C_σ is the confidence level expressed as a number of standard deviations. The error is defined as the absolute difference between the actual yield Y and the yield estimate $\overline{Y}$. That is,

$$\epsilon = \left| Y - \overline{Y} \right|$$

Therefore, the low and high values of the yield estimate are

$$\overline{Y} = Y \pm \epsilon$$

The confidence level is associated with the area under the Gaussian distribution curve for a given standard deviation. Typical values are shown in Table 2.2.

For example, for a confidence level of 68.3% (i.e. $C_\sigma = 1$) with an error of ±3% (i.e., $\epsilon = 0.03$) and a yield of 70%, the required number of trials is

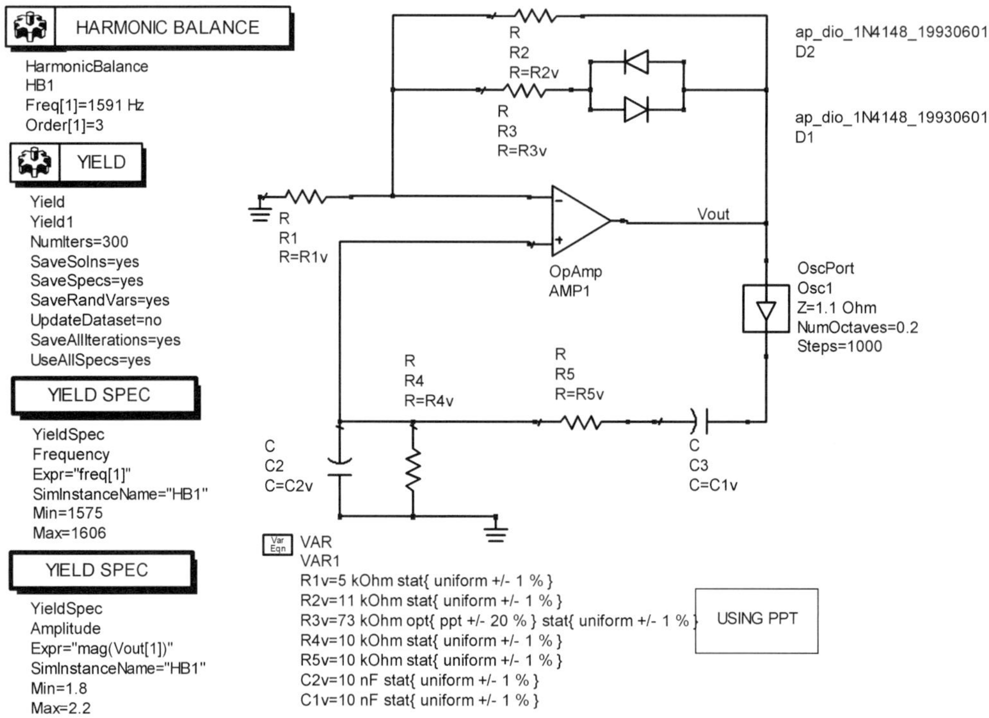

(a)

PreTuneFail	PreTunePass	PreTuneYield	PPT_Yield
218.000	82.000	27.333	79.333

(b)

Figure 2.32 (a) Postproduction tuning simulation and (b) the simulation data.

Table 2.2 Confidence Level Values

C_σ	*Confidence Level*
1	68.3%
2	95.4%
3	99.7%

$$N = \left(\frac{1}{0.03}\right)^2 0.7(1 - 0.7) = 233$$

References

[1] Leeson, D. B., "A Simple Model of a Feedback Oscillator Noise Spectrum," *Proc. IEEE*, Vol. 54, February 1966, pp. 329–330.

[2] Hajimiri, A., and T. H. Lee, "A General Theory of Phase Noise in Electrical Oscillators," *IEEE Journal of Solid-State Circuits*, Vol. 33, February 1998, pp. 179–194.

[3] Lee, T. H., and A. Hajimiri, "Oscillator Phase Noise: A Tutorial," *IEEE Journal of Solid-State Circuits*, Vol. 35, March 2006, pp. 326–336.

[4] Rohde, U. L., and D. P. Newkirk, *RF/Microwave Circuit Design for Wireless Applications*, New York: Wiley Interscience, 2000.

[5] Hewlett-Packard, *Phase Noise Characterization of Microwave Oscillators*, Product Note 11729B-1, Palo Alto, CA, 1984.

CHAPTER 3

Tuned-Circuit Oscillators

3.1 Introduction

Tuned-circuit oscillators are very popular at frequencies above 100 kHz. The tuned circuit determines the frequency of oscillation and acts as the feedback network. The use of high-Q resonant circuits produces a sinusoidal oscillator with low distortion. The amplifier section is usually designed using discrete active devices, such as BJTs and FETs.

Because of certain differences in the ac models associated with BJTs and FETs, the analysis of tuned-circuit oscillators is divided into BJT tuned-circuit oscillators and FET tuned-circuit oscillators. Tuned oscillators using op amps are also discussed.

Modern simulators programs for oscillator designs produce excellent results. Hence, the implementation of a tuned oscillator after an appropriate simulation usually requires a minimum amount of adjustment in the laboratory.

3.2 FET Tuned Oscillators

The following discussion provides an overview of FET tuned-circuit oscillators. Consider the FET tuned-circuit oscillator shown in Figure 3.1(a). Assume that there is no mutual inductance coupling between the impedances. The ac model of the oscillator is shown in Figure 3.1(b) where the RFC is open circuited and the coupling and bypass capacitors are short circuited.

From Figure 3.1(b), observing that the feedback voltage v_f is v_{gs}, the open-loop voltage gain is

$$A_v(j\omega) = \frac{v_o(j\omega)}{v_f(j\omega)} = \frac{-g_m r_d Z_L}{Z_L + r_d}$$

where Z_L is the load impedance, namely,

$$Z_L = Z_2 \,||\, (Z_1 + Z_3)$$

The resistor R_G is very large and does not load the impedance Z_1.

The feedback factor $\beta(j\omega)$ is given by

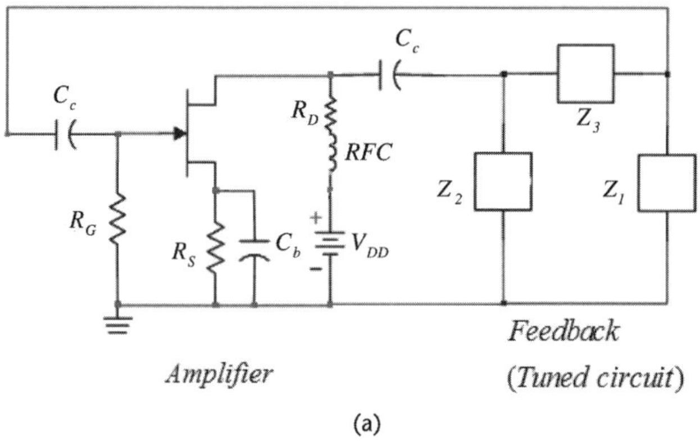

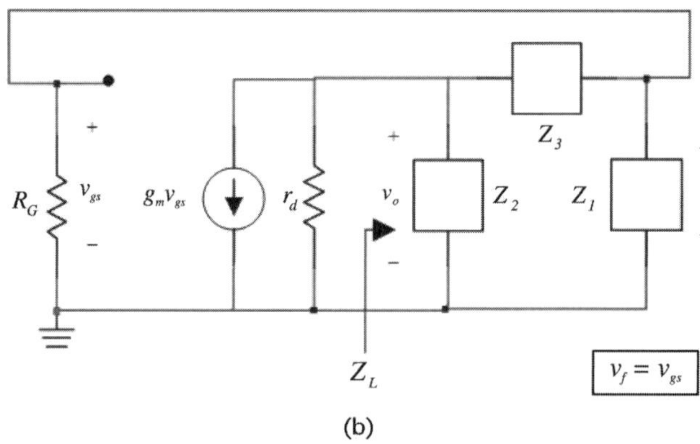

Figure 3.1 (a) An FET tuned-circuit oscillator and (b) the ac model.

$$\beta(j\omega) = \frac{v_f(j\omega)}{v_o(j\omega)} = \frac{Z_1}{Z_1 + Z_3}$$

Hence, the loop gain is given by

$$\beta(j\omega)A_v(j\omega) = \frac{-g_m r_d Z_L Z_1}{(Z_L + r_d)(Z_1 + Z_3)} = \frac{-g_m r_d Z_1 Z_2}{r_d(Z_1 + Z_2 + Z_3) + Z_2(Z_1 + Z_3)} \tag{3.1}$$

If Z_1, Z_2, and Z_3 are purely reactive impedances given by $Z_1(j\omega) = jX_1(\omega)$, $Z_2(j\omega) = jX_2(\omega)$, and $Z_3(j\omega) = jX_3(\omega)$, the loop gain in (3.1) can be expressed in the form

$$\beta(j\omega)A_v(j\omega) = \frac{g_m r_d X_1(\omega) X_2(\omega)}{jr_d[X_1(\omega) + X_2(\omega) + X_3(\omega)] - X_2(\omega)[X_1(\omega) + X_3(\omega)]} \tag{3.2}$$

The phase shift of the loop gain is zero when the imaginary part of (3.2) is zero. That is, at $\omega = \omega_o$

$$X_1(\omega_o) + X_2(\omega_o) + X_3(\omega_o) = 0 \tag{3.3}$$

Equation (3.3) determines the frequency of oscillation. At the frequency ω_o [where (3.3) is satisfied] the loop gain in (3.2) reduces to

$$\beta(j\omega_o)A_v(j\omega_o) = \frac{-g_m r_d X_1(\omega_o)}{X_1(\omega_o) + X_3(\omega_o)}$$

and using (3.3) we obtain

$$\beta(j\omega_o)A_v(j\omega_o) = \frac{g_m r_d X_1(\omega_o)}{X_2(\omega_o)} \tag{3.4}$$

For oscillations the loop gain in (3.4) must be unity at the frequency of oscillation, and greater than unity to start the oscillation. Therefore, $X_1(\omega)$ and $X_2(\omega)$ must have the same signs. That is, if Z_1 is capacitive (i.e., $X_1(\omega) =- 1/\omega C_1$), then Z_2 must also be capacitive (i.e., $X_2(\omega) =- 1/\omega C_2$). From (3.3), it also follows that $X_3(\omega_o)$, where $X_3(\omega_o) =- X_1(\omega_o) - X_2(\omega_o)$, must be inductive (i.e., $X_3(\omega) = j\omega L$).

The oscillator in Figure 3.1(a) with Z_1 and Z_2 capacitive, and Z_3 inductive is known as an FET Pierce oscillator. For such configuration, the frequency of oscillation, from (3.3), is given by

$$-\frac{1}{\omega_o C_1} - \frac{1}{\omega_o C_2} + \omega_o L = 0 \tag{3.5}$$

or

$$\omega_o = \frac{1}{\sqrt{LC_T}} \tag{3.6}$$

where

$$C_T = \frac{C_1 C_2}{C_1 + C_2}$$

At ω_o the phase shift through the amplifier is $-180°$; therefore, the phase shift through the feedback network must also be $-180°$. The phase shift due to $\beta(j\omega)$ can be seen as follows. Since

$$\beta(j\omega) = \frac{\dfrac{1}{j\omega C_1}}{\dfrac{1}{j\omega C_1} + j\omega L}$$

and observing that the denominator must have a net inductive effect at ω_o, we can express the impedance in the denominator in terms of an equivalent L'. From (3.5), the equivalent L' is

$$j\omega_o L' = j\left(\omega_o L - \frac{1}{\omega_o C_1}\right) = \frac{j}{\omega_o C_2}$$

Therefore,

$$\beta(j\omega_o) = \frac{\left(\dfrac{1}{j\omega_o C_1}\right)}{j\omega_o L'} = -\frac{C_2}{C_1}$$

which shows that the phase of $\beta(j\omega_o)$ is $-180°$.

The gain condition follows from (3.4), namely,

$$g_m r_d \geq \frac{X_2(\omega_o)}{X_1(\omega_o)} = \frac{C_1}{C_2} \tag{3.7}$$

While the analysis of tuned oscillators can be performed in terms of the loop gain, an alternate approach is in terms of the system determinant as follows. A general configuration for a common-source FET tuned oscillator is shown in Figure 3.2(a). The coupling capacitors are designed to act as short circuits at the frequency of oscillation. The drain resistor R_D is replaced in many designs by a radio frequency

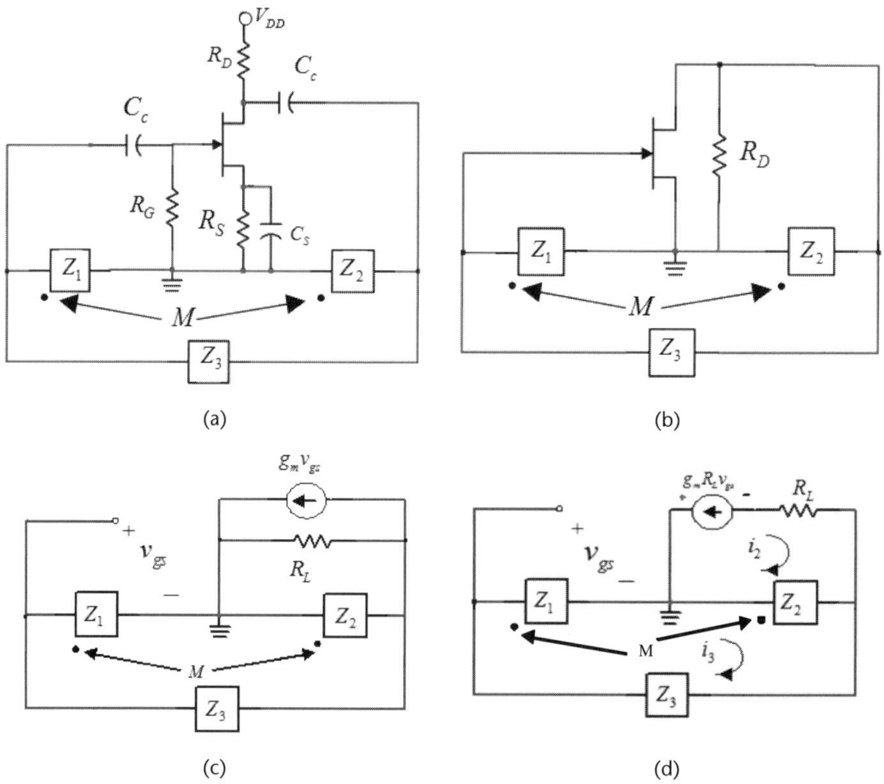

Figure 3.2 (a) General configuration of FET tuned oscillators, (b) an ac model, (c) the FET replaced by its ac model, and (d) a Thevenin's equivalent model.

coil (RFC), or an RFC in series with R_D is used. An RFC is very convenient since its inductance is selected so that its impedance is very large at the frequency of oscillation, and therefore blocks the oscillation signal from going into the dc supply. Hence, in the ac model the RFC is open circuited and the ac signal path is from the FET to the feedback network, and back to the input of the FET. The elements that determine the closed-loop path of the signal are shown in Figure 3.2(b). The elements Z_1, Z_2, and Z_3 are the reactive elements that form the tuned circuit, and provide the proper positive feedback at the frequency of oscillation. A mutual impedance is included between Z_1 and Z_2 because in some oscillators these impedances represent the inductances of a coupled coil. The mutual impedance is related to L_1 and L_2 by

$$M = \frac{k}{\sqrt{L_1 L_2}}$$

where k is the coefficient of coupling ($0 < k < 1$). If Z_1 and Z_2 are capacitive impedances, then $M = 0$. The FET provides the proper amplification and a phase shift of $-180°$. Therefore, the tuned circuit must contribute a phase shift of $-180°$ at the frequency of oscillation.

In Figure 3.2(c) the FET is replaced by its transconductance model where $R_L = r_d \| R_D$, and in Figure 3.2(d) a Thevenin's equivalent model is used. The loop equations for the ac model in Figure 3.2(d) are

$$(R_L + Z_2)i_2 - (Z_2 + Z_M)i_3 =- g_m R_L v_{gs} \tag{3.8}$$

$$-(Z_2 + Z_M)i_2 + (Z_1 + Z_2 + Z_3 + 2Z_M)i_3 = 0 \tag{3.9}$$

and

$$v_{gs} = (Z_1 + Z_M)i_3 - Z_M i_2 \tag{3.10}$$

Substituting (3.10) into (3.8) gives

$$(R_L + Z_2 - g_m R_L Z_M)i_2 - [Z_2 + Z_M - g_m R_L(Z_1 + Z_M)]i_3 = 0 \tag{3.11}$$

Equations (3.9) and (3.11) represent the system of equations for the circuit in Figure 3.1(d). The conditions for oscillations can be obtained by setting the system determinant of (3.9) and (3.11) equal to zero. That is,

$$D(j\omega) = \begin{vmatrix} -(Z_2 + Z_M) & (Z_1 + Z_2 + Z_3 + 2Z_M) \\ (R_L + Z_2 - g_m R_L Z_M) & -[Z_2 + Z_M - g_m R_L(Z_1 + Z_M)] \end{vmatrix} = 0 \tag{3.12}$$

Observe that $D(j\omega)$ is a complex quantity; that is,

$$D(j\omega) = \mathrm{Re}[D(j\omega)] + j\mathrm{Im}[D(j\omega)]$$

In order to proceed we must specify the elements Z_1, Z_2, and Z_3. Several choices of elements are possible. A simple and very practical choice is to select the three elements to be reactive. That is, let $Z_1 = jX_1$, $Z_2 = jX_2$, $Z_3 = jX_3$, and the mutual impedance is $Z_M = jX_M$. Observe that the reactances X_1, X_2, and X_3 are positive when they represent an inductor, and negative for a capacitor. With these values of impedances, the imaginary part of (3.12) is

$$\mathrm{Im}[D(j\omega)] = R_L(X_1 + X_2 + X_3 + 2X_M)$$

The frequency of oscillation is obtained by setting $\mathrm{Im}[D(j\omega)] = 0$, which gives

$$X_1(\omega_o) + X_2(\omega_o) + X_3(\omega_o) + 2X_M(\omega_o) = 0 \tag{3.13}$$

since $R_L \neq 0$. This equation determines the frequency of oscillation.

The gain condition is obtained by setting the real part of (3.12) equal to zero. Actually, we let $\mathrm{Re}[D(j\omega)] > 0$ in order to start the oscillation. Thus, the gain condition is written as

$$\mathrm{Re}[D(j\omega)] = g_m R_L(X_1 + X_M)(X_2 + X_M) - (X_2 + X_M)^2 \geq 0$$

or

$$g_m R_L \geq \frac{X_2 + X_M}{X_1 + X_M} \tag{3.14}$$

Since $g_m R_L$ is a positive quantity, and with the mutual inductance X_M smaller than X_1 or X_2, it follows that the right-hand side of (3.14) is a positive quantity only if X_1 and X_2 represent the same type of elements (i.e., either inductors or capacitors). Also, with X_1 and X_2 having the same sign, (3.13) is satisfied if X_3 has an opposite sign to that of X_1 and X_2. Hence, if Z_1 and Z_2 are capacitive, then $Z_M = 0$ and Z_3 must be inductive. If Z_1 and Z_2 are inductive with mutual inductance Z_M, then Z_3 must be capacitive.

In a design, the small-signal value of g_m is commonly used to satisfy the inequality in (3.14). Of course, as the amplitude of oscillation increases towards its stable value, the value of g_m decreases and (3.14) is approximately satisfied with the equality sign. Actually, as discussed in Chapter 1, the poles of the oscillator move between the right-half plane and the left-half plane around the frequency ω_o.

A common way of designing the circuit in Figure 3.2(a) is with an RFC in place of R_D, or with an RFC in series with R_D, Since the ac behavior of the RFC is essentially an open circuit, the previous relations apply with $R_L = r_d$. If the series resistance of the RFC is considered (i.e., a finite Q_U), it can be transformed to an equivalent parallel resistance R_p. In this case: $R_L = r_d || R_p$; which, in general, can be approximated by r_d.

When Z_1 and Z_2 are capacitive and Z_3 is an inductor, the oscillator is usually referred to as a Colpitts-type oscillator. If Z_1 and Z_2 are inductive and Z_3 is a capacitor the oscillator is called a Hartley oscillator. The oscillator in Figure 3.2 uses an FET in a CS configuration. This Colpitts-type oscillator is known as a Pierce oscillator. If the FET is used in a CD configuration, the oscillator is called a Colpitts oscillator. If the oscillator uses the FET in a CG configuration (another Colpitts-type oscillator) it is known as a Grounded-Gate oscillator.

3.2.1 FET Pierce Oscillator

An FET Pierce oscillator is shown in Figure 3.3(a). In this oscillator, Z_1 and Z_2 are capacitive impedances (i.e., $X_1 =- 1/\omega C_1$ and $X_2 =- 1/\omega C_2$), and Z_3 is an inductive impedance (i.e., $X_3 = \omega L$). There is no mutual inductance (i.e., $X_M = 0$). An RFC (or an RFC in series with a drain resistor) is used from the drain to V_{DD}. Therefore, R_L in (3.14) is simply equal to r_d. The ac model of the oscillator is shown in Figure 3.3(b). Observe that the feedback voltage v_f is v_{gs}, and that it is the same oscillator configuration as the one shown in Figure 3.1, which was analyzed using the loop gain.

From (3.13), the frequency of oscillation is obtained from

$$X_1(\omega_o) + X_2(\omega_o) + X_3(\omega_o) =- \frac{1}{\omega_o C_1} - \frac{1}{\omega_o C_2} + \omega_o L = 0$$

which gives

$$\omega_o = \frac{1}{\sqrt{LC_T}} \tag{3.15}$$

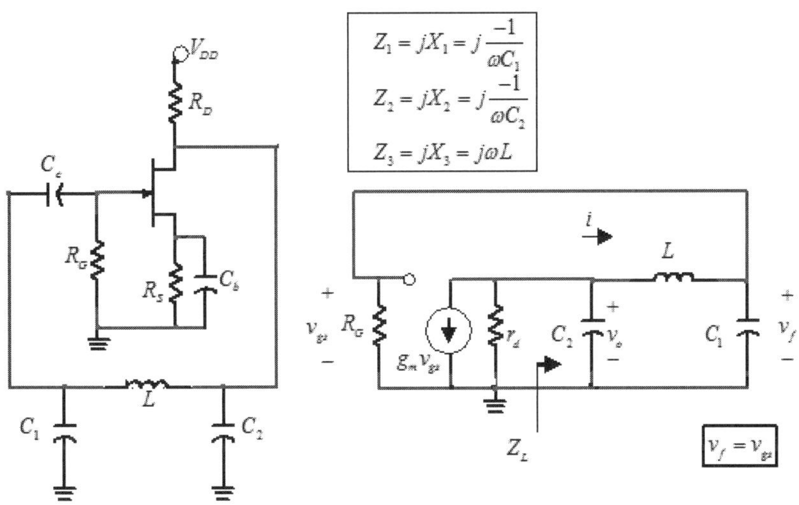

(a) (b)

Figure 3.3 (a) An FET Pierce oscillator and (b) the ac model.

where

$$C_T = \frac{C_1 C_2}{C_1 + C_2}$$

The gain condition follows from (3.14). That is,

$$g_m r_d \geq \frac{X_2(\omega_o)}{X_1(\omega_o)} = \frac{C_1}{C_2} \tag{3.16}$$

Observe that (3.15) and (3.16) are identical to (3.6) and (3.7), respectively.

To start the oscillation, the inequality in (3.16) is satisfied by making $g_m r_d$ equal to C_1/C_2 times a factor of 1.5 to 3.

In order to better understand how the Pierce oscillator works, consider the ac model shown in Figure 3.3(b). The series combination of L and C_1 must appear inductive to produce with C_2 a parallel-resonant tuned circuit. Hence, the current i lags v_o by 90°, and v_f lags i by 90°. Consequently, v_f lags v_o by 180°. Since the transistor provides a phase shift of –180°, the total phase shift around the loop is –360°, as required for oscillation.

At ω_o, the load impedance

$$Z_L = Z_2 \parallel (Z_1 + Z_3) = \frac{jX_2(X_1 + X_3)}{X_1 + X_2 + X_3}$$

is infinite. Hence, the open-loop gain at ω_o is

$$A_v(j\omega_o) = \frac{v_o}{v_f} = -g_m r_d \tag{3.17}$$

Also, the feedback factor at ω_o is given by

$$\beta(j\omega_o) = \frac{v_f}{v_o} = \frac{X_1}{X_1 + X_3} = -\frac{X_1}{X_2} = -\frac{C_2}{C_1} \tag{3.18}$$

Equations (3.17) and (3.18) show the –360° phase shift that exists around the loop at ω_o.

The gain condition is obtained from

$$\beta(j\omega_o)A_v(j\omega_o) \geq 1 \tag{3.19}$$

Substituting (3.17) and (3.18) into (3.19) gives

$$g_m r_d \geq \frac{C_1}{C_2}$$

which is recognized as the gain condition in (3.16).

The frequency of oscillation and the gain condition are affected by the series resistance of the inductor and the loading of the oscillator (which occurs when the oscillator signal is coupled to another stage). These effects can be analyzed by including appropriate resistances in (3.12). Of course, the required analysis is more involved. The series loss resistance of the inductor R_s can be included in the analysis by letting $Z_3 = R_s + j\omega L$. Then, it follows that

$$Z_L = jX_2 \,||\, (R_s + jX_3 + jX_1) = \frac{jX_2[j(X_1 + X_3) + R_s]}{j(X_1 + X_2 + X_3) + R_s} \tag{3.20}$$

and the voltage feedback factor is

$$\beta(j\omega) = \frac{v_f}{v_o} = \frac{Z_1}{Z_1 + Z_3} = \frac{jX_1}{R_s + j(X_1 + X_3)} \tag{3.21}$$

Hence, the loop gain is given by

$$\beta(j\omega)A_v(j\omega) = \beta(j\omega)\left(-\frac{g_m r_d Z_L}{Z_L + r_d}\right) = \frac{g_m r_d X_1 X_2}{jr_d(X_1 + X_2 + X_3) - X_2(X_1 + X_3) + r_d R_s} \tag{3.22}$$

where the approximation $jX_2(r_d + R_s) \approx jr_d X_2$ was used.

From (3.22), the frequency of oscillation follows from

$$X_1(\omega_o) + X_2(\omega_o) + X_3(\omega_o) = 0$$

or

$$\omega_o = \frac{1}{\sqrt{LC_T}}$$

where

$$C_T = \frac{C_1 C_2}{C_1 + C_2}$$

which is identical to (3.15).

At ω_o, the load impedance is

$$Z_L(j\omega_o) = \frac{jX_2(\omega_o)}{R_s}[-jX_2(\omega_o) + R_s]$$

Since $|X_2(\omega_o)| \gg R_s$, Z_L can be approximated by

$$Z_L \approx \frac{X_2^2(\omega_o)}{R_s} \tag{3.23}$$

The open-loop gain and feedback factors are

$$A_v(j\omega_o) \approx \frac{-g_m r_d [X_2(\omega_o)]^2}{[X_2(\omega_o)]^2 + r_d R_s} \tag{3.24}$$

and

$$\beta(j\omega_o) = \frac{jX_1}{-jX_2 + R_s} \approx - \frac{X_1}{X_2} = - \frac{C_2}{C_1} \tag{3.25}$$

At ω_o, from (3.24) and (3.25), the loop gain is given by

$$\beta(j\omega_o)A_v(j\omega_o) = \frac{g_m r_d X_1 X_2}{X_2^2(\omega_o) + r_d R_s} > 1$$

and it follows that the gain condition is

$$g_m r_d > \frac{X_2^2(\omega_o) + r_d R_s}{X_1(\omega_o) X_2(\omega_o)} \tag{3.26}$$

Observe that if $R_s \to 0$, then (3.26) is identical to (3.16).

Equation (3.26) shows how the loop gain is affected by the coil losses. In many cases $r_d R_s \gg X_2^2(\omega_o)$, and the loop gain in (3.26) is satisfied when

$$\frac{g_m}{\omega_o^2 R_s C_1 C_2} > 1 \tag{3.27}$$

This relation basically places a limit on C_1, C_2, and the required g_m. The inequality shows that small values of C_1 and C_2 are desirable. However, the values of C_1 and C_2 cannot be made too small because then the oscillator is dependent on the input and output capacitances of the transistor.

Example 3.1

Design the Pierce oscillator in Figure 3.3(a) to oscillate at 1 MHz. Use the 2N3819 n-channel JFET. This transistor has the following typical parameter values: $I_{DSS} = 12$ mA and $V_P = -3.5$V.

Solution

Selecting the Q point at $I_D = 6$ mA and $V_{DS} = 10$V, and using $V_{DD} = 20$V it follows from

$$I_D = I_{DSS}\left(1 - \frac{V_{GS}}{V_P}\right)^2$$

that $V_{GS} \approx -1$V and

$$g_m = \frac{2I_{DSS}}{V_p^2}(V_{GS} - V_P) = \frac{2(12 \times 10^{-3})}{(3.5)^2}(-1 + 3.5) \approx 5 \text{ mS}$$

The output resistance can be calculated in terms of the channel modulation voltage V_M (similar to the Early voltage) as

$$r_o = \frac{V_M}{I_D}$$

which for $I_D = 6$ mA a typical r_o is 50 kΩ.

The values of R_S and R_D are

$$R_S = \frac{-V_{GS}}{I_D} = \frac{1}{6 \times 10^{-3}} = 164\Omega$$

and

$$R_D = \frac{V_{DD} - V_{DS} - I_D R_S}{I_D} = \frac{20 - 10 - 1}{6 \times 10^{-3}} = 1.47 \text{ k}\Omega$$

The capacitor C_2 is in parallel with the transistor's output terminals. The transistor has a few picofarads of output capacitance (C_{out}) between its source and drain. This output capacitance can be neglected if $C_2 \gg C_{out}$. Several values of C_2, and the approximate value of $|X_{C_2}|$, that satisfy $C_2 \gg C_{out}$, are listed in Table 3.1.

Two design procedures are considered. First, assuming that the loss resistance of the inductor is negligible (i.e., $R_s = 0$), and second with $R_s \neq 0$.

If $R_s = 0$, (3.16) gives

$$g_m r_d = 0.05(50 \times 10^3) = 250 > \frac{C_1}{C_2}$$

Letting $C_2 = 1.6$ nF, it follows that

$$C_1 < 250(1.6 \times 10^{-9}) = 0.4\ \mu\text{F}$$

The inequality can be satisfied with $C_1 = 100$ nF.

Table 3.1 Reactance Values at 1 MHz

C_2	$\lvert X_{C_2} \rvert$
160 nF	1
16 nF	10
1.6 nF	100
160 pF	1,000

From (3.15), with $C_T \approx C_2 = 1.6$ nF, the value of the inductor is 15.8 μH.

If $R_s \neq 0$, Table 3.2 illustrates the calculation of L for $C_2 = 1.6$ nF with $C_1 = 100$ nF, and also for $C_2 = 16$ nF with $C_1 = 20$ nF. Typical values of the unloaded Q (i.e., $Q_U = \omega L/R_s$) for the inductor are listed with the associated R_s. Finally, the inequality in (3.26) is checked.

The above calculations show that the assumption $R_s = 0$ will not work in this case. Of course, when using the design procedure based on $R_s = 0$ there are values for which (3.26) is not satisfied and, therefore, oscillations will not occur.

The calculations also show that if the inductor's Q_U is 30 (i.e., $R_s = 0.6\Omega$) the oscillator will not work. However, if Q_U is increased to 50 the oscillation conditions are satisfied. Also, the inequality (3.26) is better satisfied with a higher value of g_m. This can be attained by designing the oscillator with $V_{GS} =- 0.5$V so that $g_m = 6$ mS.

The simulation of the oscillator with $C_1 = 20$ nF, $C_2 = 16$ nF, and $L = 2.86$ μH ($Q_U = 50$) is shown in Figure 3.4(a). The Q point is at $I_D = 5.68$ mA and $V_{DS} = 10.6$V. The resulting oscillation at 999.9 kHz is shown in Figure 3.4(b). Although not shown, the simulation was also performed by designing the circuit with $V_{GS} =- 0.5$V and $I_D = 6$ mA.

From a practical point of view, the above calculation (based on typical transistor parameters) shows that (3.26) is "barely" satisfied (250 > 228). This information is especially useful if many units are to be constructed since the spread of the transistor's parameters might affect the yield. Hence, the designer can use the statistical analysis tools, or might redesign the oscillator using a transistor with a higher g_m.

3.2.2 FET Colpitts Oscillator

An FET Colpitts configuration is shown in Figure 3.5(a) and its ac model in Figure 3.5(b). The capacitor C_c is a coupling capacitor. A loop-gain approach is used to analyze the oscillator. However, a system determinant approach could have been used similar to the analysis of the BJT Colpitts oscillator in Section 3.3.

From Figure 3.5(b) the load impedance [see (3.20)] is

$$Z_L = \frac{jX_2[j(X_1 + X_3) + R_s]}{j(X_1 + X_2 + X_3) + R_s}$$

At resonance

$$\omega_o = \frac{1}{\sqrt{LC_T}}$$

Table 3.2 Loop-Gain Condition [i.e., (3.26)] for Example 3.1

C_1	C_2	C_T	L	Q_U	R_s	$g_m r_d > \frac{X_2^2 + r_d R_s}{X_1 X_2}$
100 nF	1.6 nF	1.6 nF	15.8 μH	50	1.98Ω	250 > 685 (not satisfied)
20 nF	16 nF	8.9 nF	2.85 μH	30	0.6Ω	250 > 378 (not satisfied)
20 nF	16 nF	8.9 nF	2.85 μH	50	0.36Ω	250 > 228 (satisfied)

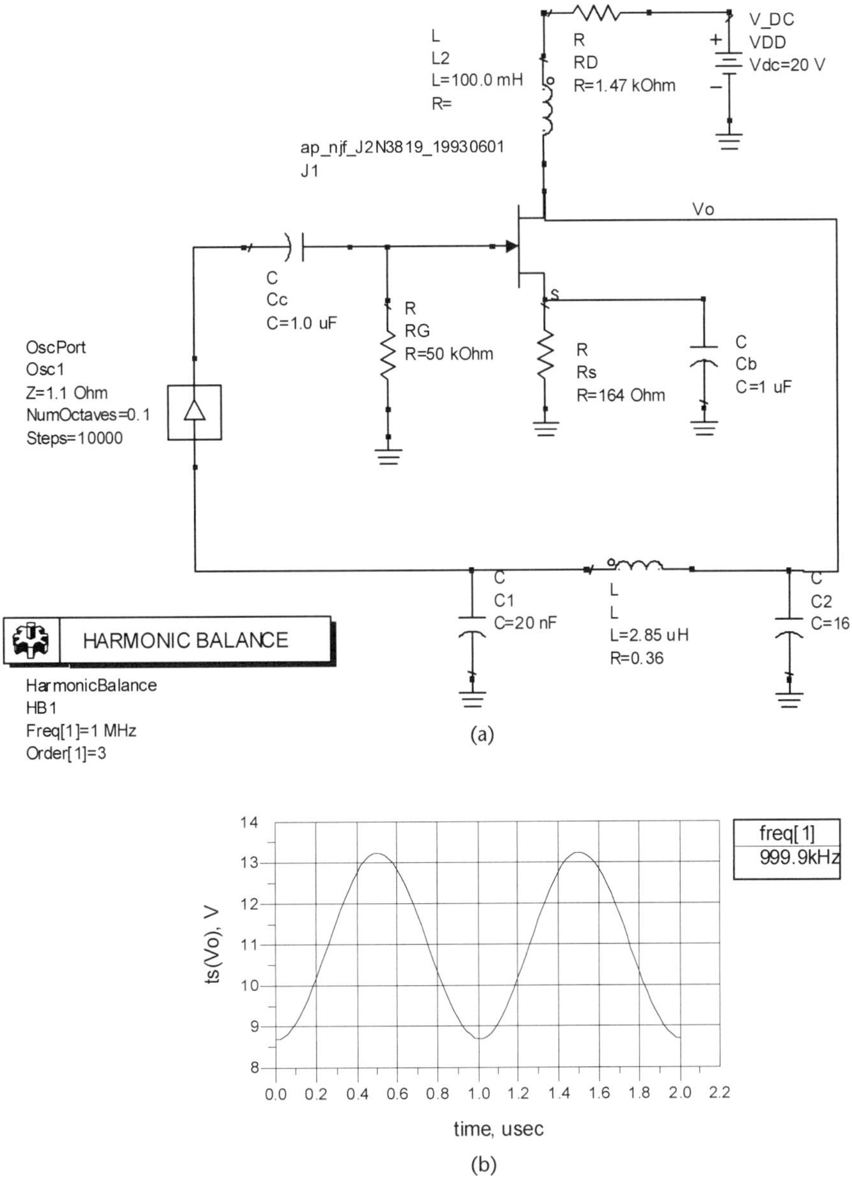

Figure 3.4 (a) Pierce oscillator for Example 3.1 and (b) the output voltage.

where

$$C_T = \frac{C_1 C_2}{C_1 + C_2}$$

and it follows that [see (3.23)]

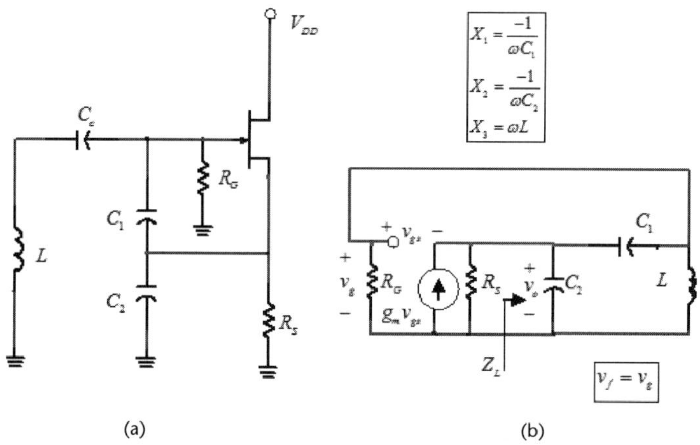

Figure 3.5 (a) An FET Colpitts oscillator and (b) its ac model.

$$Z_L(j\omega_o) \approx \frac{[X_2(\omega_o)]^2}{R_s} \tag{3.28}$$

The voltage feedback factor is

$$\beta(j\omega_o) = \frac{v_f}{v_o} = \frac{jX_3 + R_{Loss}}{R_s + j(X_1 + X_3)}$$

or

$$\beta(j\omega_o) = \frac{R_s - j(X_1 + X_2)}{R_s - jX_2}$$

If $|X_2| \gg R_s$, then

$$\beta(j\omega_o) = \frac{X_1 + X_2}{X_2} = \frac{C_1 + C_2}{C_1}$$

The gain of the source follower at ω_o (with $R_s || Z_L \approx Z_L$ and $v_{gs} = v_f - v_o$) is

$$A_v(j\omega) = \frac{v_o}{v_f} = \frac{g_m Z_L(j\omega)}{1 + g_m Z_L(j\omega)}$$

Hence, the loop gain at ω_o is

$$\beta(j\omega_o)A_v(j\omega_o) = \frac{C_1 + C_2}{C_1} \frac{g_m Z_L(j\omega_o)}{1 + g_m Z_L(j\omega_o)} > 1 \tag{3.29}$$

Substituting (3.28) into (3.29) gives

$$\frac{g_m}{\omega_o^2 R_s C_1 C_2} > 1$$

which is recognized as the gain condition.

3.2.3 FET Hartley Oscillator

In a Hartley oscillator, Z_1 and Z_2 represent a tapped inductor with mutual inductance Z_M, and Z_3 is a capacitor. A Hartley oscillator can be designed with or without mutual coupling between L_1 and L_2. If there is no mutual inductance, then $M = 0$. The general configuration of the Hartley oscillator is shown in Figure 3.6(a), and its ac model in Figure 3.6(b) where $R_L \approx r_d || R_D$.

From (3.13), the frequency of oscillation is obtained from

$$X_1(\omega_o) + X_2(\omega_o) + X_3(\omega_o) + 2X_M(\omega_o) = \omega_o L_1 + \omega_o L_2 - \frac{1}{\omega_o C} + 2\omega_o M = 0$$

which gives

$$\omega_o = \frac{1}{\sqrt{L_T C}} \tag{3.30}$$

where

$$L_T = L_1 + L_2 + 2M$$

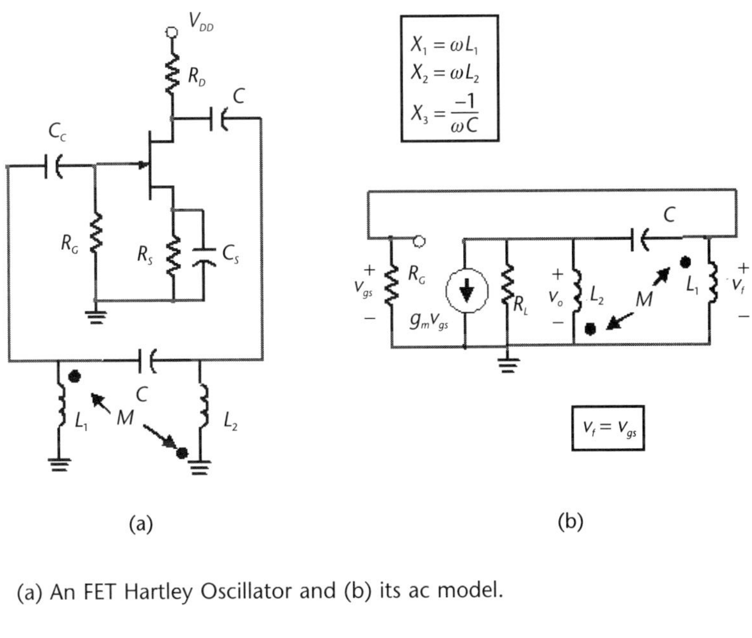

Figure 3.6 (a) An FET Hartley Oscillator and (b) its ac model.

The gain condition from (3.14) is

$$g_m R_L \geq \frac{L_2 + M}{L_1 + M} \tag{3.31}$$

A Hartley configuration that is commonly used (with $M = 0$ between L_1 and L_2) is shown in Figure 3.7. Its ac model is similar to that in Figure 3.6(b). The negative gate-to-source voltage is obtained from V_{GG}, and V_{DD} sets the drain-to-source voltage. The capacitors C_b are bypass capacitors.

The transformer $L_2 - L_3$ is used to provide the ac load to the FET. That is,

$$R_L' = n^2 R$$

where the turns ratio n is

$$n = \sqrt{\frac{L_2}{L_3}}$$

From (3.30) the frequency of oscillation is given by

$$\omega_o = \frac{1}{\sqrt{(L_1 + L_2)C}} \tag{3.32}$$

and from (3.31) the gain condition, with the load $R_L = r_d \,||\, R_L' \approx R_L'$, is

$$g_m R_L' \geq \frac{L_2}{L_1} \tag{3.33}$$

or

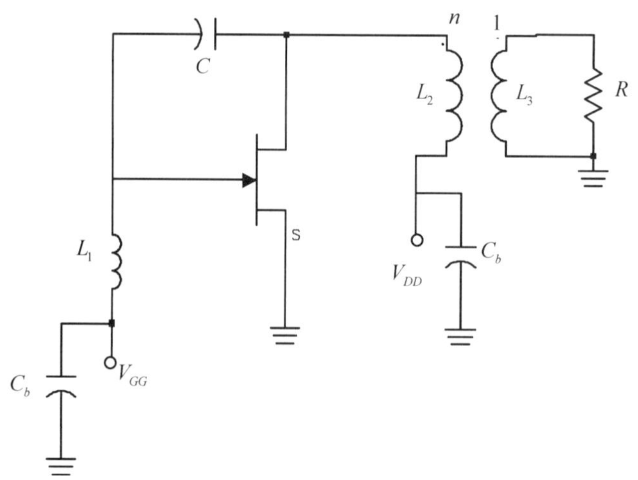

Figure 3.7 A Hartley oscillator.

$$g_m n^2 R \geq \frac{L_2}{L_1}$$

Example 3.2

Design the Hartley oscillator in Figure 3.7 to oscillate at 5 MHz with $R = 50\Omega$.

Solution

Let the supply voltage be $V_{DD} = 12\text{V}$. Letting $L_1 = L_2 = 5\ \mu\text{H}$, it follows from (3.32) that

$$C = \frac{1}{(2\pi \times 5 \times 10^6)^2 (10 \times 10^{-6})} = 101 \text{ pF}$$

For the FET, a 2N5486 JFET was selected. This JFET lists minimum and maximum values of V_P and I_{DSS} at 25°C. It also lists the typical minimum values at 25°C as $V_P =- 2.5\text{V}$ and $I_{DSS} = 7$ mA. A worst-case design requires that the minimum value of g_m be used. Hence, if the Q point is set at $V_{GS} =- 1\text{V}$, the value of $g_{m(\min)}$ is

$$g_{m(\min)} = 2\frac{I_{DSS}}{V_P^2}[V_{GS} - V_P] = 2\frac{7 \times 10^{-3}}{6.25}[-1 - (-2.5)] = 3.36 \text{ mS}$$

Let us use $g_{m(\min)} = 3$ mS. Then, from (3.33), with $L_1 = L_2$, the value of R_L' is

$$R_L' > \frac{1}{g_m} = \frac{1}{3 \times 10^{-3}} = 333\Omega$$

Selecting $R_L' > 333\Omega$, say $R_L' = 500\Omega$, the turns ratio is given by

$$n = \sqrt{\frac{R_L'}{R}} = \sqrt{\frac{500}{50}} = 3.16$$

The value of the inductor L_3 follows from

$$L_3 = \frac{L_2}{n^2} = \frac{5 \times 10^{-6}}{3.16^2} = 500 \text{ nH}$$

The oscillator was simulated in ADS using the bias circuit shown in Figure 3.8(a). The Spice parameters for the 2N5485 are $I_{DSS} = 12$ mA and $V_P =- 3.5\text{V}$. Hence, for $V_{GS} =- 1\text{V}$ the drain current is $I_D = 6.12$ mA. The Q point values calculated by the program are $V_{DS} = 12\text{V}$ and $I_D = 6.82$ mA. The resulting waveform is shown in Figure 3.8(b) where the fundamental frequency of oscillation is 4.834 MHz.

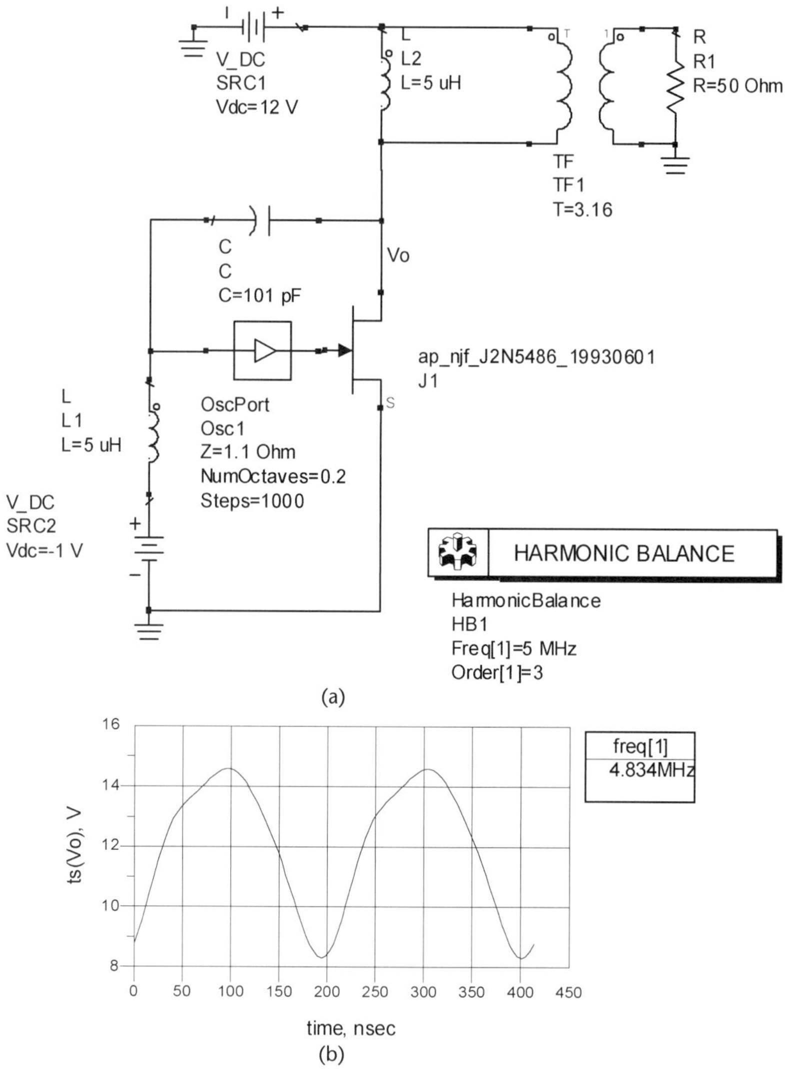

Figure 3.8 (a) The Hartley oscillator for Example 3.2 and (b) the simulation results.

Another FET Hartley type of oscillator is shown in Figure 3.9(a). This oscillator uses an FET in a *CG* configuration. The ac model is shown in Figure 3.9(b). The feedback factor is

$$\beta = \frac{v_f}{v_o} = \frac{L_2}{L_1 + L_2}$$

Using a source transformation and the substitution theorem (i.e., replacing one of the $g_m v_{gs}$ current source by the resistor $1/g_m$) produces the model in Figure 3.9(c), where R_s appears in parallel with $1/g_m$. The transformer formed by L_1 and

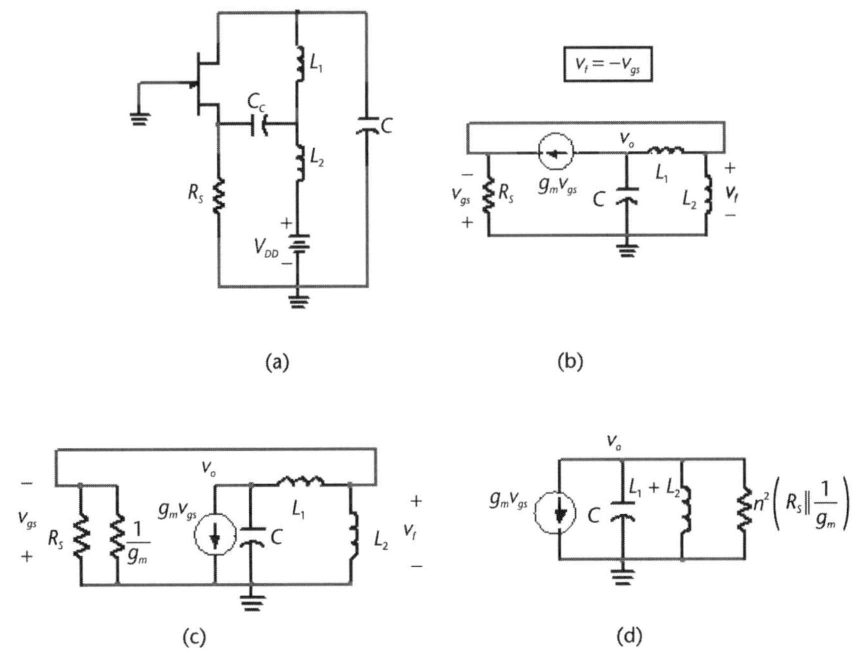

Figure 3.9 (a) A Hartley type of oscillator, (b) the ac model, (c) model for the calculation of the open-loop voltage gain, and (d) equivalent circuit across C.

L_2 convert the resistance $R_S||(1/g_m)$ to a parallel resistance across the capacitor of value

$$R_p = n^2\left(R_S \left\| \frac{1}{g_m}\right.\right)$$

where

$$n = \frac{L_1 + L_2}{L_2}$$

The equivalent circuit across C, shown in Figure 3.9(d), is convenient for the calculation of the open-loop gain.

From Figure 3.9(d), the resonant frequency is

$$\omega_o = \frac{1}{\sqrt{(L_1 + L_2)C}}$$

and the loop gain is

$$\beta(\omega_o)A_v(\omega_o) = \frac{L_2}{L_1 + L_2} g_m n^2\left(R_S \left\| \frac{1}{g_m}\right.\right) \approx g_m\left(R_S \left\| \frac{1}{g_m}\right.\right)\frac{L_1 + L_2}{L_2} \quad (3.34)$$

To start the oscillation, the loop gain is set to a value between 2 and 3. Observe that for the case that $R_S||(1/g_m) \approx 1/g_m$, (3.34) reduces to

$$\beta(\omega_o)A_v(\omega_o) = \frac{L_1 + L_2}{L_2}$$

3.2.4 FET Clapp Oscillator

The Clapp FET oscillator (also known as the Clapp-Gouriet oscillator) is shown in Figure 3.10. The configuration is somewhat similar to the Colpitts configuration. The impedance $Z_3 = j(\omega L_3 - 1/\omega C_3)$ must be inductive. The frequency of oscillation is given by

$$\omega_o = \frac{1}{\sqrt{L_3 C_T}}$$

where

$$\frac{1}{C_T} = \frac{1}{C_1} + \frac{1}{C_2} + \frac{1}{C_3}$$

The gain condition is

$$\frac{g_m}{\omega_o^2 R_s C_1 C_2} > 1$$

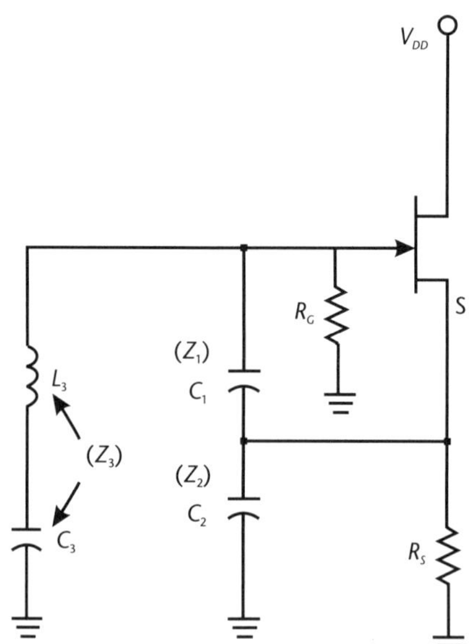

Figure 3.10 The Clapp oscillator.

where R_s represents the coil losses. Hence, C_1 and C_2 can be selected to satisfy the gain condition and C_3 to adjust the frequency of oscillation.

The addition of a tuning capacitor in series with L_3 can also be done to the Pierce oscillator. In this case the oscillator that results (i.e., a Clapp type) is shown in Figure 3.11.

The capacitor C_3 in series with L_3 provides another degree of freedom in the design, since C_1 and C_2 can be selected to satisfy the gain condition, and C_3 can be selected to set the frequency of oscillation. This improves the frequency stability, making the frequency stability of the Clapp oscillator better than that of the Colpitts. The frequency stability of the Clapp oscillator due to a change in capacitance is

$$\Delta\omega_o = -\frac{1}{2}\omega_o\frac{\Delta C_T}{C_T}$$

The changes to the total capacitance usually come from the active device. If C_1 and C_2 are designed with large values, then $C_T \approx C_3$, making $\Delta\omega_o$ independent of the active device.

3.2.5 The Grounded-Gate Oscillator

A Grounded-Gate (GG) oscillator is shown in Figure 3.12(a) and its ac model in Figure 3.12(b), where R_p represents the equivalent parallel resistance of the inductor. The GG is in the form of a Colpitts oscillator but with the gate grounded. The analysis of this oscillator is similar to that of the grounded-base oscillator, which is done in Section 3.3.

Using a source transformation and the substitution theorem (i.e., replacing one of the $g_m v_{gs}$ current sources by the resistor $1/g_m$) produces the model in Figure

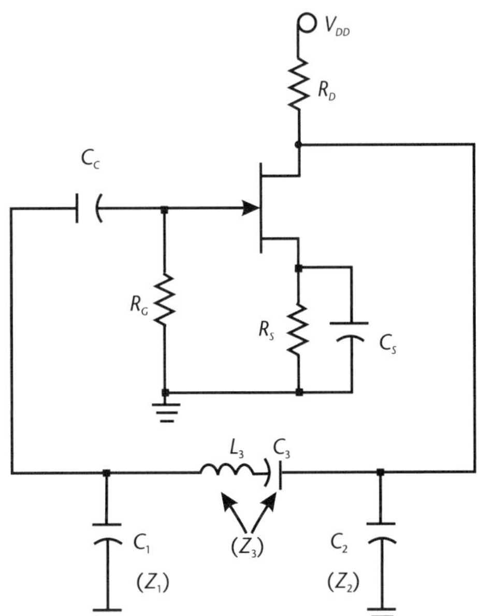

Figure 3.11 A Clapp-type oscillator.

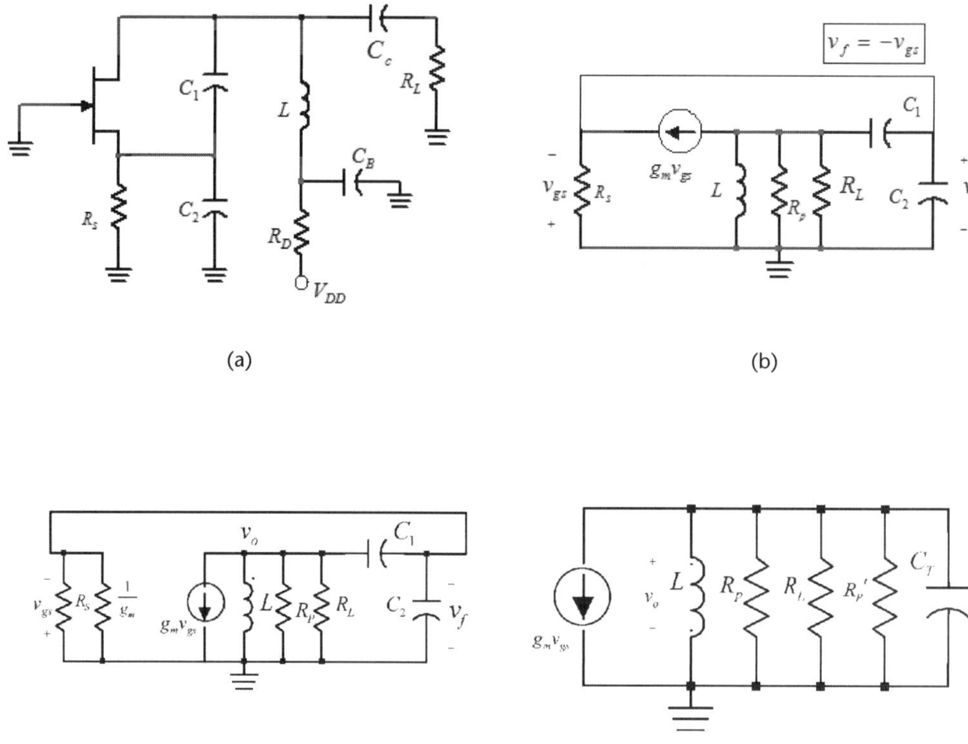

Figure 3.12 (a) The GG oscillator, (b) an ac model, (c) an alternative ac model, and (d) equivalent circuit across the inductor.

3.12(c), where R_S appears in parallel with $1/g_m$. The capacitive transformer formed by C_1 and C_2 transforms $(R_S \| 1/g_m)$ to a parallel resistance across the inductor with value

$$R'_P = n^2\left(R_S \left\| \frac{1}{g_m}\right.\right)$$

where

$$n = \frac{C_1 + C_2}{C_1}$$

The equivalent ac circuit across the inductor is shown in Figure 3.12(d). It is seen that the resonant frequency occurs at

$$\omega_o = \frac{1}{\sqrt{LC_T}} \tag{3.35}$$

where

$$C_T = \frac{C_1 C_2}{C_1 + C_2} \tag{3.36}$$

At ω_o, the gain is

$$A_v(\omega_o) = \frac{v_o}{v_f} = g_m(R_P \parallel R_L \parallel R'_P)$$

and the feedback factor is

$$\beta = \frac{v_f}{v_o} = \frac{C_1}{C_1 + C_2}$$

In the case that $R_P \parallel R_L \parallel R'_P$ can be approximated by R'_P, then

$$A_v(\omega_o) \approx g_m n^2 \left(R_S \middle\| \frac{1}{g_m} \right) \tag{3.37}$$

Furthermore, if $R_S \| (1/g_m) \approx 1/g_m$, (3.37) reduces to

$$A_{vo} = n^2 = \left(\frac{C_1 + C_2}{C_1} \right)^2$$

and the loop gain is

$$\beta(\omega_o) A_v(\omega_o) = \frac{C_1 + C_2}{C_1} \tag{3.38}$$

To start the oscillation, the loop gain can be set to a value between 2 and 3.

Example 3.3

Design the GG oscillator in Figure 3.12(a) to oscillate at 20 MHz using the 2N5486 JFET. Assume that R_L = 10 kΩ, and an RFC is in series with R_S.

Solution

Let the Q point be at I_D = 4.5 mA, and V_{DS} = 5V. For this transistor, typical values are V_P =– 3.5V and I_{DSS} = 12 mA. Therefore, at I_D = 4.5 mA the value of V_{GS} is approximately –1.36V and g_m = 4.2 mS. The Q point design requires that

$$R_S = \frac{-V_{GS}}{I_{DS}} = \frac{1.36}{4.5 \times 10^{-3}} = 302\Omega$$

and

$$R_D = \frac{V_{DD} - V_{DS} - V_S}{I_{DS}} = \frac{12 - 5 - 1.36}{4.5 \times 10^{-3}} = 1.25 \text{ k}\Omega$$

For the inductor, select $L = 1\ \mu$H having $Q_U = 80$ (or $R_p = 10$ kΩ). From (3.35)

$$C_T = \frac{1}{\omega_o^2 L} = 63.4 \text{ pF}$$

Letting the loop gain be 3, it follows from (3.36) and (3.38) that

$$C_2 = 3C_T = 3(63.4 \times 10^{-12}) = 190 \text{ pF}$$

and

$$C_1 = \frac{C_2}{2} = \frac{190 \times 10^{-12}}{2} = 95 \text{ pF}$$

Since

$$n = \frac{C_1 + C_2}{C_1} = 3$$

and with $R_S||(1/g_m) \approx 1/g_m$, it follows that

$$R'_P = \frac{n^2}{g_m} = 238(9) = 2.14 \text{ k}\Omega$$

Hence, the equivalent resistance of the tuned circuit is

$$R_P \,||\, R_L \,||\, R'_P = 10 \times 10^3 \,||\, 10 \times 10^3 \,||\, 2.14 \times 10^3 = 1.5 \text{ k}\Omega$$

which shows that the approximation (3.37) is not exactly valid. However, making $n = 3$ should be sufficient to start the oscillation.

The simulation of this oscillator is shown in Figure 3.13. The Q point is at $V_{GS} =-1.41$V and $I_D = 4.7$ mA, and the fundamental frequency of oscillation is 19.95 MHz.

3.2.6 Tuned-Drain Oscillator

Another FET oscillator is the tuned-drain oscillator shown in Figure 3.14(a). The transistor produces a phase shift of –180° and the transformer the other –180°. In this oscillator, in terms of the notation employed in Figure 3.2(a), Z_1 is the impedance of the gate coil, Z_2 is the impedance of the tuned circuit at the drain terminal, Z_3 represents the small impedance coupling through C_{gd}, (which at the frequency of oscillation can be neglected), and Z_M is the mutual coupling of the coils. However, instead of using the general equations for the calculation of the frequency of oscillation and the gain conditions, it is worthwhile to analyze

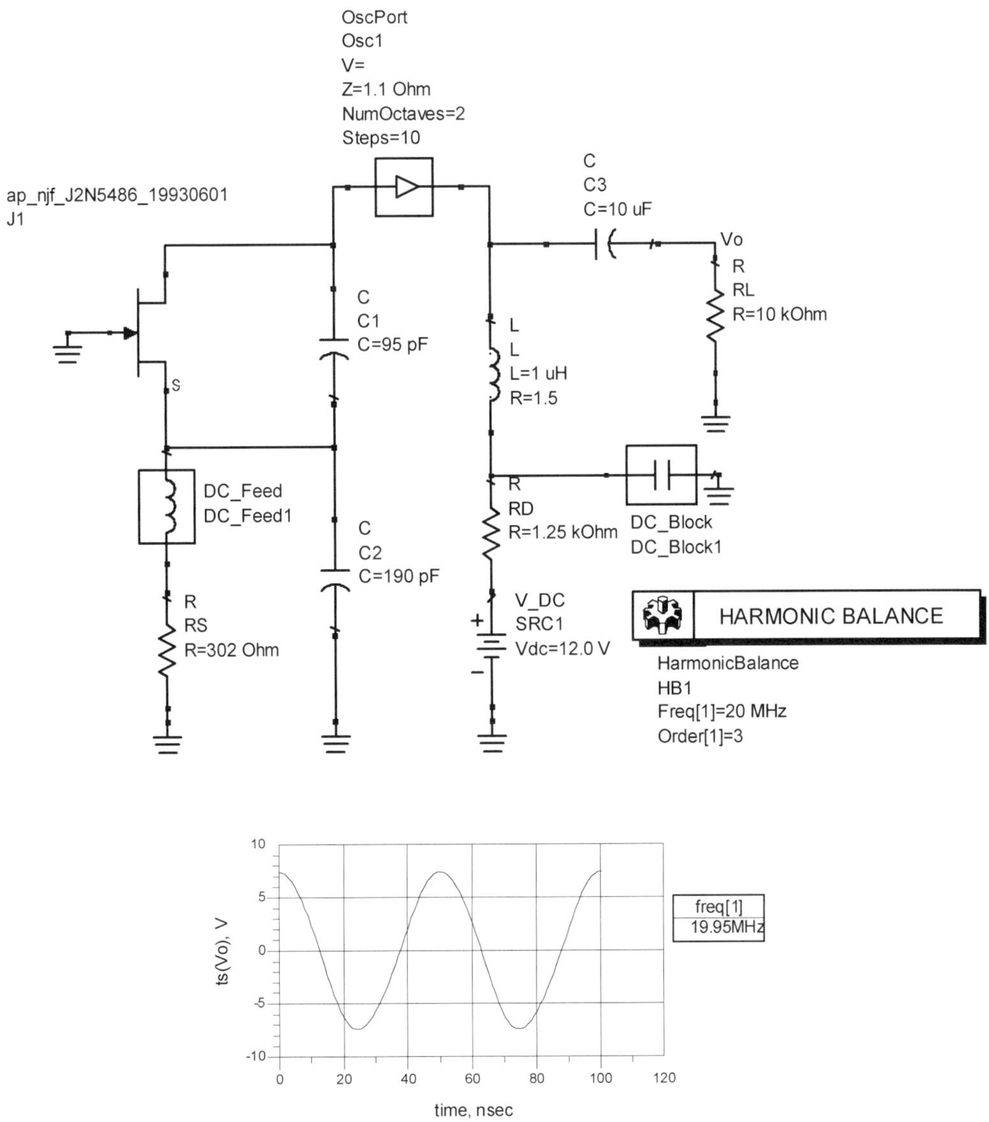

Figure 3.13 Simulation of the oscillator in Example 3.3.

this circuit using the ac model shown in Figure 3.14(b). The resistor R_p represents the equivalent parallel resistance of the coil losses. Also, since $i_2 = 0$ there is no voltage induced in the primary of the transformer due to i_2. The open-loop gain is given by

$$A_v(j\omega) = \frac{v_o}{v_{gs}} = -g_m\left(r_d \,\Big\|\, R_p \,\Big\|\, \frac{1}{j\omega C} \,\Big\|\, j\omega L_2\right) \tag{3.39}$$

and the voltage-feedback factor is

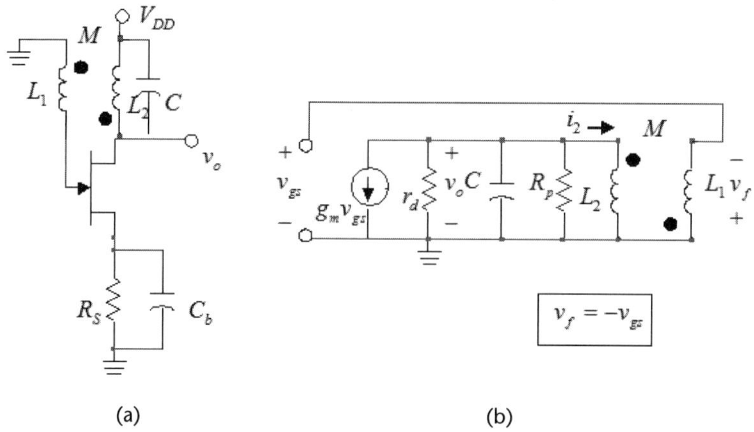

Figure 3.14 (a) A tuned-drain oscillator and (b) the ac model.

$$\beta(j\omega) = \frac{v_{gs}}{v_o} = \frac{-M\dfrac{di_2}{dt}}{L_2\dfrac{di_2}{dt}} = -\frac{M}{L_2} \tag{3.40}$$

Consequently, using (3.39) and (3.40), the loop gain can be expressed in the form

$$\beta(j\omega)A_v(j\omega) = \frac{M}{L_2}g_m\left(r_d \left\| R_p \right\| \frac{1}{j\omega C} \left\| j\omega L_2\right.\right) \tag{3.41}$$

Obviously, the transistor sees a parallel RLC circuit whose resonant frequency [i.e., the frequency at which the imaginary part of (3.41) vanishes] is

$$\omega_o = \frac{1}{\sqrt{L_2 C}}$$

At ω_o, the gain condition [from (3.41)] is

$$\beta(j\omega_o)A_v(j\omega_o) = \frac{M}{L_2}g_m\left(r_d \| R_p\right) \geq 1$$

or

$$g_m\left(r_d \| R_p\right) \geq \frac{L_2}{M}$$

3.2.7 Cross-Coupled Tuned Oscillator

A configuration that is used in IC oscillators is shown in Figure 3.15(a), and redrawn in Figure 3.15(b). This oscillator is known as a cross-coupled tuned

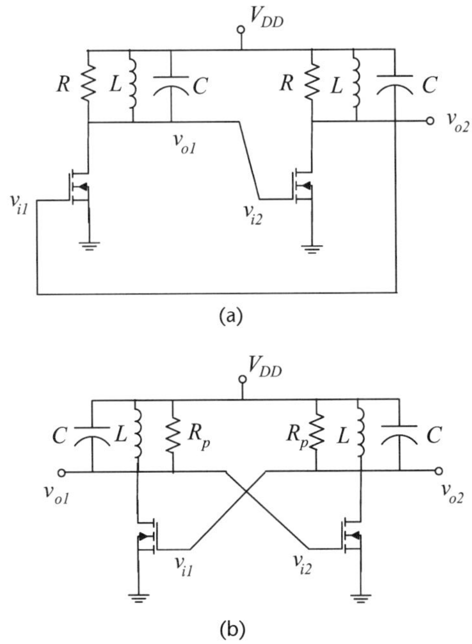

Figure 3.15 (a) A cross-coupled tuned oscillator and (b) alternate drawing.

oscillator. In Figure 3.15, R_p represents the equivalent parallel resistance of the tuned circuit.

Each stage provides a phase shift of $-180°$ at its resonant frequency. The gain of the first stage is

$$A_{v1}(j\omega) = \frac{v_{o1}}{v_{i1}} = \frac{-g_m}{\dfrac{1}{R_p} + j\left(\omega C - \dfrac{1}{\omega L}\right)}$$

where $v_{i1} = v_{o2}$ and $v_{i2} = v_{o1}$. The gain of the second stage is identical to that of the first stage. Hence, the loop gain is given by

$$\frac{v_{o1}}{v_{i1}}\frac{v_{o2}}{v_{i2}} = \left[\frac{-g_m}{\dfrac{1}{R_p} + j\left(\omega C - \dfrac{1}{\omega L}\right)}\right]^2$$

From (3.42) the frequency of oscillation is given by

$$\omega_o \frac{1}{\sqrt{LC}}$$

and the gain condition is

$$g_m R_p = 1$$

For IC implementation a current bias is used as shown in Figure 3.16(a). Another IC implementation [similar to that in Figure 3.16(a)] is shown in Figure 3.16(b).

In this section various FET tuned amplifiers have been designed, and the resulting waveforms obtained using the harmonic balance simulator. The transient simulator could have also been used. However, in some cases the transient simulator can take a significant amount of time. For example, the transient simulation of the oscillator in Figure 3.13 can take up to 5 minutes in order for the oscillator to reach steady state.

In the analysis of the initial response of an oscillator a transient simulator is required. Of particular interest is the oscillator behavior of when the power supply is turned on. Figure 3.17 shows the transient response of the oscillator in Figure 3.13 when a step voltage source is used to represent the power supply turn-on condition. Observe that the steady state response is identical to the one obtained in Figure 3.13.

An envelope simulator is also provided in ADS. It can be used to obtain the envelope information associated with the transient response of an oscillator. For the purposes of this book the envelope simulator was not needed. The reader is referred to the ADS documentation from Agilent for specific details about the use of the envelope simulator in oscillator systems.

3.3 BJT Tuned Oscillators

The general configuration of a BJT tuned oscillator is shown in Figure 3.18(a). In some designs RFCs are used in series with R_C, R_1, and R_2. The elements that

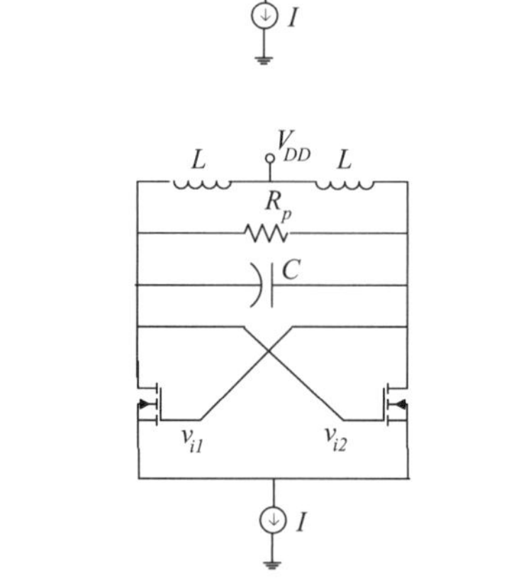

Figure 3.16 (a) A cross-coupled tuned oscillator with current bias and (b) another cross-coupled oscillator with current bias.

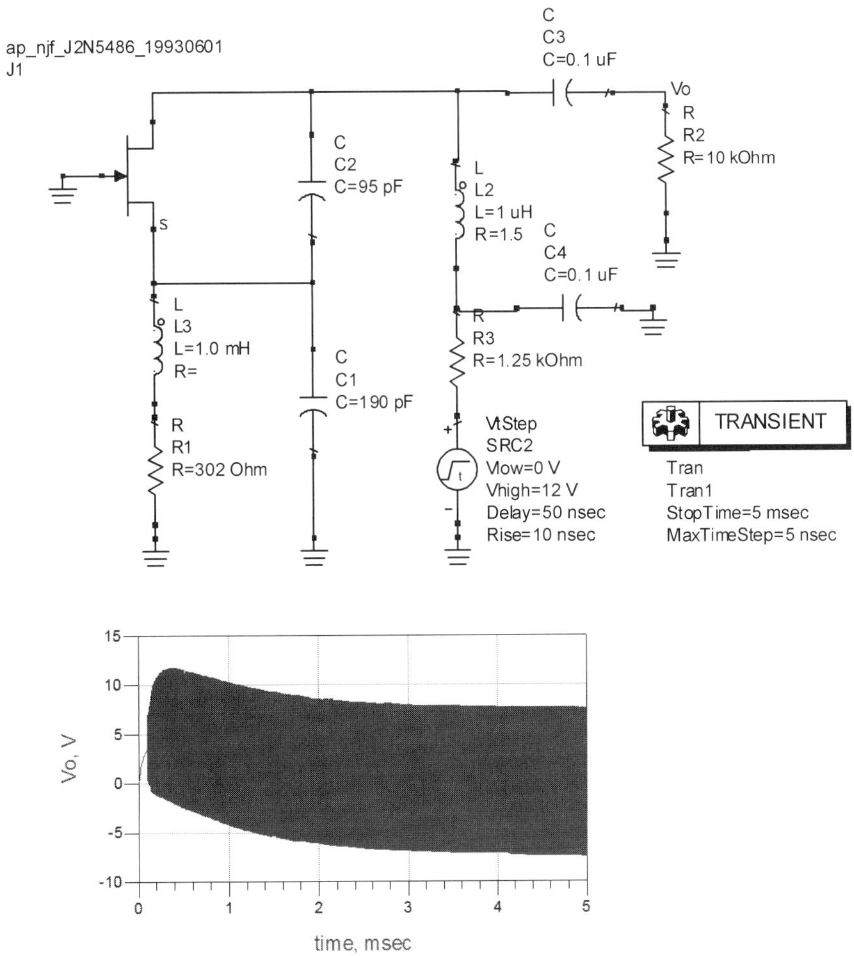

Figure 3.17 Transient simulation for the turn-on condition of the oscillator in Figure 3.13.

determine the path of the ac signal are shown in Figure 3.18(b). The BJT provides a voltage gain with a phase shift of –180°. The tuned circuit introduces another –180° phase shift at the frequency of oscillation.

In Figure 3.18(c) the BJT is replaced by its hybrid model. The resistor R_L represents R_C in parallel with $1/h_{oe}$. If an RFC is used in series with R_C, then $R_L = 1/h_{oe}$. If the resistance of the RFC is considered, then $R_L = (1/h_{oe}) || R_p$ where R_p is the equivalent parallel loss resistance of the coil. Using a Thevenin's equivalent circuit across R_L, the model in Figure 3.18(d) is obtained. The loop equations are

$$i_1(h_{ie} + Z_1) + i_2 Z_M - i_3(Z_1 + Z_M) = 0$$

$$i_1(h_{fe} R_L + Z_M) + i_2(R_L + Z_2) - i_3(Z_2 + Z_M) = 0$$

and

$$-i_1(Z_1 + Z_M) - i_2(Z_2 + Z_M) + i_3(Z_1 + Z_2 + Z_3 + 2Z_M) = 0$$

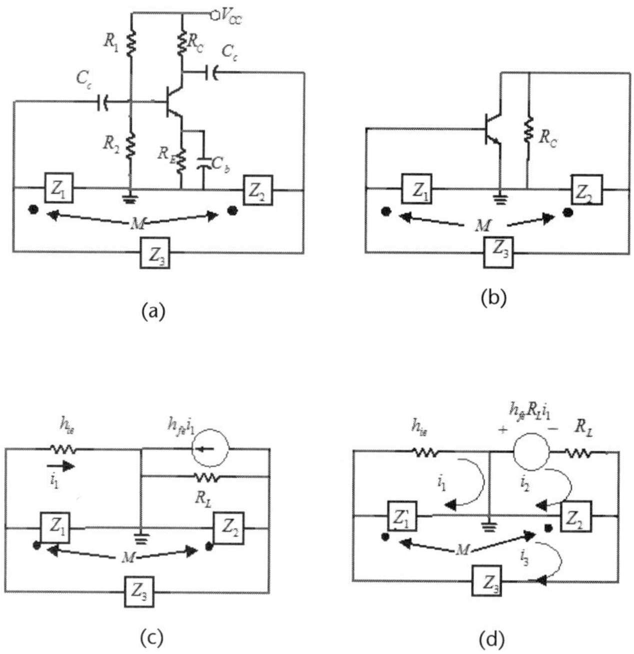

Figure 3.18 (a) General configuration of BJTs tuned oscillators, (b) ac model, (c) the BJT replaced by its ac model, and (d) using a Thevenin's model across R_L.

Hence, the system determinant is

$$D(j\omega) = \begin{vmatrix} h_{ie} + Z_1 & Z_M & -(Z_1 + Z_M) \\ h_{fe}R_L + Z_M & R_L + Z_2 & -(Z_2 + Z_M) \\ -(Z_1 + Z_M) & -(Z_2 + Z_M) & Z_1 + Z_2 + Z_3 + 2Z_M \end{vmatrix} \tag{3.42}$$

The hybrid parameters h_{fe} and h_{ie} in the previous equations are also denoted by β and r_π, respectively. In what follows, both notations are used.

3.3.1 BJT Pierce Oscillator

The schematic of a BJT Pierce oscillator (i.e., a Colpitts-type oscillator) is shown in Figure 3.19(a), and its ac model is shown in Figure 3.19(b). Setting the imaginary part of (3.42) equal to zero, it follows that the frequency of oscillation is given by

$$\omega_o = \sqrt{\frac{1}{LC_T} + \frac{1}{C_1C_2R_Lh_{ie}}} \approx \sqrt{\frac{1}{LC_T}} \tag{3.43}$$

where

$$C_T = \frac{C_1C_2}{C_1 + C_2}$$

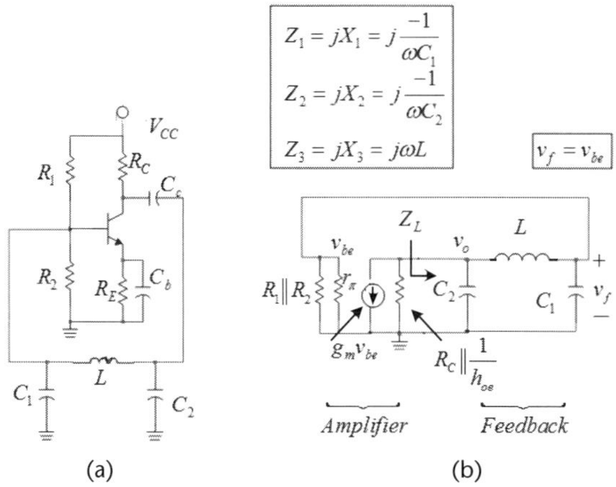

Figure 3.19 (a) A BJT Pierce oscillator and (b) its ac model.

The gain condition follows from the real part of (3.42), namely,

$$\left[g_m - \frac{C_2}{C_1 h_{ie}}\right] R_L \geq \frac{C_1}{C_2}$$

where $g_m = h_{fe}/h_{ie}$ and the load resistance is $R_L = R_c \,\|\, (1/h_{oe})$. If $h_{fe} \gg C_2/C_1$ (which is simple to satisfy), the gain condition reduces to

$$g_m R_L \geq \frac{C_1}{C_2} \tag{3.44}$$

Observing that the voltage feedback factor in the Pierce oscillator is $\beta(j\omega_o) = -C_2/C_1$, and that the open-loop gain of the amplifier is $A_v(j\omega_o) = -h_{fe} R_L / h_{ie} = -g_m R_L$, (3.44) is simply recognized as the loop-gain condition $\beta(j\omega_o) A_v(j\omega_o) \geq 1$.

If the coil losses are included in the analysis (i.e., with $Z_3 = R_s + jX_3$) the impedance $Z_L(j\omega)$ is given by

$$Z_L = jX_2 \,\|\, \left[R_s + jX_3 + \left(jX_1 \,\|\, r_\pi\right)\right]$$

where the approximation $R_1 \,\|\, R_2 \,\|\, r_\pi \approx r_\pi$ was used. Furthermore, letting $|X_1| \ll r_\pi$, we can write

$$Z_L = \frac{jX_2[j(X_1 + X_3) + R_s]}{j(X_1 + X_2 + X_3) + R_s}$$

The voltage feedback factor is

$$\beta(j\omega) = \frac{v_f}{v_o} = \frac{jX_1}{R_s + j(X_1 + X_3)}$$

Hence, the loop gain is

$$\beta(j\omega)A_v(j\omega) = \beta(j\omega)\left[-g_m\left(R_L \parallel Z_L\right)\right] \tag{3.45}$$
$$= \frac{g_m R_L X_1 X_2}{jR_L(X_1 + X_2 + X_3) - X_2(X_1 + X_3) + R_L R_s}$$

which is similar to (3.22).

From (3.45) it follows that the frequency of oscillation is

$$\omega_o = \sqrt{\frac{1}{LC_T}}$$

where

$$C_T = \frac{C_1 C_2}{C_1 + C_2}$$

and the gain condition is

$$g_m R_L > \frac{X_2^2(\omega_o) + R_L R_s}{X_1(\omega_o)X_2(\omega_o)} \tag{3.46}$$

which is similar to (3.26). Observe that if $R_s \to 0$ then (3.46) reduces to (3.44). Also, in the cases that $R_L R_s \gg X_2^2$, (3.46) gives

$$\frac{g_m}{\omega_o^2 R_s C_1 C_2} > 1 \tag{3.47}$$

Example 3.4

Design the BJT Pierce oscillator in Figure 3.19(a) to oscillate at 800 kHz.

Solution

A 2N2222 BJT can be used at a bias point of $V_{CE} = 10\text{V}$ and $I_C = 1$ mA, where the minimum beta is 50 and $f_T = 300$ MHz. Letting $V_{CC} = 20\text{V}$ and using a typical beta of 100, the resistors are designed as follows:

$$R_E = \frac{10\% V_{CC}}{I_C} = \frac{0.1(20)}{10^{-3}} = 2 \text{ k}\Omega$$

$$R_C = \frac{V_{CC} - V_{CE} - V_{R_E}}{I_C} = \frac{20 - 10 - 2}{10^{-3}} = 8 \text{ k}\Omega$$

$$R_{TH} = \frac{\beta R_E}{10} = \frac{100(2 \times 10^3)}{10} = 20 \text{ k}\Omega$$

$$V_{TH} = I_B R_{TH} + 0.7 + I_E R_E = \frac{20}{100} + 0.7 + 10^{-3}(2 \times 10^3) = 2.9\text{V}$$

$$R_1 = R_{TH} \frac{V_{CC}}{V_{TH}} = 20 \times 10^3 \frac{20}{2.9} = 138 \text{ k}\Omega$$

and

$$R_2 = \frac{R_{TH}}{1 - \dfrac{V_{TH}}{V_{CC}}} = \frac{20 \times 10^3}{1 - \dfrac{2.9}{20}} = 23.4 \text{ k}\Omega$$

At $I_C = 1$ mA: $g_m = 10^{-3}/25 \times 10^{-3} = 40$ mS and $h_{ie} > \beta_{min}/g_m = 50/0.04 = 1.25$ kΩ. From (3.44), with $R_L = R_C$, the gain condition is

$$(0.04)(8 \times 10^3) = 320 \geq \frac{C_1}{C_2} \tag{3.48}$$

The input resistance is

$$R_{IN} = R_1 \,||\, R_2 \,||\, h_{ie} = 1.18 \text{ k}\Omega$$

The reactance of C_1 should be much smaller than 1.18 kΩ, say, 50 times smaller. Using a factor of 50 gives

$$|X_{C_1}| = \frac{1}{2\pi 800 \times 10^3 C_1} = \frac{1.18 \times 10^3}{50}$$

or $C_1 = 8.43$ nF. Then, from (3.48),

$$C_2 > \frac{8.42 \times 10^{-9}}{320} = 26.3 \text{ pF}$$

Let $C_2 = 50$ pF.

Finally, from (3.43), with $C_T = 49.7$ pF we obtain $L = 797$ μH. The magnitude of the reactance of the bypass capacitor is selected to be negligible with respect to R_E, say $C_b = 0.1$ μF. A coupling capacitor can be placed from the collector to C_2. Hence, its reactance (magnitude) must be much smaller than that of C_2, say $C_2 = 100$ nF. It can also be connected in series with the inductor, in which case its reactance should be much smaller than that of the inductor.

The simulation and resulting waveform are shown in Figure 3.20. The fundamental frequency of oscillation is calculated to be $f_o = 783$ kHz.

Example 3.5

The BJT Pierce oscillator shown in Figure 3.19(a) was designed using a 2N2907 BJT with $V_{CC} = 15$V, $R_1 = 12$ kΩ, $R_2 = 6$ kΩ, $R_E = 500\Omega$, $C_e = C_{c2} = 0.1$ μF,

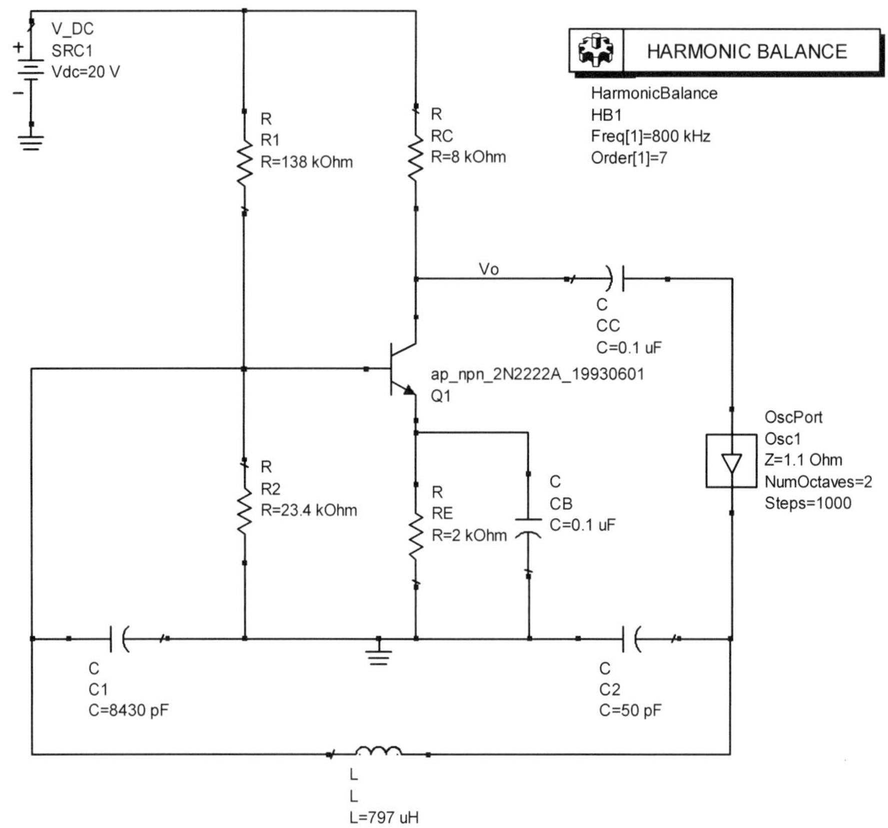

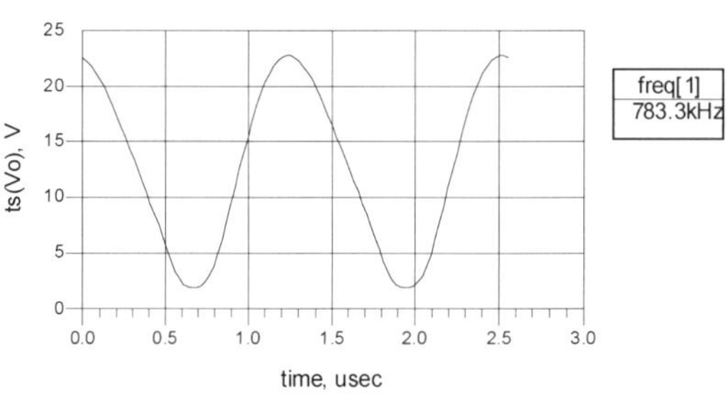

Figure 3.20 Simulation of the Pierce oscillator in Example 3.4.

$C_1 = 50$ nF, $C_2 = 500$ pF, $L = 300$ μH, and the collector resistor R_c was replaced by an RFC whose inductance is 20 mH. Determine the frequency of oscillation and check if the RFC, the bypass capacitor, and the coupling capacitor were properly designed.

Solution

The 2N2907 transistor lists $h_{FE}(\text{min}) = 30$, $I_C(\text{max}) = 800$ mA, and $f_T = 300$ MHz. An analysis of the circuit with the values given shows that the Q point is located at $I_C = 8.16$ mA and $V_{CE} = 11$V. Hence, $g_m = 326$ mS.

The total capacitance across L is

$$C_T = \frac{C_1 C_2}{C_1 + C_2} = \frac{50 \times 10^{-9}(500 \times 10^{-12})}{50 \times 10^{-9} + 500 \times 10^{-12}} = 495 \text{ pF}$$

The frequency of oscillation is

$$f_o = \frac{1}{2\pi\sqrt{LC_T}} = \frac{1}{2\pi\sqrt{300 \times 10^{-6}(495 \times 10^{-12})}} = 413 \text{ kHz}$$

The voltage feedback is

$$\beta(j\omega_o) =- \frac{C_2}{C_1} =- \frac{0.5 \times 10^{-9}}{50 \times 10^{-9}} =- 0.01$$

Hence, the gain for oscillation is

$$A_v(j\omega_o) \geq \frac{1}{\beta(j\omega_o)} =- \frac{1}{0.01} =- 100$$

Since $h_{fe}(\text{min}) = 30$, and with $R_L = 1/h_{oe} > 50 \text{ k}\Omega$, the gain condition in (3.44) is readily satisfied.

The resistance $R_1 \| R_2 \| h_{ie} \approx h_{ie}$ appears in parallel with the reactance of C_1, which is $X_{C_1} =- 7.7\Omega$ at 413 kHz. Using a typical value of 100 for h_{fe}, the typical value of h_{ie} is 306Ω. Hence, the loading is small.

At f_o the reactances of the bypass and coupling capacitors are

$$X_{C_b} = X_{C_c} = \frac{-1}{2\pi f_o C_b} = \frac{-1}{2\pi(413 \times 10^3)(0.1 \times 10^{-6})} =- 3.85\Omega$$

and the reactance of the RFC is

$$X_{RFC} = 2\pi f_o L = 2\pi(413 \times 10^3)(20 \times 10^{-3}) = 51.9 \text{ k}\Omega$$

Also, C_c acts as a short circuit to the ac signal, thus properly coupling the ac signal to the resonant circuit. Since $R_E \gg |X_{C_b}|$, the emitter capacitor acts like a short circuit at f_o. A finite loss of 5Ω was assumed for the RFC. The reactance of the coil is large and appears in parallel with r_d. Hence, the gain condition in (3.47) is satisfied with $R_L \approx r_d$.

The simulation of the oscillator is shown in Figure 3.21(b). The fundamental frequency of oscillation is calculated to be $f_o = 424.5$ kHz.

3.3.2 BJT Colpitts Oscillator

The schematic of a BJT Colpitts oscillator is shown in Figure 3.22(a), and its ac model in Figure 3.22(b). If the RFC is not used in series with R_E, then for proper operation

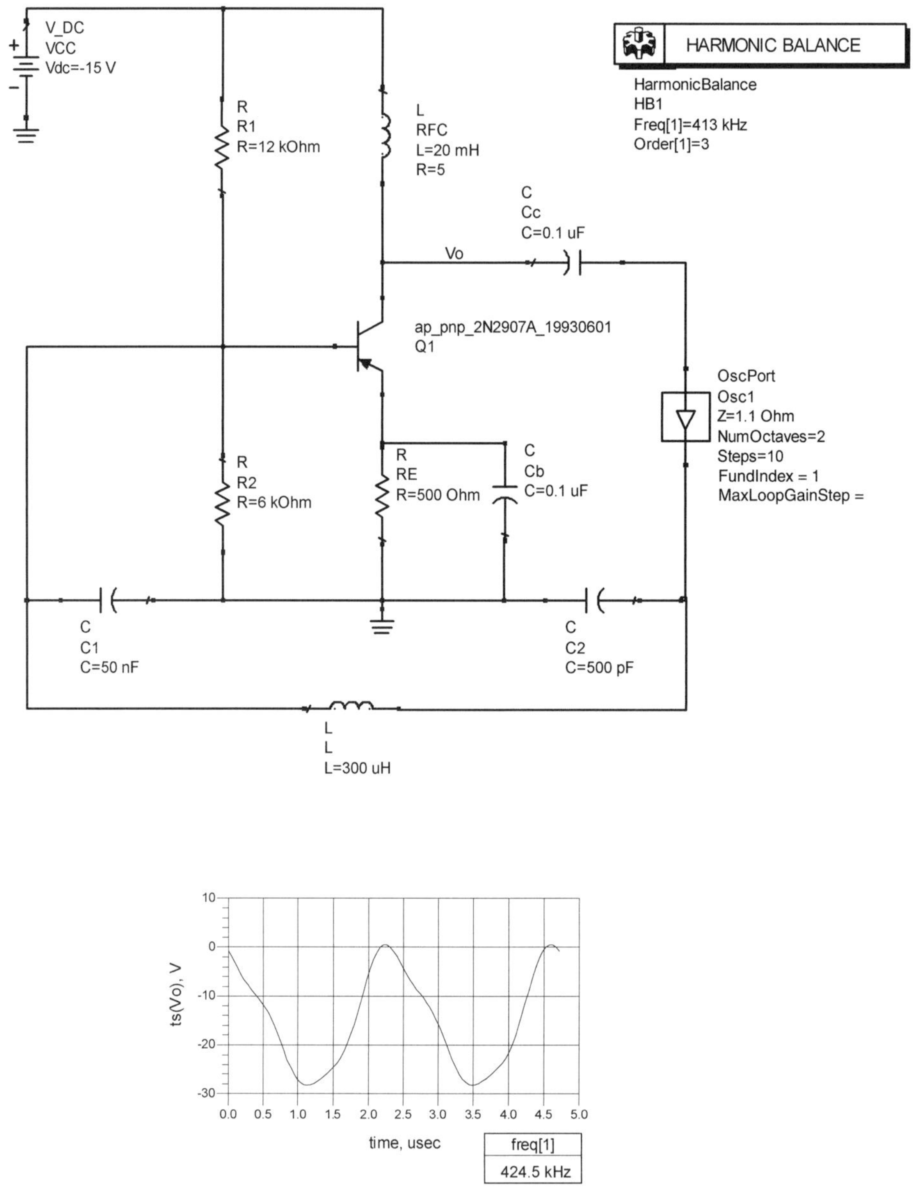

Figure 3.21 Simulation of the oscillator in Example 3.5.

$$R_E \gg \frac{1}{\omega C_2}$$

Also, $C_1 \gg C_{be}$ must be satisfied, and C_C is a coupling capacitor. Either RFCs are used or the input resistance $R_1 || R_2 || \beta R_E$ must be high in order not to load the resonant circuit.

The analysis of the BJT Colpitts oscillator can be performed like the one for the FET Colpitts oscillator in Section 3.2. The analysis can also be performed in

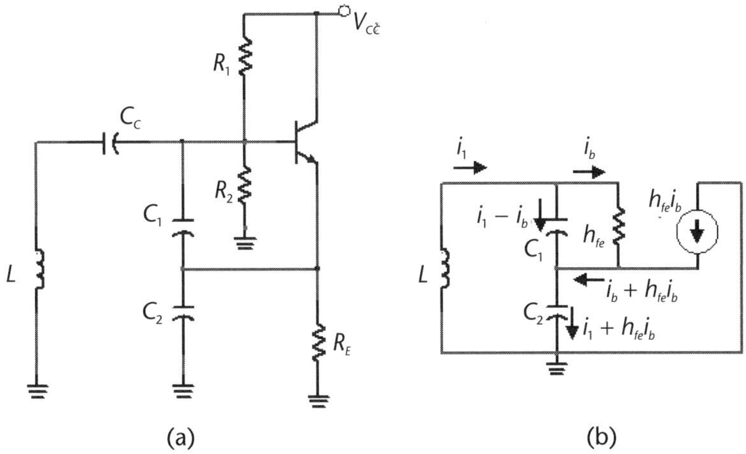

Figure 3.22 (a) A BJT Colpitts oscillator and (b) its ac model.

terms of the system determinant using (3.42). However, Figure 3.22(b) shows that in the Colpitts configuration the current in $R_L = 1/h_{oe}$ can be neglected and only two loop equations need to be written. That is, with $Z_1 = 1/j\omega C_1$, $Z_2 = 1/j\omega C_2$, and $Z_3 = j\omega L$, we obtain

$$0 = i_1(Z_1 + Z_2 + Z_3) + i_b(h_{fe}Z_2 - Z_1) \tag{3.49}$$

and

$$0 = i_1 Z_1 - i_b(Z_1 + h_{ie}) \tag{3.50}$$

Oscillation occurs when the determinant of (3.49) and (3.50) is set equal to zero. That is, when

$$\begin{vmatrix} Z_1 + Z_2 + Z_3 & h_{fe}Z_2 - Z_1 \\ Z_1 & -(Z_1 + h_{ie}) \end{vmatrix} = 0$$

or

$$(Z_1 + Z_2 + Z_3)h_{ie} + h_{fe}Z_1Z_2 + Z_1(Z_2 + Z_3) = 0 \tag{3.51}$$

Setting the imaginary part of (3.51) equal to zero gives

$$Z_1 + Z_2 + Z_3 = 0$$

or

$$-\frac{1}{\omega_o C_1} - \frac{1}{\omega_o C_2} + \omega_o L = 0$$

Hence, the frequency of oscillation is

$$\omega_o = \frac{1}{\sqrt{LC_T}}$$

where

$$C_T = \frac{C_1 C_2}{C_1 + C_2}$$

Setting the real part of (3.51) equal to zero gives

$$Z_1 Z_2 (h_{fe} + 1) + Z_1 Z_3 = 0$$

which can be expressed as

$$h_{fe} + 1 >- \frac{Z_3}{Z_2} = 1 + \frac{Z_1}{Z_2}$$

or

$$h_{fe} > \frac{C_2}{C_1} \tag{3.52}$$

which is the start of oscillation condition.

If coil losses are included in the analysis (i.e., $Z_3 = R_s + j\omega L$) the real part of (3.51) gives

$$\left(-\frac{1}{\omega_o C_1} - \frac{1}{\omega_o C_2} + \omega_o L\right) h_{ie} - \frac{R_s}{\omega_o C_1} = 0 \tag{3.53}$$

Equation (3.53) gives the frequency of oscillation as

$$\omega_o = \frac{1}{\sqrt{LC_T}} \tag{3.54}$$

where

$$C_T = \frac{C_1' C_2}{C_1' + C_2}$$

and

$$C_1' = \frac{C_1}{1 + \dfrac{R_s}{h_{ie}}} \approx C_1$$

The gain condition follows from the imaginary part of (3.51), which is,

$$R_s h_{ie} - \frac{h_{fe}}{\omega_o^2 C_1 C_2} + \frac{L}{C_1} = 0 \tag{3.55}$$

Since the second term is much larger than the third term (i.e., $h_{fe} X_{C_2} \gg X_L$), (3.55) can be expressed as

$$\frac{g_m}{\omega_o^2 R_s C_1 C_2} > 1 \tag{3.56}$$

It is also observed from (3.55) that if $R_s \to 0$, it reduces to

$$h_{fe} > \frac{C_2}{C_1}$$

as in (3.52).

Example 3.6

Design the Colpitts oscillator in Figure 3.22(a) to oscillate at 1 MHz.

Solution

Let $C_1 = C_2 = 10$ nF, so that $C_1 \gg C_{be}$ and $C_T = 5$ nF. It follows that $|X_{C_2}| = |X_{C_1}| = 15.9\Omega$. Then, from (3.54)

$$L = \frac{1}{(2\pi \times 10^6)^2\ 5 \times 10^{-9}} = 5.07\ \mu\text{H}$$

If the coil Q_U is 60, then $R_s = 0.64\Omega$.

The condition (3.56) requires that

$$g_m > (2\pi \times 10^6)^2 (0.64)(10 \times 10^{-9})^2 = 2.52\ \text{mS}$$

which is simple to attain.

A 2N2222 BJT can be selected and the bias point set at 8V, 10 mA, with $V_{CC} = 15$V. Then,

$$R_E = \frac{7}{10 \times 10^{-3}} = 700\Omega$$

$$I_B = \frac{10 \times 10^{-3}}{100} = 100\ \mu\text{A}$$

Let $I_{R_1} \approx I_{R_2} = 10 I_B = 1$ mA. Then,

$$R_1 = \frac{15 - 7.7}{10^{-3}} = 7.3\ \text{k}\Omega$$

$$R_2 = \frac{7 + 0.7}{10^{-3}} = 7.7\ \text{k}\Omega$$

The value of g_m is

$$g_m = \frac{10 \times 10^{-3}}{25 \times 10^{-3}} = 400\ \text{mS}$$

which certainly satisfies the gain condition.

The coupling capacitor C_c is selected so that its reactance is negligible to that of the inductor at 1 MHz. Hence, let $C_c = 0.1\ \mu$F.

The simulation of the oscillator is shown in Figure 3.23(a). The Q point values are: $V_{CE} = 8.13$V and $I_C = 9.78$ mA. The oscillation signal is viewed across the emitter resistor. The fundamental frequency of oscillation is 1.023 MHz.

A modification to the configuration in Figure 3.23(a) is shown in Figure 3.23(c). The collector-tuned circuit at 1 MHz is used to couple the output signal and provide further filtering at the fundamental frequency. The simulation results are shown in Figure 3.23(d) where v_{BE} and i_C are shown. The plots of v_{BE} and i_C are useful with phase noise considerations.

The pulse width of the collector current affects the phase noise of the oscillator (see Chapter 2). Further considerations associated with controlling the collector pulse width and the phase noise of the oscillator are discussed in Section 3.7. The SSB phase noise of the oscillator is shown in Figure 3.23(e). The harmonic balance controller used for the calculation of the phase noise is shown. ADS recommend that for phase noise calculation in the harmonic balance controller use: *Order* = 7 and *Oversample* = 4.

3.3.3 BJT Hartley Oscillator

The schematic of a BJT Hartley oscillator is shown in Figure 3.24(a), and its ac model is shown in Figure 3.24(b). For this oscillator the frequency of oscillation is

$$\omega_o = \frac{1}{\sqrt{L_T C - \dfrac{L_1 L_2 - M^2}{\left(\dfrac{h_{ie}}{R_L}\right)}}} \approx \frac{1}{\sqrt{L_T C}}$$

where

$$L_T = L_1 + L_2 + 2M$$

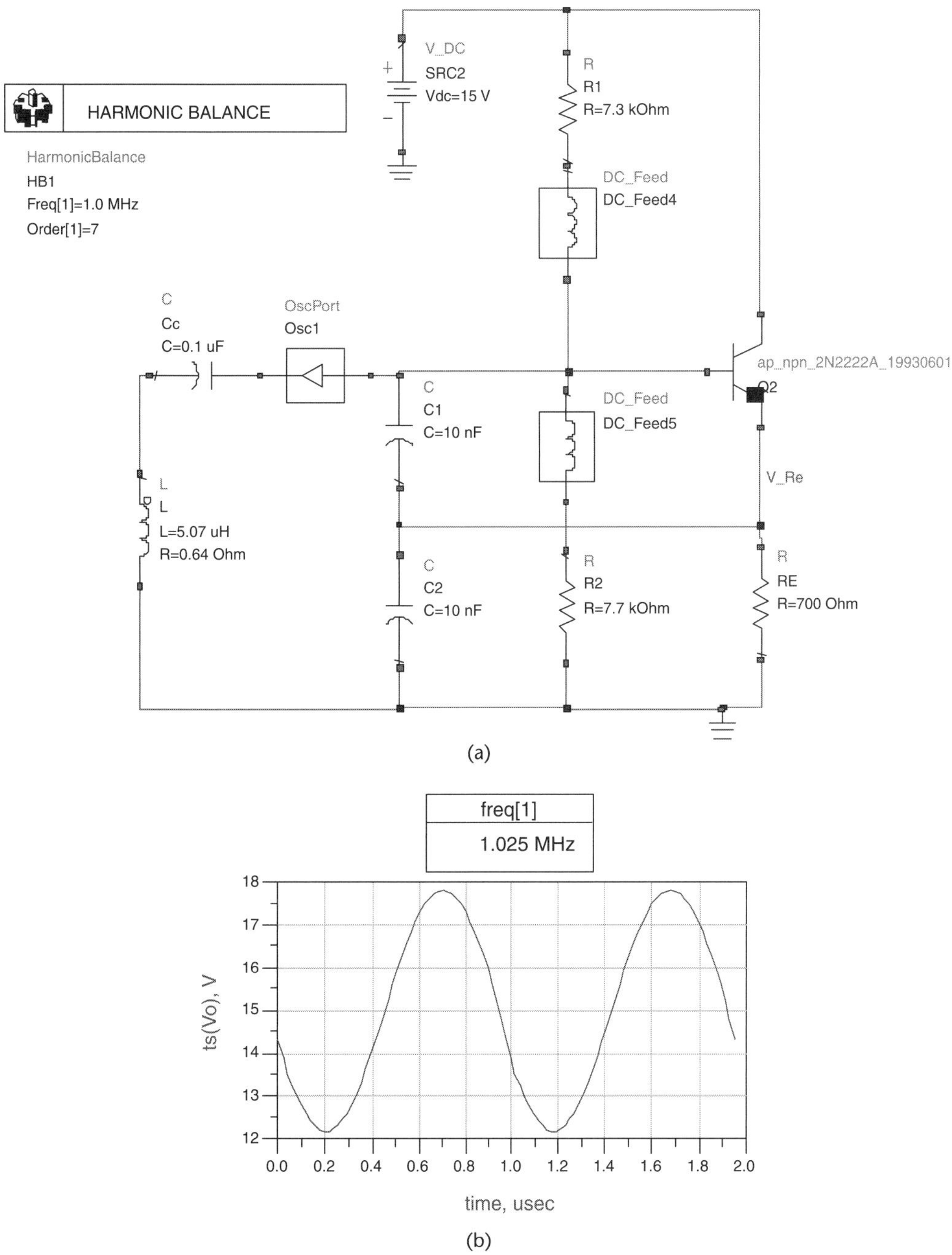

Figure 3.23 (a) Colpitts oscillator simulation for Example 3.6, (b) the oscillation waveform, (c) a collector coupling circuit for the Colpitts oscillator, (d) the waveforms of v_{BE} and i_C, and (e) the SSB phase noise.

With an RFC from the collector to V_{CC}, we have $R_L = 1/h_{oe}$ and the gain condition is

$$g_m R_L \geq \frac{L_2 + M}{L_1 + M}$$

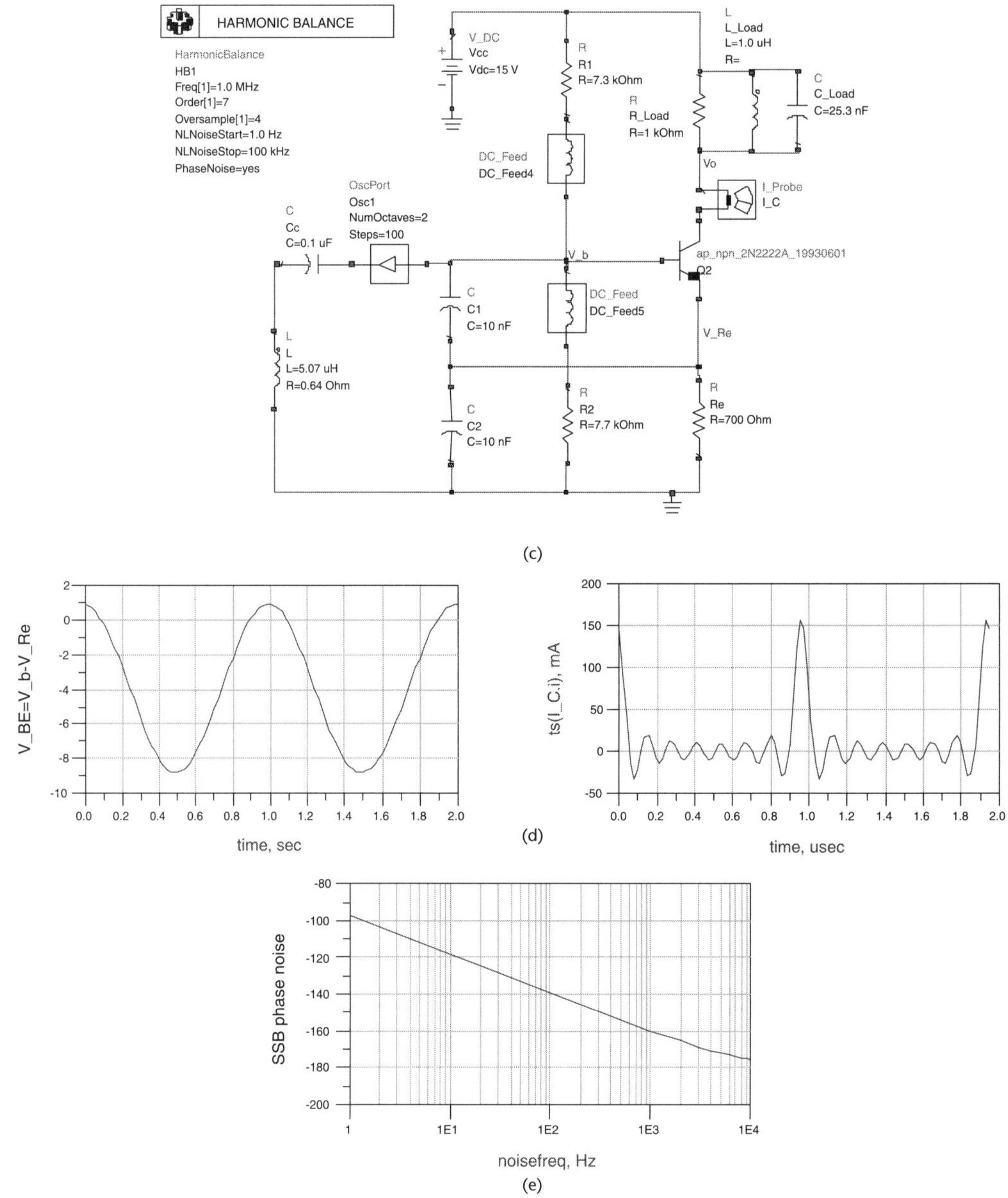

Figure 3.23 (Continued).

Another BJT Hartley type of oscillator is shown in Figure 3.25(a). In this oscillator the BJT is in a CB configuration. The biasing could also be done as shown in Figure 3.25(b). The ac model is shown in Figure 3.25(c).

From Figure 3.25(c) the feedback factor is

$$\beta = \frac{v_f}{v_o} = \frac{L_2}{L_1 + L_2}$$

The transformer formed by L_1 and L_2 converts the resistance $R_E || r_e \approx r_e$ to a parallel resistance across the capacitor of value

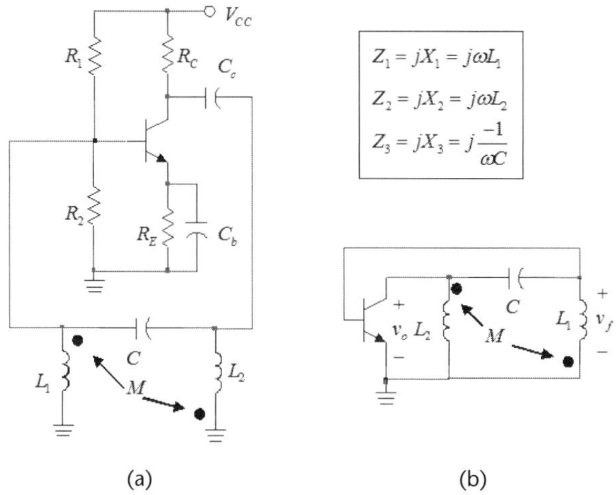

Figure 3.24 (a) A BJT Hartley oscillator and (b) its ac model.

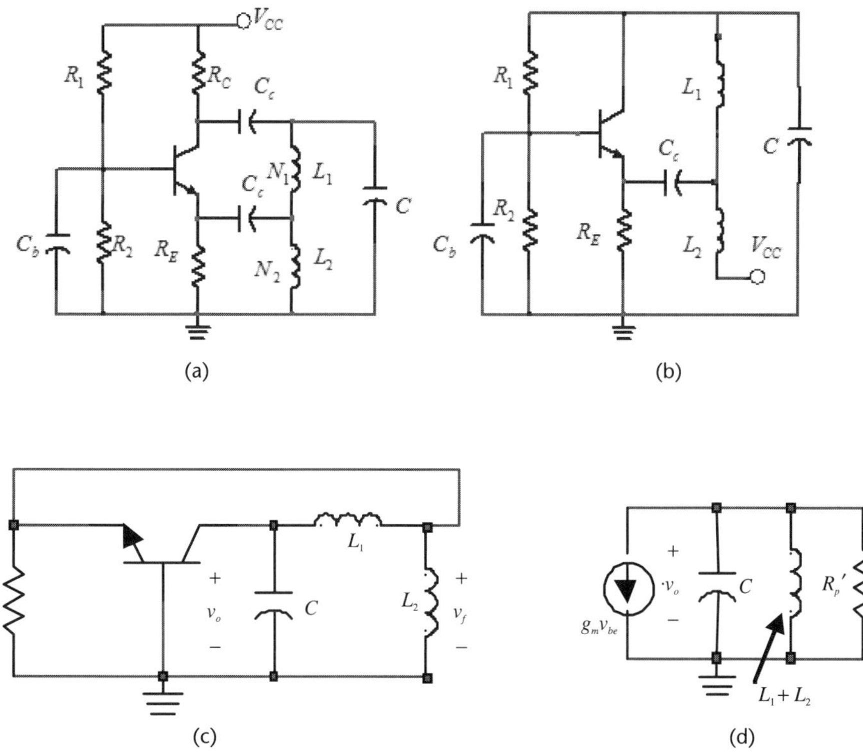

Figure 3.25 (a) A Hartley type of oscillator, (b) a dc bias variation, (c) the ac model, and (d) equivalent circuit across the capacitor.

$$R'_P = n^2 r_e$$

where

$$n = \frac{L_1 + L_2}{L_2}$$

The equivalent model of the tuned circuit is shown in Figure 3.25(d).

From Figure 3.25(d), the resonant frequency is

$$\omega_o = \frac{1}{\sqrt{(L_1 + L_2)C}} \tag{3.57}$$

At ω_o, the loop gain is

$$\beta(\omega_o)A_v(\omega_o) = \frac{L_2}{L_1 + L_2} g_m n^2 r_e \approx \frac{L_1 + L_2}{L_2} \tag{3.58}$$

where $g_m r_e = 1$. To start the oscillation, the loop gain is set to be between 2 and 3.

Example 3.7

Design the BJT Hartley oscillator in Figure 3.25(b) to oscillate at 1 MHz using the 2N3904, a general purpose BJT. The Q point is to be at V_{CE} = 6V and I_C = 3 mA. The 2N3904 lists $f_T > 300$ MHz and a typical beta of 100.

Solution

Let V_{CC} = 12V, then

$$R_E = \frac{12 - 6}{3 \times 10^{-3}} = 2\ \text{k}\Omega$$

Let

$$R_{TH} = \beta \frac{R_E}{10} = 100 \frac{2{,}000}{10} = 20\ \text{k}\Omega$$

and it follows that

$$V_{TH} \approx 6 + 0.7 = 6.7\text{V}$$

$$R_1 = R_{TH} \frac{V_{CC}}{V_{TH}} = (20 \times 10^3) \frac{12}{6.7} = 20\ \text{k}\Omega$$

and

$$R_2 = \frac{R_{TH}}{1 - \dfrac{V_{TH}}{V_{CC}}} = \frac{(20 \times 10^3)}{1 - \dfrac{6.7}{20}} = 45.3 \text{ k}\Omega$$

Let $C = 5$ nF, then from (3.57)

$$L_1 + L_2 = \frac{1}{\omega_o^2 C} = \frac{1}{(2\pi 10^6)^2 \, 5 \times 10^{-9}} = 5.1 \ \mu\text{H}$$

Letting the value of the loop gain be 3, it follows from (3.58) that

$$L_2 = \frac{L_1 + L_2}{3} = \frac{5.1 \times 10^{-6}}{3} = 1.7 \ \mu\text{H}$$

and

$$L_1 = 2L_2 = 3.4 \ \mu\text{H}$$

The simulation of the oscillator is shown in Figure 3.26. The fundamental frequency of oscillation is 999.5 kHz.

3.3.4 The Grounded-Base Oscillator

Two grounded-base (GB) oscillators are shown in Figure 3.27(a, b). In Figure 3.27(b), R_L also serves as the collector resistor. The GB oscillator is in the form of a Colpitts oscillator but with the base grounded. An ac model is shown in Figure 3.27(c). Replacing the BJT by its ac model we obtain the ac model in Figure 3.27(d), where

$$h_{ib} = \frac{h_{ie}}{h_{fe}}$$

The parameter h_{ib} is also denoted by r_e in the literature. Observe that $r_e = 1/g_m$.

It is of interest to analyze this oscillator by calculating the loop gain. To this end, the model in Figure 3.27(d) is drawn in the form shown in Figure 3.28(a) where it is assumed that $C_2 \gg C_{be}$ and $C_1 \gg C_{ce}$. Observe that the impedance across C_2 is $R_E \| h_{ib}$. Since h_{ib} is small it follows that $R_E \| h_{ib} \approx h_{ib}$. Therefore, if

$$\frac{1}{\omega C_2} \ll h_{ib} \tag{3.59}$$

the resistor h_{ib} does not load the reactance of the capacitor C_2. However, the resistor h_{ib} transforms to a parallel resistor across L, which must be considered in the calculations of the voltage gain. In Figure 3.28(a) the losses of the inductor are represented by R_p.

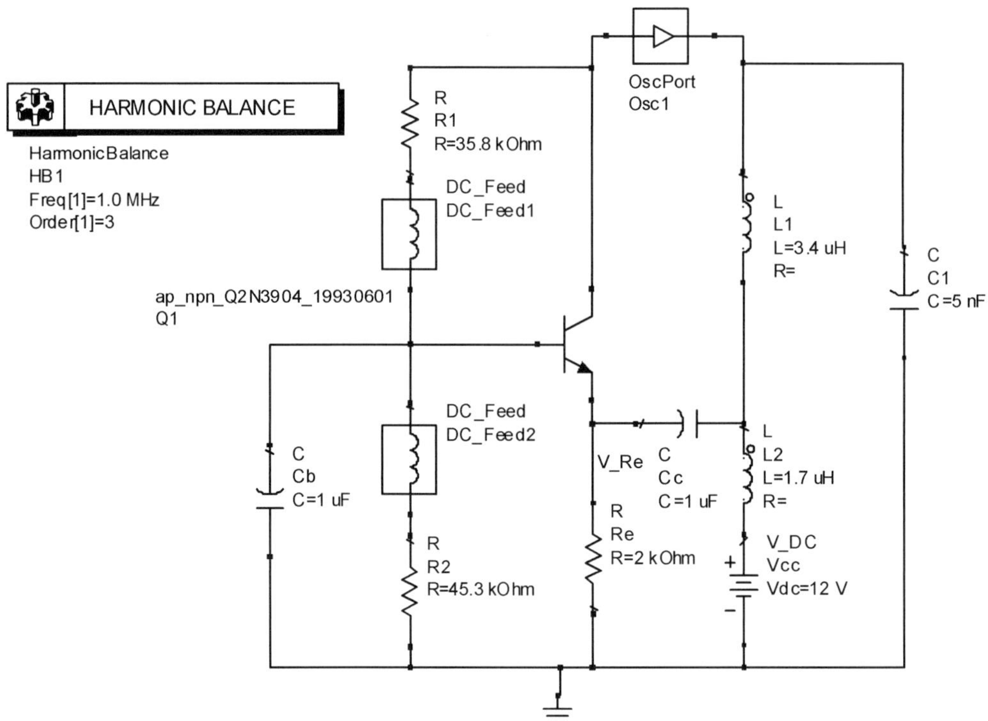

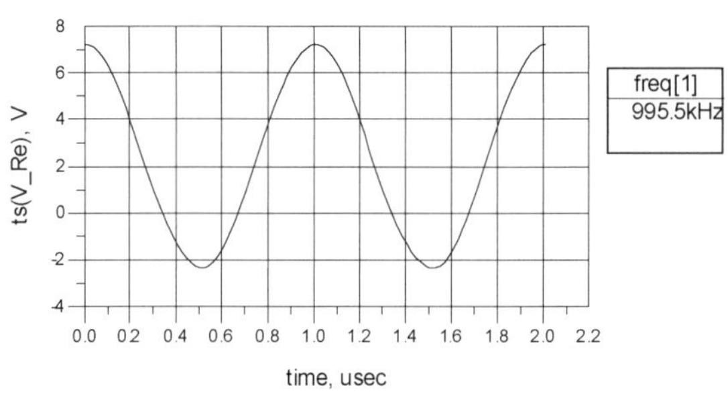

Figure 3.26 Simulation of Example 3.7.

Assuming that (3.59) is satisfied, the voltage feedback factor is

$$\beta(\omega) = \frac{v_f}{v_o} = \frac{X_{C_2}}{X_{C_2} + X_{C_1}} = \frac{C_1}{C_1 + C_2} \tag{3.60}$$

Defining

$$n = \frac{C_1 + C_2}{C_1}$$

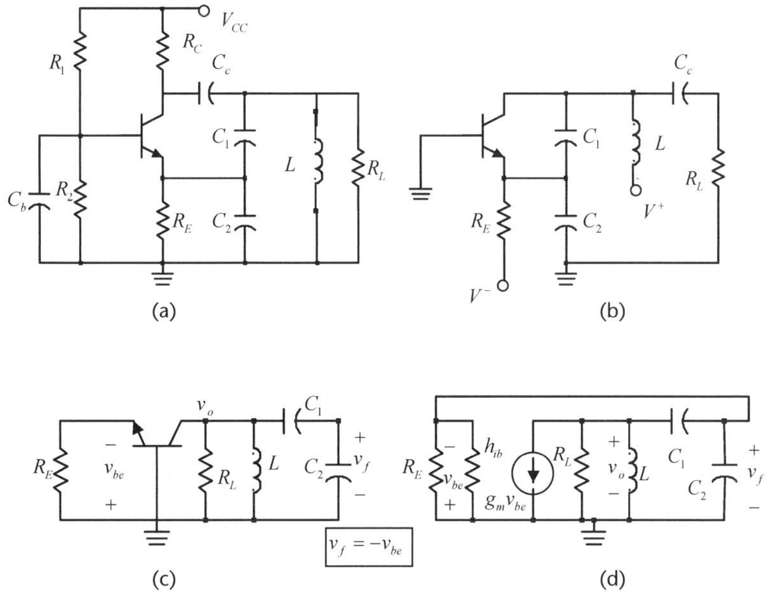

Figure 3.27 (a) A GB oscillator, (b) a GB oscillator with a different dc bias network, (c) an ac model, and (d) using the ac model of the transistor.

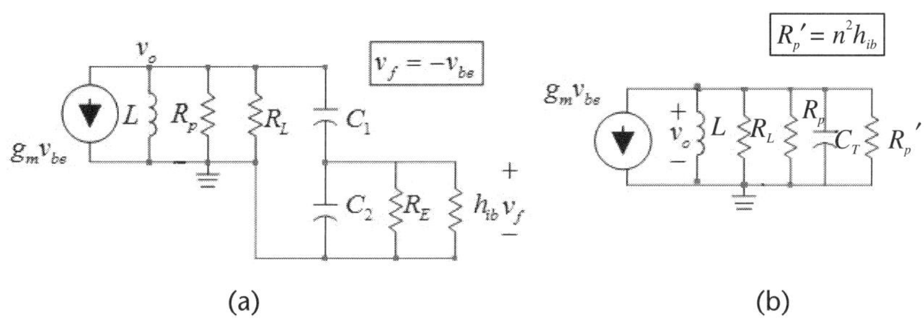

Figure 3.28 (a) Model to calculate the loop gain and (b) an equivalent ac model.

it follows that for a high-Q tuned circuit the capacitive transformer formed by C_1 and C_2 transform the resistor $R_E || h_{ib}$ to a parallel resistance across the coil given by

$$R_P' = (h_{ib} \parallel R_E) n^2 \approx h_{ib}\left(\frac{C_1 + C_2}{C_1}\right)^2$$

The total parallel resistance is

$$R_T = R_p \parallel R_L \parallel R_P'$$

The equivalent ac model across L is shown in Figure 3.28(b). Therefore, the open-loop gain is

$$A_v(j\omega) = \frac{v_o}{v_f} = g_m Z_L(j\omega) \tag{3.61}$$

where

$$Z_L(j\omega) = R_T \left\| j\omega L \right\| \frac{1}{j\omega C_T} \tag{3.62}$$

and

$$C_T = \frac{C_1 C_2}{C_1 + C_2}$$

The phase shift through the feedback network is 0°. Therefore, the phase from $A_v(j\omega)$ in (3.61) must also be 0°. This occurs at the resonant frequency of the tuned circuit, which is the frequency of oscillation. From (3.62), the imaginary part is zero at

$$\omega_o = \frac{1}{\sqrt{LC_T}}$$

Then, observing that $R_T \approx R_P'$, it follows that

$$A_v(j\omega_o) \approx g_m R_P' = g_m h_{ib}\left(\frac{C_1 + C_2}{C_1}\right)^2 \approx \left(\frac{C_1 + C_2}{C_1}\right)^2 \tag{3.63}$$

since $g_m h_{ib} = \alpha \approx 1$.

From (3.60) and (3.63), the loop-gain requirement to start the oscillation is

$$\beta(j\omega_o) A_v(j\omega_o) = \frac{C_1 + C_2}{C_1} \tag{3.64}$$

A design procedure is to use a factor of 10 to satisfy the inequality (3.59), or

$$C_2 \geq \frac{10}{\omega h_{ib}} \tag{3.65}$$

and in (3.64) to make the loop gain a value between 2 and 3.

Example 3.8

Design the GB oscillator in Figure 3.29(a) to oscillate at 10 MHz. The transistor (a 2N3904) has $\beta = 100$ and it is to be biased at $I_C = 1$ mA.

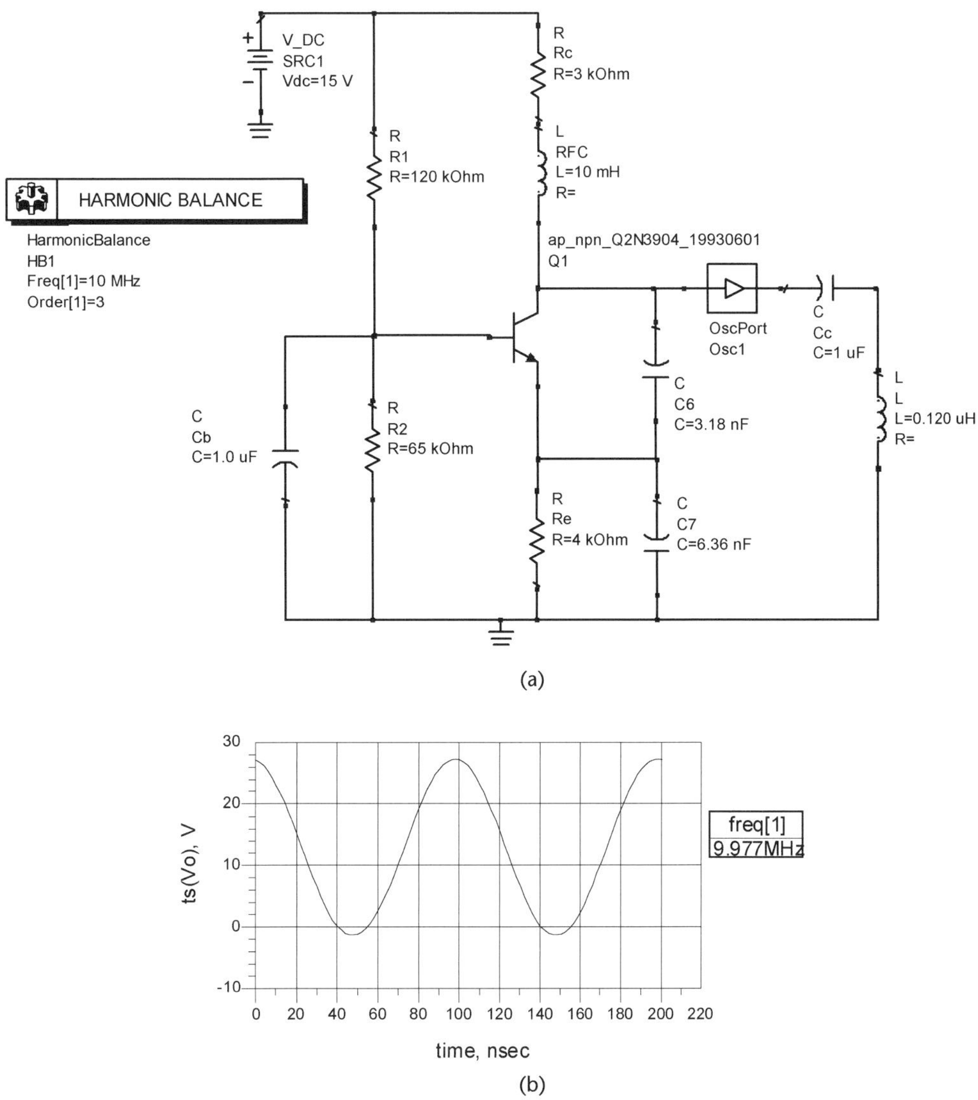

Figure 3.29 (a) GB oscillator for Example 3.8 and (b) the simulation results.

Solution

With $I_C = 1$ mA, it follows that

$$h_{ib} = \frac{V_T}{I_C} = \frac{25 \times 10^{-3}}{1 \times 10^{-3}} = 25\Omega$$

and

$$g_m = \frac{I_C}{V_T} = \frac{1}{h_{ib}} = \frac{1}{25} = 40 \text{ mS}$$

From (3.65)

$$C_2 \geq \frac{10}{\omega h_{ib}} = \frac{10}{2\pi \times 10^7 \, (25)} = 6.36 \text{ nF}$$

and from (3.64) letting the loop gain be 3, we obtain

$$C_1 = \frac{C_2}{2} = \frac{6.36 \times 10^{-9}}{2} = 3.18 \text{ nF}$$

Then,

$$C_T = C_1 \parallel C_2 = 2.12 \text{ nF}$$

The value of L is

$$L = \frac{1}{\omega_o^2 C_T} = \frac{1}{(2\pi \times 10^7)^2 (2.12 \times 10^{-9})} = 0.12 \; \mu\text{H}$$

and

$$R_P' \approx h_{ib}\left(\frac{C_1 + C_2}{C_1}\right)^2 = 25(3)^2 = 225\Omega$$

Since $R_P' = 225\Omega$, any R_L such that $R_L \gg 225\Omega$ can be used.

The simulation results are shown in Figure 3.29(b) where the fundamental frequency of oscillation is 9.977 MHz. The actual Q point values are: $V_{CE} = 7.5$V and $I_C = 1.07$ mA.

3.3.5 BJT Clapp Oscillator

Two Clapp oscillators are shown in Figure 3.30. For the oscillator in Figure 3.30(a), the frequency of oscillation occurs when

$$X_L(\omega_o) =- X_{C_1}(\omega_o) - X_{C_2}(\omega_o) - X_{C_3}(\omega_o)$$

or

$$\omega_o = \frac{1}{\sqrt{LC_T}} \tag{3.66}$$

where

$$\frac{1}{C_T} = \frac{1}{C_1} + \frac{1}{C_2} + \frac{1}{C_3}$$

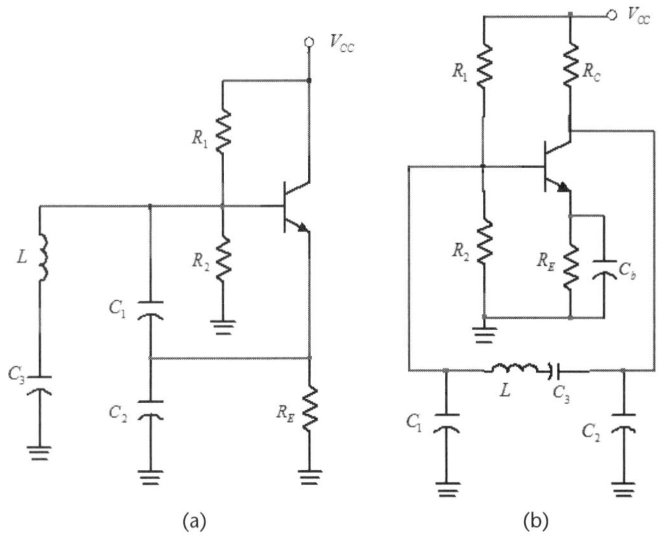

Figure 3.30 (a) A BJT Clapp oscillator and (b) another Clapp-type oscillator.

The gain condition is

$$\frac{g_m}{\omega_o^2 C_1 C_2 R_s} > 1 \tag{3.67}$$

Example 3.9

Design the Clapp oscillator in Figure 3.30a to oscillate at 10 MHz.

Solution

The BFS17P BJT was selected for this design. This RF transistor lists a maximum f_T value of 2.5 GHz. At a typical Q point of V_{CE} = 5V and I_C = 2 mA, the value of f_T is 1.5 GHz.

The Q point is obtained with V_{CC} = 10V, R_E = 2.5 kΩ, R_1 = 4.3 kΩ and R_2 = 5.7 kΩ. The value of g_m at I_C = 2 mA is 80 mS.

Let L = 25 μH with Q_U = 80. Then, X_L = 1,571Ω at 10 MHz and

$$R_s = \frac{\omega L}{Q_U} = \frac{2\pi \times 10^7 (25 \times 10^{-6})}{80} = 19.6\Omega$$

From (3.67), with $C_1 = C_2$, we obtain

$$C_1^2 < \frac{g_m}{\omega_o^2 R_s} = \frac{0.08}{(2\pi \times 10^7)^2 19.6} = 1.04 \times 10^{-18}$$

or C_1 < 1 nF, which is satisfied with $C_1 = C_2$ = 100 pF (or X_{C_1} =– 159Ω).

The value of C_3 follows from

$$X_{C_3} = - X_L - X_{C_1} - X_{C_2} = - 1{,}571 + 159 + 159 = - 1{,}253\Omega$$

or $C_3 = 12.7$ pF. The total capacitance across L is $C_T = 10$ pF.

The simulation of the oscillator is shown in Figure 3.31. The fundamental frequency of oscillation is 10.01 MHz.

3.3.6 Tuned-Collector Oscillator

A tuned-collector oscillator is shown in Figure 3.32. The ac model is similar to the one shown in Figure 3.14(b), except that the resistor h_{ie} must be included.

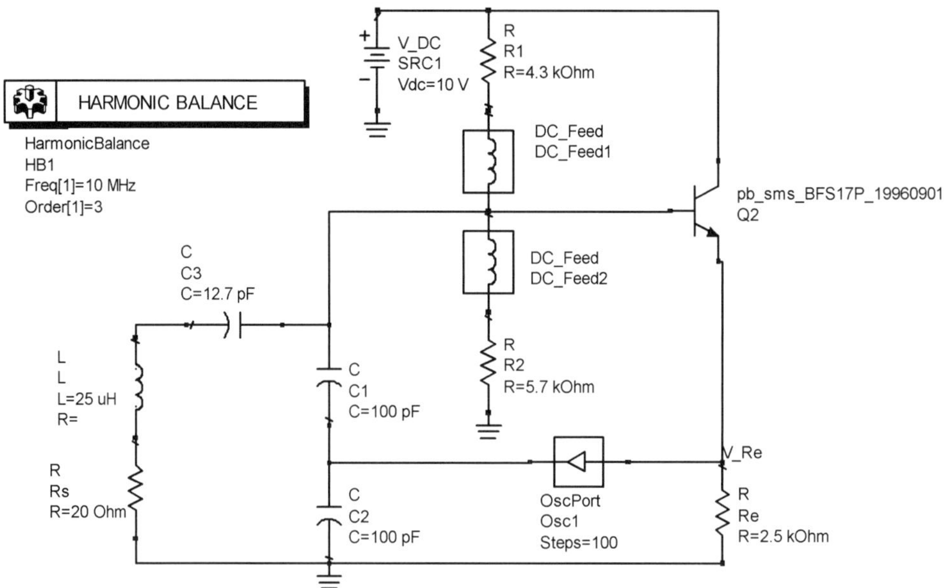

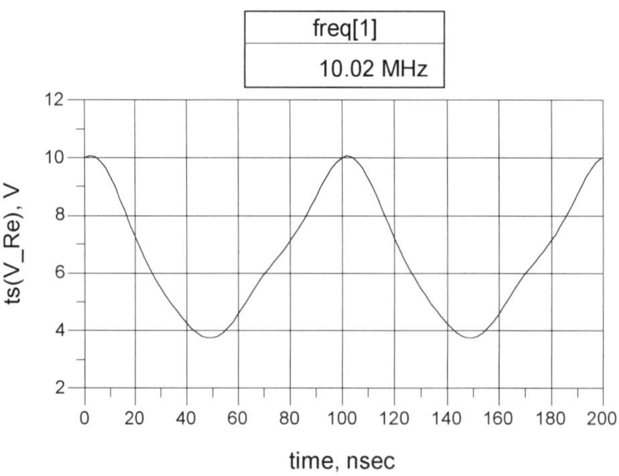

Figure 3.31 Simulation for Example 3.9.

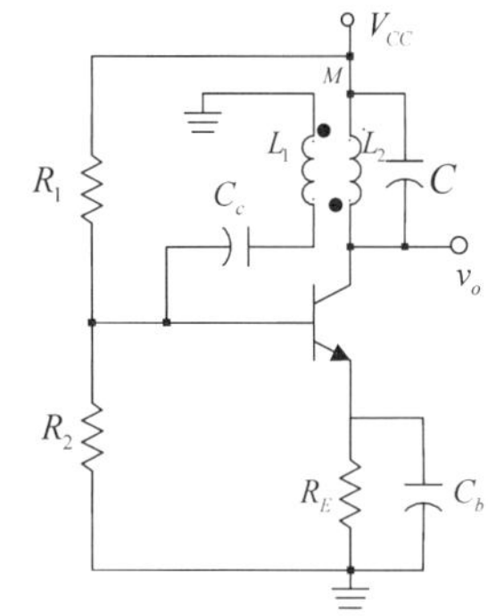

Figure 3.32 A tuned-collector oscillator.

Hence, the analysis of this oscillator is somewhat similar to the one used for the FET tuned-drain oscillator.

3.4 Op-Amp Tuned Oscillators

Tuned oscillators can also be designed using an op amp. Figure 3.33 illustrates a Pierce oscillator using an op amp. The analysis is very similar to that of the FET Pierce oscillators. The frequency of oscillation is given by [see (3.12)]

$$\omega_O = \frac{1}{\sqrt{LC_T}} \tag{3.68}$$

where

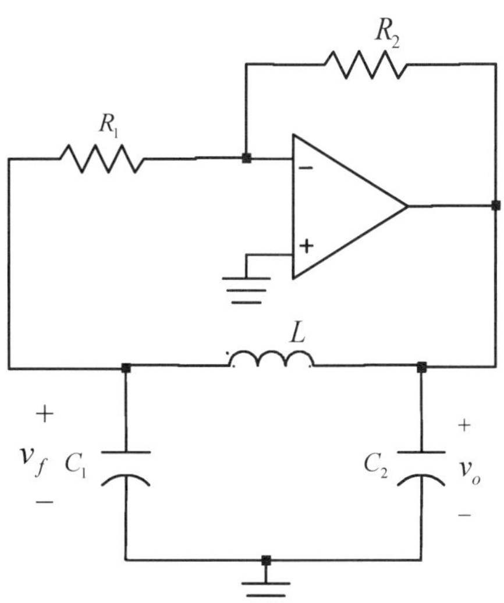

Figure 3.33 A Pierce oscillator using an op amp.

$$C_T = \frac{C_1 C_2}{C_1 + C_2}$$

The gain of the inverter configuration is $A_{vo} =- R_2/R_1$ and the feedback factor is

$$\beta(\omega_o) =- \frac{C_2}{C_1}$$

Therefore, the gain condition is

$$|A_{vo}| = \frac{R_2}{R_1} \geq \frac{C_1}{C_2} \tag{3.69}$$

Although the previous results are straightforward, it is of interest to analyze the op-amp oscillator ac model shown in Figure 3.34 where the inverter gain is $A_{vo} =- R_2/R_1$, and its output resistance is R_o. From Figure 3.34, assuming that $R_1 \gg 1/\omega C_1$, the open-loop gain is

$$A_v = \frac{v_o}{v_f} = \frac{A_{vo} Z_R}{Z_R + R_o}$$

where

$$Z_R = \frac{Z_2(Z_1 + Z_3)}{Z_1 + Z_2 + Z_3}$$

The feedback factor is

$$\beta = \frac{v_f}{v_o} = \frac{Z_1}{Z_1 + Z_3}$$

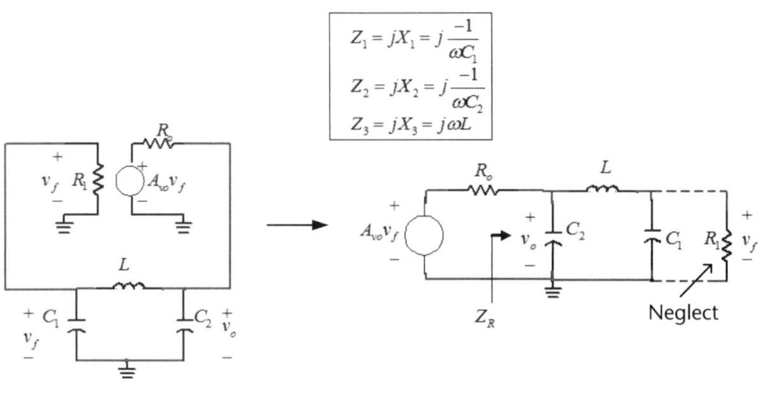

Figure 3.34 An ac model for the op-amp oscillator.

Hence, the loop gain is

$$\beta A_v = \frac{A_{vo} Z_1 Z_2}{Z_2(Z_1 + Z_3) + R_o(Z_1 + Z_2 + Z_3)}$$

$$\frac{-A_{vo} X_1 X_2}{-X_2(X_1 + X_3) + jR_o(X_1 + X_2 + X_3)}$$

Setting the loop gain equal to one shows that

$$X_1 + X_2 + X_3 = 0 \tag{3.70}$$

and the loop-gain condition is

$$\frac{A_{vo} X_1}{X_1 + X_3} =- \frac{A_{vo} X_1}{X_2} = 1$$

or

$$A_{vo} =- \frac{X_2}{X_1} =- \frac{C_1}{C_2} \tag{3.71}$$

which of course, is identical to (3.69).

If the inductor losses are considered in the analysis, it follows from (3.23) that at ω_o

$$Z_R \approx \frac{X_2^2(\omega_o)}{R_s}$$

and the gain condition is

$$A_{vo} > \frac{X_2^2(\omega_o) + R_s R}{X_1(\omega_o) X_2(\omega_o)}$$

Figure 3.35 illustrates a Hartley oscillator using an op amp. The frequency of oscillation is given by

$$\omega_o = \frac{1}{\sqrt{L_T C}}$$

where $L_T = L_1 + L_2$ if $M = 0$.

The feedback factor is

$$\beta(\omega_o) =- \frac{L_1}{L_2}$$

and the required gain is

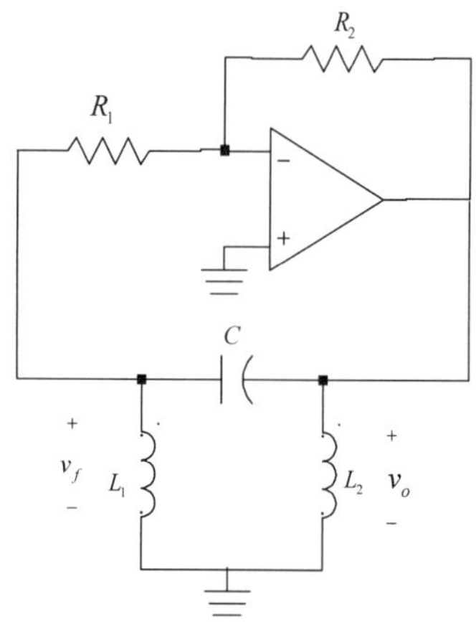

Figure 3.35 A Hartley oscillator using an op amp.

$$|A_{vo}| = \frac{R_2}{R_1} > \frac{L_2}{L_1}$$

Example 3.10

Design the op-amp Pierce oscillator in Figure 3.33 to oscillate at 10 kHz.

Solution

The op amp used is a general-purpose op amp, such as the 1458 or 741. General-purpose op amps have typical slew rates of 0.5 V/μs and gain-bandwidth products (f_T) of 1 MHz. These parameters limit the frequency of oscillation. Of course, there are many special purpose op amps with high slew-rate values and high f_T values, which can be used to attain higher frequency of oscillation.

Using supply voltage values of $V^+ = 5$V and $V^- = -5$V, the op amp output is limited to its saturation values (i.e., approximately ±5V). In this case the *SR* limits the frequency of oscillation to

$$f_o < \frac{SR}{2\pi|v_o|} = \frac{0.5 \times 10^6}{2\pi(5)} = 15.9 \text{ kHz}$$

Hence, the oscillator should perform well at 10 kHz.

Letting $L = 100\ \mu$H, the required total capacitance, from (3.68), is

$$C_T = \frac{1}{(2\pi f_o)^2 L} = \frac{1}{(2\pi 10^4)^2(100 \times 10^{-6})} = 2.53\ \mu\text{F}$$

Letting $C_1 = 10\ \mu$F, then

$$C_2 = \frac{C_1 C_T}{C_1 - C_T} = \frac{10 \times 10^{-6}(2.53 \times 10^{-6})}{10 \times 10^{-6} - 2.53 \times 10^{-6}} = 3.39\ \mu\text{F}$$

From (3.69), the required gain is

$$|A_{vo}| \geq \frac{C_1}{C_2} = \frac{10 \times 10^{-6}}{3.39 \times 10^{-6}} = 2.9$$

The gain condition can be satisfied with $R_2 = 200$ kΩ and $R_1 = 50$ kΩ (i.e., $|A_{vo}| = 4$).

The simulation is shown in Figure 3.36. The output is shown across C_1. The fundamental frequency of oscillation is 10.79 kHz. Although not shown, the waveform across C_2 has a peak value of approximately 5V.

A circuit that limits the amplitude of v_o in an op-amp tuned oscillator is shown in Figure 3.37. Other limiting circuits that can be used are those shown in Figures 1.18, 1.19, and 1.21.

3.5 Delay-Line Oscillators

A delay-line oscillator uses a transmission line as a delay line in the feedback path to accomplish the necessary phase shift for oscillation. The oscillator is shown in Figure 3.38(a). The amplifier is assumed to have a constant voltage gain with phase shift θ, namely,

$$A_v = A_o e^{j\theta}$$

The transmission line is assumed to be ideal and to be described by the feedback factor

$$\beta(j\omega) = e^{-j\omega\tau}$$

where τ is the delay time. The phase shift introduced by the transmission line (i.e., $\phi =- \omega t$) is a linear function of frequency, as shown in Figure 3.38(b). The slope of the phase shift is

$$\frac{d\phi}{d\omega} =- \tau$$

The phase shift associated with the loop gain is $\theta + \phi$ (or $\theta - \omega t$). Oscillations occur at the frequency ω_o when $\theta + \phi = 0$.

In an RLC tuned circuit, Q_L and the phase shift are related by

$$Q_L =- \frac{\omega_o}{2} \frac{d\phi}{d\omega}$$

Defining the Q of the delay-line oscillator in a similar way gives

$$Q_L = \frac{\omega_o \tau}{2}$$

which shows that high-Q values are obtained with long time delays.

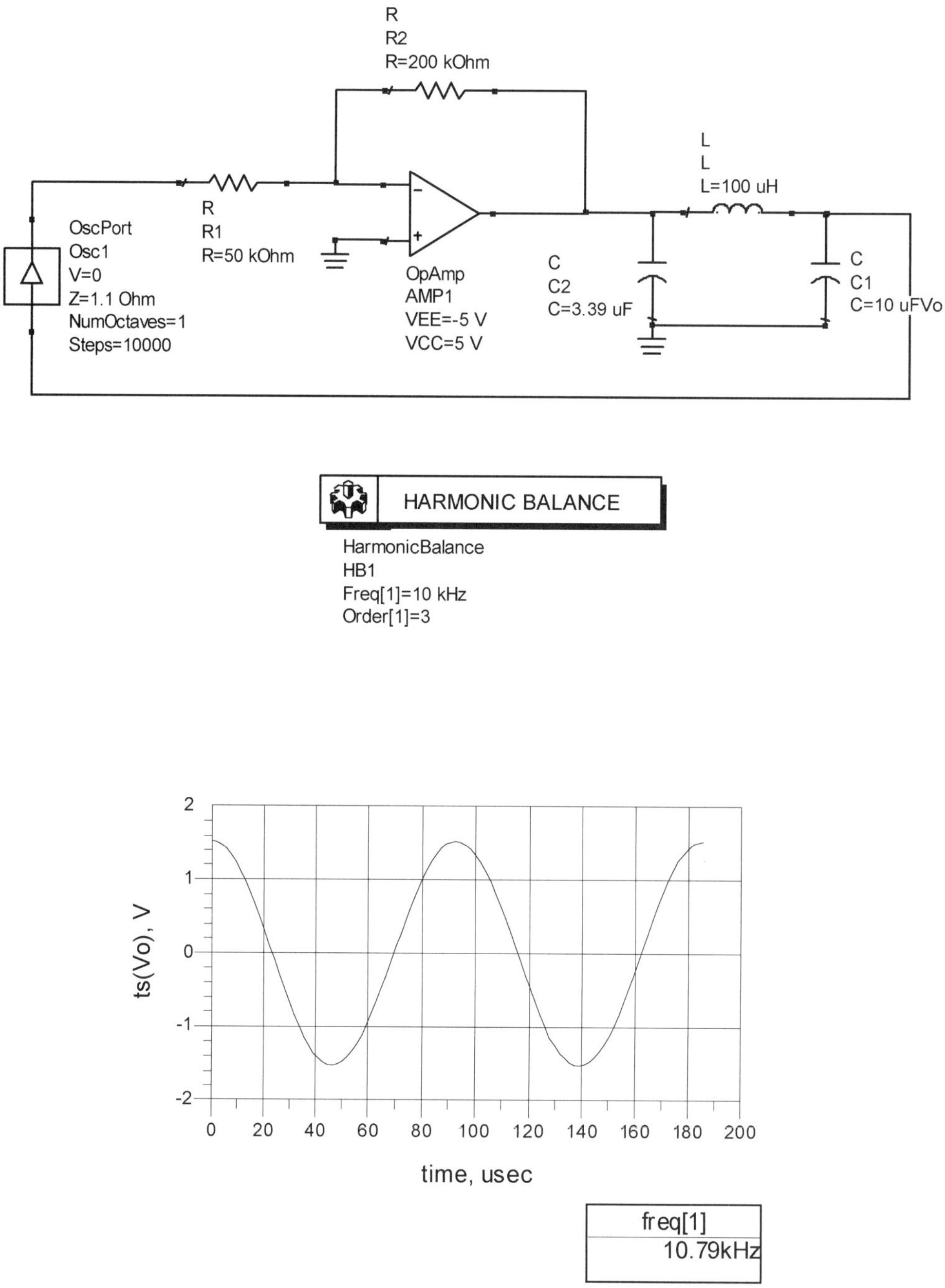

Figure 3.36 Simulation for the oscillator in Example 3.10.

Delay lines can be implemented with coaxial cable, although the resultant length can be long. Surface acoustic delay lines are more practical, but more expensive.

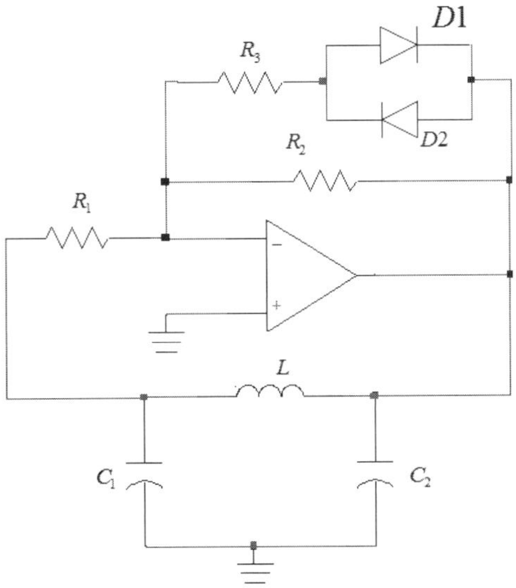

Figure 3.37 An amplitude-limiting mechanisms for op-amps tuned oscillators.

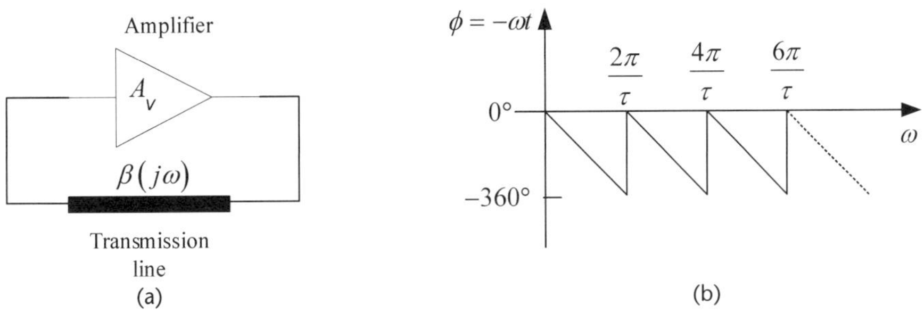

Figure 3.38 (a) Delay-line oscillator and (b) its phase shift.

3.6 Voltage-Controlled Tuned Oscillators

In a voltage-controlled oscillator (VCO) either the capacitance or inductance of the tuned circuit is varied to change the frequency of oscillation. VCOs are used in many electronic applications, such as AM and FM modulation, TV tuners, and frequency synthesizers.

A varactor-tuned oscillator uses the voltage-controlled capacitance of a varactor diode (also known as varicaps) to accomplish the electronic tuning. Varactor diodes of different types (i.e., abrupt and hyper-abrupt) having a wide range of capacitances are available. Abrupt varactor diodes operate over a wide range of tuning voltages, and the phase noise performance is lower than the one obtained with hyper-abrupt varactor diodes. The phase noise performance of a varactor diode is determined by its series resistance.

Hyper-abrupt varactor diodes have a more linear characteristic than the abrupt diode. Hence, they are a good choice for wide-band VCOs. For example, it is possible to cover a range of one octave with a tuning voltage of 1V to 20V.

The varactor diode symbol is shown in Figure 3.39(a), and its circuit model in Figure 3.39(b). The varactor diode junction capacitance (C) is given by the formula

$$C = \frac{C_o}{\left(1 + \frac{V_R}{\phi}\right)^n}$$

where C_o is the value of the capacitance at zero voltage, V_R represents the reverse bias voltage, ϕ is the junction contact potential ($\phi \approx 0.7$V for Si and 1.2V for GaAs), and n is a number from 0.3 to 2. Lower values of n are associated with abrupt junction diodes, and larger values with hyper-abrupt junction diodes.

The total varactor capacitance C_T is composed of the sum of the junction capacitance C plus the case capacitance C_C. The resistance R_s represents the series-loss resistance of the diode. The unloaded Q of a varactor diode is

$$Q_U = \frac{1}{\omega C_T R_s}$$

For example, if at 100 MHz the capacitance C_1 is 100 pF and $Q_U = 80$, then $R_s = 0.2\Omega$. The Q of many varactor diodes is specified at 50 MHz, and for varactor diodes used at microwave frequencies the Q_U is usually specified at 1 GHz.

Figure 3.40 shows typical characteristic of a varactor diode. For this diode a bias voltage of 4V produces a capacitance of 9.5 pF.

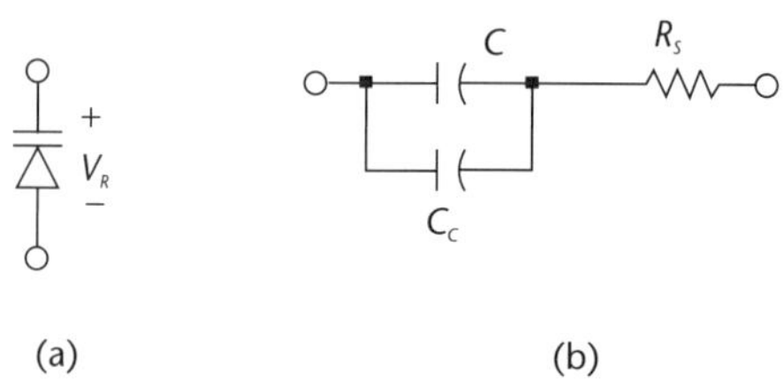

Figure 3.39 (a) Varactor diode symbol and (b) model.

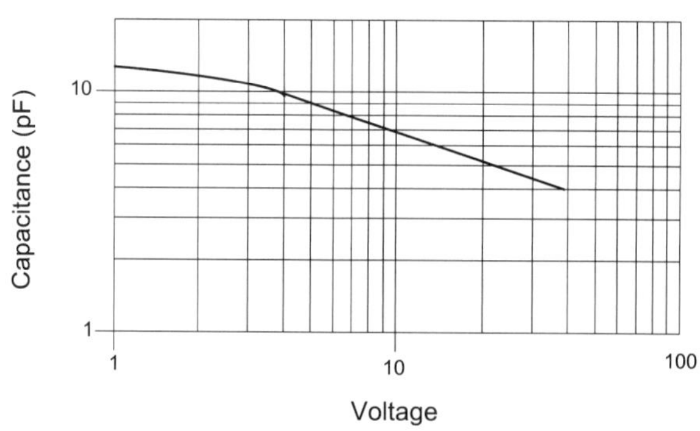

Figure 3.40 Typical varactor's characteristics.

A basic schematic of a varactor-tuned oscillator is shown in Figure 3.41(a). The capacitance of the varactor is determined by the voltage V, which is set by V^+, R_a, and R_b. The RFC is used to isolate the dc signal from the ac signal. Another varactor-tuned oscillator is shown in Figure 3.41(b). This configuration is recognized as a Clapp oscillator with the varactor diode used in place of C_3 (see Figure 3.30). The relations (3.66) and (3.67) apply to the VCO with Rs being the sum of the series-loss resistance of the inductor and the varactor equivalent series resistance.

Another important parameter is the tuning ratio TR (or the capacitance ratio). For example, if a varactor specifies

$$TR = \frac{C_T(2\text{V})}{C_T(30\text{V})} = 3$$

and $C_T(2\text{V}) = 15$ pF (i.e., the total capacitance at 2V), then $C_T(30\text{V}) = 5$ pF.

In the previous oscillators, at the frequency of oscillation, the resonant circuit can be represented by a parallel tuned circuit with an equivalent inductance L, a fixed value of capacitance C_f, and a variable varactor diode capacitance C_v. Therefore, the frequency of oscillation is

$$\omega_o = \frac{1}{\sqrt{L(C_f + C_v)}} \tag{3.72}$$

For simplicity, write C_v in the form

$$C_v = \frac{C_o}{V_b^n}$$

where C_o now represents the capacitance at $V_b = 1\text{V}$. Then, (3.72) can be written as

$$L = \frac{1}{\omega^2\left(C_f + C_o V_b^{-n}\right)}$$

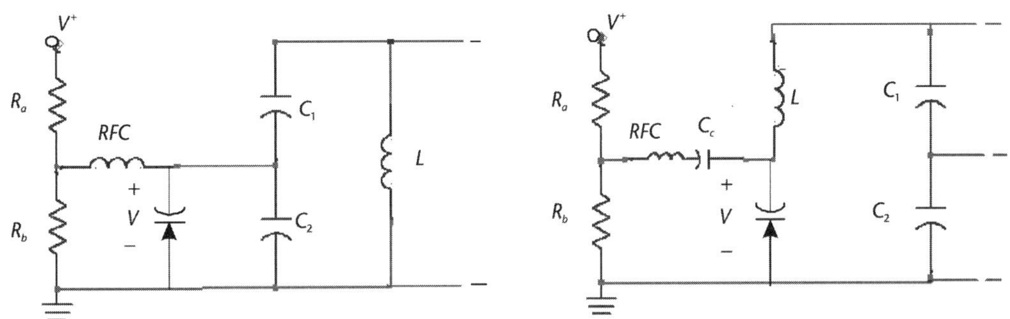

Figure 3.41 (a) A varactor-tuned oscillator and (b) a Clapp oscillator using a varactor diode.

The unmodulated frequency of oscillation (i.e., $\omega = \omega_o$) is obtained when $V_b = V_o$ and $C_v = C_{vo}$, where

$$C_{vo} = C_o V_o^{-n}$$

If V_o is varied by ΔV, the unmodulated frequency is varied by a small $\Delta\omega$ given by

$$L = \frac{1}{(\omega_o + \Delta\omega)^2 [C_f + C_o (V_o + \Delta V)^{-n}]}$$

or

$$\frac{1}{\omega_o^2 \left(C_f + C_o V_o^{-n}\right)} = \frac{1}{(\omega_o + \Delta\omega)^2 [C_f + C_o (V_o + \Delta V)^{-n}]}$$

The previous relation gives

$$\left(1 + \frac{\Delta\omega}{\omega_o}\right)^2 = \frac{C_f + C_o V_o^{-n}}{C_f + C_o (V_o + \Delta V)^{-n}}$$

Finally, the tuning sensitivity (K) in hertz/volt is

$$K = \frac{d\omega}{dV} = \frac{n\omega_o}{2V_o}\left(\frac{C_{v,o}}{C_1 + C_{v,o}}\right)$$

The use of a varactor diode in a crystal oscillator is illustrated in Figure 4.40.

3.7 Large-Signal Analysis of Oscillators

A large-signal analysis is required to analytically determine the amplitude of the oscillation. In tuned oscillators the mechanism that stabilizes the oscillation amplitude is the nonlinear behavior of the active device. Basically, the transistor's transconductance decreases as the oscillation amplitude grows. To analyze the transconductance behavior of a BJT under large-signal conditions we consider the circuit in Figure 3.42 where

$$\begin{aligned} v_{BE} &= V_{BEQ} + v_i(t) \\ &= V_{BEQ} + V_1 \cos \omega t \end{aligned} \tag{3.73}$$

Therefore, i_C is given by

$$i_C = I_s e^{v_{BE}/V_T} = I_s e^{V_{BEQ}/V_T} e^{x \cos \omega t} \tag{3.74}$$

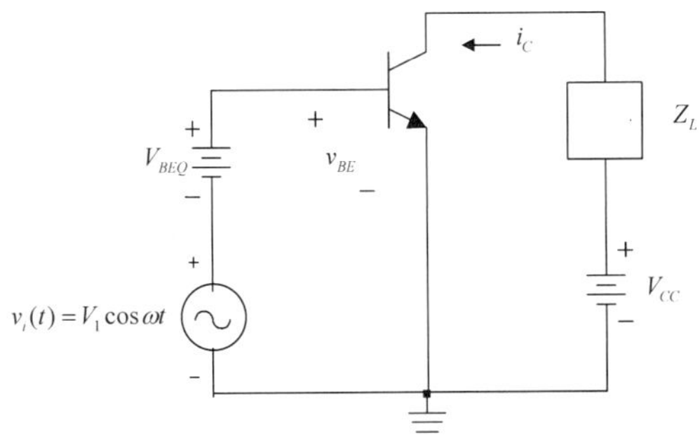

Figure 3.42 Large-signal analysis of a BJT.

where

$$V_T = \frac{nq}{kT}$$

and

$$x = \frac{V_1}{V_T}$$

where I_s is the saturation current, V_T is the thermal voltage, q is the electron charge (1.6×10^{-19} C), k is Boltmann's constant (1.38×10^{-23} J/K), and T is the temperature in degrees Kelvin. At a room temperature of 300°K, it follows that $V_T = 26$ mV.

For small-signal excitation $x \ll 1$ (or $V_1 \ll 26$ mV), and (3.74) can be approximated by

$$i_C = I_s e^{V_{BEQ}/V_T}[1 + x \cos \omega t]$$

or

$$i_C = I_{CQ} + i_c(t)$$

where the dc value of i_C is

$$I_{CQ} = I_s e^{V_{BEQ}/V_T}$$

and its ac value is

$$i_c(t) = I_{CQ} x \cos \omega t = \frac{I_{CQ}}{V_T} V_1 \cos \omega t = g_m V_1 \cos \omega t$$

where g_m is the small-signal transconductance, namely,

$$g_m = \left.\frac{\partial i_C}{\partial v_{BE}}\right|_{i_C = I_{CQ}} = \frac{i_c(t)}{v_i(t)} = \frac{I_C}{V_T} \tag{3.75}$$

For large-signals, the function $e^{x \cos \omega t}$ in (3.74) can be expanded in a Fourier series as

$$e^{x \cos \omega t} = I_o(x) + 2\sum_{n=1}^{\infty} I_n(x) \cos n\omega t$$

where $I_n(x)$ ($n = 0, 1, \ldots,$) are the modified Bessel functions of the first kind of order n and argument x. Hence, (3.74) can be expressed in the form

$$i_C = I_{CQ}\left[I_o(x) + 2\sum_{n=1}^{\infty} I_n(x) \cos n\omega t\right] \tag{3.76}$$

From (3.75) it is seen that the dc value of i_C is affected by the amplitude of the sinusoidal signal and it is given by

$$I_{C,dc} = I_{CQ} I_o(x)$$

If the load circuit Z_L in Figure 3.42 is a tuned circuit with a high value of Q_L, only the fundamental component of i_C contributes to the output voltage v_o. From (3.76) the fundamental component of i_C is $2I_{CQ}I_1(x) \cos \omega t$. The large-signal transconductance $G_m(x)$ is defined as the ratio of the fundamental component of i_C to the voltage V_1. That is,

$$G_m(x) = \frac{2I_{CQ}I_1(x)}{V_1} = \frac{2I_{C,dc}I_1(x)}{V_1 I_o(x)} = \frac{2I_{C,dc}I_1(x)}{V_T x I_o(x)}$$

or, using (3.75),

$$G_m(x) = g_m \frac{2I_1(x)}{xI_o(x)} \tag{3.77}$$

A plot of (3.77) is shown in Figure 3.43. For small argument ($x \to 0$) the modified Bessel function can be approximated by $I_o(x) = 1$ and $I_1(x) = x/2$. Hence, it follows that for small signals (i.e., for $x \to 0$) (3.77) reduces to $G_m(x) = g_m$, as shown in Figure 3.43.

It also follows from (3.76) that a calculation of $2I_n(x)/I_o(x)$ for a given x provides information about the amplitude of the harmonics. For example, for $x = 10$ it follows that $2I_1(x)/I_o(x) \approx 1.9$, $2I_2(x)/I_o(x) \approx 1.6$, and $2I_3(x)/I_o(x) \approx 1.25$. Hence, if the dc collector current is 1 mA, the amplitude of the first harmonic is 1.9 mA, for the second harmonic it is 1.6 mA, and for the third harmonic it is 1.25 mA, showing that a drive of $x = 10$ produces a significant number of harmonics.

Low values of x, say $x = 1$, results in low harmonic distortion, but can also lead to problems with the oscillator design.

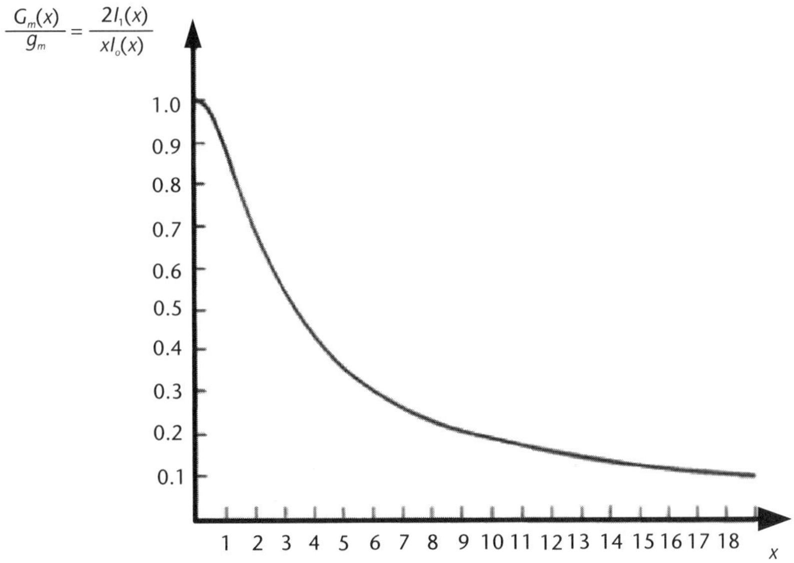

Figure 3.43 Plot of the function $\frac{G_m(x)}{g_m} = \frac{2I_1(x)}{xI_o(x)}$. (*From:* [1]. © 1971. Reprinted with permission of Pearson Education, Inc.)

The exponential relationship between the collector current and the base-to-emitter voltage shows that a large amplitude v_{be} voltage produces a series of periodic pulses in the collector current. The smaller the width of these current pulses, the higher the harmonic content. However, from the phase noise discussion in Chapter 2 there is a relation between the collector pulse width and the resulting phase noise.

In tuned *oscillators*, capacitors are used to provide feedback. In addition, capacitors might be needed for coupling and bypass purposes in order to ensure that the feedback oscillator signal appears across the base to emitter junction of the transistor. For example, consider the GB oscillator shown in Figure 3.44. If there is no oscillation the voltage difference between the capacitors C_b and C_2 sets the Q point value of v_{BE} (i.e., $V_{BEQ} \approx 0.7\text{V}$), and I_{CQ} is given by

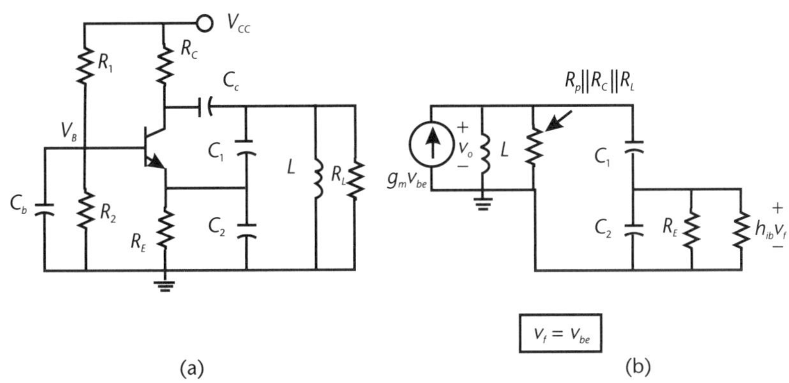

Figure 3.44 (a) A GB oscillator and (b) the small-signal model.

$$I_{CQ} = \frac{V_{TH} - 0.7}{R_E + \frac{R_{TH}}{\beta}}$$

where

$$V_{TH} = \frac{V_{CC} R_2}{R_1 + R_2}$$

and

$$R_{TH} = R_1 \parallel R_2$$

An oscillation signal will produce an ac variation in v_{BE} [see $v_i(t)$ in Figure 3.42] that affects the resulting average value of v_{BE} and i_C. However, this is usually a second-order effect and the large-signal transconductance can be calculated using (3.77) with g_m given by (3.75). The analysis of this effect is discussed in Clarke and Hess [1] where the resulting $G_m(x)$ was found to be given by

$$G_m(x) = g_m \left[1 + \frac{\ln I_o(x)}{V_\lambda V_T} \right] \frac{2I_1(x)}{x I_o(x)} \tag{3.78}$$

where

$$V_\lambda = I_{CQ}\left(R_E + \frac{R_{TH}}{\beta} \right)$$

The parameter V_λ represents the sum of the quiescent voltages across R_E and R_B. It is also equal to $V_{TH} - 0.7$.

If

$$V_\lambda \gg \frac{\ln I_o(x)}{V_T} \tag{3.79}$$

which is satisfied in most circuits with good beta stability, the value of $G_m(x)$ in (3.78) reduces to that in (3.77).

The small-signal model of the GB oscillator in Figure 3.44(a) is shown in Figure 3.44(b). For this oscillator

$$\beta = \frac{v_f}{v_o} = \frac{C_1}{C_1 + C_2}$$

and

$$A_v = \frac{v_o}{v_f} = \frac{g_m \frac{s}{C}}{s^2 + s\frac{1}{R_T C_T} + \frac{1}{LC_T}} = \frac{g_m R_T (2\alpha s)}{s^2 + 2\alpha s + \omega_o^2}$$

where

$$R_T = (R_p \parallel R_C \parallel R_L) \parallel \left[n^2(R_E \parallel h_{ib})\right]$$

$$n = \frac{C_1 + C_2}{C_1}$$

$$\alpha = \frac{1}{R_T C_T}$$

and

$$\omega_o = \sqrt{\frac{1}{LC_T}}$$

Therefore, it follows that the loop gain is

$$\beta A_v = \frac{1}{n} \frac{g_m R_T(2\alpha s)}{s^2 + 2\alpha s + \omega_o^2} \tag{3.80}$$

At the resonant frequency ω_o the oscillation will grow if $\beta A_v > 1$, or from (3.80) when

$$\frac{g_m R_T}{n} > 1$$

or

$$g_m > \frac{n}{R_T} \tag{3.81}$$

The small-signal g_m decreases to its large-signal value $G_m(x)$ and the small-signal model in Figure 3.44(b) is replaced by the large-signal model shown in Figure 3.45. Only the fundamental components of $v_{be} = V_1 \cos \omega_o t$ are considered since the tuned-circuit shorts all other harmonics to ground. The amplitude of $v_{be}(t)$

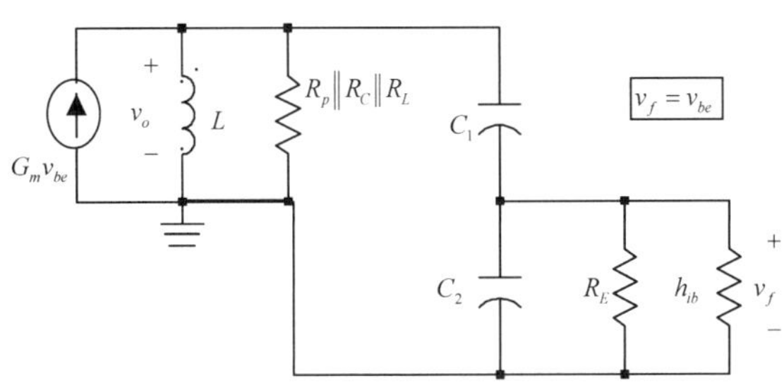

Figure 3.45 Large-signal model of the GB oscillator shown in Figure 3.44.

(i.e., V_1) at which the oscillator stabilizes can be calculated assuming that the transistor does not saturate. For the large-signal model in Figure 3.45 the loop gain is identical to (3.80) but with g_m replaced by $G_m(x)$, namely,

$$\beta A_v = \frac{1}{n}\frac{G_m(x)R_T(2\alpha s)}{s^2 + 2\alpha s + \omega_o^2}$$

At the resonant frequency the large-signal loop gain is unity when

$$G_m(x) = \frac{n}{R_T} \tag{3.82}$$

In order to use the plot in Figure 3.43, (3.82) is written in the form

$$\frac{G_m(x)}{g_m} = \frac{n}{g_m R_T} \tag{3.83}$$

Equation (3.83) gives the value of $G_m(x)/g_m$ that sets the loop-gain equal to unity. Then, from the plot in Figure 3.43 the value of x is obtained (or $V_1 = xV_T$). The amplitude V_1 is the amplitude of v_{be} at which the oscillation amplitude stabilizes.

Example 3.11 illustrates the previous large-signal analysis in a GB oscillator.

Example 3.11

(a) A design for the GB oscillator in Figure 3.27(b) at $f_o = 1.5$ MHz is shown in Figure 3.46. For this design, calculate the oscillation waveform of $v_o(t)$.
(b) Calculate $v_o(t)$ if the load resistor is changed to 20 kΩ.

Solution
(a) For this oscillator the feedback factor is

$$\beta = \frac{C_1}{C_1 + C_2} = \frac{1.142 \times 10^{-9}}{1.142 \times 10^{-9} + 80 \times 10^{-9}} = 0.0141$$

and

$$n = \frac{C_1 + C_2}{C_1} = \frac{1}{0.0141} = 71$$

The total capacitance C_T is

$$C_T = \frac{C_1 C_2}{C_1 + C_2} = 1.126 \text{ nF}$$

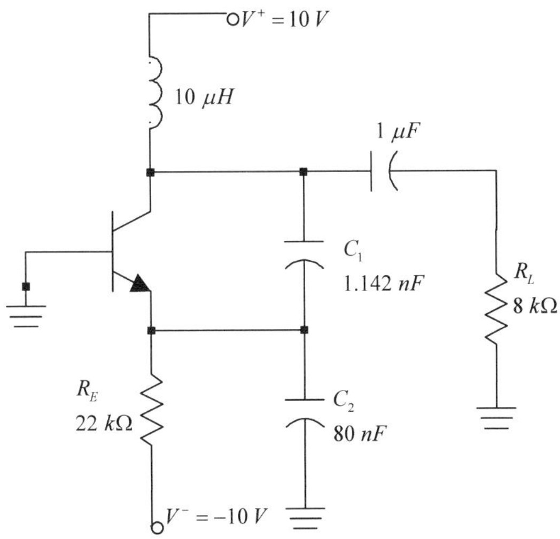

Figure 3.46 GB oscillator for Example 3.11.

and the resonant frequency is

$$f_o = \frac{1}{2\pi}\sqrt{\frac{1}{LC_T}} = \frac{1}{2\pi}\sqrt{\frac{1}{10 \times 10^{-6}(1 \times 10^{-9})}} = 1.5 \text{ MHz}$$

The Q point value of I_C is

$$I_{CQ} \approx I_{EQ} = \frac{10 - 0.7}{22 \times 10^3} = 0.423 \text{ mA}$$

and the small-signal g_m is

$$g_m = \frac{I_{CQ}}{V_T} = \frac{0.423 \times 10^{-3}}{26 \times 10^{-3}} = 16.27 \text{ mS}$$

From (3.82) the amplitude V_1 stabilizes when

$$\frac{G_m(x)}{g_m} = \frac{n}{g_m R_T} = \frac{71}{(16.27 \times 10^{-3})(8 \times 10^3)} = 0.545$$

and it follows from Figure 3.43 that $x = 3$. Therefore,

$$V_1 = xV_T = 3(26 \times 10^{-3}) = 78 \text{ mV}$$

$$V_o = nV_1 = 71(78 \times 10^{-3}) = 5.54\text{V}$$

and the output voltage is

$$v_o = 5.54 \cos \omega_o t \tag{3.84}$$

It is of interest to observe that for this example the value of V_1 (i.e., the dc value across R_E) is

$$V_\lambda = 10 - 0.7 = 9.3\text{V}$$

and therefore, (3.79) is satisfied.

The simulation of this oscillator is shown in Figure 3.47(a). The Q point is at $V_{CE} = 10.6\text{V}$ and $I_C = 0.426$ mA. The output waveform, shown in Figure 3.47(b), oscillates between ±5.4V at a fundamental frequency of 1.498 MHz in fairly good agreement with (3.84). The base-to-emitter voltage and collector current waveforms are shown in Figure 3.47(c). The peak value of v_{be} shows that $V_1 \approx 78$ mV. From the discussion in Chapter 2, the pulse width and amplitude of the i_C waveform when the BJT conducts affects significantly the phase noise of the oscillator. The SSB phase noise for this oscillator is shown in Figure 3.47(d).

(b) If $R_L = 20\ \text{k}\Omega$, then

$$\frac{G_m(x)}{g_m} = \frac{n}{g_m R_T} = \frac{71}{(16.27 \times 10^{-3})(20 \times 10^3)} = 0.218$$

In this case, from Figure 3.43, the value of x is 8.8, and it follows that

$$V_1 = xV_T = 8.8(26 \times 10^{-3}) = 228\text{ mV}$$

and

$$V_o = nV_1 = 71(0.228) = 16.1\text{V}$$

But V_o cannot be 16.1V. In this case the BJT saturates and the amplitude V_o is limited to approximately $V_{CC} = 10\text{V}$. With $V_o = 10\text{V}$, it follows that $V_1 = 141$ mV and $x = 5.42$. This value of x implies that the large-signal value of $G_m(x)$ is

$$\frac{G_m(x)}{g_m} = 0.33$$

or

$$G_m(x) = 16.27 \times 10^{-3}(0.33) = 5.37\text{ mS}$$

The reason that G_m settles at 5.37 mS is due to the saturation of the BJT. The corresponding v_o is

$$v_o(t) = 10 \cos \omega_o t \tag{3.85}$$

The simulation of the oscillator with $R_T = 20\ \text{k}\Omega$ is shown in Figure 3.48(a). The output waveform is limited by saturation and in agreement with (3.85). The

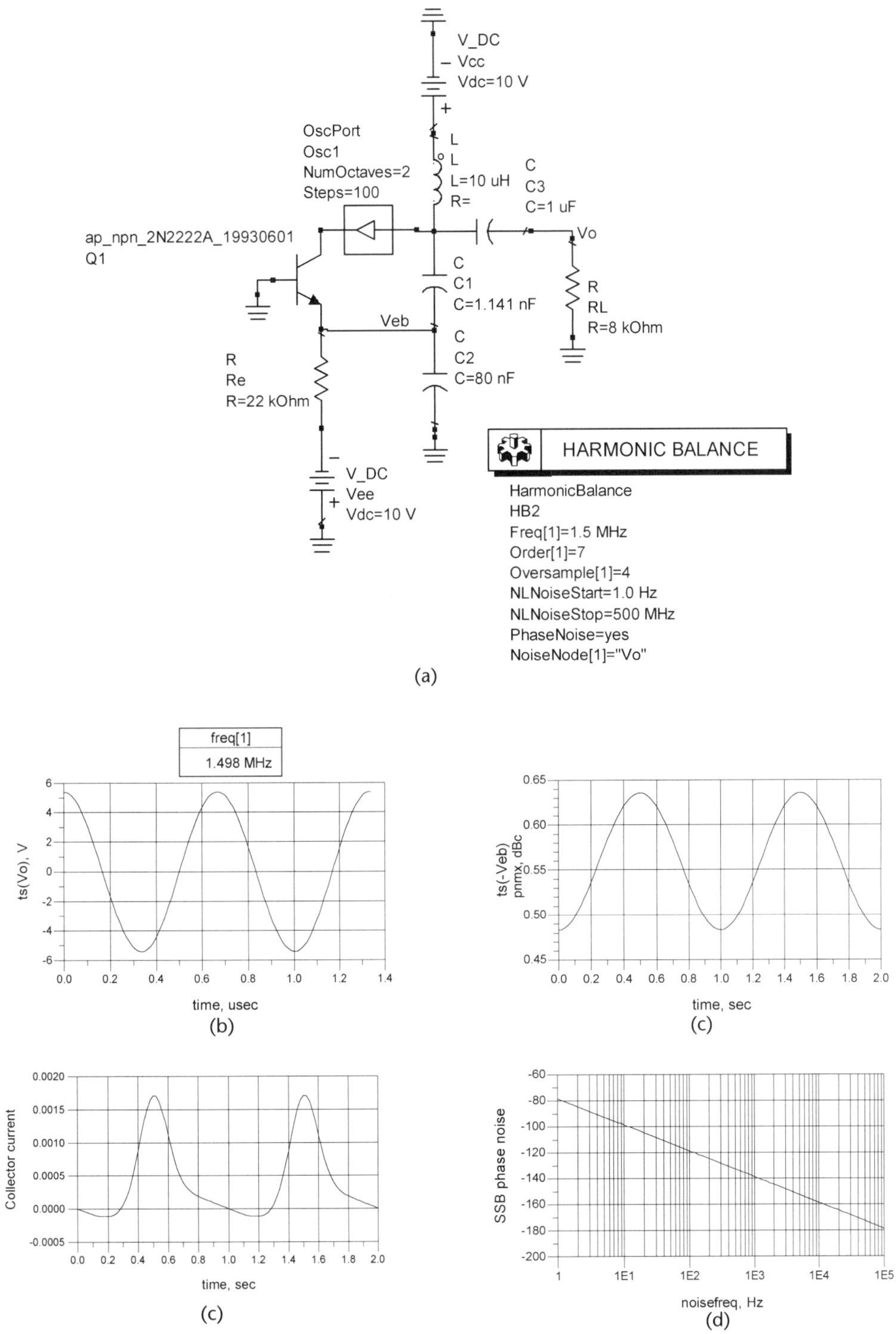

Figure 3.47 (a) The GB oscillator for Example 3.11, (b) the output waveform, (c) the v_{BE} and i_C waveforms, and (d) the SSB phase noise.

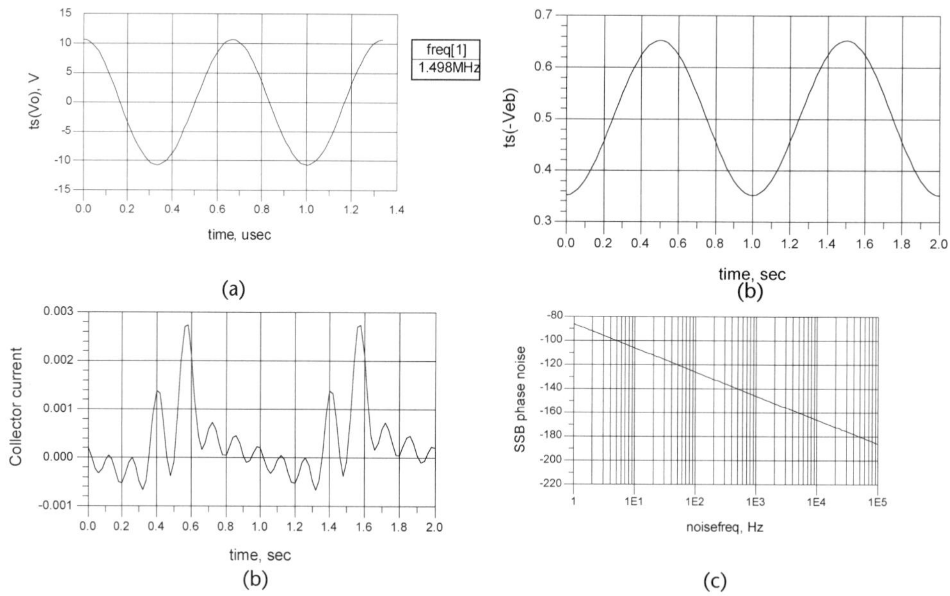

Figure 3.48 (a) Simulation data for the GB oscillator in Example 3.11 with $R_L = 20$ kΩ, (b) the v_{BE} and i_C waveforms, and (c) the SSB phase noise.

base-to-emitter voltage and the collector current waveforms are shown in Figure 3.48(b), and the SSB phase noise is shown in Figure 3.48(c).

This example shows how the SSB phase noise of the oscillator is affected by the collector current pulse width and amplitude. In this oscillator, a higher value of x decreased the current pulse width with a corresponding decrease in the SSB phase noise. This is one type of analysis that can be performed to reduce the phase noise.

For n-channel FETs if the device is operating in the square-law region the drain current is given by

$$i_D = I_{DSS}\left(1 - \frac{v_{GS}}{V_P}\right)^2 \text{ for } V_P < v_{GS} < 0 \tag{3.86}$$

where v_{GS} and V_P are negative numbers.

To analyze the transconductance of such FETs under large-signal conditions, we let

$$v_{GS} = V_{GSQ} + V_1 \cos \omega t \tag{3.87}$$

Substituting (3.87) into (3.86) produces

$$i_D = \frac{I_{DSS}}{V_p^2}\left[V_x^2 - 2V_xV_1 \cos \omega t + V_1^2 \cos^2 \omega t\right] \tag{3.88}$$

$$\frac{I_{DSS}}{V_p^2}\left[\left(V_x^2 + \frac{V_1^2}{2}\right) - 2V_xV_1 \cos \omega t + \frac{V_1^2}{2} \cos 2\omega t\right]$$

where

$$V_x = V_P - V_{GSQ}$$

The voltage V_x is negative if the FET is biased in the square-law region.

Equation (3.88) shows that i_D contains three terms. The first term is a dc term (called I_o) of value

$$I_o = \frac{I_{DSS}}{V_p^2}\left(V_x^2 + \frac{V_1^2}{2}\right)$$

The second term, whose amplitude is denoted by I_1, represents the fundamental current,

$$I_1 =- 2\frac{I_{DSS}}{V_p^2}V_xV_1 \tag{3.89}$$

and the third term, with amplitude denoted by I_2, represents the second harmonic,

$$I_2 = \frac{I_{DSS}}{V_p^2}\frac{V_1^2}{2}$$

From (3.89) a large-signal transconductance can be defined as

$$G_m = \frac{I_1}{V_1} =- 2\frac{I_{DSS}}{V_p^2}V_x \tag{3.90}$$

which happens to be identical to the small-signal transconductance g_m, since

$$g_m =- 2\frac{I_{DSS}}{V_p^2}(V_p - v_{GS})\bigg|_{v_{GS}=V_{GSQ}} =- 2\frac{I_{DSS}}{V_p^2}V_x$$

A plot of (3.90) is shown in Figure 3.49. The transconductance is a linearly decreasing function of v_{GS}, where the largest value, denoted by g_{mo}, occurs at $v_{GS} = 0$ and it is given by

$$g_{mo} =- 2\frac{I_{DSS}}{V_P}$$

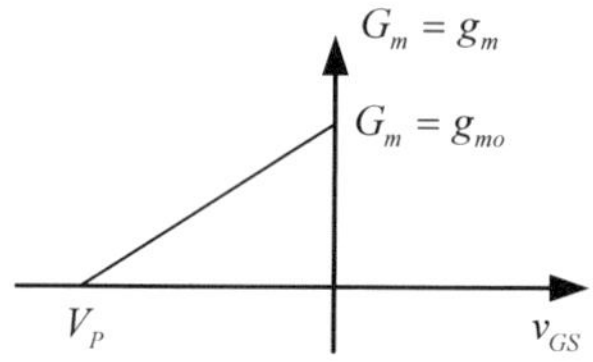

Figure 3.49 Transconductance variation for operation in the square-law region.

If the value of v_{GS} in (3.88) is such that it exceeds V_P during some part of the cycle, as shown in Figure 3.50, then the corresponding i_D can no longer be expressed by (3.88). The resulting i_D contains many Fourier terms. The Fourier analysis of the signal shown in Figure 3.50 can be found in Clarke and Hess [1]. The amplitude of the fundamental component is given by

$$I_1 = \frac{2I_p}{\pi} \frac{\frac{3}{4}\sin\phi + \frac{1}{12}\sin 3\phi - \phi\cos\phi}{(1-\cos\phi)^2} \tag{3.91}$$

where ϕ is the conduction angle, namely,

$$\cos\phi = \frac{V_x}{V_1}$$

and I_p is the peak value of i_D, namely,

$$I_p = \frac{I_{DSS}}{V_p^2}(V_1 - V_x)^2$$

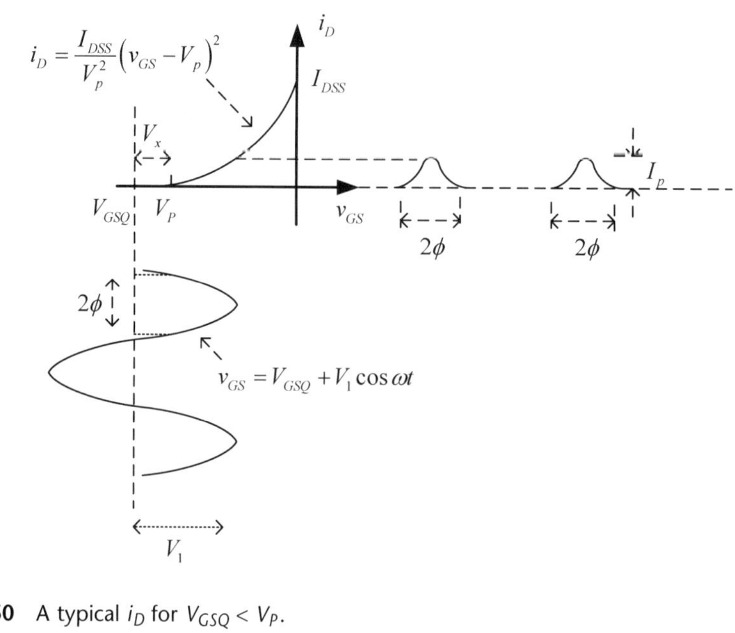

Figure 3.50 A typical i_D for $V_{GSQ} < V_P$.

Figure 3.51 shows a plot of the normalized fundamental component I_1/I_p versus V_x/V_1. Negative values of V_x occur when V_{GSQ} is selected to bias the FET in the square-law region (i.e., $V_P \ll V_{GSQ}$).

For operation in the square-law region, the graph of I_1/I_p agrees with (3.89). This can be seen better if (3.89) is written in the form

$$\frac{I_1}{I_p} = \frac{-2\left(\frac{V_x}{V_1}\right)}{\left(1 - \frac{V_x}{V_1}\right)^2} \tag{3.92}$$

and observing that for $V_x/V_1 = -1$, (3.92) gives $I_1/I_p = 0.5$ in agreement with Figure 3.51.

If the load circuit of the FET is a tuned circuit with a high value of Q_L, only the fundamental component of i_D contributes to the output voltage. A large-signal transconductance G_m is defined as the ratio of I_1 to V_1. That is,

$$G_m = \frac{I_1}{V_1}$$

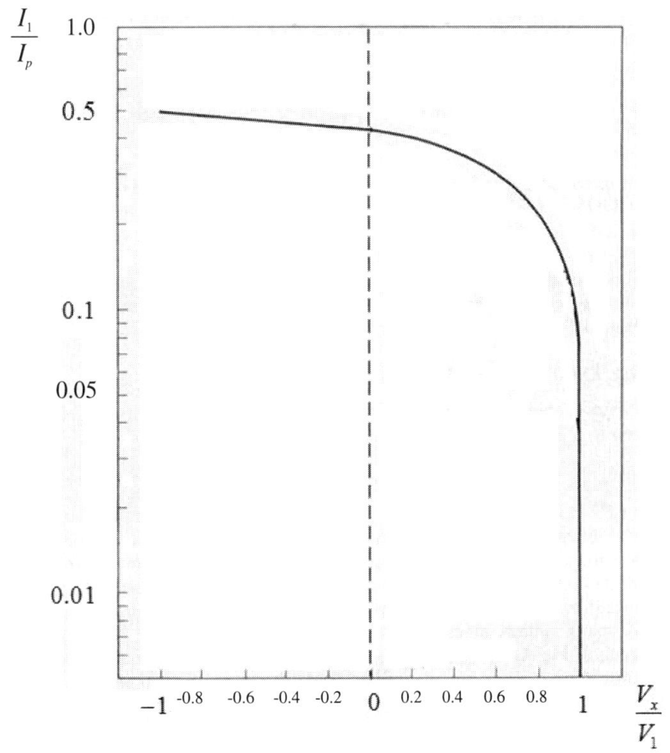

Figure 3.51 Plot of the normalized fundamental component vs. V_x/V_1. (*From:* [1]. © 1971. Reprinted with permission of Pearson Education, Inc.)

If the peak value of i_D is I_{DSS}, G_m can be expressed in the form

$$G_m = \frac{2I_{DSS}\left(\frac{I_1}{I_p}\right)}{2\left(-\frac{V_1}{V_p}\right)} = \frac{g_{mo}\left(\frac{I_1}{I_p}\right)}{2\left(-\frac{V_1}{V_p}\right)}$$

A plot of G_m/g_{mo} is shown in Figure 3.52.

Just as in the discussion of BJTs, the small-signal g_m decreases to its large-signal value G_m at which point a large-signal model of the oscillator can be developed for the calculation of the loop gain and the value of G_m at which the oscillator voltage stabilizes.

For example, consider the grounded-gate oscillator shown in Figure 3.12(a). It follows from the ac model in Figure 3.12(d), with g_m replaced by G_m, that the large-signal loop gain is

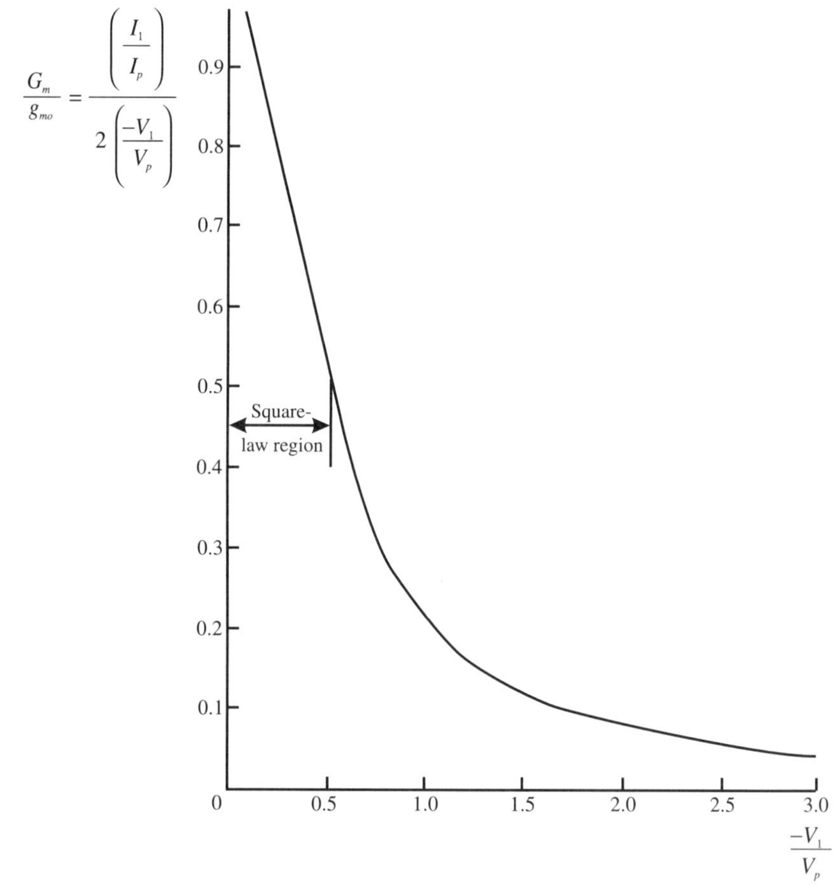

Figure 3.52 Plot of G_m/g_{mo} versus $-V_1/V_p$. (*From:* [1]. © 1971. Reprinted with permission of Pearson Education, Inc.)

$$\beta A_v = \frac{C_1}{C_1 + C_2} \frac{\frac{G_m}{C_T}}{s^2 + s\frac{1}{R_T C_T} + \frac{1}{LC_T}} \tag{3.93}$$

where

$$R_T = R_p \parallel R_L \parallel R'_P \approx R'_P \approx n^2\left(R_S \middle\| \frac{1}{g_m}\right)$$

and

$$C_T = \frac{C_1 C_2}{C_1 + C_2}$$

Setting the loop gain in (3.93) equal to 1 gives the frequency of oscillation as

$$\omega_o = \frac{1}{\sqrt{LC_T}}$$

and the gain condition is

$$\frac{C_1}{C_1 + C_2} G_m R_T \geq 1$$

which can be expressed as

$$G_m \geq \frac{C_1}{C_1 + C_2} \frac{1}{\left(R_s \middle\| \frac{1}{g_m}\right)} \tag{3.94}$$

Equation (3.94) gives the value of G_m that sets the large-loop gain equal to unity. Then, from Figure 3.52 the value of the oscillation amplitude V_1 follows. Of course, V_1 is the amplitude of the oscillator voltage between the gate and source. The voltage v_o is

$$v_o = \left(\frac{C_1 + C_2}{C_1}\right) V_1 \cos \omega_o t \tag{3.95}$$

Example 3.12

Calculate $v_o(t)$ for the GG oscillator designed in Example 3.3.

Solution

In the oscillator of Example 3.3 (refer to Figure 3.13): $C_1 = 95$ pF, $C_2 = 190$ pF, $L = 1\ \mu$H, $R_S = 302\Omega$, $R_D = 1.25$ kΩ, $I_D = 4.5$ mA, $g_{mo} = 6.86$ mS, and $g_m = 4.2$ mS. Therefore, from (3.95),

$$G_m = \frac{C_1}{C_1 + C_2} \frac{1}{\left(R_s + \frac{1}{g_m}\right)} = \frac{(95 \times 10^{-12})}{95 \times 10^{-12} + 190 \times 10^{-12}} \frac{1}{(302 + 238)} = 2.5 \text{ mS}$$

From Figure 3.52, with

$$\frac{G_m}{g_{mo}} = \frac{2.5 \times 10^{-3}}{6.86 \times 10^{-3}} = 0.364$$

it follows that

$$V_1 \approx 0.7(-V_P) = 0.7(3.5) = 2.45\text{V}$$

Then,

$$V_o = nV_1 = 3(2.45) = 7.3\text{V}$$

or

$$v_o(t) = 7.3 \cos \omega_o t \tag{3.96}$$

where the frequency of oscillation is 19.75 MHz.

The simulation for this example is shown in Figure 3.13. The output waveform varies between ±7.2V with a fundamental frequency of 19.95 Hz in good agreement with (3.96).

Reference

[1] Clarke, K. K., and D. T. Hess, *Communications Circuits: Analysis and Design*, Upper Saddle River, NJ: Pearson Education, Inc., 1971.

CHAPTER 4

Crystal Oscillators

4.1 Introduction

As their name implies, crystal-controlled oscillators use a crystal usually made from quartz for their operation. Quartz exhibits a very interesting property: if a mechanical stress is applied to the crystal along a certain direction or axis, a voltage is generated along a perpendicular direction. Conversely, an ac voltage applied to the crystal changes its physical shape, producing mechanical vibration. This phenomenon between mechanical and electrical effects is called the *piezoelectric effect*. Other crystalline materials exhibit piezoelectric effects, but quartz is the most suitable material for oscillators. If the frequency of an applied signal is equal to the mechanical resonant frequency of the crystal, the crystal will vibrate and only a small voltage is required to keep it vibrating. In fact, the main reason for using a crystal oscillator instead of an LC tuned oscillator is that the Q associated with crystal oscillators are 1,000 to 10,000 times greater than those associated with LC tuned oscillators. The crystal will act like a very-high Q tuned circuit, permitting oscillations (i.e., vibrations) only at its resonant frequency. The resonant frequency of the crystal is very important since at frequencies slightly above and below its resonant frequency the amplitude of the crystal vibrations are essentially zero. The oscillation frequency of a crystal oscillator is the same at that of the crystal.

4.2 Crystal Characteristics

The crystal symbol is shown in Figure 4.1(a). A typical quartz crystal and mount are shown in Figure 4.1(b). The crystal plate is supported by springs or wires attached to points that do not attenuate the mechanical vibrations. The wires are then connected to the pins for external electrical connections. The electrical contacts to the crystals are made with special metal plate contacts. Typical crystal-holder (HC) dimensions are shown in Figure 4.1(c) for an HC-49/U holder. The HC-49/U holder is the resistance-weld version of the older solder-seal HC-18/U holder. Many types of holders (i.e., metal, plastic, ceramic) are available for crystals.

The quartz crystals used in oscillators must be cut and polished to extremely accurate dimensions. Crystals are classified according to the manner in which the crystal was cut from the original quartz material. A variety of cuts are used, among them are the X, Y, AT, BT, GT, and SC cuts. For example, in an AT cut the crystal

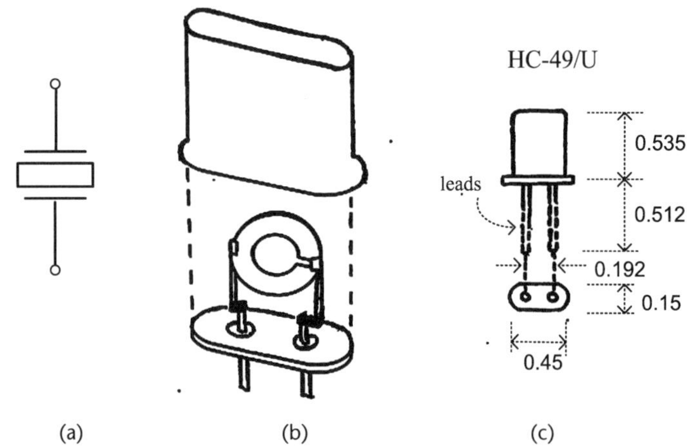

Figure 4.1 (a) Crystal symbol, (b) crystal and mount, and (c) crystal-holder dimensions (in inches) for the HC-49/U.

is a cut along its x-axis at an angle of 35° with respect to its z-axis. Each cut exhibits different characteristics, especially temperature effects. Temperature stability in crystals refers to the variation of its nominal resonant frequency as a function of temperature. It is the allowable frequency drift in ppm (parts per million) over a specified temperature range. Typical temperature stability specifications are from 10 to 100 ppm in hertz per degrees Celsius. The AT-cut crystals are very popular because of their small temperature variation. For example, an AT-cut crystal can be specified with a temperature range of 30 ppm (or 0.0003%) over a temperature range of –55°C to 105°C. The X-cut crystals have negative temperature coefficient. That is, the crystal frequency decreases with increases in temperature. The Y-cut crystals have positive temperature coefficients.

Frequency tolerance (or accuracy), also expressed in ppm, describes how close the actual crystal frequency is to its specified frequency at a reference temperature (usually 25°C). For example, a 3.24500-MHz crystal with a frequency tolerance of 10 ppm is cut differently from one with a specified frequency of 3.24555 MHz.

The sum of the frequency tolerance and the temperature tolerance is known as the overall tolerance. The overall tolerance is a measure of the frequency deviation over a temperature range of operation with respect to the nominal frequency.

Example 4.1

The frequency tolerance of a 32.768-kHz crystal is listed as 20 ppm at 25°C. Calculate the frequency of operation of the crystal and the associated timing error.

Solution

The 32.768-kHz crystal is used in digital watches. Since $2^{15} = 32{,}765$, it follows that division by 15 stages of division 2 provides a 1-Hz pulse.

The 20-ppm accuracy means that the crystal frequency of operation can vary by

$$\Delta f = \frac{\pm 20}{10^6}(32{,}768) = \pm\ 0.6554 \text{ Hz}$$

or the frequency of operation can be between $32{,}767.3446 < f < 32{,}768.6554$.

In a week, a watch using such crystal can vary by

$$\frac{\pm 20}{10^6}\left(604{,}800\ \frac{\text{seconds}}{\text{week}}\right) \approx \pm\ 12 \text{ seconds/week}$$

or 50 seconds/month, or 10 minutes/year, which is not bad for an inexpensive watch.

The 32.768-kHz watch crystals are of the tuning-fork variety. It is an NT-cut crystal that is shaped like a tuning fork. The crystal is commonly manufactured in the NC-38 holder, which is shaped like a tube (about 3 mm by 8 mm).

Crystals are affected by temperature variations, vibrations, shock, and aging. Aging is a systematic crystal effect that results in a change in the operational frequency of the crystal. It is observed over a period of time (days to years). The main causes are associated with a mass transfer due to internal contamination, excessive drive level, or a chemical reaction in the crystal structure. For example, a crystal can specify a maximum aging per year of 5 ppm.

In an AT-cut crystal the fundamental resonant frequency is related to the crystal thickness by the formula

$$f_o \approx \frac{167.64 \times 10^3}{t}$$

where t is expressed in centimeters. For example, for f_o = 1 MHz the required thickness is $t = 0.16764$ cm. As the fundamental frequency increases, the required thickness becomes smaller. The small thickness associated with higher f_o makes the crystal very fragile and susceptible to fracture. This places a practical limitation on the highest f_o possible. Many manufacturers provide crystals with fundamental frequencies from a few kilohertz to about 30 MHz. Fundamental frequencies higher than 30 MHz are associated with crystals fabricated using the *inverted mesa* technology. Crystals can also be operated at their higher mechanical overtones, especially the third and fifth overtone. Crystals operated at their overtones are normally used above 20 MHz. For example, typical frequency ranges for AT crystals are: 30 to 90 MHz at the third overtone; 50 to 150 MHz at the fifth overtone; and 100 to 180 MHz at the seventh overtone.

There are several classifications of crystal oscillators. Standard crystal oscillators (XOs) have typical accuracy of 20 to 50 ppm. These crystals are suitable for many digital timing applications. Temperature-controlled crystal oscillators (TCXOs) use the output signal from a temperature sensor to generate a correction voltage that depends on temperature. This voltage is applied to a varactor diode whose reactance variation compensates for the crystal frequency variation due to temperature. TCXOs have typical aging rate of 0.5 ppm/year and typical temperature stability of 0.5 ppm. Accuracy is of the order of 1 ppm.

Oven-controlled crystal oscillators (OCXOs) use a crystal placed in a stable oven. The oven temperature is adjusted so that the frequency versus temperature slope is zero. OCXOs have typical aging rate of 0.005 ppm/year and typical temperature stability of 0.001 ppm. Accuracy is of the order of 0.01 ppm. These devices are costly.

An important application of crystals is in voltage-controlled crystal oscillators (VCXOs). This type of oscillator commonly uses a crystal in series with a voltage-controlled capacitor to vary the frequency of oscillation.

The reactance characteristics of a crystal are illustrated in Figure 4.2(a). An expanded view of the characteristics of the fundamental mode is shown in Figure 4.2(b). The crystal exhibits several series resonances, which are modeled by several series-resonant branches, as shown in Figure 4.2(c). The first RLC branch (i.e., R_1, C_1, and L_1) models the fundamental mode of oscillation, and the other branches the odd overtones. The overtone frequencies are approximately odd multiples of the fundamental frequency. Spurious resonances appear between the dominant modes. For operation around a certain resonant frequency the crystal can be modeled by the circuit of a single branch, simply denoted by L, C, and R, as shown

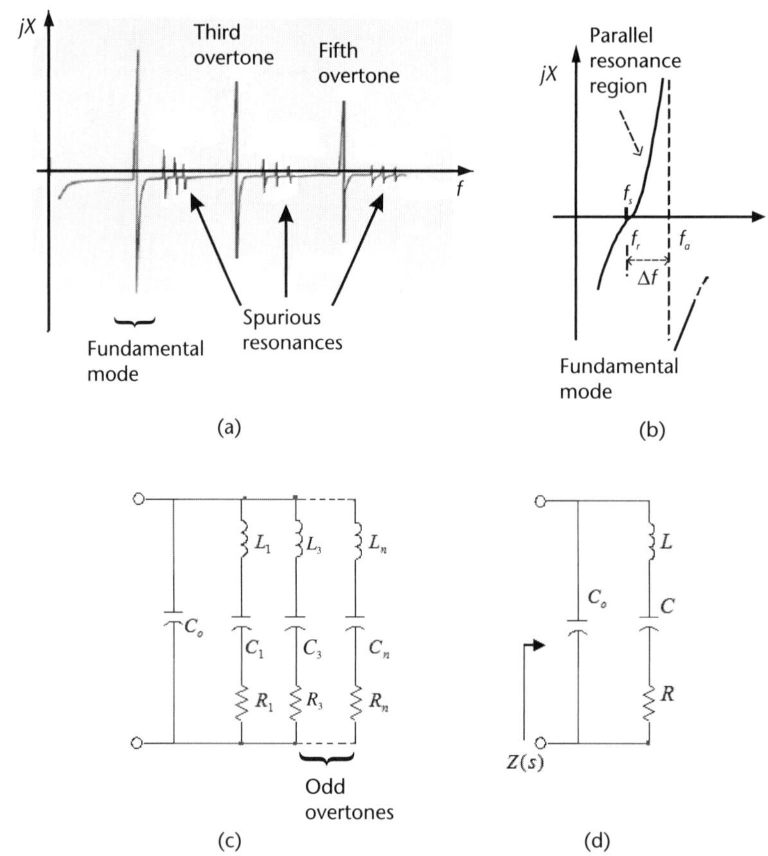

Figure 4.2 (a) Typical reactance characteristics of a crystal, (b) reactance characteristics of the fundamental mode, (c) equivalent circuit of a crystal where *n* represents the *n*th odd overtone, and (d) model for operation at a certain resonance.

in Figure 4.2(d). The inductor L is the electrical equivalent of the crystal mass, the capacitor C represents the crystal stiffness or elasticity, and R represents the heat losses due to mechanical friction in the crystal. The series L, C, R portion of the circuit is called the motional arm of the crystal (or circuit). The capacitor C_o represents the shunt capacitance of the electrodes in parallel with the holder capacitance. It is called the static arm of the crystal (or circuit).

The overtone element values are calculated using

$$L_n = L$$

$$C_n = \frac{C}{n^2}$$

and

$$R_n = n^2 R$$

where n is the nth overtone ($n = 3, 5, 7, \ldots$).

The impedance of the equivalent circuit in Figure 4.2(d) is given by

$$Z(s) = \frac{s^2 + \left(\frac{R}{L}\right)s + \omega_s^2}{sC_o\left[s^2 + \left(\frac{R}{L}\right)s + \left(1 + \frac{C}{C_o}\right)\omega_s^2\right]} \tag{4.1}$$

$$= \frac{s^2 + \left(\frac{\omega_s}{Q_U}\right)s + \omega_s^2}{sC_o\left[s^2 + \left(\frac{\omega_s}{Q_U}\right)s + \left(1 + \frac{C}{C_o}\right)\omega_s^2\right]}$$

where

$$\omega_s = 2\pi f_s = \frac{1}{\sqrt{LC}} \tag{4.2}$$

and

$$Q_U = \frac{\omega_s L}{R} \tag{4.3}$$

If R is neglected (due to a high-Q_U value) the impedance reduces to

$$Z(s) = \frac{s^2 + \omega_s^2}{sC_o\left[s^2 + \left(1 + \frac{C}{C_o}\right)\omega_s^2\right]}$$

which shows that the crystal exhibits a series resonance at ω_s and a parallel resonance (i.e., an antiresonance) at

$$\omega_a = 2\pi f_a = \omega_s\left(1 + \frac{C}{C_o}\right)^{1/2} = \frac{1}{\sqrt{LC_T}} \tag{4.4}$$

where

$$C_T = \frac{CC_o}{C + C_o}$$

Typical magnitude and phase plots of (4.1) are shown in Figure 4.3. The impedance plot in Figure 4.3 [or reactance plot in Figure 4.2(b)] shows that the circuit has a series and a parallel resonance. At the series resonance frequency ω_s, $X_L = -X_C$ and the crystal branch impedance is simply R. The resistor R is also called the equivalent series resistance (ESR) at series resonance. At series resonance, the resistor R appears in parallel with the reactance C_o. However, $|X_{C_o}| \gg R$ so the crystal essentially appears resistive. Between f_s and f_a the impedance is inductive with the phase being 90°. This is an important region of operation, which is called the *parallel-resonance region* (or the region of usual parallel resonance). In Figure 4.2(b) the frequency f_r is the frequency where the reactance is zero. The difference between f_r and f_s is due to C_o. Hence, for practical purposes $f_s \approx f_r$ and no distinction is made between these two frequencies.

The parallel-resonant frequency is given by (4.4). From (4.4), since $C_o \gg C$, the relation between f_a and f_s is approximately given by

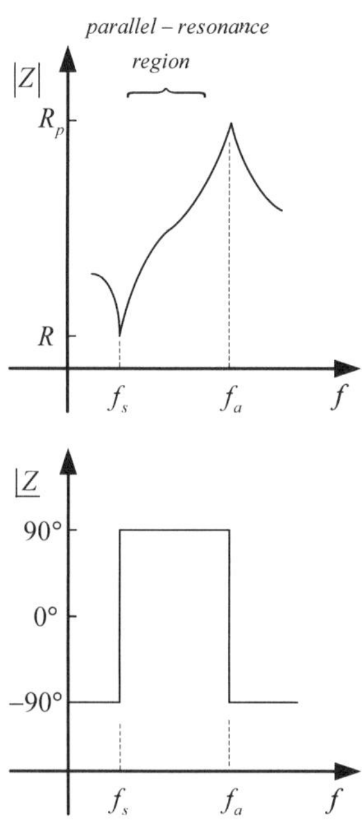

Figure 4.3 Impedance plot of the crystal.

$$f_a = f_s\left(1 + \frac{C}{C_o}\right)^{1/2} \approx f_s\left(1 + \frac{C}{2C_o}\right) \tag{4.5}$$

Hence, the frequency separation (Δf) between f_a and f_s is

$$\Delta f = f_a - f_s \approx f_s\left(\frac{C}{2C_o}\right)$$

This frequency separation is very small. For example, for a crystal with $f_s = 2$ MHz and $C_O = 300C$, it follows that $\Delta f = 3.33$ kHz and $f_a = 2003.33$ kHz. The frequency range Δf is known as the pulling range.

The crystal parameters can be measured very accurately using a variety of equipment, such as an impedance analyzer, a capacitance meter, and a crystal impedance meter. It is important to know if a crystal is specified to operate at its fundamental or at an overtone mode. The reason is that crystal overtones are not exactly integers multiples of the fundamental. Hence, a crystal manufacturer will calibrate a third overtone crystal at the third overtone and not at the fundamental.

Two figures of merit used in the specifications of crystals are r and M. The parameter r is defined as

$$r = \frac{C_o}{C}$$

It provides a measure of the separation between f_s and f_a. The parameter M is the ratio of the impedance of the static arm of the crystal to the impedance of the motional arm at series resonance, which is R. That is,

$$M = \frac{1}{\omega_s C_o R} = Q_U \frac{C}{C_o} = Q_U \frac{2(f_a - f_s)}{f_s}$$

Hence, the parameter M also provides a measure of the separation between f_a and f_s. In fact, M and r are related by $Q_U = Mr$. Crystals operating in the fundamental mode have high values of M (see Example 4.2). Crystals operating at an overtone have smaller values of M. It can be shown that an inductive region exists for values of M greater than 2.

From Figure 4.2(b) it is seen that for frequencies between f_s and f_a the impedance is inductive (i.e., the crystal appears inductive), and for $f < f_s$ or $f > f_a$, the impedance is capacitive. The region above series resonance, where $f_s < f < f_a$, is known as the parallel-resonance region. In oscillator applications the crystal is operated either at series resonance or in the parallel resonance region. Crystals are not operated at the anti-resonant frequency.

Next, consider the crystal circuit shown in Figure 4.4(a). At frequencies lower than f_s the motional arm impedance is high and the current increases as the frequency increases due to the decreasing reactance of C_o. As the frequency reaches the series resonant frequency ($f = f_s \approx f_r$), the current increases significantly since it is only limited by the crystal resistance R [see Figure 4.4(b)]. As the frequency increases

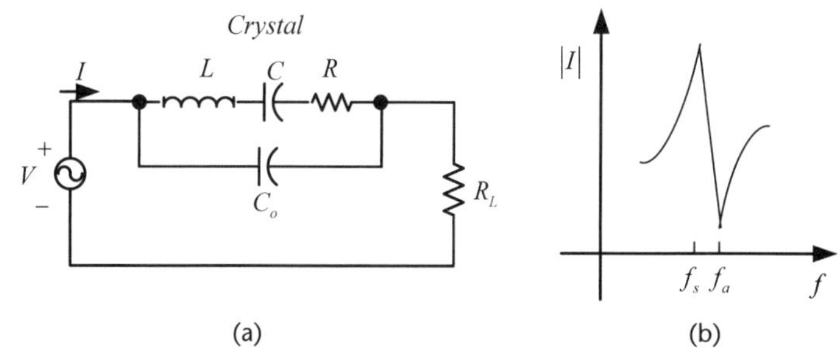

Figure 4.4 (a) A crystal in series with a load R_L and (b) the current response.

toward f_a the motional arm exhibits a net inductive reactance, which is in parallel resonance at f_a with C_o. At parallel resonance the circuit exhibits a high resistance and the resulting current is low. At series and parallel resonance the crystal appears resistive and the applied voltage and current are in phase. Figure 4.4(b) also shows that at frequencies higher than f_a the circuit is again controlled by C_o.

An equivalent circuit for the crystal for frequencies between f_s and f_a can be obtained as follows. Consider Figure 4.5(a) where

$$jX_m = j\left(\omega L - \frac{1}{\omega C}\right)$$

The input impedance of the circuit is given by

$$Z_{IN} = \frac{jX_{C_o}(R + jX_m)}{R + j\left(X_m + X_{C_o}\right)}$$

or

$$Z_{IN} = \frac{RX_{C_o}^2}{R^2 + \left(X_m + X_{C_o}\right)^2} + j\frac{X_{C_o}\left[R^2 + X_m\left(X_m + X_{C_o}\right)\right]}{R^2 + \left(X_m + X_{C_o}\right)^2} \tag{4.6}$$

The ESR, denoted by R_e, is given by the real part of (4.6). The imaginary part of (4.6) is denoted by X_e. A typical plot of R_e and X_e is shown in Figure 4.5(b).

From Figure 4.5(b) it is observed that the function X_e is zero at f_r. At $f = f_r \approx f_s$, the equivalent resistance is approximately R, since $\left|X_{C_o}\right| \gg R$. From (4.6), in the cases that

$$\left|X_{C_o} + X_m\right| \gg R \tag{4.7}$$

the following relation approximates the behavior of the crystal impedance in the parallel-resonant region:

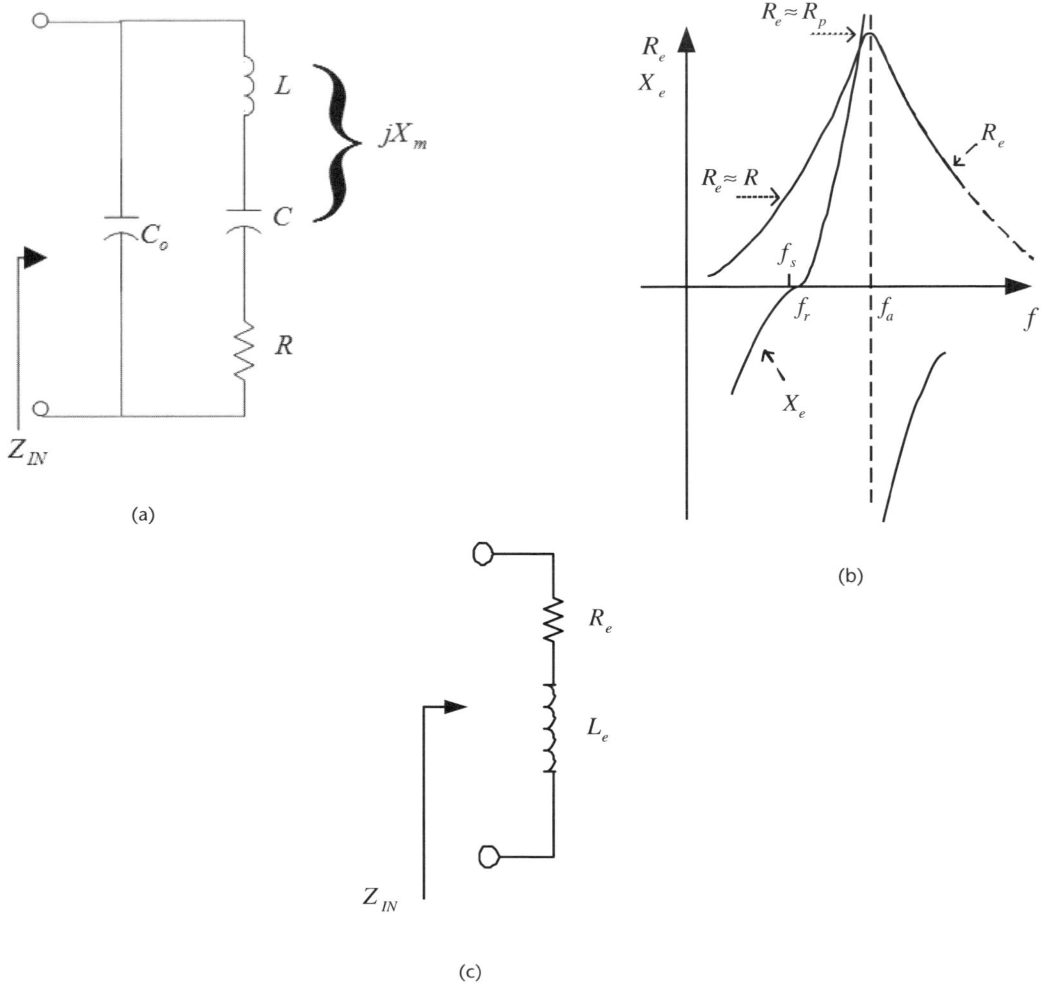

Figure 4.5 (a) Crystal model, (b) typical plot of R_e and X_e, and (c) equivalent model of the crystal in the parallel-resonance region.

$$Z_{IN} = \frac{R}{\left(1 + \frac{X_m}{X_{C_o}}\right)^2} + j\frac{X_m X_{C_o}}{X_m + X_{C_o}}$$

or

$$R_e \approx \frac{R}{\left(1 + \frac{X_m}{X_{C_o}}\right)^2} \tag{4.8}$$

and

$$X_e = \omega L_e \approx \frac{X_m X_{C_o}}{X_m X_{C_o}} \tag{4.9}$$

The equivalent model for the crystal in the parallel-resonance region is shown in Figure 4.5(c).

For frequencies around the series resonance of the crystal we let $\omega = \omega_s + \delta\omega$, where $\delta\omega$ represents a small frequency increment. Hence, the impedance behavior of X_m can be approximated by

$$X_m = \omega L - \frac{1}{\omega C} = \frac{1}{\omega C}[\omega^2 LC - 1]$$

$$\approx \frac{1}{\omega_s C}\left[\frac{2\delta\omega}{\omega_s} + \frac{(\delta\omega)^2}{\omega_s^2}\right] \approx \frac{1}{\omega_s C}\left(\frac{2\delta\omega}{\omega_s}\right)$$

and

$$\frac{X_m}{X_{C_o}} \approx - \frac{2C_o}{C}\frac{(\omega - \omega_s)}{\omega_s}$$

Therefore, (4.8) and (4.9) are approximated as

$$R_e \approx \frac{R}{\left[1 - \frac{2C_o(\omega - \omega_s)}{C\omega_s}\right]^2} \approx R$$

and

$$X_e \approx \frac{\left(\frac{1}{\omega_s C}\right)\frac{2(\omega - \omega_s)}{\omega_s}}{\left[1 - \frac{2C_o(\omega - \omega_s)}{C\omega_s}\right]^2} \approx \left(\frac{1}{\omega_s C}\right)\frac{2(\omega - \omega_s)}{\omega_s}$$

and it follows that

$$Z_{IN} \approx R + j\left(\frac{1}{\omega_s C}\right)\left[\frac{2(\omega - \omega_s)}{\omega_s}\right] \qquad (4.10)$$
$$= R\left[1 + jQ_U \frac{2(\omega - \omega_s)}{\omega_s}\right]$$

which shows that the equivalent circuit is composed of a resistor of value R in series with a reactance whose sign depends on whether ω is greater or less than ω_s. An inductive reactance results for $\omega > \omega_s$. It is also observed that the series resonance does not depend on C_o.

The previous approximations are based on the assumption in (4.7). It is observed that if $\delta\omega \to 0$ (i.e., $X_m \to 0$), the assumption is identical to the figure of merit requirement that

$$M = \frac{1}{\omega_s C_o R} \gg 1$$

which is valid for crystals operating in the fundamental mode. Also, if $\delta\omega \to \Delta\omega$ (i.e., for operation close to f_a) the assumption (4.7) is not valid.

An equivalent circuit can also be obtained for the behavior of the impedance around f_a, namely

$$Z_{IN} \approx \frac{R_p}{1 + jQ_U \frac{2(\omega - \omega_a)}{\omega_s}} \qquad (4.11)$$

where the equivalent parallel resistance R_p is given by

$$R_p = \frac{1}{R(\omega_s C_o)^2} \qquad (4.12)$$

This relation for Z_{IN} is recognized as the behavior of a high-Q parallel-resonant circuit.

Equations (4.10) and (4.11) are used to develop the narrowband equivalent circuits shown in Figure 4.6 for the crystal behavior around the series resonance and the parallel resonance.

Crystal oscillator design is either based on a design at series resonance or in the parallel-resonance region where the crystal behavior is inductive.

When the crystal is operated with a parallel load capacitance C_L, as shown in Figure 4.7(a), its parallel resonance, denoted as the load-resonant frequency f_L, is given by

$$f_L = f_s\left(1 + \frac{C}{C_o + C_L}\right)^{1/2} \approx f_s\left[1 + \frac{C}{2(C_o + C_L)}\right] \qquad (4.13)$$

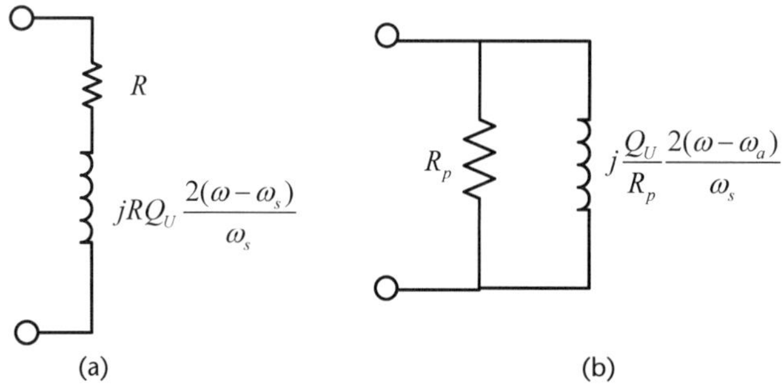

Figure 4.6 (a) Narrowband equivalent circuit for the crystal around the series resonance for $\omega > \omega_s$ and (b) narrowband equivalent circuit for the crystal around the parallel resonance for $\omega < \omega_a$.

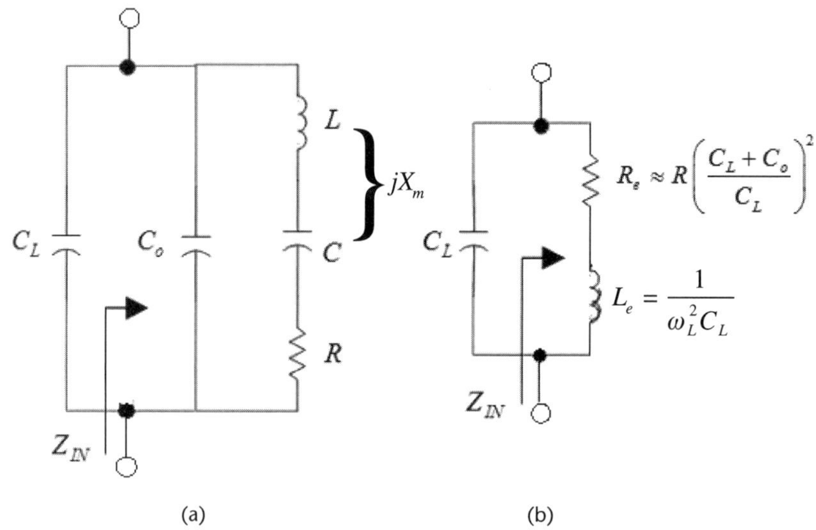

Figure 4.7 (a) Crystal with external load capacitance C_L and (b) equivalent model at f_L.

where $f_s < f_L < f_a$. Equation (4.13) follows from (4.5) with C_o replaced by $C_o + C_L$. The frequency separation (Δf) between f_L and f_s due to C_L is given by

$$\Delta f = \frac{f_s C}{2(C_o + C_L)}$$

For the circuit in Figure 4.7(a), for frequencies around its parallel resonance, the input impedance [see (4.6)], is

$$Z_{IN} = R_e + jX_e \tag{4.14}$$

where

$$R_e = \frac{RX_{C_o}^2}{R^2 + \left[\frac{1}{\omega(C_L + C_o)} - \frac{1}{\omega C_o}\right]^2} = \frac{RX_{C_o}^2}{R^2 + X_{C_o}^2\left(\frac{C_L}{C_L + C_o}\right)^2} \tag{4.15}$$

and, since $X_{C_L} = |X_{C_o} \parallel X_m|$ at f_L, we obtain

$$X_e = \omega_L L_e = |X_{C_L}| = \frac{1}{\omega_L C_L} \tag{4.16}$$

If

$$X_{C_o}\left(\frac{C_L}{C_L + C_o}\right) \gg R$$

(4.15) can be further approximated by

$$R_e \approx R\left(\frac{C_L + C_o}{C_L}\right)^2 \tag{4.17}$$

Equations (4.16) and (4.17) permit the crystal with the external load capacitance C_L to be represented by the model in Figure 4.7(b).

Example 4.2

(a) Consider a 2-MHz series resonant crystal described by L = 0.528H, C = 0.011993 pF, R = 100Ω, and C_o = 4 pF. Calculate the series-resonance frequency f_s, the parallel-resonant frequency f_a, and the crystal Q_U. Also, calculate the figure of merits r and M.
(b) Consider the 2-MHz crystal in part (a) with the value of C specified as C = 0.012 pF instead of 0.011993 pF. Calculate f_s and f_a.

Solution
From (4.2), f_s is

$$f_s = \frac{1}{2\pi\sqrt{LC}} = \frac{1}{2\pi\sqrt{0.528(11.993 \times 10^{-15})}} = 2.000042 \text{ MHz}$$

and from (4.4), with C_T = 0.011957 pF, f_a is

$$f_a = \frac{1}{2\pi\sqrt{LC_T}} = \frac{1}{2\pi\sqrt{0.528(11.957 \times 10^{-15})}} = 2.003051 \text{ MHz}$$

The frequency separation is

$$\Delta f = f_a - f_s = 3 \text{ kHz}$$

Also, note that (4.5) provides a good approximation for f_a since it gives

$$f_a = f_s\left(1 + \frac{C}{2C_o}\right) = 2.000042 \times 10^6\left[1 + \frac{11.993 \times 10^{-15}}{2(4 \times 10^{-12})}\right] = 2.003049 \text{ MHz}$$

The unloaded Q_U of the 2-MHz crystal is

$$Q_U = \frac{\omega_s L}{R} = \frac{2\pi(2 \times 10^6)(0.528)}{100} = 66{,}350$$

The reactances of L and C at 2 MHz are

$$X_L =- X_C = \omega L = 2\pi(2 \times 10^6)(0.528) = 6.635 \text{ M}\Omega$$

In the parallel-resonance region, say at 2.001 MHz, the reactances of the crystal elements are

$$X_L = \omega L = 2\pi(2.001 \times 10^6)(0.528) = 6.6384 \text{ M}\Omega$$

$$X_C =- \frac{1}{\omega C} =- \frac{1}{2\pi(2.001 \times 10^6)(11.993 \times 10^{-15})} =- 6.632 \text{ M}\Omega$$

and

$$X_{C_o} =- \frac{1}{\omega C_o} =- \frac{1}{2\pi(2.001 \times 10^6)(4 \times 10^{-12})} =- 19.9 \text{ k}\Omega$$

Hence, $jX_m = j(6.6384 - 6.632) = j6.4 \text{ k}\Omega$.

The parameter r is

$$r = \frac{C_o}{C} = \frac{4 \times 10^{-12}}{11.993 \times 10^{-15}} = 333.5$$

and the parameter M is

$$M = \frac{1}{\omega_s C_o R} = \frac{1}{2\pi(2 \times 10^6)4 \times 10^{-12}(100)} = 199$$

(b) With C = 0.012 pF, it follows that f_s = 1.999459 MHz and f_a = 2.002456 MHz. In this case the specified crystal frequency of operation (i.e., 2 MHz) is between f_s and f_a. This example shows how small variations in the value of C (and also on L) can make a difference in the resulting f_s and f_a values.

To further understand the electrical properties of a crystal consider its pole-zero diagram, shown in Figure 4.8(a), where (4.1) is expanded in the form

$$Z_{IN} = \frac{(s - s_{z1})(s - s_{z2})}{sC_o(s - s_{p1})(s - s_{p2})} \tag{4.18}$$

For example, for a 2-MHz crystal with $L = 0.528$H, $C = 0.011993$ pF, $R = 100\Omega$, and $C_o = 4$ pF it follows that

$$Q_U = \frac{\omega_s L}{R} = \frac{2\pi(2 \times 10^6)(0.528)}{100} = 66{,}350$$

$$\omega_s = 12.56663 \text{ Mrad/s}$$

$$\Delta\omega = \omega_s\left(\frac{C}{2C_o}\right) = 12.56663 \times 10^6\left[\frac{0.011993 \times 10^{-12}}{2(4 \times 10^{-12})}\right] = 18.839 \text{ krad/s}$$

and

$$\frac{\omega_s}{2Q_U} = \frac{12.56663 \times 10^6}{2(66{,}350)} = 94.7$$

Hence, there is a pole at $s = 0$ and the conjugate poles and zeroes are located at

$$s_{z1} = -94.7 + j12.56663 \times 10^6$$

$$s_{z2} = -94.7 - j12.56663 \times 10^6$$

$$s_{p1} = -94.7 + j12.58547 \times 10^6$$

$$s_{p2} = -94.7 - j12.58547 \times 10^6$$

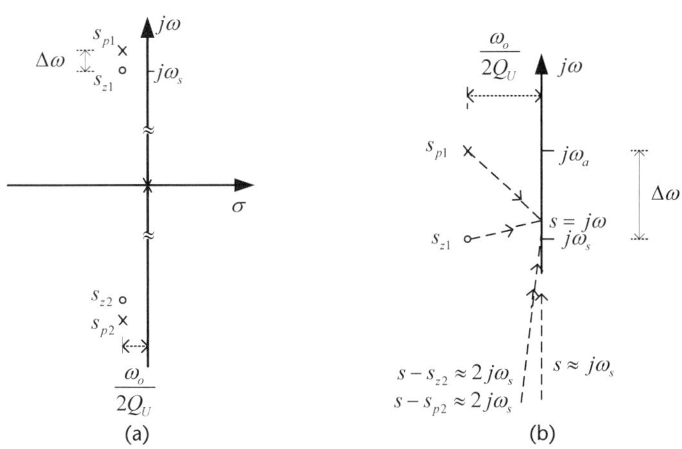

Figure 4.8 (a) Typical pole-zero diagram of a crystal and (b) pole-zero diagram in the vicinity of the series resonance.

These values show that the vertical spacing is about 10^5 times larger than the horizontal spacing. Hence, near a complex zero the impedance behaves as an isolated zero with a very high Q_U, and near a complex pole it behaves like an isolated complex pole with, also, a very high Q_U. The high Q_U values associated with a crystal oscillator provide excellent frequency stability and reduction of the harmonics.

From Figure 4.8(b), in the vicinity of the series resonant frequency ω_s, we can make the following approximations in (4.18):

$$s \approx j\omega_s$$

$$(s - s_{z1}) = \frac{\omega_s}{2Q_U} + j(\omega - \omega_s)$$

$$(s - s_{p1}) \approx -j\Delta\omega = -j\omega_s \frac{C}{2C_o}$$

$$(s - s_{z2}) \approx (s - s_{p2}) \approx 2j\omega_s$$

Hence, Z_{IN} is approximated by

$$Z_{IN} \approx \frac{\dfrac{\omega_s}{2Q_U} + j(\omega - \omega_s)}{j\omega_s C_o\left(-j\omega_s \dfrac{C}{2C_o}\right)} = R\left[1 + jQ_U \frac{2(\omega - \omega_s)}{\omega_s}\right]$$

which is recognized as the behavior of a high-Q series resonant circuit.

In the vicinity of the parallel-resonant frequency ω_a, the following approximations apply:

$$s \approx j\omega_s$$

$$(s - s_{z1}) \approx j\Delta\omega = j\omega_s \frac{C}{2C_o}$$

$$(s - s_{p1}) = \frac{\omega_s}{2Q_U} + j(\omega - \omega_a)$$

$$(s - s_{z2}) \approx (s - s_{p2}) \approx 2j\omega_a$$

Hence, Z_{IN} is approximated by

$$Z_{IN} \approx \frac{R_p}{1 + jQ_U \dfrac{2(\omega - \omega_a)}{\omega_s}}$$

where R_p is given by (4.12).

When a crystal is to be operated in the parallel-resonance region. the manufacturer constructs the crystal to resonate at the listed frequency when an external load capacitance is connected in series with the crystal. The load resonant frequency, denoted by f_L, is the frequency at which the crystal is resonant when a given external load capacitance C_L is connected in series with the crystal. This is illustrated in Figure 4.9(a) where the basis of a crystal impedance bridge measuring system is illustrated. In Figure 4.9(a) a tuned oscillator is completed by either a variable resistor or the crystal (see switch S1). The tuned oscillator is designed to operate at a frequency close to the desired crystal frequency. Hence, it plays a role in selecting the desired mode of operation for the crystal. From a pole-zero point of view, the tuned circuit adds canceling poles and zeroes at the operating frequency of the crystal. The oscillator is adjusted so the frequency and amplitude of oscillation is the same through both paths. Hence, the crystal operates at f_s (zero-phase shift) and the value of the variable resistor is equal to R. If switch S2 connects a load capacitor in series with the crystal, the circuit will oscillate at the frequency f_L.

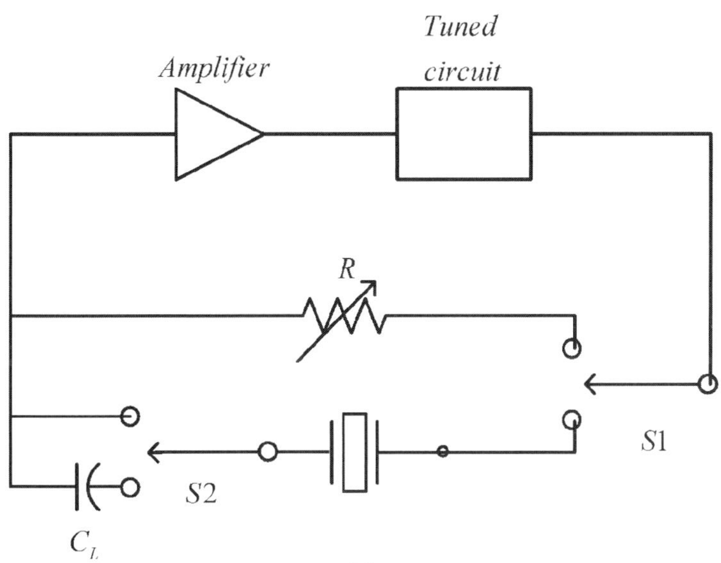

(a)

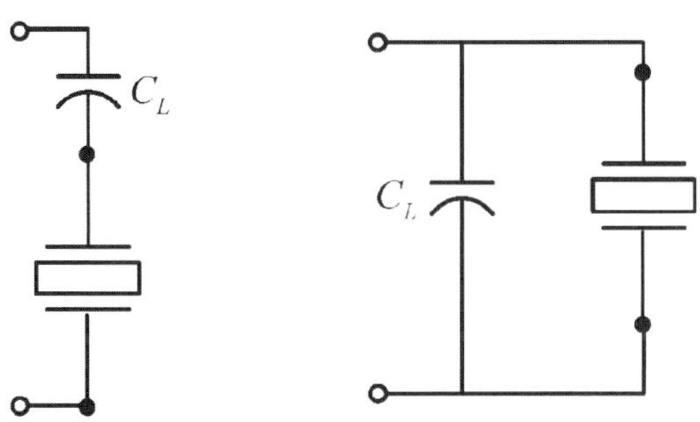

(b)

Figure 4.9 (a) Crystal impedance bridge setup and (b) equivalent crystal load circuits.

The value of the variable resistor that produces the same oscillation is equal to R_e.

At f_L, the series combination of the crystal reactance (X_{XTAL}) and the reactance of the load capacitor C_L (i.e., $X_{C_L} =- 1/\omega_L C_L$ where $\omega_L = 2\pi f_L$) is zero (i.e., zero-phase shift). Hence, the crystal reactance is

$$X_{XTAL} = \frac{1}{\omega_L C_L} \tag{4.19}$$

For example, specifying a 10-MHz crystal at a load capacitance of 30 pF is the same as specifying a crystal reactance of 530.5Ω at 10 MHz. Typical values of the load capacitance are from 10 to 32 pF.

Crystals operating in the parallel-resonance region operate between f_s and f_a with a certain load capacitance. Series resonant crystals do not have a load capacitance specification.

From the previous discussion it is seen that a crystal has a frequency specification of f_L at a load capacitance C_L when its reactance at f_L is given by (4.19). There is a single frequency f_L for each value of C_L. In a parallel-resonance oscillator configuration, such as the tuned Colpitts oscillators in Chapter 3, the crystal replaces the inductor and the load capacitance specification is the total capacitance across the (inductive) crystal. This capacitance value, equal to C_L, produces parallel resonance at f_L. In Section 4.4 various oscillator configurations that take advantage of this effect are discussed.

A crystal is usually specified by giving the maximum value of R, the maximum value of C_o, and the minimum value of C. The values of C_o and C fix the maximum value of the ratio $r = C_o/C$. For computer simulation purposes once the values of R, C_o, and C are specified, the value of L follows from (4.2) for a crystal operating at series resonance. For a crystal operating in the parallel-resonance region at a given load capacitance, the value of f_s is calculated using (4.13), and then, (4.2) is used to calculate L.

Some typical crystal parameters are given in Table 4.1. Since there are many varieties of crystals, the designer should obtain from the manufacturer the specific data for the desired crystal.

It is common for manufacturers to list for a given crystal holder the maximum shunt capacitance, the maximum ESR, the maximum load capacitance, the typical value of C, and the maximum drive level.

For example, for a crystal frequency in the 10- to 20-MHz range in an HC-49/U holder, in the −10°C to 60°C temperature range, a manufacturer specifies a temperature stability of 10 ppm, series C_L between 10 and 40 pF, a calibration tolerance of 10 ppm, a maximum ESR of 25Ω, a maximum shunt capacitance of $C_o = 7$ pF, a value of C in the 10- to 20-fF range, and a maximum drive level of 0.5 mW.

The crystal maximum drive level is the maximum dissipation allowable in the crystal. The oscillator circuit determines the drive level in the crystal. An overdriven crystal will deteriorate fast. For example, a manufacturer lists the maximum drive level for a crystal in an HC-49/U holder at 1 mW. Other maximum drive levels listed by manufacturers for crystals operating at frequencies below 100 kHz is in the range of 5 μW, in the frequency range of 1 to 30 MHz maximum drive levels

Table 4.1 Typical Crystal Parameters

	L (H)	C (pF)	R (Ω)	C_o (pF)	Q_U
20 kHz	4,000	0.005	40,000	7	12,500
200 kHz	27	0.024	2,000	10	17,000
450 kHz	8.5	0.015	1,050	5	23,200
1 MHz	3.16	0.008	400	3.2	50,000
2 MHz	0.528	0.012	100	4	66,000
4 MHz	0.21	0.007	22	2.4	240,990
8 MHz	0.014	0.027	8	5.6	88,680
10 MHz	0.0101	0.025	5	5.5	126,000
15 MHz	0.00417	0.028	5	4	78,500
30 MHz (3rd overtone)	0.0101	0.0027	45	6	42,300
150 MHz (7th overtone)	0.002814	0.0004	70	6	37,887

of 10 mW are found, with 1 mW being typical. Overtone crystals are usually rated at 1 to 2 mW.

The drive level can be calculated using

$$P_{XTAL} = I_{\text{rms}}^2 R_e$$

where I_{rms} is the rms current through the crystal. A current probe can be used to measure I_{rms}.

Example 4.3

Calculate Q_U, f_p, Δf, and L for the seventh overtone crystal operating at 150 MHz listed in Table 4.1.

Solution

From (4.5), with $f_s = 150$ MHz, we obtain

$$f_p = 150 \times 10^6 \left[1 + \frac{0.0004 \times 10^{-12}}{2(6 \times 10^{-12})} \right] = 150.005000 \text{ MHz}$$

and it follows that $\Delta f = f_p - f_s = 5$ kHz. This result also follows from

$$\Delta f = f_s \frac{C_1}{2C_o} = 150 \times 10^6 \frac{0.0004 \times 10^{-12}}{2(6 \times 10^{-12})} = 5 \text{ kHz}$$

The value of L is calculated using (4.2):

$$L = \frac{1}{(2\pi \times 150 \times 10^6)^2 (0.0004 \times 10^{-12})} = 2.814 \text{ mH}$$

The value of Q_U, from (4.3), is

$$Q_U = \frac{2\pi \times 150 \times 10^6 (0.002814)}{70} = 37.887$$

and the value of M is

$$M = \frac{1}{\omega_s C_o R} = \frac{1}{(2\pi \times 150 \times 10^6)(0.0004 \times 10^{-12})70} = 2.52$$

High overtone crystals have low values of M. In fact, values of M less than 1 are found.

Example 4.4

A 1-MHz crystal is specified to operate with a load capacitance of 32 pF. The crystal manufacturer specifies $C_o = 5$ pF, $R = 400\Omega$ and $C = 10$ fF for the 1-MHz crystal with $C_L = 32$ pF. Calculate L and Q_U for the crystal.

Solution

The load resonant frequency is 1 MHz. Hence, from (4.13)

$$10^6 = f_s \left[1 + \frac{10 \times 10^{-15}}{2(2 \times 10^{-12} + 32 \times 10^{-12})} \right] \Rightarrow f_s = 0.999865 \text{ MHz}$$

and it follows that

$$L = \frac{1}{(2\pi f_s)^2 C} = \frac{1}{(2\pi 10^6)^2 10 \times 10^{-15}} = 2.539\text{H}$$

The value of Q_U follows from

$$Q_U = \frac{\omega_s L}{R} = \frac{2\pi 10^6 (2.539)}{400} = 39{,}828$$

Because of its resonant characteristics, a crystal is operated either as a series resonant circuit or in the parallel resonant region. A series-mode oscillator uses a crystal in a series-resonant configuration where it appears as a pure resistor of value R at the frequency f_s. A parallel-mode oscillator uses a crystal in the parallel resonance region, with C_L specified. In the parallel mode of operation the crystal appears inductive. Crystals operated in the parallel mode are specified by the manufacturer to resonate at the frequency f_L with a specific load capacitance. The frequency listed on a crystal package is the desired frequency of operation. However, it depends on the mode of operation. That is, it can be for a series-mode oscillator or for a parallel-mode oscillator with a specific load capacitance C_L.

The basic parallel-resonant crystal oscillator is the Pierce oscillator shown in Figure 4.10(a) and discussed in detail in Section 4.4. This oscillator consists of an inverter amplifier, the inductive crystal and two capacitors. These capacitors and the circuit stray capacitance determine the value of the load capacitance. The resistor limits the drive current in the crystal. The crystal operates at a certain load frequency in the parallel-resonant region at the given load capacitance. If the crystal fails the oscillator will not oscillate.

The basic series oscillator consists of two inverters with the crystal connected in series, as shown in Figure 4.10(b). The crystal operates at its series resonant frequency. If the crystal fails, the oscillator usually continues to oscillate at a higher frequency due to the high-frequency feedback paths through the ICs.

In the series-resonance oscillator the frequency of oscillation is determined by the crystal with no easy means for frequency adjustment. The capacitor C_c is a coupling capacitor, and the resistor R_1 limits the drive current in the crystal.

A popular, low-cost clock oscillator is shown in Figure 4.11. It consists of two CMOS inverters (7404 type) and a series resonant crystal. The feedback resistors R_f bias the inverters in the linear region.

4.3 Frequency Pulling in a Crystal Oscillator

The frequency of resonance of a crystal can be adjusted by using a load capacitor C_L in series with the crystal, as shown in Figure 4.12.

The input impedance in Figure 4.12 is

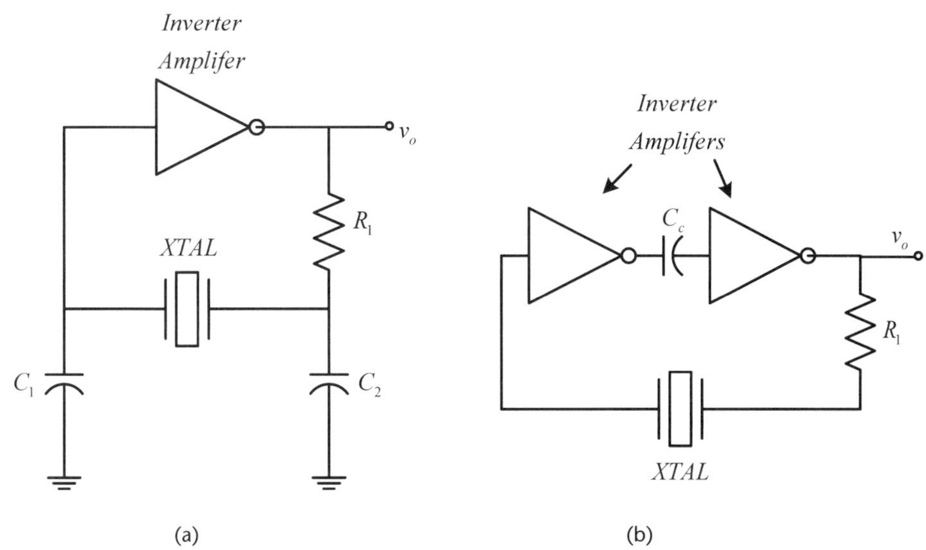

Figure 4.10 (a) A basic parallel-resonant configuration and (b) a basic series-resonant configuration.

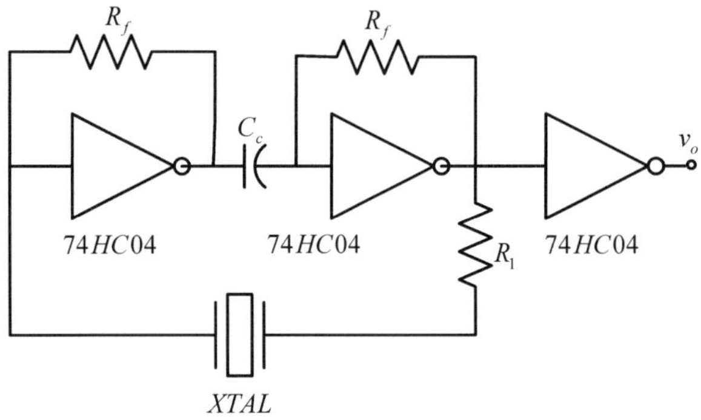

Figure 4.11 A clock oscillator using a series-resonant crystal.

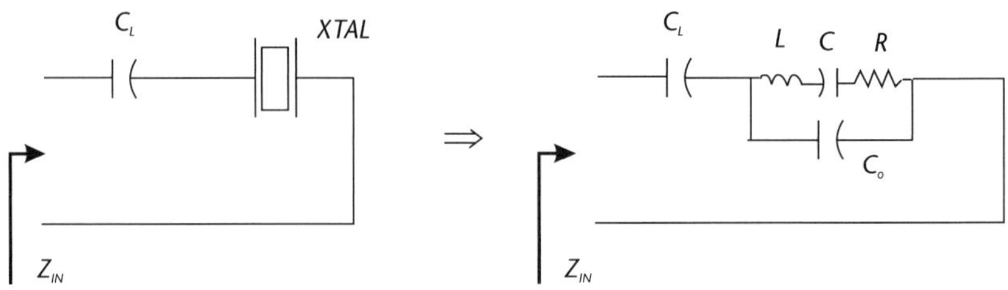

Figure 4.12 Capacitor C_L in series with the crystal.

$$Z_{IN}(j\omega) = jX_{C_L} + \frac{jX_{C_o}(R + jX_L + jX_C)}{jX_{C_o} + R + jX_L + jX_C} \tag{4.20}$$

$$= \frac{j(X_{C_o} + X_{C_L})\left[R + jX_L + j\left(X_C + \dfrac{X_{C_o}X_{C_L}}{X_{C_o} + X_{C_L}}\right)\right]}{R + jX_L + j(X_{C_o} + X_C)}$$

where X_{C_L}, X_{C_o}, X_L, and X_C are the reactances of C_L, C_o, L, and C, respectively. Let

$$X'_C = X_C + \frac{X_{C_o}X_{C_L}}{X_{C_o} + X_{C_L}}$$

and observing that

$$X_{C_o} + X_C = X'_C + \frac{X^2_{C_o}}{X_{C_o} + X_{C_L}}$$

we can write (4.20) in the form

$$Z_{IN}(j\omega) = \frac{j(X_{C_o} + X_{C_L})(R + jX_L + jX'_C)}{R + jX_L + j\left(X'_C + \dfrac{X^2_{C_o}}{X_{C_o} + X_{C_L}}\right)} \tag{4.21}$$

Letting

$$n = \frac{X_{C_o} + X_{C_L}}{X_{C_o}} = \frac{C_o + C_L}{C_L}$$

(4.21) can be expressed in the form

$$Z_{IN}(j\omega) = \frac{jnX_{C_o}(R + jX_L + jX'_C)}{R + jX_L + j\left(X'_C + \dfrac{X_{C_o}}{n}\right)}$$

or

$$Z_{IN}(j\omega) = \frac{jnX_{C_o}(n^2R + jn^2X_L + jn^2X'_C)}{jnX_{C_o} + n^2R + jn^2X_L + jn^2X'_C} \tag{4.22}$$

which is recognized as the parallel combination of a capacitor C_o/n with three elements in series consisting of an inductor n^2L, a resistor n^2R, and a capacitor C'/n^2, as shown in Figure 4.13. The capacitor C' is given by

$$C' = \frac{C(C_o + C_L)}{C + C_o + C_L}$$

Since the circuit in Figure 4.13 is similar to that in Figure 4.2(d) it follows that (4.22) is similar in form to (4.1), and therefore, the series resonance is given by

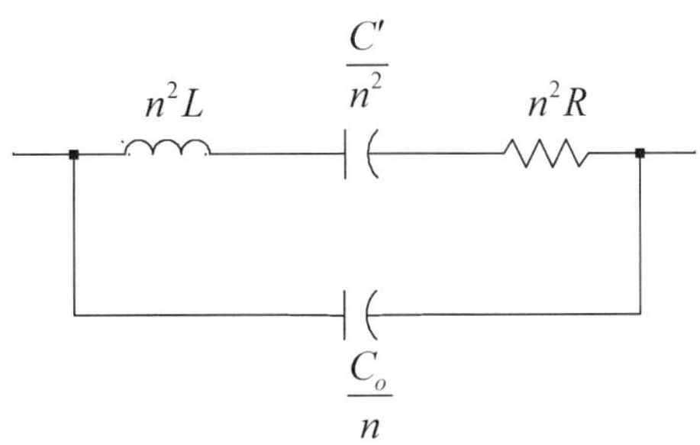

Figure 4.13 Equivalent circuit of the circuit in Figure 4.12.

$$\omega_s' = \sqrt{\frac{1}{(n^2L)\left(\frac{C'}{n^2}\right)}} = \frac{1}{\sqrt{LC'}} \tag{4.23}$$

and the parallel-resonant frequency is

$$\omega_a = \frac{1}{\sqrt{(n^2L)\dfrac{\left(\frac{C'}{n^2}\right)\left(\frac{C_o}{n}\right)}{\left(\frac{C'}{n^2}\right)+\left(\frac{C_o}{n}\right)}}} = \sqrt{\frac{1}{L\left(\frac{CC_o}{C+C_o}\right)}} = \omega_s\sqrt{1+\frac{C}{C_o}}$$

or

$$f_a \approx f_s\left(1+\frac{C}{2C_o}\right)$$

The parallel-resonant frequency of the crystal with a series load capacitor is identical to that of the crystal by itself. Hence, the series load capacitor only affects the series-resonant frequency. In terms of poles and zeroes, the poles of Z_{IN} are not affected by the addition of the series capacitor, only the zeroes are affected.

Equation (4.23) can also be written in the form

$$\omega_s' = \omega_s\sqrt{1+\frac{C}{C_o+C_L}}$$

or

$$f_s' \approx f_s\left[1+\frac{C}{2(C_o+C_L)}\right] \tag{4.24}$$

This relation shows that $f_s' \to f_a$ as $C_L \to 0$ and $f_s' \to f_s$ as $C_L \to \infty$. The pulling range (i.e., the interval between f_s' and f_s) is

$$\Delta f = \frac{f_s C}{2(C_o+C_L)} \tag{4.25}$$

Equation (4.24) shows that the series load capacitor can be used to pull the series resonant frequency.

Further analysis of the circuit in Figure 4.13 shows that its impedance is given by

$$Z_{IN} = \frac{s^2 + \frac{\omega_s}{Q_U}s + \omega_s^2\left(1 + \frac{C}{C_o + C_L}\right)}{s\frac{C_o}{n}\left[s^2 + \frac{\omega_s}{Q_U}s + \omega_s^2\left(1 + \frac{C}{C_o}\right)\right]} \tag{4.26}$$

A comparison of (4.26) with (4.1) shows that the poles of the two functions are identical. The zeroes of (4.26) are located at

$$s = -\frac{\omega_s}{Q_U} \pm j\omega_s\sqrt{1 + \frac{C}{C_o + C_L} - \frac{1}{4Q_U^2}} \approx -\frac{\omega_s}{Q_U} \pm j\omega_s\left[1 + \frac{C}{2(C_o + C_L)}\right]$$

Since the real part of the zeroes in (4.26) and (4.1) are the same, the circuit Q is not affected by C_L. Narrowband equivalent circuits are similar to those in Figure 4.6, with R replaced by nR.

The capacitor C_o has no effect on the series resonance of the crystal. However, from (4.23) it is seen that f_s' can be changed by varying C_o. Of course, the value of C_o also affects the parallel-resonant frequency f_a.

The parallel-resonant frequency can be changed by adding a load capacitor in parallel with the crystal (i.e., with C_o). In this case, as in (4.13), the load capacitor adds to C_o and the parallel-resonant frequency, previously denoted by f_L, is given by

$$f_L \approx f_s\left[1 + \frac{C}{2(C_o + C_L)}\right] \tag{4.27}$$

which shows that $f_L \to f_a$ as $C_L \to 0$ and $f_L \to f_s$ as $C_L \to \infty$. Hence, in this case the pulling range (i.e., the interval between f_L and fs) is

$$\Delta f = \frac{f_s C}{2(C_o + C_L)} \tag{4.28}$$

The frequency f_L is the specified frequency of operation for the crystal at a given C_L. At this frequency the crystal reactance is inductive and given by (4.19). In an appropriate configuration, such that the crystal sees a load capacitance of C_L, the circuit will oscillate at the frequency f_L.

The *pullability* of a crystal is an important parameter to the designer that wants to obtain several operating frequencies by changing the load capacitance. The pullability, which is given by (4.27), provides a measure of the frequency shift that can be obtained for a crystal operating in the parallel mode. It can be expressed in ppm as

$$\frac{\Delta f}{f_s} = \frac{10^6 C}{2(C_O + C_L)} \text{ (ppm)} \tag{4.29}$$

The pullability, expressed in ppm/pF, is known as the pullability sensitivity (PS) or trimming sensitivity (TS). It represents the average pulling per pF around a certain value of C_L, and it can be obtained by taking the derivative of (4.10) with respect to C_L, namely,

$$PS = \frac{-10^6 C}{2(C_0 + C_L)^2} \left(\frac{\text{ppm}}{\text{pF}}\right) \tag{4.30}$$

Hence, the PS is calculated at a certain value of load capacitance.

Another useful relation is the pullability between two load capacitances, denoted by C_{L1} and C_{L2}. It is given by

$$\Delta f = \frac{C(C_{L2} - C_{L1})10^6}{2(C_0 + C_{L1})(C_0 + C_{L2})} \text{ (ppm)} \tag{4.31}$$

A typical pullability curve (or frequency shift curve) is shown in Figure 4.14. It is seen from Figure 4.14, or from (4.30), that small values of C_L results are large values of PS, while larger values of C_L are associated with smaller PS values. In general, in a fixed frequency oscillator a high value of C_L is desired. In a VCXO a low value of C_L is desired so that an appropriate pulling range is obtained.

Example 4.5

(a) Calculate the pullability for a crystal that lists C_o = 5 pF and C = 0.015 pF if a load capacitance of 20 pF is used.
(b) Repeat part (a) if C_L = 30 pF.
(c) Calculate the pullability (or timing range) between C_L = 20 pF and C_L = 30 pF.
(d) Calculate the pullability sensitivity for the crystal.

Solution
(a) Using (4.29), the shift from f_s with C_L = 20 pF is

$$\frac{\Delta f}{f_s} = \frac{10^6(0.015 \times 10^{-12})}{2(25 \times 10^{-12})} = 300 \text{ (ppm)}$$

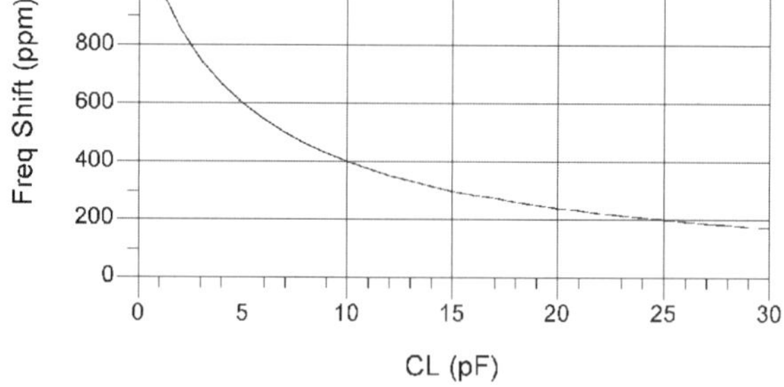

Figure 4.14 A pullability curve for C_o = 4 pF and C = 0.012 pF.

(b) Similarly, using $C_L = 30$ pF the shift from f_s is 214 ppm.
(c) The pullability (or timing range) is (300 – 214) ppm = 86 ppm. The pullability between two values of load capacitances can also be calculated using (4.31), namely,

$$\Delta f = \frac{0.015 \times 10^{-12}(10 \times 10^{-12})10^6}{2(25 \times 10^{-12})(35 \times 10^{-12})} = 86 \text{ (ppm)}$$

(d) From (4.30), the pulling sensitivity around $C_L = 20$ pF is

$$PS = \frac{-10^6(0.015 \times 10^{-12})}{2(35 \times 10^{-12})^2} = 6.12 \left(\frac{\text{ppm}}{\text{pF}}\right)$$

Example 4.6

Discuss the selection of a crystal to oscillate at 10 MHz when the circuit load capacitance is 15 pF. A series pulling capacitor is used to vary the load capacitance between 10 and 25 pF.

Solution

Applications that use a pulling capacitor usually require the selection of an AT-cut crystal due to its good temperature stability. When compared to a BT-cut crystal (which has poor temperature stability), the required pulling in a given system with an AT-cut crystal is less.

The power dissipation that the crystal will experience in the oscillator determines the type of crystal holder and the maximum power dissipation of the crystal. Let us select an HC-49 holder that can dissipate 1 mW.

Some typical specifications for the desired AT-cut crystal are a calibration tolerance of 25 ppm at 25°C, a temperature stability of 30 ppm in the range of –10°C to 60°C, and an aging specification of 5 ppm in the first year, and 2 ppm/year thereafter.

Assume that the oscillator is to be operated for 5 years. Therefore, the total tolerance for the crystal is 25 + 30 + (5 + 10) = 70 ppm.

For the package selected, the maximum value of C_o is listed as 5 pF. Then, from (4.31) the value of C that produces a pulling of –70 ppm when C_L varies from 15 to 10 pF is

$$\frac{10^6 C(10 \times 10^{-12} - 15 \times 10^{-12})}{2(20 \times 10^{-12})(15 \times 10^{-12})} =- 70 \Rightarrow C = 8.4 \text{ fF}$$

Next, we check that the value $C = 8.4$ pF can pull +70 ppm as C_L varies from 15 to 25 pF. Using (4.31), it follows that $C = 8.4$ pF can attain such a pulling value. To allow for some margin of error, let us specify a larger value of C, say $C = 11$ fF.

The value of R_e can be specified to be between 25 and 50Ω, say $R_e = 40\Omega$. If $R_e = 40\Omega$ it follows from (4.19) that R is

$$R = \frac{R_e}{\left(1 + \frac{C_o}{C_L}\right)^2} = \frac{40}{\left(1 + \frac{5 \times 10^{-12}}{15 \times 10^{-12}}\right)^2} = 22.5\Omega$$

The value of L does not need to be specified, since the crystal is specified to operate at 10 MHz. However, for simulation purposes the value of L is needed. From (4.14) the value of f_s is

$$f_s = \frac{10 \times 10^6}{1 + \frac{11 \times 10^{-15}}{2(20 \times 10^{-12})}} = 9.997251 \text{ MHz}$$

and from (4.2) we obtain

$$L = \frac{1}{(2\pi \times 9.997251 \times 10^6)^2 (11 \times 10^{-15})} = 23.04 \text{ mH}$$

The pullability can also be controlled with the use of an inductor L_L, as shown in Figure 4.15. For this circuit the pulling range is given by

$$\Delta f = \frac{f_s C}{2\left[C_o + C_L\left(\frac{1}{1 - \omega_s^2 L_L C_L}\right)\right]}$$

where $\omega_s^2 L_L C_L < 1$.

4.4 The Pierce, Colpitts, and Clapp Crystal Oscillators

Tuned-circuit oscillator configurations were discussed in Chapter 3. If the inductor in these oscillators is replaced by a crystal operated in the parallel mode, the result is a crystal oscillator. Depending where the ground is placed, the oscillator is known as either a Pierce, Colpitts, or Clapp crystal oscillator.

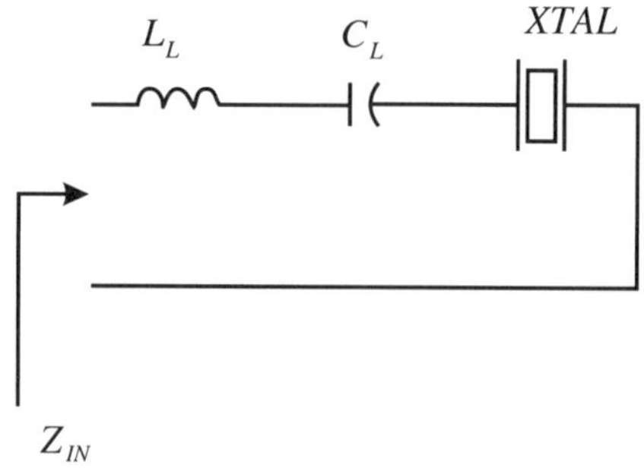

Figure 4.15 Using an inductor to increase the pulling range.

The three resulting ac configurations for crystal oscillator circuits are shown in Figure 4.16. Figure 4.16(a) illustrates the Pierce oscillator. In the Pierce oscillator the emitter is grounded. The crystal impedance must appear inductive since the crystal is operated at f_L. Pierce oscillators work fine and are simple to design in the 100-kHz to 20-MHz range. They are very stable and output powers of 3 to 5 mW are typical.

The Colpitts crystal oscillator is shown in Figure 4.16(b) where the crystal behaves inductively. This oscillator is grounded at the collector. It operates well in the 1- to 20-MHz region. Typical output powers are in the low milliwatt range.

In crystal oscillator nomenclature the Colpitts type of configuration with the base grounded is usually called the Clapp crystal oscillator. The Clapp crystal oscillator is shown in Figure 4.16(c) where the crystal must appear inductive. Its frequency of operation and power characteristics are similar to those of the Colpitts crystal oscillator.

The Pierce, Colpitts, and Clapp crystal oscillators are basically the same circuit (i.e., a Colpitts type configuration), but with the transistor grounded at its different terminals. The circuit biasing resistors and stray capacitances appear in shunt with different elements in each of the three configurations. This makes the performance of the three configurations different. The Pierce configuration is usually the most desirable. The analysis of the Pierce oscillator circuit follows.

The ideal operation of the Pierce oscillator is illustrated in Figure 4.17. The crystal at series resonance is resistive, and the RC_1 circuit provides a phase shift of 90°. In the parallel-resonance region the crystal appears as an L_eR_e circuit. If R_e is neglected the phase shift provided by the tuned circuit is −180°. Therefore, it is seen that if the crystal operates in the parallel resonance region, the total phase shift is −360°. More specifically, if the crystal operates in the parallel resonance

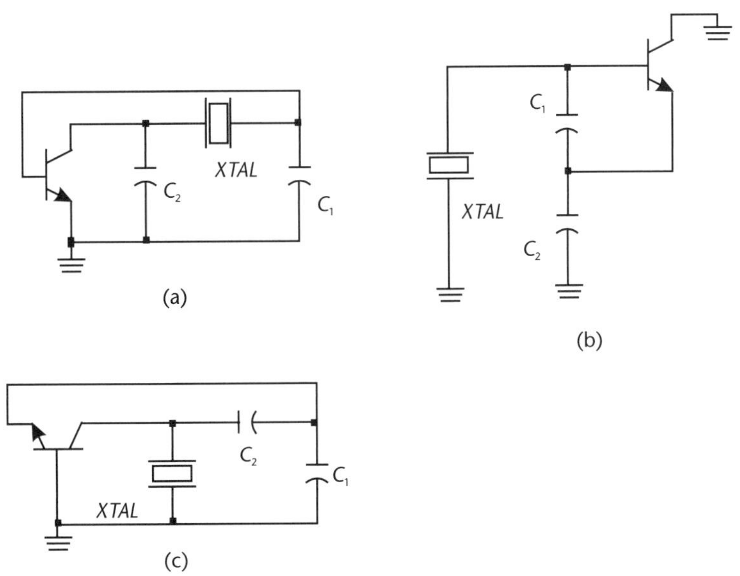

Figure 4.16 (a) The Pierce crystal oscillator, (b) the Colpitts crystal oscillator, and (c) the Clapp crystal oscillator.

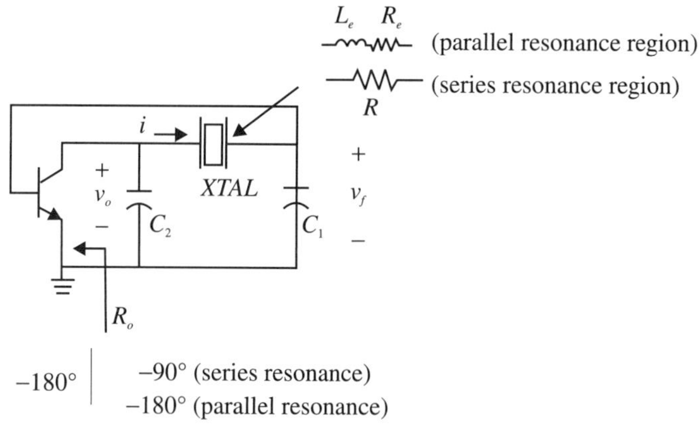

Figure 4.17 Ideal operation of the Pierce crystal oscillator.

region, the current i lags v_o by 90°, and v_f lags i by 90°. Thus, the base-to-ground voltage v_f lags the collector-to-ground voltage by 180°. The total phase shift around the loop is −360°.

In an actual circuit the phase shift through the transistor is more than −180° due to the junction capacitance and delay time. Let us assume that the phase shift is −181°. For the total phase shift to be −360°, the crystal must appear inductive (i.e., operate in the parallel resonance region) and the tuned circuit must provide a phase shift of −179°.

The Pierce oscillator circuit is shown in Figure 4.18(a). For proper ac operation the impedances of C_1 and C_2 must not load the input and output of the transistor, respectively. Figure 4.18(b) shows the load impedance Z_L seen by the transistor. The crystal is replaced by its equivalent crystal resistance R_e in series with its equivalent inductive reactance jX_e, where $X_e = \omega L_e$. The impedance Z_L, assuming that $R_1 \parallel R_2 \parallel h_{ie} \approx h_{ie}$, is given by

$$Z_L = jX_2 \parallel \left[R_e + jX_e + j\left(X_1 \parallel h_{ie}\right)\right]$$

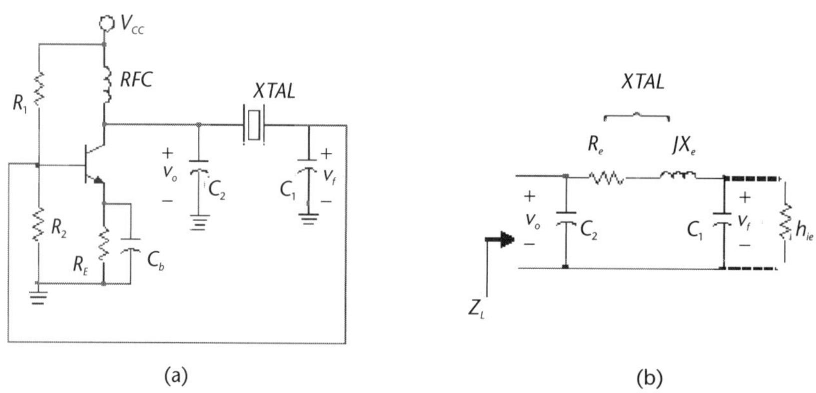

Figure 4.18 (a) A Pierce oscillator and (b) the load impedance Z_L seen by the transistor.

where $X_1 = -1/\omega C_1$ and $X_2 = -1/\omega C_2$. Also, if $|X_1| \ll h_{ie}$, we can approximate Z_L by

$$Z_L = \frac{jX_2[R_e + j(X_1 + X_e)]}{R_e + j(X_1 + X_2 + X_e)} \tag{4.32}$$

At resonance

$$X_1(\omega_o) + X_2(\omega_o) + X_e(\omega_o) = 0 \tag{4.33}$$

which shows that the frequency of oscillation is

$$\omega_o = \frac{1}{\sqrt{L_e C_T}} \tag{4.34}$$

where

$$C_T = \frac{C_1 C_2}{C_1 + C_2} \tag{4.35}$$

Note that if the crystal is specified to operate in the parallel-resonance region at f_L with an external load capacitance C_L, then f_o in (4.34) is equal to f_L provided that C_T in (4.34) is equal to C_L. The frequency of operation in a Pierce oscillator is usually 5 to 50 ppm above its series resonance.

Using (4.33), (4.32) reduces to

$$Z_L(j\omega_o) = \frac{X_2(\omega_o)}{R_e}[X_2(\omega_o) + jR_e]$$

Furthermore, since $|X_2(\omega_o)| \gg R_e$, we obtain

$$Z_L(j\omega_o) \approx \frac{[X_2(\omega_o)]^2}{R_e} \tag{4.36}$$

The open-loop gain at resonance is

$$A_v(j\omega_o) = \frac{v_o}{v_f} = -g_m Z_L(j\omega_o) = -g_m \frac{[X_2(\omega_o)]^2}{R_e} \tag{4.37}$$

The voltage-feedback factor is

$$\beta(j\omega) = \frac{v_f}{v_o} = \frac{jX_1}{R_e + j(X_1 + X_e)} \tag{4.38}$$

Substituting (4.33) into (4.38) gives the voltage-feedback factor as

$$\beta(j\omega_o) = \frac{jX_1}{R_e - jX_2} \approx -\frac{X_1}{X_2} = -\frac{C_2}{C_1} \quad (4.39)$$

which shows that v_1 lags v_2 by $-180°$.

The loop-gain condition for oscillation is

$$\beta(j\omega_o)A_v(j\omega_o) > 1 \quad (4.40)$$

Substituting (4.37) and (4.39) into (4.40) gives the gain condition, namely,

$$g_m \frac{X_2^2}{R_e} > \frac{C_1}{C_2}$$

or

$$\frac{g_m}{\omega_o^2 R_e C_1 C_2} > 1 \quad (4.41)$$

To summarize, (4.34) gives the frequency of oscillation, and (4.41) gives the gain condition.

The Pierce crystal oscillator is an attractive configuration since the stray capacitances appear across the large capacitors C_1 and C_2. Furthermore, the resistors R_1 and R_2 do not appear in parallel with the crystal. The effect of the transistor input and output impedances are minimized by selecting an appropriate transistor and the values of C_1 and C_2.

Some observations are in order. First, observe that larger output powers can be obtained by making X_2 large (i.e., small values of C_2). Another observation is that the crystal power dissipation [see Figure 4.18(b)] can be approximated by

$$P = i_{1(\mathrm{rms})}^2 R_e \approx \frac{v_{2(\mathrm{rms})}^2 R_e}{X_2^2}$$

which shows that it can be reduced by making $|X_2|$ large. A final observation is that in a Pierce oscillator the crystal is not grounded. Some applications that require the interchanging of crystals might be unsuitable for the Pierce configuration.

Example 4.7

Design a 2-MHz Pierce crystal oscillator. The load capacitance of the 2-MHz crystal is specified to be 32 pF.

Solution

The oscillator configuration is shown in Figure 4.19(a). It uses a variable series capacitor C_s with the crystal (shown as 40 pF) to adjust the frequency of oscillation. Using an MPS5179 BJT with $V_{CC} = 12$V, $R_1 = 26.5$ kΩ, $R_2 = 33.5$ kΩ and $R_E = 6$ kΩ places the Q point at $I_C = 1$ mA and $V_{CE} = 6$V. The RFC is implemented

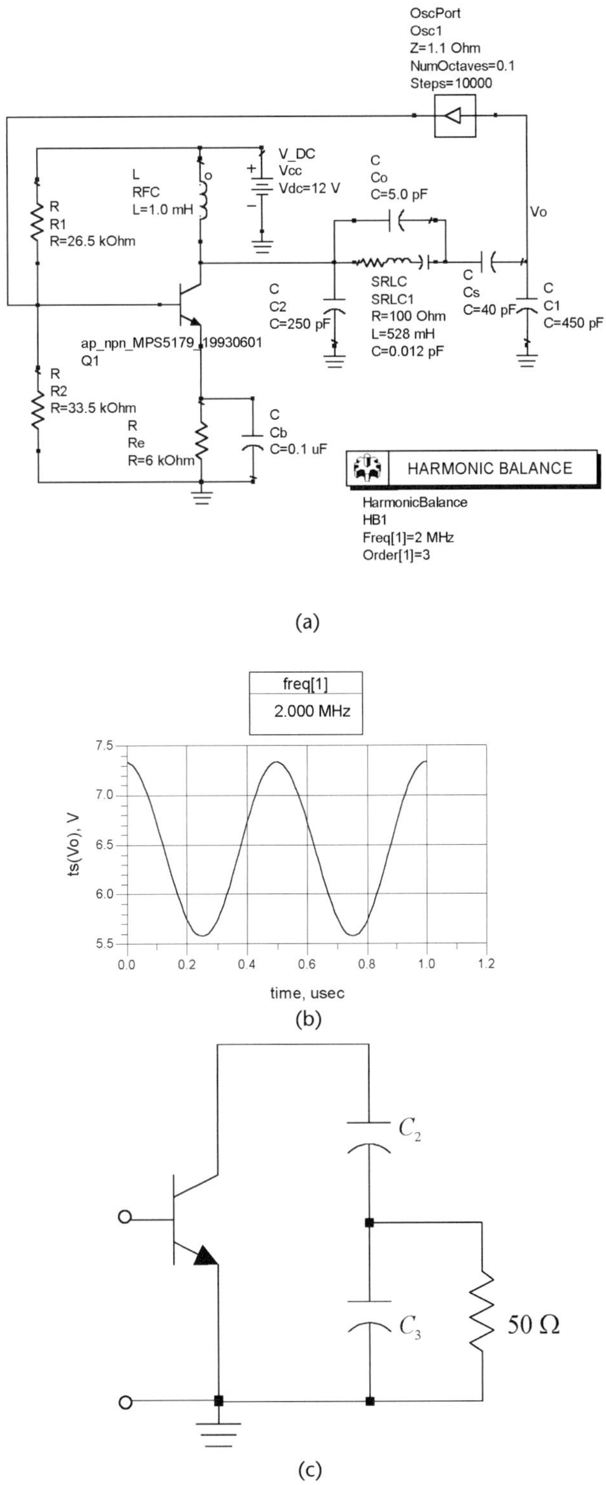

Figure 4.19 (a) A 2-MHz Pierce crystal oscillator, (b) the resulting waveform, and (c) a capacitive transformer.

with a 1-mH inductor ($X_{L_1} = 12.6$ kΩ). The crystal sets the frequency of oscillation at 2 MHz when the load capacitance is 32 pF. The crystal parameters listed in Table 4.1 were used. The value of $C_1 = 450$ pF ($X_{C_1} =- 177\Omega$) was selected so that its reactance is much smaller than h_{ie} ($h_{ie} = 2.5$ kΩ). The ratio C_1/C_2 was selected to be between 1.5 and 2.

The total capacitance seen by the crystal is the series combination of C_1, C_2, and C_s. The series capacitance of C_1 and C_2 is 161 pF. The trimmer capacitor C_s is set so that the series capacitance of C_s and 161 pF is 32 pF.

The inequality (4.41), with $R_e \approx R = 100\Omega$, and $C_1' = 36.7$ pF being the series capacitance of C_s and C_1, requires that

$$g_m > \omega_o^2 R_e C_1' C_2 = [2\pi(2 \times 10^6)]^2 100(36.7 \times 10^{-12}) 250 \times 10^{-12} = 0.14 \text{ mS}$$

which is easily satisfied since at $I_C = 1$ mA the value of g_m is 40 mS.

The simulation data is shown in Figure 4.19(b). The fundamental frequency of oscillation is 2 MHz. There are several ways that the oscillator signal can be coupled to a load. For example, a capacitive coupled resistor connected to the collector can be used. Such a resistor should not load the capacitor C_2. One can also use an amplifier connected to the collector to extract the oscillation signal. Another way is to connect a low-value resistive load using a capacitive transformer, as shown in Figure 4.19(c). The oscillator signal is coupled to the 50Ω resistor through the coupling network formed by C_2 and C_3. When viewed from the collector, the load resistor R_L appears as

$$R_L\left(\frac{C_2 + C_3}{C_2}\right)^2 = 50\left(\frac{2{,}250}{250}\right)^2 = 4.05 \text{ k}\Omega$$

which does not load C_1.

In the Pierce configuration a trimmer capacitor for frequency adjustment can also be placed in parallel with C_2.

Another dc bias configuration that works well with a Pierce oscillator is shown in Figure 4.20. The design shown produces a Q point at $V_{CE} = 7.7$V and $I_C =$ 3.6 mA. The load capacitance is 21 pF. The data shows that the fundamental frequency of oscillation is 2 MHz.

An FET Pierce crystal oscillator is shown in Figure 4.21(a), and its ac model in Figure 4.21(b). Figure 4.21(b) shows that C_1 should be larger than C_{gs} and the output resistance of the FET can be neglected if

$$r_d \gg |Z_L| = \frac{X_2^2(\omega_o)}{R_e} \tag{4.42}$$

It also follows from the gain condition in (4.41) that

$$\frac{g_m}{\omega_o^2 R_e C_1 C_2} > 1 \tag{4.43}$$

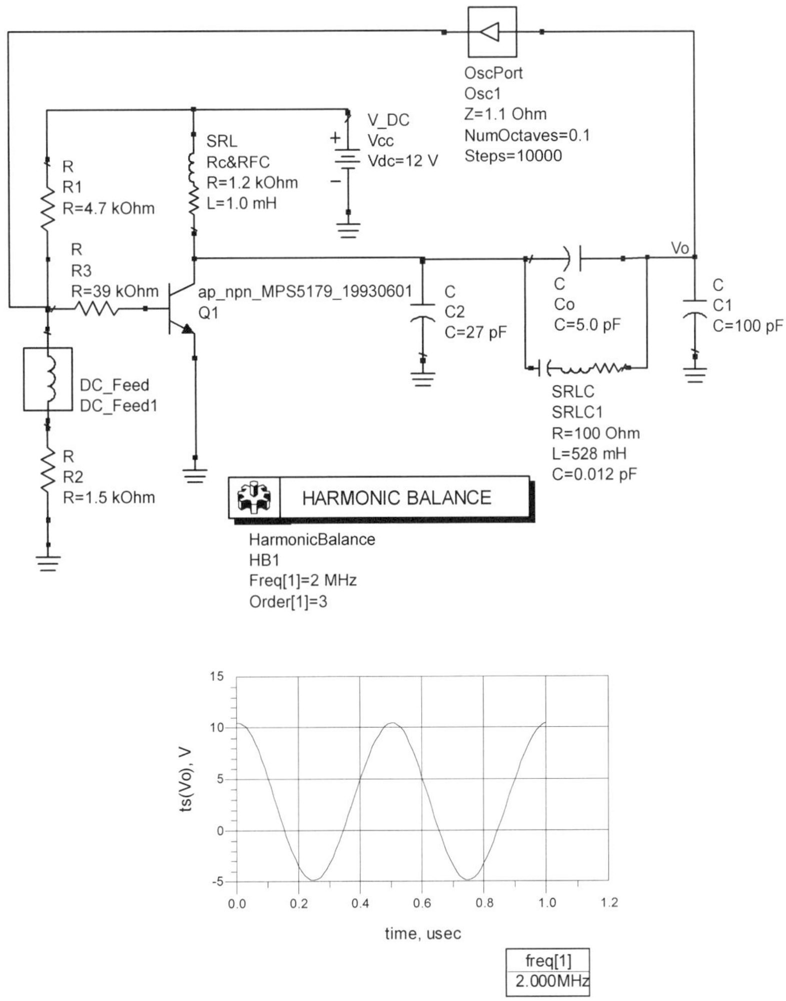

Figure 4.20 A Pierce oscillator at 2 MHz.

Example 4.8

Design the FET Pierce crystal oscillator in Figure 4.22(a) to oscillate at 10 MHz. The load capacitance of the crystal is specified as 32 pF.

Solution

The 10-MHz crystal parameters are those listed in Table 4.1. Using a typical value for r_d of 20 kΩ, it follows from (4.42) that $C_2 \gg 5$ pF. Hence, let $C_2 = 19$ pF.

Next, we let $C_1 = 100$ pF, and from

$$\frac{C_1 C_2}{C_1 + C_2} = 32 \text{ pF}$$

it follows that the value of C_2 is 47 pF.

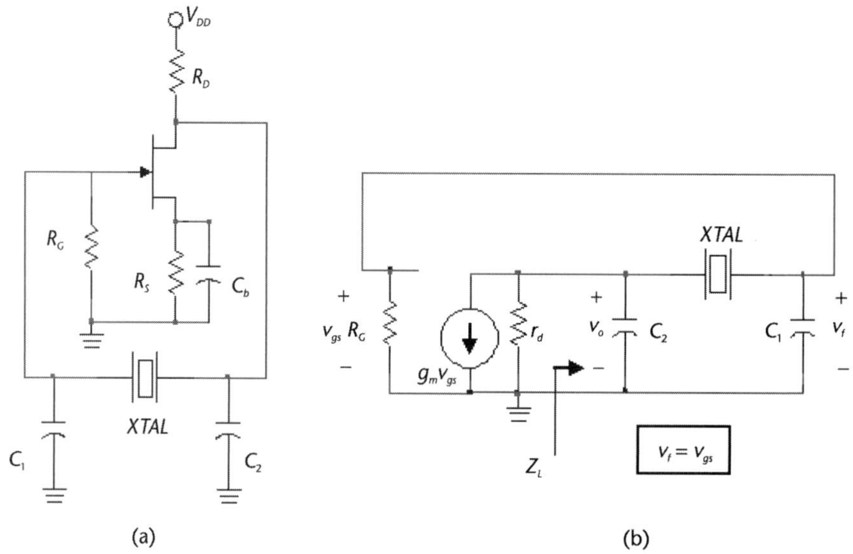

Figure 4.21 (a) An FET Pierce crystal oscillator and (b) its ac model.

From (4.43), using $R_e \approx R = 65\Omega$, it follows that

$$g_m > (2\pi 10^7)^2 65(100 \times 10^{-12})(47 \times 10^{-12}) = 0.1 \text{ mS}$$

which is simple to satisfy.

The oscillator was designed using the 2N5459 JFET at $V_{DS} = 5\text{V}$ and $I_D = 2$ mA. This JFET has the following typical parameters: $V_P = -3\text{V}$ and $I_{DSS} = 4.5$ mA. Then, for $I_D = 2$ mA it follows that $V_{GS} = -1\text{V}$, $g_m = 2$ mS,

$$R_s = \frac{-V_{GS}}{I_D} = \frac{1}{2 \times 10^{-3}} = 500\Omega$$

and

$$R_D = \frac{V_{DD} - V_{DS} - I_D R_S}{I_D} = \frac{12 - 5 - 1}{2 \times 10^{-3}} = 3 \text{ k}\Omega$$

An RFC ($L = 3$ mH) was used in series with R_D.

The simulation of the oscillator is shown in Figure 4.22. The resulting Q point is at $I_D = 1.93$ mA and $V_{DS} = 5.2\text{V}$. The resulting oscillation is at 10.02 MHz.

A popular Pierce configuration is shown in Figure 4.23. It uses an unbuffered IC inverter (such as the 4049 CMOS, the 7SU04 CMOS, the 74HCU04 CMOS, or the HA7210) as the active linear device to provide a phase shift of approximately −180°, and the other 180° are provided by the crystal and capacitors circuit. The second inverter is used as a buffer to couple the oscillation signal.

The inverter is biased in the linear region by the feedback resistor R_f whose value is in the 500-kΩ to 10-MΩ range. The resistor R_1 is employed in certain

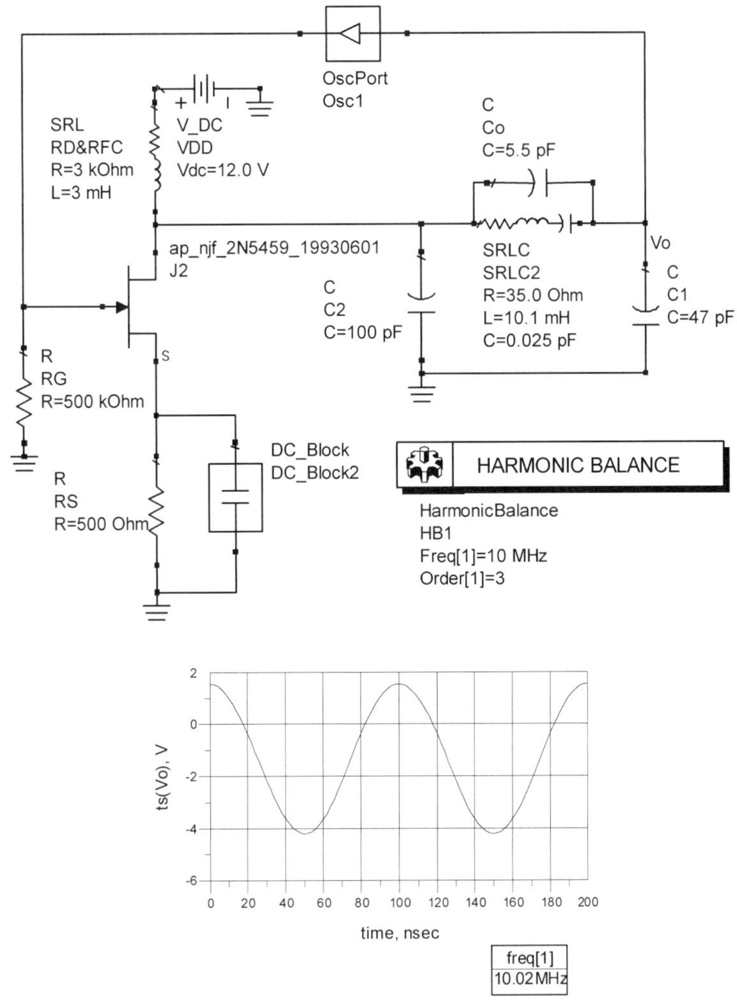

Figure 4.22 The FET Pierce crystal oscillator for Example 4.8.

situations to reduce the crystal drive level. The crystal behaves inductively since it operates in the parallel-resonance region. The load capacitance in the ideal case is given by

$$C_L = \frac{C_1 C_2}{C_1 + C_2}$$

In the actual oscillator the input and output capacitances of the inverter, as well as the board parasitic capacitances, affect the capacitance seen by the crystal at its terminals (i.e., the crystal load capacitance). In practice, the actual load capacitance is estimated using

$$C_L = \frac{C_1 C_2}{C_1 + C_2} + C_s$$

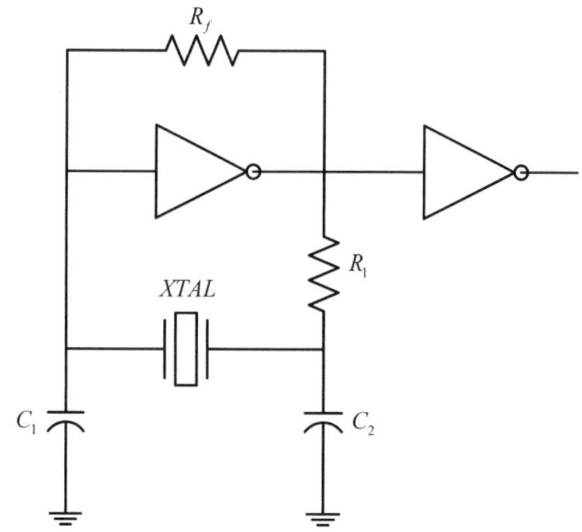

Figure 4.23 A Pierce configuration using an IC inverter.

where C_s is the total effective stray capacitance. Typical values of C_s are from 2 to 7 pF.

Table 4.2 shows typical component values for operation of the oscillator in the 1- to 25-MHz range using an appropriate crystal at the given load capacitance.

The design of the 10-MHz oscillator is now discussed. The loop gain is first analyzed to check that the oscillation conditions are satisfied. To this end, the inverter is simulated using a typical MOSFET configuration and biased using ±5V. The 10-MHz crystal is represented with $R = 15\Omega$, $L = 10.1$ mH, $C = 25$ fF, and $C_o = 6$ pF. The stray capacitance across the crystal is assumed to be 5 pF. In Figure 4.24(a) the loop gain is calculated by representing the circuit as a continuous circuit. That is, the feedback loop is broken in Figure 4.23 and the crystal is loaded with the input impedance of the inverter. A current signal of 1A is placed at the input and the loop gain [v_o/v_i in Figure 4.24(a)] is calculated. The results of the simulation are shown in Figure 4.24(b). The gain at 10.022 MHz where the phase is 0° is 15.85 dB, which shows that with the loop closed the circuit should oscillate at 10.022 MHz. This loop-gain simulation can be used to analyze the effects of the loading capacitance and the crystal resistance on the loop gain. For example, the maximum values of the load capacitance and of R can be determined.

The simulation of the 10-MHz oscillator is shown in Figure 4.24(c). The resulting output signal oscillates at 10.02 MHz with an amplitude of approximately 5V.

Table 4.2 Component Values for Oscillation in the 5- to 25-MHz Range Using $R_f = 1$ MΩ and a Crystal at Its Fundamental Frequency

f (MHz)	$C_1 = C_2$	R_1	C_L
5	33 pF	5 kΩ	20 pF
10	22 pF	1 kΩ	16 pF
15	15 pF	500Ω	12 pF
25	10 pF	330Ω	10 pF

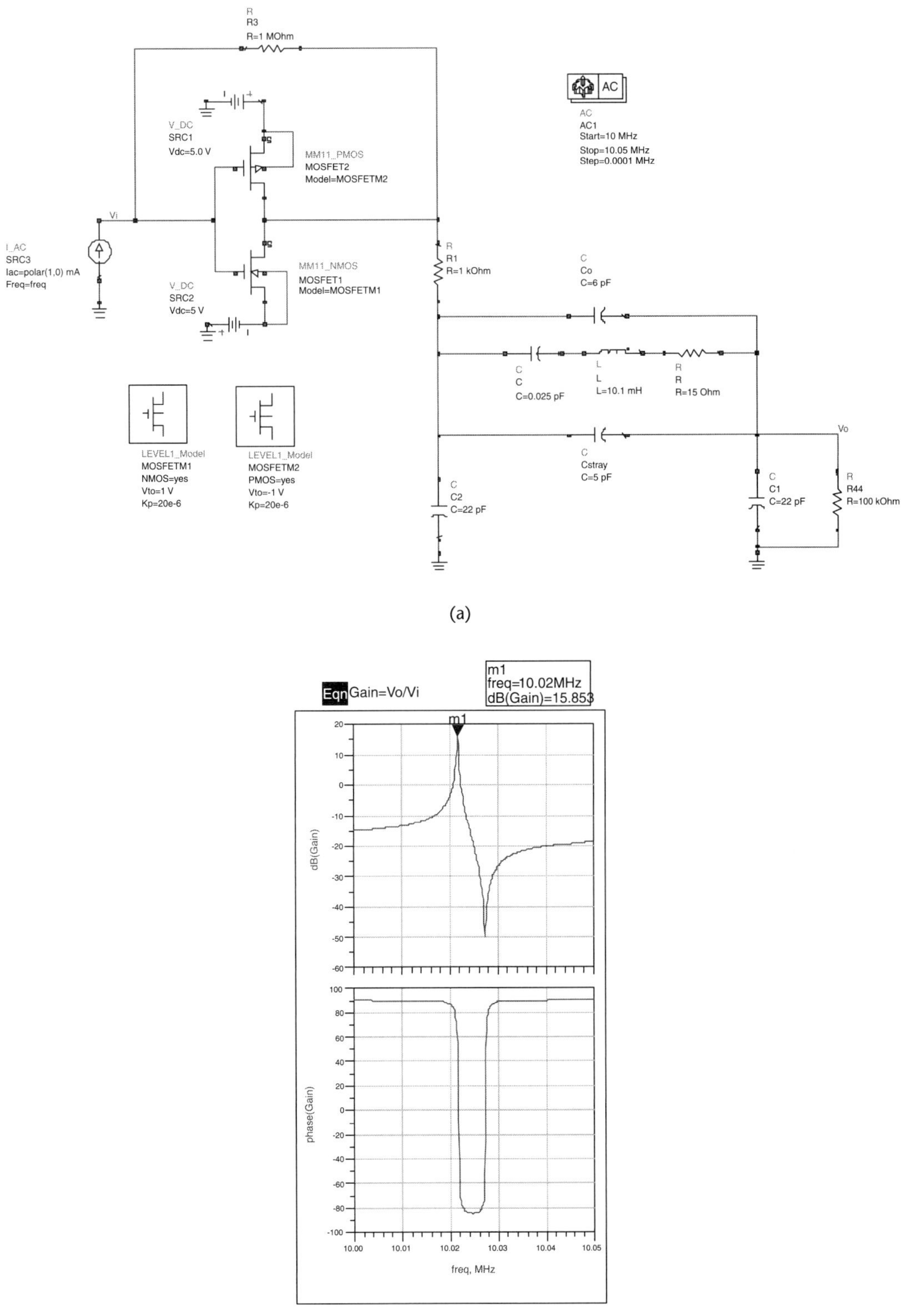

Figure 4.24 (a) Loop-gain simulation for the 10-MHz oscillator, (b) magnitude and phase of the loop gain, and (c) simulation of the 10-MHz oscillator.

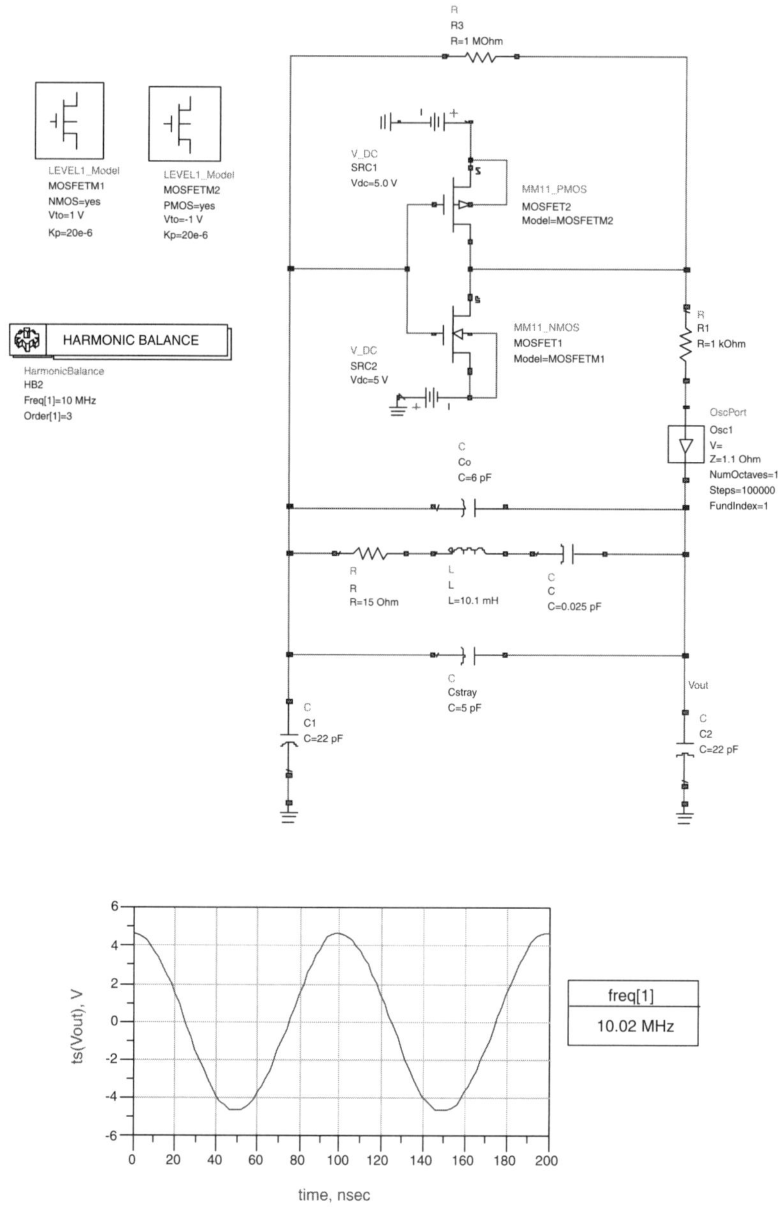

(c)

Figure 4.24 (Continued.)

A design for a 32.768-kHz oscillator (i.e., a watch oscillator) is shown in Figure 4.25. This oscillator uses a 4049 CMOS, and the crystal load capacitance is specified as 20 pF. The stray capacitance is estimated to be 7 pF.

Another design for a 32.768-kHz oscillator, using the Harris HA7210, is shown in Figure 4.26. This IC can be externally programmed to operate between 10 kHz and 10 MHz. In the 10- to 100-kHz range with the enable pin high and with a high in the FREQ1 and FREQ2 pins, the IC places internal 15-pF capacitors from

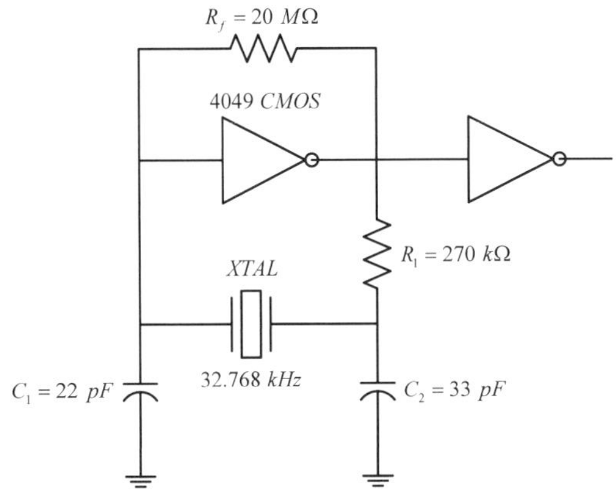

Figure 4.25 A Pierce configuration for a watch oscillator.

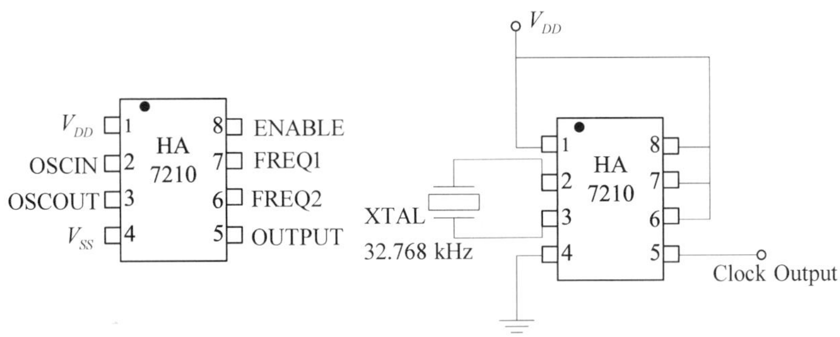

Figure 4.26 A 32.768-kHz Pierce oscillator using the HA7210.

pin 2 and pin 3 to ground (i.e., grounding pin 4). The two 15-pF capacitors make the loading capacitance of the crystal to be 7.5 pF.

In the frequency range above 100 kHz the oscillator circuit in Figure 4.27 can be used where C_3 is a trimming capacitor. For example, with a 1-MHz crystal at a load capacitance of 2.5 pF, the values of the components can be $C_1 = C_2 =$ 33 pF and $C_3 = 5$ pF to 20 pF trimmer.

Two Colpitts crystal oscillators are shown in Figure 4.28. The bias resistors R_1 and R_2 appear in parallel with the crystal and, therefore, reduce the circuit Q. The transistor's input impedance should not load C_1, and its output impedance should not load C_2.

Although the basic results in (4.32) to (4.41) apply to the Colpitts crystal oscillator, it is of interest to perform the following analysis of the circuit. Consider the ac model shown in Figure 4.28(c) where the crystal is represented by $R_e + jX_e$ The ac model assumes that the input and output impedances of the transistor do not load the resonator. Hence, Z_L is given by (4.32), and at resonance (4.33) to (4.36) are also satisfied. Hence,

$$Z_L(j\omega_o) \approx \frac{[X_2(\omega_o)]^2}{R_e} \tag{4.44}$$

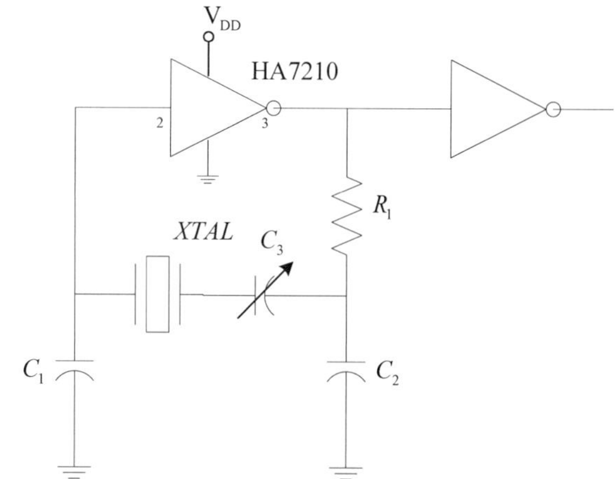

Figure 4.27 A Pierce oscillator using the HA7210.

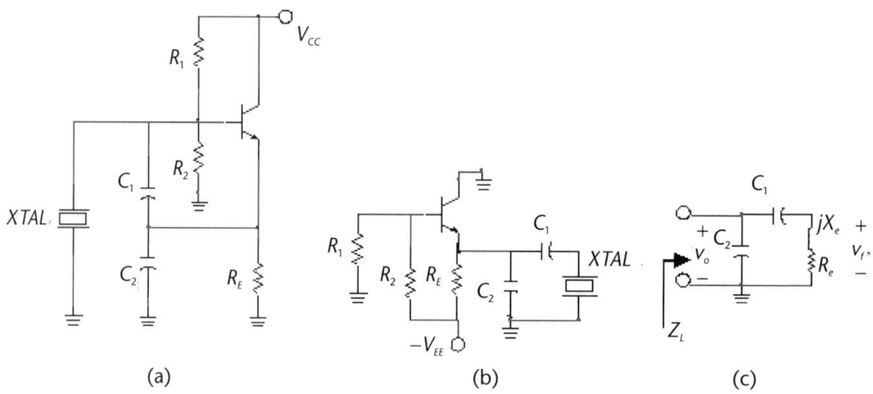

Figure 4.28 (a) A Colpitts crystal oscillator, (b) a Colpitts crystal oscillator using a negative supply voltage, and (c) the ac model.

The voltage v_f is given by

$$v_f = v_o \frac{R_e + jX_e}{R_e + j(X_1 + X_e)} \tag{4.45}$$

Using (4.33) in (4.45), we can write the voltage feedback factor as

$$\beta(j\omega_o) = \frac{v_f}{v_o} = \frac{R_e - j(X_1 + X_2)}{R_e - jX_2}$$

if $|X_2| \gg R_e$ then

$$\beta(j\omega_o) \approx \frac{X_1 + X_2}{X_2} = \frac{C_1 + C_2}{C_1}$$

The gain of the emitter follower is

$$A_v(j\omega) = \frac{v_o}{v_f} = \frac{g_m Z_L}{1 + g_m Z_L}$$

Hence, the required loop gain is

$$\beta(j\omega_o)A_v(j\omega_o) = \frac{X_1 + X_2}{X_2}\left(\frac{g_m Z_L}{1 + g_m Z_L}\right) > 1$$

Using (4.44), the gain condition is

$$\frac{g_m}{\omega_o^2 R_e C_1 C_2} \geq 1 \quad (4.46)$$

which agrees with (4.41). Observe that in this configuration the phase shift through the emitter follower is 0°, and also the phase shift through the feedback network is 0°. A trimmer capacitor can be placed in series with the crystal to adjust the frequency of oscillation. One way of pulling the frequency and coupling the oscillation signal is shown in Figure 4.29(a). Another way of coupling the signal from

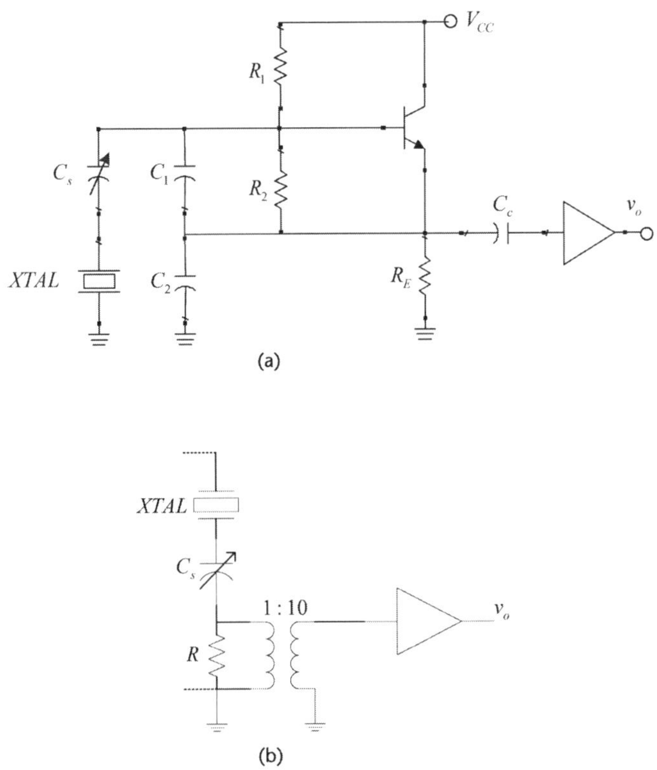

Figure 4.29 (a) A Colpitts crystal oscillator with a frequency pulling capacitor C_s and coupling circuit and (b) another coupling circuit.

the oscillator is shown in Figure 4.29(b). A small resistance in series with the crystal permits the extraction of the signal. The filtering produced by the crystal results in an output signal with very low distortion.

An FET Colpitts crystal oscillator is shown in Figure 4.30. This oscillator works better than the BJT version at the lower frequencies (down to 10 kHz) because of the high value of R_G. However, at the higher frequencies the low values of the FET's transconductance might present a problem.

Example 4.9

Design the Colpitts crystal oscillator in Figure 4.31(a) using the 10-MHz crystal described in Table 4.1. The crystal load capacitance is specified by the manufacturer to be 30 pF.

Solution

Let V_{DD} = 10V and the Q point be at V_{CE} = 5V and I_C = 2 mA. The resistor values for the desired Q point are shown in Figure 4.31(a). The capacitor C_1 is selected at 200 pF (or $X_{L_1} =- 79\Omega$), and C_2 = 270 pF. Then, the value of C_s is 40.6 pF for a total load capacitance of 30 pF. The trimming capacitor C_s is implemented with a variable capacitor (say, a 20- to 100-pF trimmer).

From (4.46), with $R_e \approx R = 5\Omega$, the required loop gain is satisfied if

$$g_m > \omega^2 C_1 C_2 R_e = (2\pi 10 \times 10^6)^2 (200 \times 10^{-12})(270 \times 10^{-12}) 5 = 1.1 \text{ mS}$$

The transistor's g_m value satisfies the above inequality since

$$g_m = \frac{I_C}{V_T} = \frac{2 \times 10^{-3}}{25 \times 10^{-3}} = 80 \text{ mS}$$

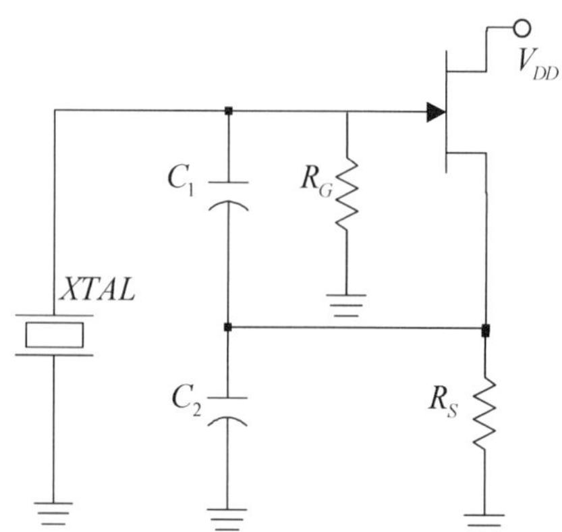

Figure 4.30 An FET Colpitts crystal oscillator.

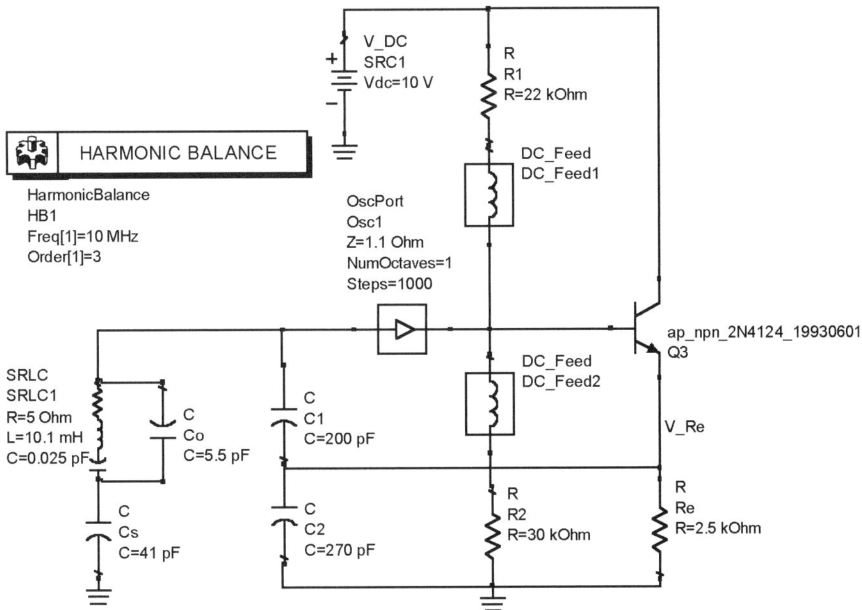

(a)

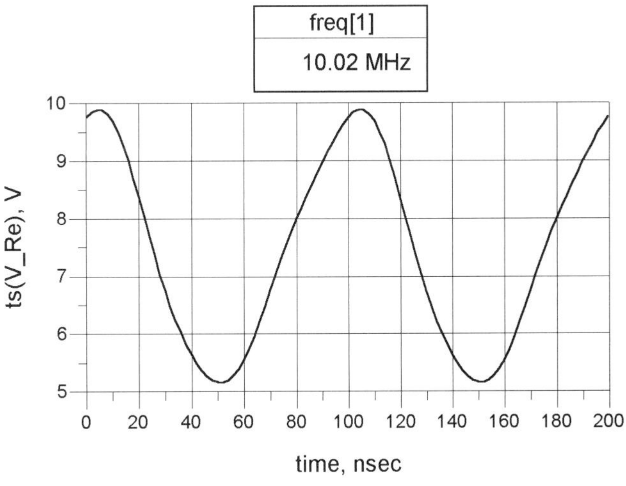

(b)

Figure 4.31 (a) Simulation of the Colpitts crystal oscillator in Example 4.9 and (b) the simulation data.

The simulation data is shown in Figure 4.31(b) with C_s set at 41 pF. The resulting Q point is at $V_{CE} = 5.03$V and $I_C = 1.98$ mA. The frequency of oscillation is at 10.02 MHz.

The following example illustrates the design of an oscillator with an overtone crystal. For example, when using a third-overtone crystal, care must be exercise that the oscillator does not also oscillate at the fundamental frequency of the crystal. Two methods that can be used to prevent the oscillation at the fundamental frequency are shown in Figure 4.32. In Figure 4.32(a) a trap (i.e., a parallel LC network) at the fundamental frequency is placed in the feedback path, and in Figure 4.32(b) the L in parallel with C_2 is resonant at $2f_o$; this produces a net inductive reactance at f_o and a net capacitive reactance at $3f_o$. Hence, only at $3f_o$ are X_1 and X_2 capacitive for the oscillation condition to be satisfied.

Example 4.10

Design a 30-MHz Colpitts crystal oscillator. A third-overtone crystal is available with a frequency stability of 20 ppm from –25°C to 50°C. The load capacitance is specified as 27 pF, and the supply voltage is a 9V battery.

Solution

The Colpitts crystal oscillator is shown in Figure 4.33(a). The bias point for the MPS5179 BJT was selected at V_{CE} = 5V and I_C = 500 μA. The Q point was designed with R_E = 8 kΩ, R_1 = 46 kΩ, and R_2 = 44 kΩ.

The load capacitance of 27 pF is obtained with C_1 = 43 pF and C_2 = 68 pF. Observe that $|X_{C_2}| \ll R_E$. Also, in order to provide for some tuning, a variable capacitor C_v is placed in parallel with the crystal, as shown in Figure 4.33(a). Using a 2- to 20-pF variable capacitor, it follows that the total capacitance seen by the crystal when C_v is at a mid-value of 10 pF is

$$C_T + C_v = 27 \times 10^{-12} + 10 \times 10^{-12} = 37 \text{ pF}$$

In order to bring the load capacitance seen by the crystal back to 27 pF when C_v = 10 pF, a series capacitor, as shown in Figure 4.33(a) is used. Its value is

$$\frac{(27 \times 10^{-12})C_x}{27 \times 10^{-12} + C_x} = 27 \times 10^{-12}$$

or C_x = 46 pF.

The value of g_m is 20 mS and satisfies (4.46).

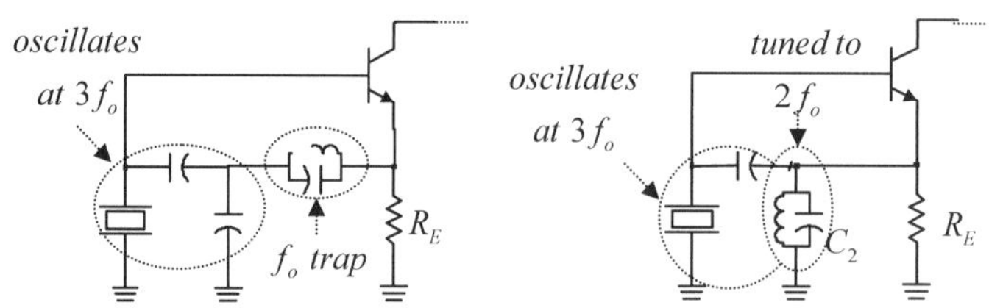

Figure 4.32 Two methods for oscillation at the third overtone of the crystal.

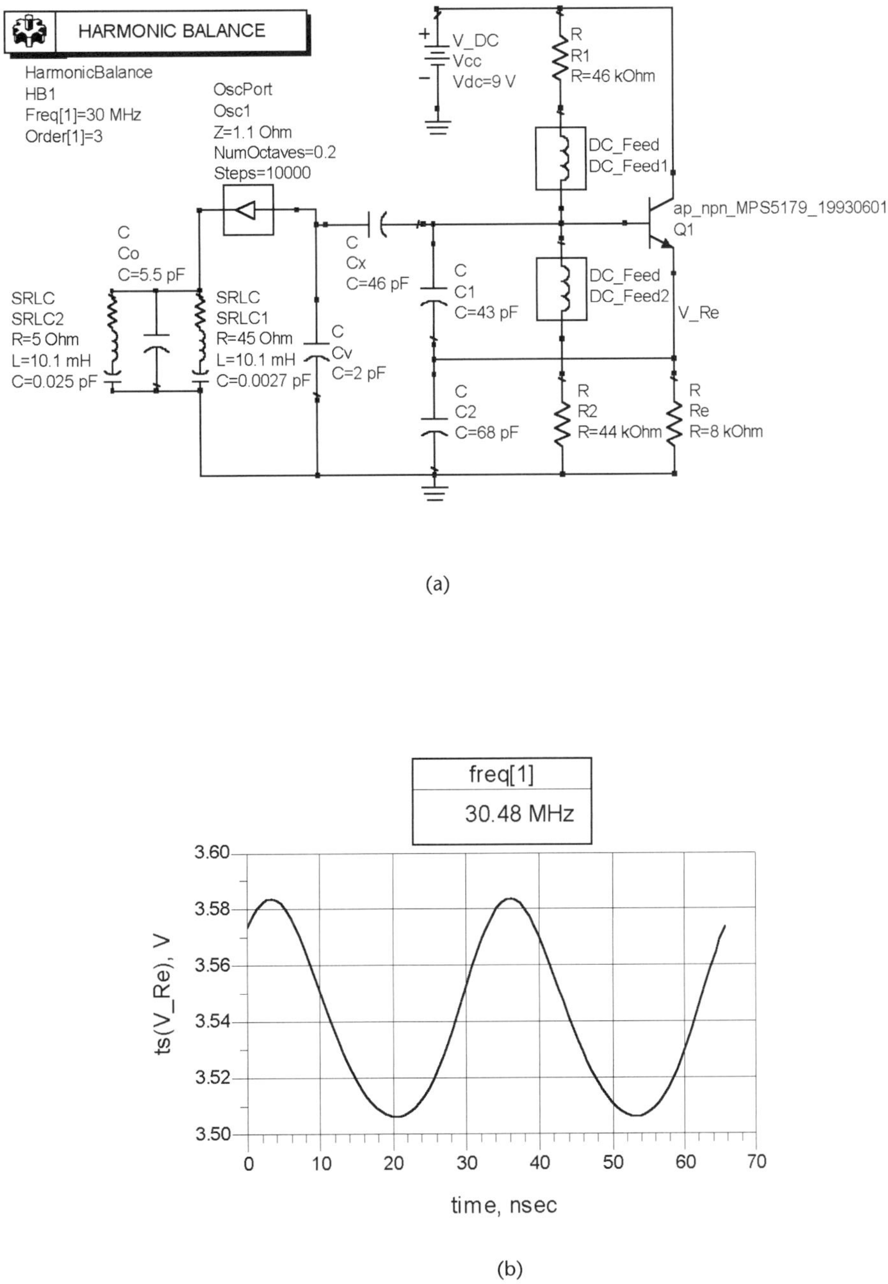

Figure 4.33 (a) The simulation of the 30-MHz Colpitts oscillator for Example 4.10 and (b) the oscillator waveform.

The ADS simulation data for the oscillator is shown in Figure 4.33(b). The Q point is at $V_{CE} = 5.47\text{V}$ and $I_C = 0.435$ mA. The frequency of oscillation is at the third overtone, or 30.32 MHz. No oscillation occurred at the fundamental.

An overtone oscillator that uses the 7SHU04 inverter (a very-high speed unbuffered inverter) is shown in Figure 4.34. The values listed are for a crystal oscillator

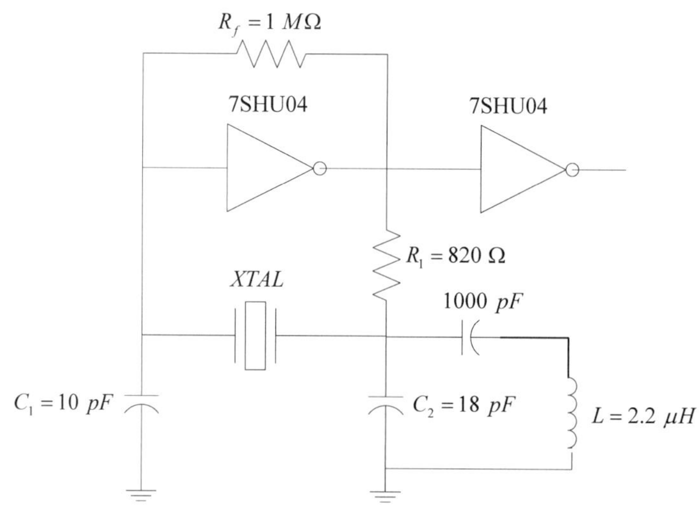

Figure 4.34 A 30-MHz oscillator using a third-overtone crystal.

at 30 MHz with a load capacitance of 10 pF. The 18-pF and 2.2-μH inductor are designed to be parallel resonant at 25 MHz. Hence, at 10 MHz the resonant circuit appears inductive, and at 30 MHz it appears capacitive. Hence, only at 30 MHz are the conditions for oscillations satisfied.

A Clapp crystal oscillator is shown in Figure 4.35(a). The ac model is similar to that in Figure 3.27(c, d) (with no R_L). If $R_E \parallel h_{ib} \approx h_{ib}$, $C_2 \gg C_{be}$ and $C_1 \gg C_{ce}$, the ac model in Figure 4.35(b) is obtained. The transistor's output capacitance appears across the crystal, thus limiting the frequency of operation.

From Figure 4.35(b) the open-loop voltage gain is

$$A_v = \frac{v_o}{v_f} = g_m Z_L \tag{4.47}$$

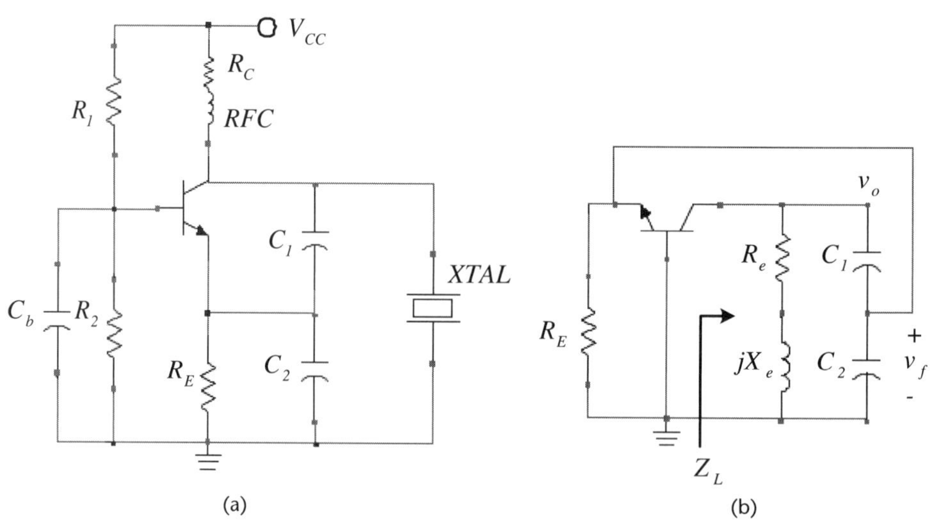

Figure 4.35 (a) A Clapp crystal oscillator and (b) its ac model.

where

$$Z_L = \frac{(R_e + jX_e)\left[jX_1 + \left(h_{ib} \parallel jX_2\right)\right]}{R_e + jX_e + jX_1 + \left(h_{ib} \parallel jX_2\right)} \tag{4.48}$$

$$= \frac{(R_e + jX_e)\left[j(X_1 + X_2) - \dfrac{X_1 X_2}{h_{ib}}\right]}{R_e - \dfrac{X_2 X_e}{h_{ib}} - \dfrac{X_1 X_2}{h_{ib}} + j(X_1 + X_2 + X_e) + j\dfrac{R_e X_2}{h_{ib}}}$$

If $R_e \ll X_e$, and

$$X_1 + X_2 \gg X_1 X_2 / h_{ib} \tag{4.49}$$

then with $X_1 + X_2 + X_e = 0$ it follows that (4.48) reduces to

$$Z_L = \frac{-X_e(X_1 + X_2)}{R_e - \dfrac{X_2(X_e + X_2)}{h_{ib}}} \tag{4.50}$$

$$= \frac{(X_1 + X_2)^2}{R_e + \dfrac{X_2^2}{h_{ib}}}$$

Substituting (4.50) into (4.32) gives

$$A_v = \frac{g_m(X_1 + X_2)^2}{R_e + g_m X_2^2} \tag{4.51}$$

Observe that if $g_m X_2^2 \gg R_e$, (4.51) reduces to

$$A_v = \left(\frac{X_1 + X_2}{X_2}\right)^2 = \left(\frac{C_2 + C_1}{C_1}\right)^2$$

which is identical to (3.63).

The feedback factor is

$$\beta = \frac{v_f}{v_o} = \frac{h_{ib} \parallel jX_2}{jX_1 + \left(h_{ib} \parallel jX_2\right)}$$

$$= \frac{X_2}{X_1 + X_2 + j\dfrac{X_1 X_2}{h_{ib}}}$$

Using (4.49), it follows that

$$\beta = \frac{X_2}{X_1 + X_2} = \frac{C_1}{C_1 + C_2} \tag{4.52}$$

From (4.51) and (4.52) the loop gain condition is

$$\beta A_v = \frac{X_2}{X_1 + X_2} \frac{g_m (X_1 + X_2)^2}{R_e + g_m X_2^2} > 1$$

which simplifies to

$$\frac{g_m}{\omega_o^2 R_e C_1 C_2} > 1 \tag{4.53}$$

The oscillator signal can be coupled to a high resistance load R_L connected to the collector or to a resistive load capacitively coupled to the emitter.

Example 4.11

Design the Clapp crystal oscillator shown in Figure 4.36(a) to oscillate at 10 MHz.

Solution

A 2N4124 BJT was selected and biased at $V_{CE} = 5V$ and $I_C = 1.5$ mA, as shown in Figure 4.36(a). This is a general purpose transistor that lists: $f_T = 300$ MHz and $h_{fe}(\text{min}) = 120$.

Assume that the 10-MHz crystal described in Table 4.1 is used with a specified load capacitance of 28 pF. This load capacitance can be obtained with $C_1 =$ 40 pF and $C_2 = 100$ pF $\left(X_{C_2} =- 158\Omega\right)$. The condition in (4.53) is satisfied since $g_m = 60$ mS.

The simulation data for the oscillator is shown in Figure 4.36(b), and the resulting output is shown in Figure 4.36(b). The SSB phase noise of the oscillator is shown in Figure 4.36(c).

4.5 The Grounded-Base Crystal Oscillator

The transistor's f_T value limits its frequency of operation, creating a problem when the values of the capacitors needed to satisfy the oscillation condition in the Pierce, Colpitts, and Clap crystal oscillators are comparable to the transistor's terminal capacitances. One way to solve this problem is to use a transistor with a higher f_T value or to use the crystal in a series resonance configuration. The grounded-base (GB) crystal oscillator, shown in Figure 4.37(a), uses the crystal as a series resonator.

The GB crystal oscillator works well at frequencies up to about 150 MHz. Usually, third-overtone crystals are used up to about 70 MHz, and fifth-overtone crystals up to about 150 MHz. The ac model is shown in Figure 4.37(b). Observe

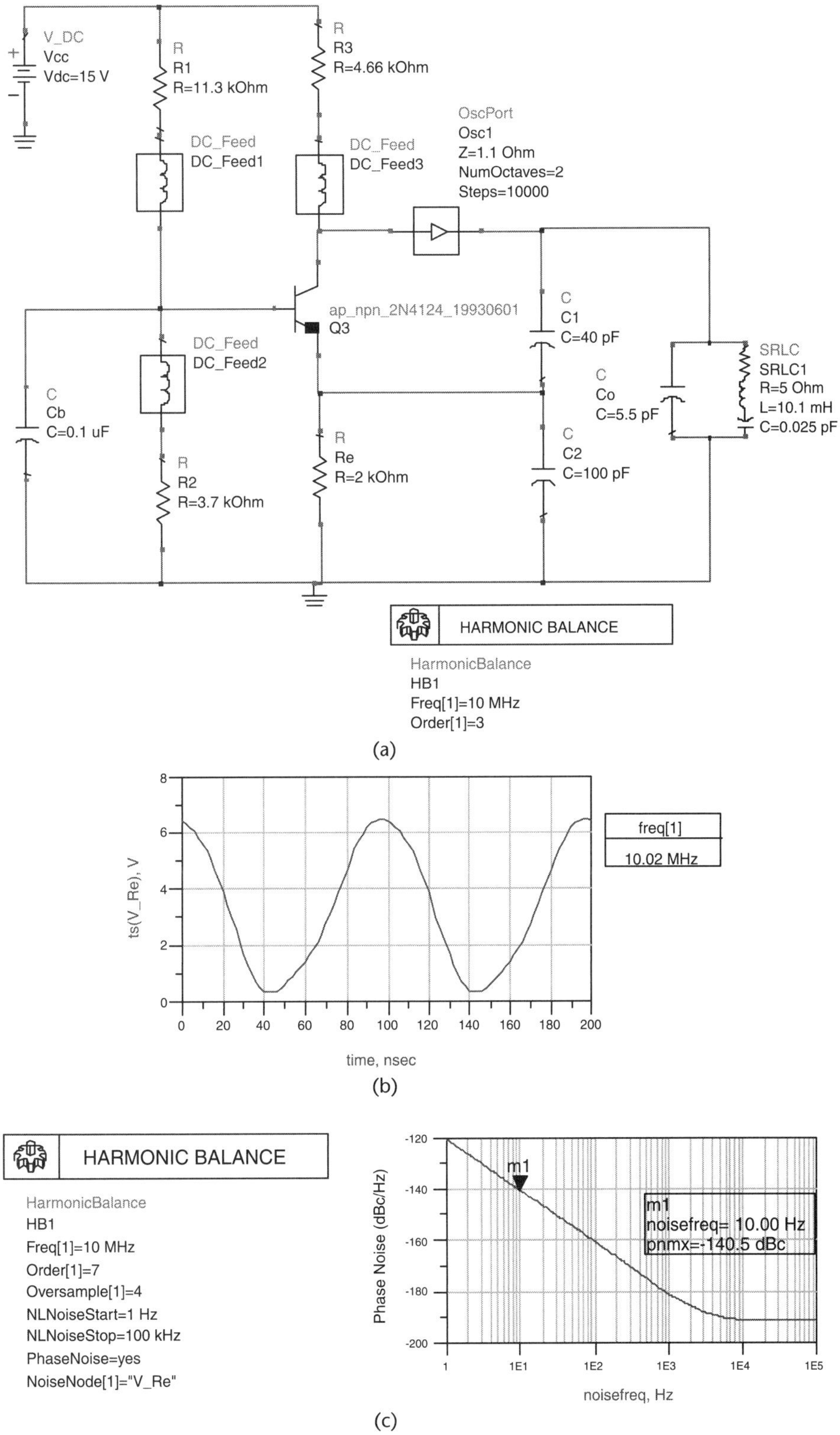

Figure 4.36 (a) Simulation of the Clapp oscillator in Example 4.11, (b) the simulation data, and (c) the SSB phase noise.

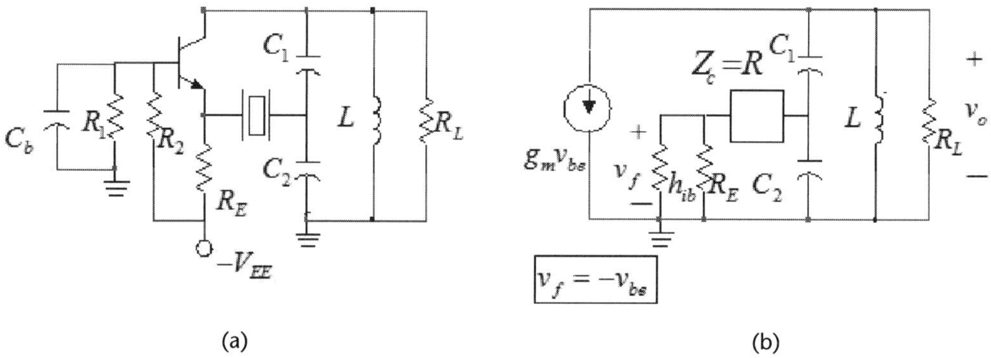

Figure 4.37 (a) The GB crystal oscillator and (b) its ac model.

the series connection of the crystal. At the frequency of oscillation the crystal operates in series resonance, the transistor provides a phase shift of 0° and the feedback-network phase shift is also 0°. The frequency of oscillation is

$$f_o = \frac{1}{2\pi\sqrt{LC_T}}$$

where

$$C_T = \frac{C_1 C_2}{C_1 + C_2} \tag{4.54}$$

A simple way to check if the crystal is operating in the series mode is to replace the crystal by a short circuit and observe if oscillations can occur. Obviously, with the crystal replaced by a short circuit in Figure 4.37(a) the circuit is recognized as a GB configuration (see Figure 3.27).

A more detailed analysis of the operation of the circuit follows. In Figure 4.37(b) a reasonable approximation is

$$h_{ib} \parallel R_E \approx h_{ib}$$

Also, at the crystal series resonance, the resistance $h_{ib} + Z_c \approx h_{ib} + R$ does not load C_2, where R is the crystal resistance. That is,

$$h_{ib} + R \gg \frac{1}{\omega C_2}$$

Then, the relation between v_o and v_f can be approximated by

$$v_f \approx \frac{v_o C_1}{C_1 + C_2}\left(\frac{h_{ib}}{h_{ib} + R}\right)$$

or

$$\beta(j\omega) = \frac{v_f}{v_o} = \frac{C_1}{C_1 + C_2}\left(\frac{h_{ib}}{h_{ib} + R}\right) \tag{4.55}$$

Equation (4.55) expresses the feedback factor. Observe that the phase shift from the feedback network, at the crystal series resonance, is 0°.

Since $h_{ib} + R$ transforms to an equivalent resistance across the inductor L given by $n^2(h_{ie} + R)$, where

$$n = \frac{C_1 + C_2}{C_1}$$

the load impedance seen by the transistor is

$$Z_L(j\omega) = R_p \left\| j\omega L \right\| \frac{1}{j\omega C_T}$$

where

$$R_p = R_L \parallel n^2(h_{ib} + R)$$

and C_T is given by (4.54).

The open-loop voltage gain is given by

$$A_v(j\omega) = \frac{v_o}{v_f} = g_m Z_L(j\omega) \tag{4.56}$$

Since the phase shift through the feedback network is 0°, the phase shift from (4.56) must also be zero at the frequency of oscillation. This occurs at the resonant frequency of the tank circuit. That is, at

$$\omega_o = \frac{1}{\sqrt{LC_T}}$$

which is designed to be the same as the series resonant frequency of the crystal. Hence, from (4.56)

$$A_v(j\omega_o) = g_m\left[R_L \parallel n^2(h_{ib} + R)\right]$$

At ω_o, the loop-gain condition is

$$\beta(j\omega_o)A_v(j\omega_o) = \frac{C_1}{C_1 + C_2}\left(\frac{h_{ib}}{h_{ib} + R_c}\right)g_m\left[R_L \parallel n^2(h_{ib} + R_c)\right] > 1$$

where $g_m h_{ib} \approx 1$.

In the case that $R_L \gg n^2(h_{ib} + R)$ the loop gain reduces to

$$\beta(j\omega)A_v(j\omega_o) > \frac{C_1 + C_2}{C_1}$$

In the tuned GB oscillator the loaded Q_L of the circuit was low due to the small resistive loading across the inductor L. In the GB crystal oscillator the Q_U of the crystal determines the Q_L. This can be deduced from the analysis of the loop gain of the circuit, namely,

$$\beta(j\omega)A_v(j\omega) = \frac{C_1}{(C_1 + C_2)} \frac{Z_L(j\omega)}{(h_{ib} + Z_c)} \tag{4.57}$$

The frequency stability can be evaluated from (4.57). The phase of (4.57) is

$$\phi(j\omega) = \phi_1(j\omega) - \phi_2(j\omega) \tag{4.58}$$

where $\phi_1(j\omega)$ is the phase of $Z_L(j\omega)$ and $\phi_2(j\omega)$ is the phase of $h_{ib} + Z_c$. Near the resonant frequency ω_o, $Z_L(j\omega)$ can be expressed in the form

$$Z_L(j\omega) \approx \frac{R_p}{1 + jQ_L\left(\dfrac{\omega}{\omega_o} - \dfrac{\omega_o}{\omega}\right)}$$

where Q_L is the loaded Q of the parallel tuned circuit. Also, at series resonance the crystal impedance can be approximated by the series RLC branch of the crystal. Hence, we can write

$$Z_c \approx R + j\omega_o L\left(\frac{\omega}{\omega_o} - \frac{\omega_o}{\omega}\right)$$

and it follows that

$$\frac{1}{h_{ib} + Z_c} = \frac{1}{(h_{ib} + R)\left[1 + jQ_x\left(\dfrac{\omega}{\omega_o} - \dfrac{\omega_o}{\omega}\right)\right]}$$

where

$$Q_x = \frac{\omega_o L}{R + h_{ib}} = \frac{Q_U}{1 + \dfrac{h_{ib}}{R}}$$

Q_U is the unloaded Q of the crystal.

From (2.15) and (4.58) it follows that

$$\frac{d\phi_1}{d\omega} = \frac{-2Q_L}{\omega_o}$$

and

$$\frac{d\phi_2}{d\omega} = \frac{-2Q_x}{\omega_o}$$

Hence, the frequency stability factor S_F [see (2.1)] is given by

$$S_F =- 2(Q_L - Q_x)$$

Since $Q_x \gg Q_L$, it follows that

$$S_F \approx 2Q_x = \frac{2Q_U}{1 + \dfrac{h_{ib}}{R}}$$

Observing that h_{ib} is of the same magnitude as R, the Q of the series resonant circuit in Figure 4.37 is controlled by Q_U, which is very high.

Example 4.12

Design the GB crystal oscillator shown in Figure 4.37 to oscillate at 45 MHz using a series-resonant third-overtone crystal.

Solution

From Table 4.1, using the third overtone of the 15-MHz crystal, the parameters are: $R = 45\Omega$, $C = 0.0031$ pF, $C_o = 4$ pF, and $L = 4.17$ mH. The tuned circuit is designed with $C_1 = 50$ pF, $C_2 = 270$ pF, and $L = 0.293$ μH.

The transistor selected was the BFR92 at a Q point of $V_{CE} = 6$V and $I_C = 1$ mA. This transistor lists $f_T = 5$ GHz.

The simulation is shown in Figure 4.38, as well as the resulting oscillation at 44.95 MHz.

4.6 The PI-Network Crystal Oscillator

The PI-network crystal oscillator is shown in Figure 4.39(a), and the ac model is shown in Figure 4.39(b). This oscillator uses the series-resonant properties of the crystal. It is used in the 1- to 100-kHz range. The output power is low with 0.5 to 1 mW being typical values. This oscillator is simple to design, but its frequency stability is not as good as the Pierce oscillator.

At the frequency of oscillation, which is the series-resonance frequency of the crystal, the crystal behaves like a short circuit. Hence, the transistor provides a phase shift of −180° and the PI network the other −180° at

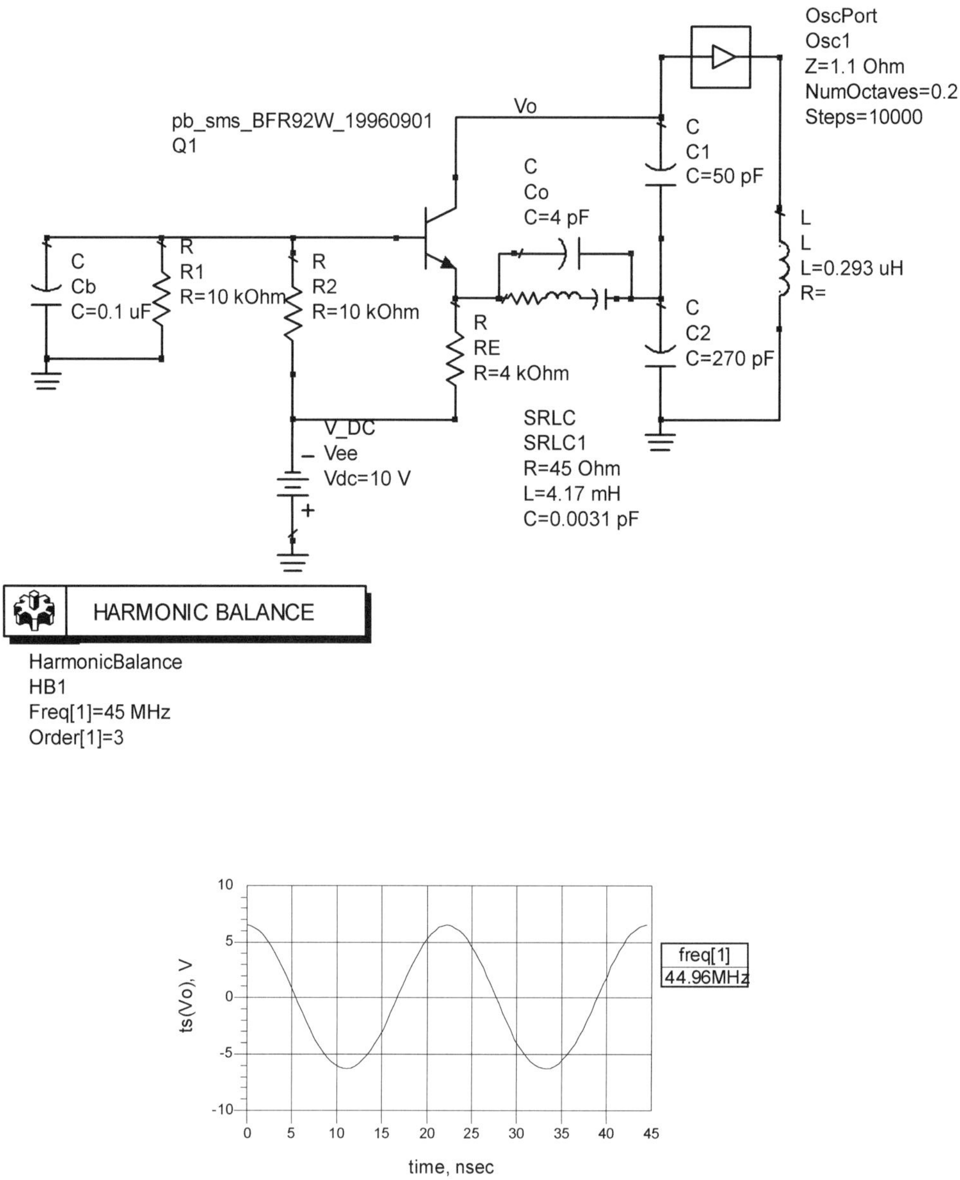

Figure 4.38 The GB crystal oscillator for Example 4.12.

$$f_o = \frac{1}{2\pi\sqrt{LC_T}}$$

where

$$C_T = \frac{C_1 C_2}{C_1 + C_2}$$

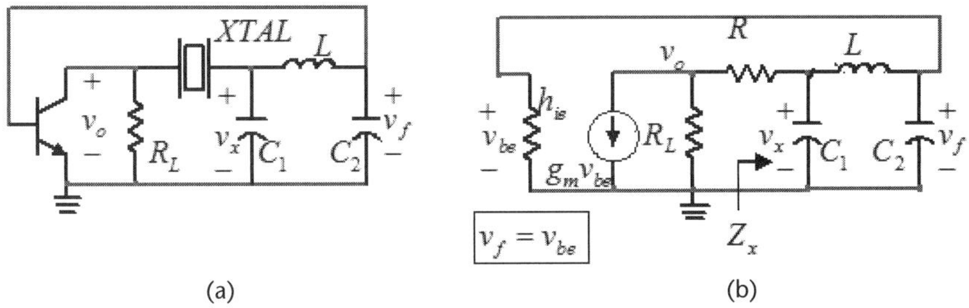

Figure 4.39 (a) The PI-network crystal oscillator and (b) the ac model.

The analysis of the oscillator in Figure 4.39(a) is a bit involved. Figure 4.39(b) shows the PI-network loaded by the input resistance of the transistor, which is h_{ie}. The impedance Z_x is given by

$$Z_x = \frac{jX_1\left(jX_L + \dfrac{jh_{ie}X_2}{h_{ie} + jX_2}\right)}{jX_1 + jX_L + \dfrac{jh_{ie}X_2}{h_{ie} + jX_2}} = \frac{-h_{ie}X_1(X_2 + X_L) - jX_1X_2X_L}{-X_2(X_1 + X_L) + jh_{ie}(X_1 + X_2 + X_L)} \tag{4.59}$$

At the frequency of oscillation

$$X_1(\omega_o) + X_2(\omega_o) + X_L(\omega_o) = 0 \tag{4.60}$$

Then, (4.59) simplifies to

$$Z_x(j\omega_o) = h_{ie}\left(\frac{X_1}{X_2}\right)^2 - jX_L\left(\frac{X_1}{X_2}\right) = h_{ie}\left(\frac{C_2}{C_1}\right)^2 - jX_L\left(\frac{C_2}{C_1}\right)$$

Since

$$h_{ie}\left(\frac{C_2}{C_1}\right) \gg |X_L| = \omega L \tag{4.61}$$

it follows that the real part of Z_x is much greater than the imaginary part, or

$$Z_x(j\omega_o) \approx R_x = h_{ie}\left(\frac{C_2}{C_1}\right)^2$$

The voltage v_f is given by

$$v_f = v_x \frac{jh_{ie}X_2}{jh_{ie}(X_2 + X_L) - X_2X_L} \tag{4.62}$$

Using (4.60), we can express (4.62) in the form

$$\frac{v_f}{v_x} = \frac{-h_{ie}}{h_{ie}\left(\frac{C_2}{C_1}\right) + j\omega L} \tag{4.63}$$

Using (4.61), (4.63) reduces to

$$\frac{v_f}{v_x} \approx -\left(\frac{C_1}{C_2}\right) \tag{4.64}$$

Also, at ω_o

$$v_x = v_o\left(\frac{R_x}{R_x + R}\right) \tag{4.65}$$

where R is the crystal resistance. Therefore, from (4.64) and (4.65) the feedback factor is

$$\beta(j\omega_o) = \frac{v_f}{v_o} =- \frac{C_1}{C_2}\left(\frac{R_x}{R_x + R}\right)$$

The open-loop voltage gain of the transistor is given by

$$A_v(j\omega_o) = \frac{v_o}{v_f} =- g_m\left[R_x \parallel (R_{IN} + R)\right]$$

and the loop gain condition is

$$\beta(j\omega_o)A_v(j\omega_o) = \frac{C_1}{C_2}\left(\frac{R_x}{R_x + R}\right)g_m\left[R_L \parallel (R_x + R)\right] > 1$$

4.7 Voltage-Controlled Crystal Oscillators

Voltage-controlled crystal oscillators find applications in many circuits, just like those using a voltage-controlled tuned oscillator. The main difference is that with a crystal the selectivity is high and the phase noise is low. Also, the range of frequencies that can be controlled with a single crystal is small due to the limited pulling range. For this reason some VCXO designs commonly use several crystals to cover different frequency ranges.

The parallel-resonant frequency of crystals used in the parallel mode can be varied using a capacitor in parallel with the crystal. For a series mode crystal a series capacitor can be used to vary the series resonant frequency. A varactor diode can be used to implement and vary the series capacitance, as shown in Figure 4.40.

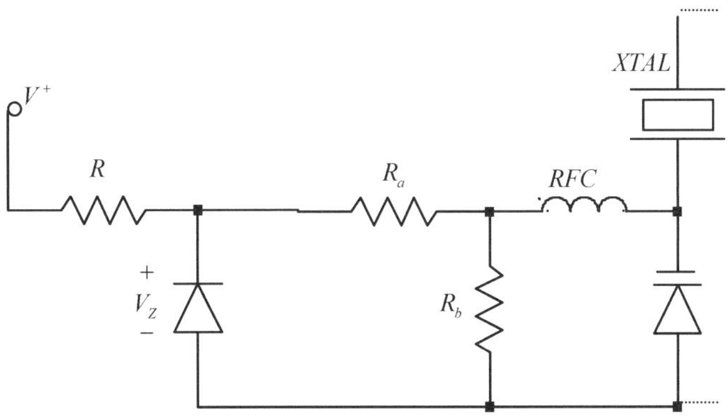

Figure 4.40 A VCXO configuration using a varactor diode as the voltage control device.

4.8 Ceramic-Resonator Oscillators

Ceramic resonators (CRs) make use of the mechanical resonances that occur in piezoelectric ceramics (usually PZT–lead zirconium titanate). As a resonator, the ceramic resonators' performance is better than those of the *LC* tuned circuits, and less than that of a crystal resonator. CRs are smaller in size and less expensive than crystals; also, they are half the size of a comparable crystal. However, crystals have typical temperature coefficients of 10^{-6}/°C, and CRs have 10^{-5}/°C. Figure 4.41 shows the typical shapes of two and three lead CRs. Murata has a good application manual for their CRs [1]. Those with three leads have an internal built-in load capacitor. CRs are used in many digital-clock applications, just like crystal oscillators.

The symbol and equivalent circuit of the CR (shown in Figure 4.42) are identical to those of the crystal. The impedance characteristics are similar to the one shown in Figure 4.3 [or Figure 4.2(b)] for the fundamental vibration. Between the series (f_s) and parallel (f_a) resonance of the CR, the resonator is inductive. In this frequency range the equivalent circuit is shown in Figure 4.42(c), where R_e and L_e are the equivalent resistance and inductance for $f_s < f < f_a$.

The series resonant frequencies is

$$f_s = \frac{1}{2\pi\sqrt{LC}}$$

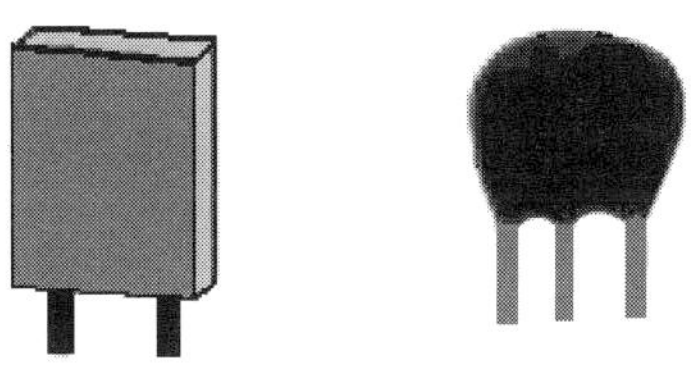

Figure 4.41 Typical shapes of CRs.

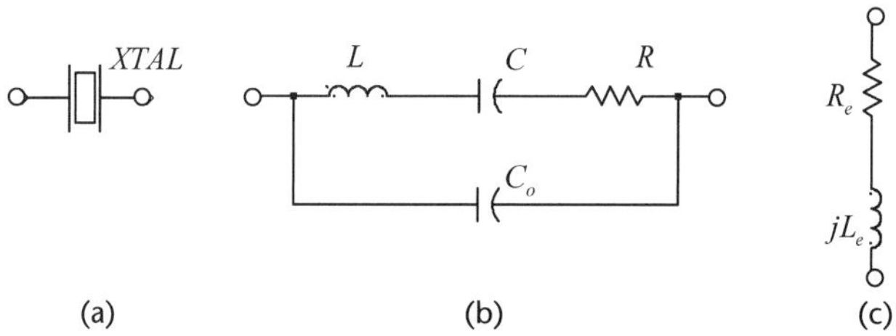

Figure 4.42 (a) CR symbol, (b) equivalent circuit, and (c) equivalent circuit for $f_s < f < f_a$.

and the parallel-resonant frequency is

$$f_a = f_s\sqrt{1 + \frac{C}{C_o}}$$

Since the operation is similar to that of a crystal, it can be used in the same oscillator circuit configurations used with crystals. Typical element values for some CRs are listed in Table 4.3. These values are, of course, for the fundamental mode. The resulting Q_U values are lower than those of the crystal. Higher harmonic modes exist due to the mechanical nature of the resonances.

A common way to construct a square-wave oscillator using a CR is shown in Figure 4.43. The amplifier is a digital inverter gate; a second inverter is used as a buffer stage. The feedback resistor R_f provides negative feedback to place the inverter in the linear amplification region. Usually, a gain of about 10 is obtained from a CMOS inverter. An R_f value of 1 MΩ is typical. In addition, the oscillator might require a resistor R_d. The R_dC_2 combination is used to add some additional phase shift and to reduce the gain at high frequencies, thus preventing undesirable high-frequency oscillations. The resistor R_d is sometimes referred to as a damping resistor.

The frequency of oscillator of the circuit in Figure 4.43 is given by

$$f_o = f_s\sqrt{1 + \frac{C}{C_o + C_L}}$$

where

$$C_L = \frac{C_1 C_2}{C_1 + C_2}$$

Table 4.3 Typical Values of the Components in the Equivalent Circuit of Some Ceralock CRs (*From:* [1])

f_o (MHz)	L (μH)	C (pF)	R (Ω)	C_o (pF)	Q_U
0.455	$7.68(10^3)$	16.7	10.1	272.8	2,136
2.0	$1.71(10^3)$	4.0	43.9	20.8	475
4.0	$0.46(10^3)$	3.8	9.0	19.8	1,220
8.0	$0.13(10^3)$	3.5	8.0	19.9	775

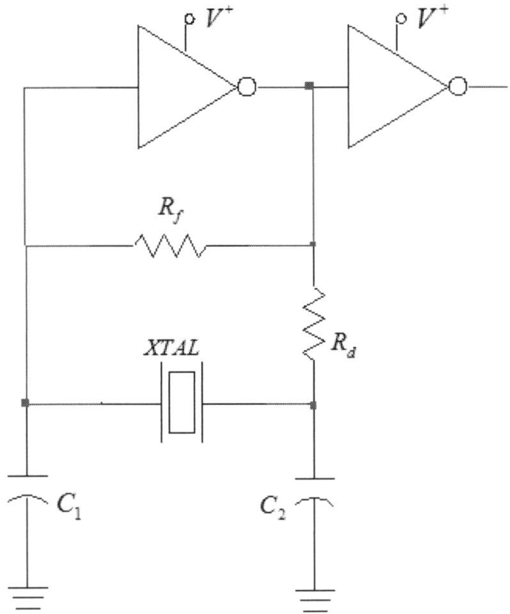

Figure 4.43 A CR square-wave oscillator.

The design of square-wave oscillators in the frequency range of 375 kHz to 13 MHz is illustrated in Figure 4.44. The values listed in Table 4.4 are typical for this oscillator. The RCA (HARRIS) CD4069UBE is a C-MOS that is commonly used as the amplifier. Some C-MOS inverters ICs can have too much gain, which can cause RC oscillations and ringing.

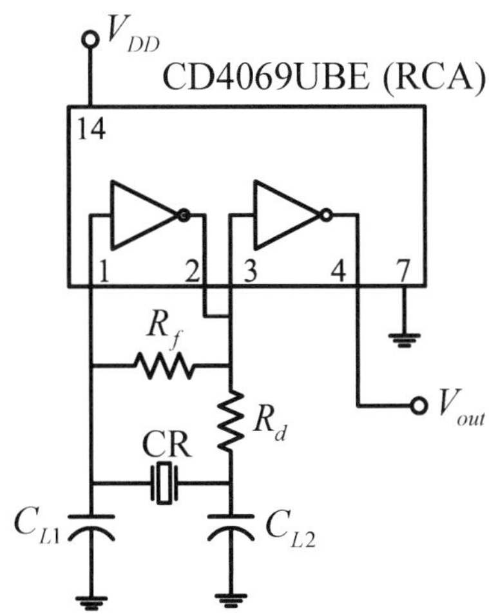

Figure 4.44 A C-MOS square-wave oscillator for use with CRs.

Table 4.4 Typical Values for the Oscillator in Figure 4.44

Frequency (kHz)	V_{DD} (V)	C_{L1} (pF)	C_{L2} (pF)	R_f (MΩ)	R_d (kΩ)
375–429	5	120	470	1	0
429–699	5	100	100	1	0
700–1,250	5	100	100	1	5.6
2,000–6,000	5	30	30	1	0
6,000–13,000	12	30	30	1	0

An example of a Colpitts CR oscillator at 455 kHz is shown in Figure 4.45. This oscillator uses the Murata CSBLA455KC8-B0 ceramic resonator. This CR has a fundamental frequency of oscillator at 455 kHz.

4.9 SAW Oscillators

Surface acoustic wave (SAW) resonators are available as one-port and two-port resonators. The structure of the one-port resonator chip is shown in Figure 4.46. It consists of an interdigitated transducer (IDT) and two grating reflectors. The structure is fabricated on a quartz crystal substrate using a photolithographic process. The chip is usually encapsulated in a ceramic package, and the resulting component is small in size. The SAW resonators use the mechanical vibration of

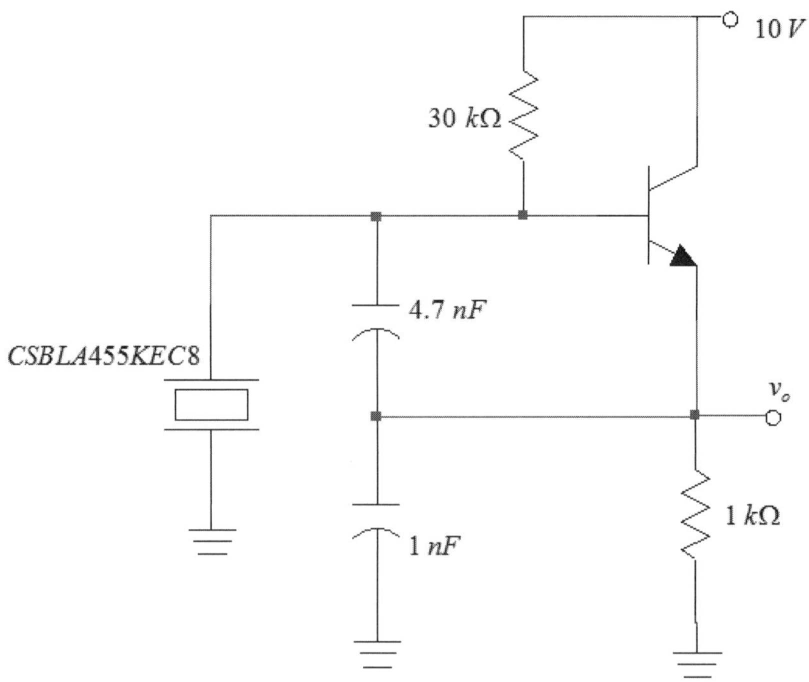

Figure 4.45 A 455-kHz CR oscillator.

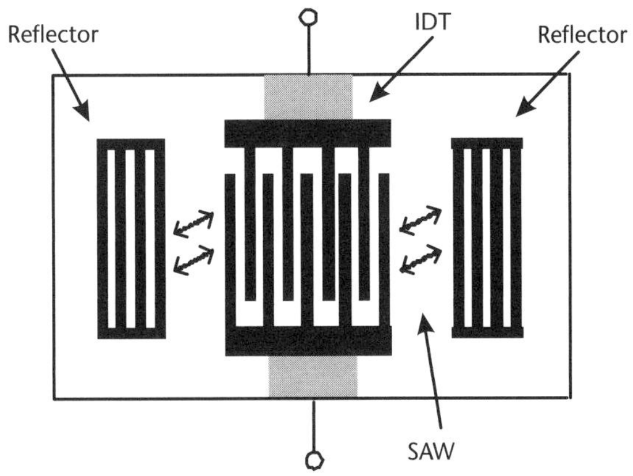

Figure 4.46 Structure of a one-port resonator. (*After:* The Murata SAW Resonator Manual [2].)

the piezoelectric material (quartz) to generate the surface acoustic wave between the reflectors.

SAW resonators can be used in oscillators just like quartz crystals or ceramic resonators. In fact, they are used in many HF and RF applications where crystals do not work at their fundamental frequency, or when better performance than that obtained from a ceramic oscillator is needed. The fundamental frequency of operation of SAW resonators is from 50 MHz to 1 GHz.

The equivalent circuit for a one-port SAW resonator is shown in Figure 4.47, and typical values for the electrical parameters are shown in Table 4.5. The model is identical to that of a quart crystal or ceramic resonator. Hence, its impedance characteristic is similar to the one shown in Figure 4.2(a). The SAW resonator exhibits a series resonance (f_s) and a parallel resonance (f_a). The oscillator configurations used with quartz crystals or ceramic resonators are also used for SAW resonators.

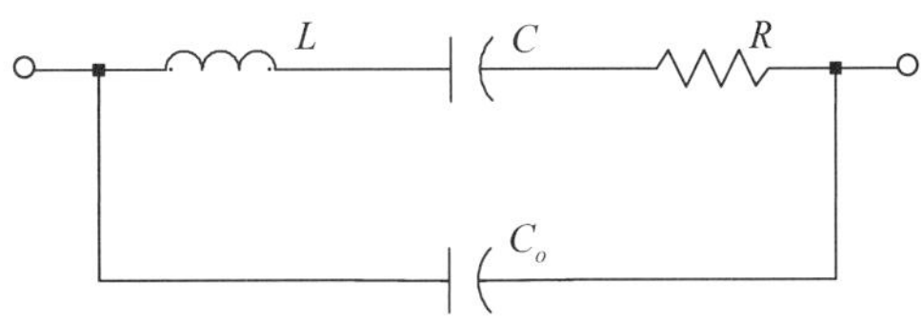

Figure 4.47 SAW resonator equivalent circuit.

Table 4.5 Typical Element Values for Some Murata SAW Resonators (*From:* [2])

Part Number	f_o (MHz)	L (μH)	C (pF)	R (Ω)	C_o (pF)
SARCC304M30BX	304.30	164.495	0.001663	22.0	2.37
SARCC315M00BX	315.00	159.331	0.001602	22.0	2.25
SARCC423M22BX	423.22	110.088	0.001284	22.2	2.00
SARCC433M30BX	433.87	92.747	0.001451	20.2	2.00
SARCC433M92BX	433.92	96.529	0.001394	22.1	2.11
SARCC434M15BX	434.15	95.288	0.001410	20.0	1.97

A Colpitts SAW oscillator configuration is shown in Figure 4.48. The SAW resonator behaves inductively in the region $f_s < f < f_a$. The BJT must have an f_T value of a few gigahertz.

Example 4.13

Design the SAW oscillator in Figure 4.48 to oscillate at 315 MHz using the Murata SARCC31500BLO.

Solution

The Murata SARCC31500BLO lists the following parameter values: $L = 159.33$ μH, $C = 0.001602$ pF, $R = 22\Omega$ and $C_o = 2.25$ pF. The transistor selected is the surface mount NEC 2SC4228 BJT, which has $f_T = 9$ GHz. The Q point selected is at $V_{CE} = 2.9$V and $I_C = 7$ mA, using $V_{CC} = 4$V. Good beta stability is obtained by designing the bias circuit with $R_E = 160\Omega$, $R_1 = 2$ kΩ and $R_2 = 2$ kΩ. Letting $C_1 = 12$ pF and $C_2 = 8$ pF, it follows that $C_T = 4.8$ pF. This SAW resonator has a series resonance at

$$f_s = \frac{1}{2\pi\sqrt{LC}} = \frac{1}{2\pi\sqrt{159.33 \times 10^{-6}(0.001602 \times 10^{-12})}} = 315.021 \text{ MHz}$$

and the parallel-resonant frequency occurs at

$$f_a = f_s\sqrt{1 + \frac{C}{C_o + C_T}} = 315 \times 10^6\sqrt{1 + \frac{0.001602 \times 10^{-12}}{2.25 \times 10^{-12} + 4.8 \times 10^{-12}}} = 315.055 \text{ MHz}$$

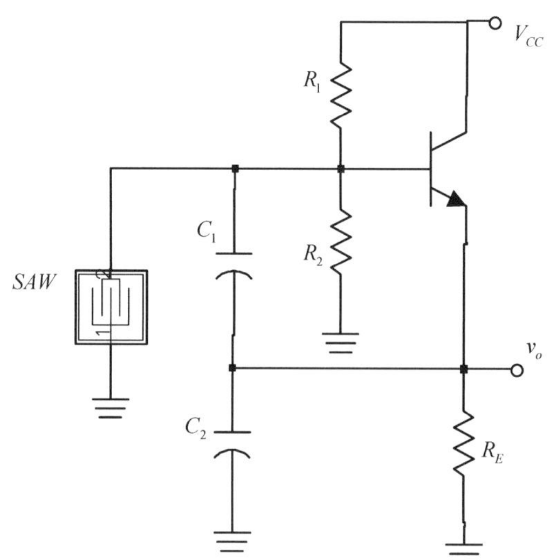

Figure 4.48 A Colpitts SAW oscillator.

The simulation is shown in Figure 4.49(a). The resulting Q point is at V_{CE} = 2.7V and I_C = 8 mA, and the frequency of oscillation is at 315.1 MHz.

The phase noise of the oscillator is shown in Figure 4.49(b). The harmonic balance controller used with the oscillator in Figure 4.49(a) for the calculation of the SSB phase noise is shown. The marker shows that $\mathscr{L}(f_m)$ =– 105.1 dBc/Hz at an offset of 10 kHz from the carrier.

In the Colpitts configuration shown in Figure 4.48 at frequencies above 500 MHz the parasitic capacitances begin to affect the design. For this reason, a preferred configuration at higher frequencies is the Pierce configuration, since the

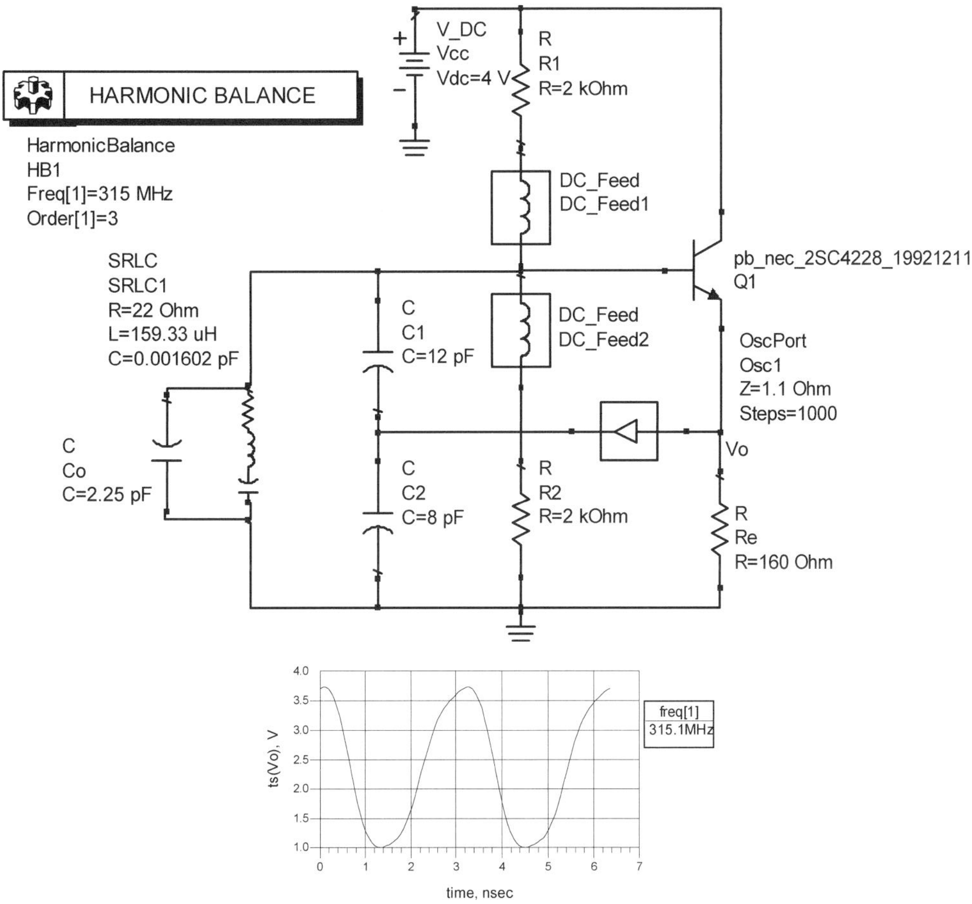

(a)

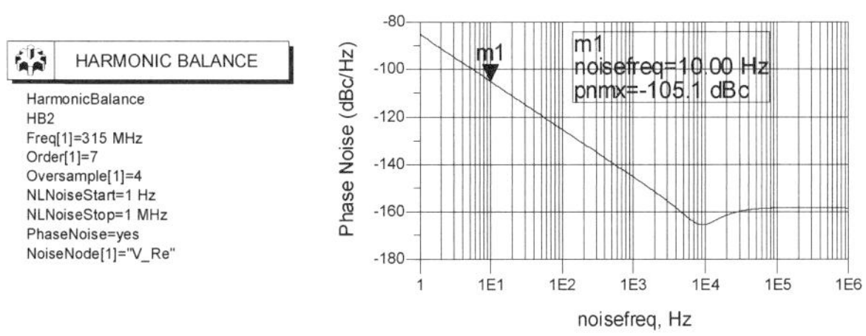

(b)

Figure 4.49 (a) ADS simulation for the oscillator in Example 4.13 and (b) the SSB phase noise.

parasitic effects on the resonator are minimized. The Pierce oscillator requires the use of a two-port SAW resonator in the feedback path to attain the required 180° phase shift.

A two-port SAW resonator is shown in Figure 4.50(a) and its equivalent circuit model in Figure 4.50(b). The coupling from port 1 to port 2 results in a phase shift of 180°. Internal capacitances from ports 1 and 2 to ground can be included in the model. However, these are usually small in comparison with the matching circuit capacitances.

Figure 4.51 illustrates a Pierce SAW oscillator using a two-port SAW resonator. The matching networks might be necessary to properly match the SAW resonator. The collector matching network is used to transform the high collector resistance to a lower resistance, and the base matching network is used to transform the match the base resistance to the output resistance of the SAW resonator. The total phase shift in the feedback path is due to the phase shift from the matching networks plus the phase shift of the SAW resonator.

SAW oscillators are a good choice for remote keyless entry (RKE) transmitters. These transmitters find extensive applications in the automobile industry due to their simple design, low current consumption, long battery life, low phase noise characteristics, and resistance to shock and vibration.

The following example illustrates a Colpitts SAW oscillator at 315 MHz, which is one of the frequencies used in the United States for RFK transmitters.

A 315-MHz SAW resonator from RF Monolithics, Inc (i.e., the RO2073E) is illustrated in Figure 4.52, as well as a bottom view showing the connections, and typical values for the equivalent model. This is a 3 × 3 mm device. Some nominal dimensions are $E = 3$ mm and $F = 1.6$ mm. Terminals 2 and 5 are interchangeable and NC means no connection. The NC pads are used for mechanical support of the SAW resonator. The manufacturer recommends grounding the support NC pads in order to reduce the parasitic capacitance.

Two oscillator configurations, adapted from the RF Monolithics application notes, are illustrated in Figure 4.53. These configurations are recognized as a GB

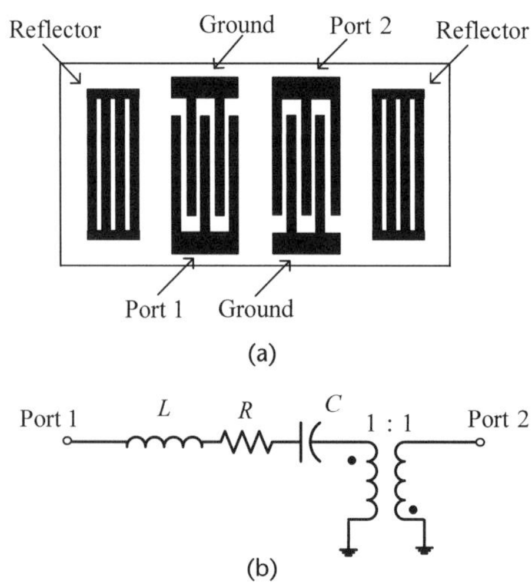

Figure 4.50 (a) A two-port SAW resonator and (b) equivalent model.

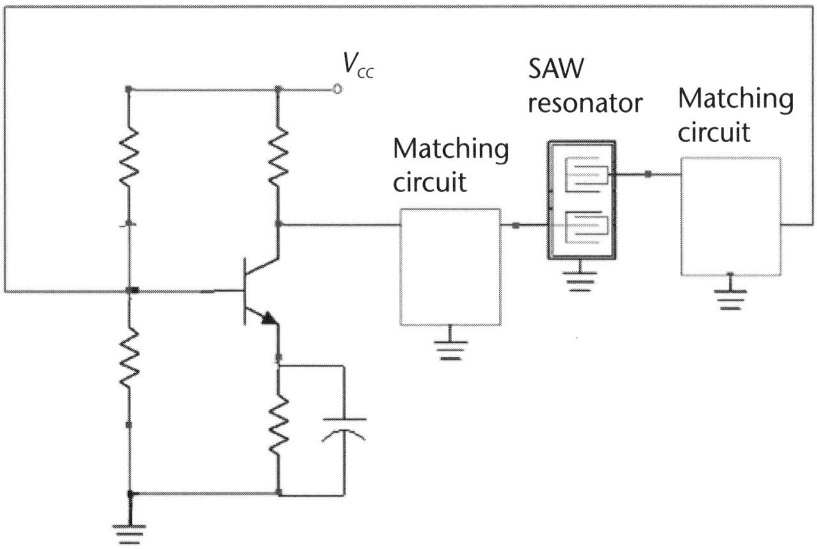

Figure 4.51 A Pierce oscillator using a two-port SAW resonator.

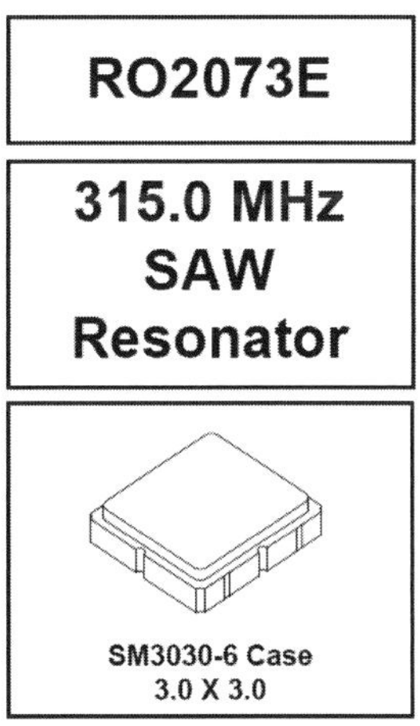

(a)

Pin	Connection
1	NC
2	Terminal
3	NC
4	NC
5	Terminal
6	NC

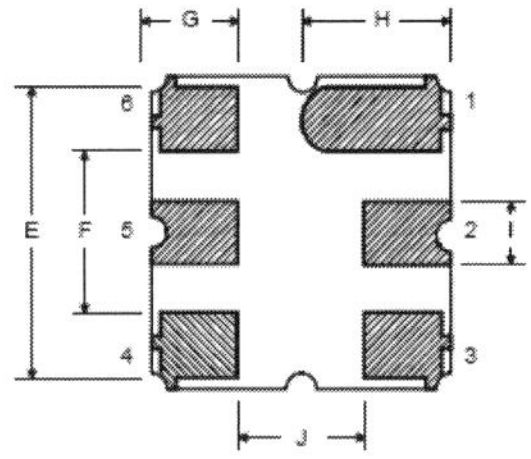

(b)

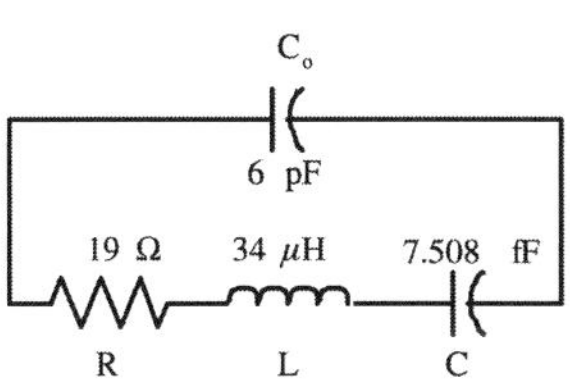

(c)

Figure 4.52 (a) The RO2073E SAW resonator from RF Monolithics, (b) the electrical connections, and (c) equivalent model of the resonator. (*From:* [3]. Reproduced with permission of RF Monolithics, Inc.)

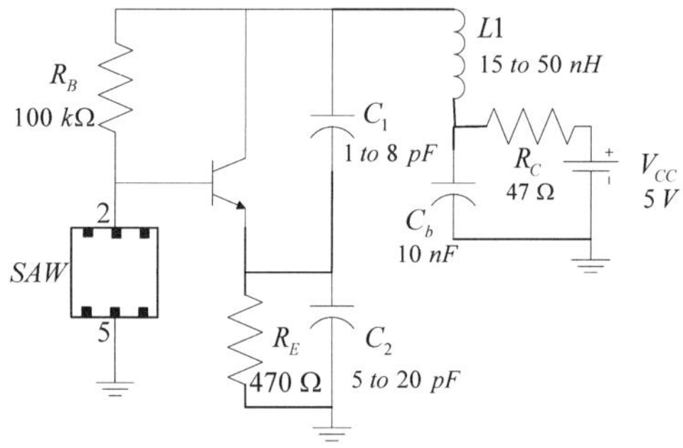

(a)

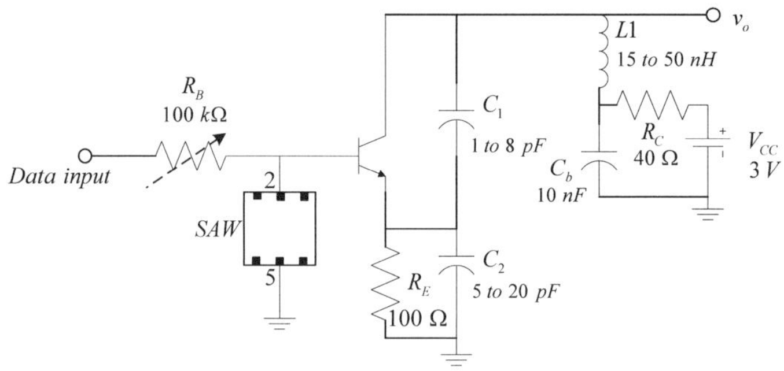

(b)

Figure 4.53 (a) A SAW oscillator configuration for use in the 300- to 500-MHz range and (b) a low-voltage design for RFK applications.

configuration with the SAW resonator acting as a high-*Q* bandpass filter between base and ground. The inductor $L1$ can represent the inductance of a transmitter antenna. Typical values for the oscillator components used in the 300- to 500-MHz range are shown.

Figure 4.53(b) shows a design for RKE applications using a 3V battery. The data input represents the modulation input signal. The C_b capacitors are RF bypass capacitors and $L1$ represents the transmitter antenna.

Another popular band for remote control applications is the 418-MHz band. The RO2103 SAW resonator from Monolithics operates at this frequency. This SAW resonator is constructed in a TO39-3 case. A design using this resonator follows.

Example 4.14

Design a 418-MHz SAW oscillator.

Solution

The oscillator configuration is shown in Figure 4.54. The SAW resonator used is the Monolithics R02103. Typical values for the equivalent model of this SAR

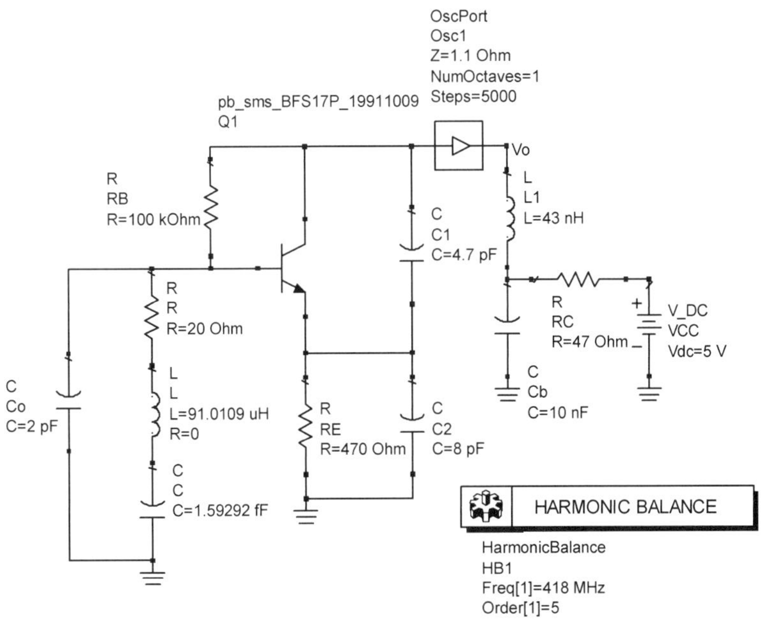

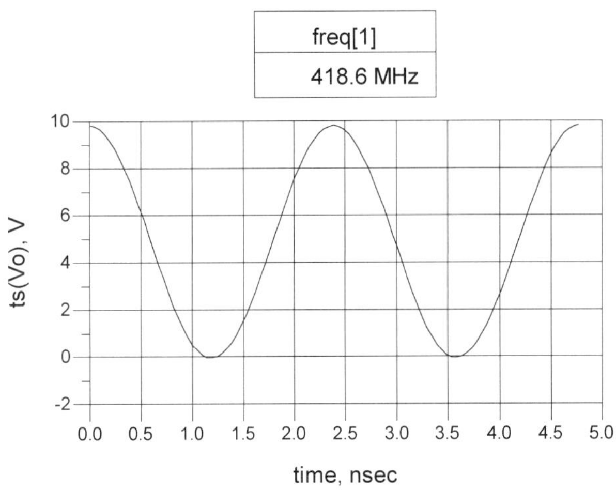

Figure 4.54 Design of the 418-MHz SAW oscillator for Example 4.14.

resonator are: $R = 20\Omega$, $L = 91.0109$ mH, $C = 1.59292$ fF, and $C_o = 2$ pF. The transistor selected is the BFS17, which has an f_T value of 2.5 GHz. The resistor values in Figure 4.54 set the Q point at $V_{CE} = 3.75$V and $I_C = 2.4$ mA.

The design method for C_1, C_2, and $L1$ is the same as the method used for the GB tuned oscillator. If $L1 = 43$ nH, it follows that letting $C_1 = 5$ pF and $C_2 = 10$ pF produces a resonant frequency of 418 MHz, with proper values for the associated reactances.

The simulation data is shown in Figure 4.54, where the frequency of oscillation is 418.6 MHz with $C_1 = 4.7$ pF and $C_2 = 8$ pF.

References

[1] Murata Manufacturing Co., Ltd, "Ceramic Resonators (CERALOCK®) Applications Manual," Catalog P17E-14, Kyoto, Japan, 2004.

[2] Murata Manufacturing Co., Ltd, "SAW Resonators Applications Manual," Catalog P36E, Kyoto, Japan, 2001.

[3] RF Monolithics, Inc., "RO2073E SAW Resonator Datasheet," Dallas, Texas.

CHAPTER 5

Negative-Resistance Oscillators

5.1 Introduction

In this chapter the techniques that are used in the design of negative-resistance oscillators (NROs) are discussed. Tuned-circuit oscillators and crystal oscillators are usually designed using a feedback approach. In fact, whenever external lumped elements are used for the feedback network, the feedback approach is the method of choice. In oscillators operating in the microwave region the transistor's internal capacitances play an important part in the design. In this range of frequencies it is more convenient to represent the transistor in terms of its scattering parameters (i.e., S parameters). The matching network at one of the two ports of the transistor is called the *terminating matching network* and the one at the other port is called the *load-matching network*. The terminating matching network is used to provide the proper termination so that the transistor presents a negative-resistance at the other port (i.e., the load port), which is designed to satisfy the oscillation conditions.

Some of the material in Sections 5.2 to 5.7, and in the TE-Mode DRs in Section 5.12 were adapted or reproduced from the book by Gonzalez [1] (© 1997. Reprinted by permission of Pearson Education, Inc., Upper Saddle River, NJ.)

5.2 Negative-Resistance Method

The theory and design of oscillators can also be approached using the negative-resistance method. The method is used extensively in the design of RF and microwave frequency oscillators. If an active device is used to supply an amount of energy equal to the energy dissipated, the circuit can sustain oscillations. The behavior of the active device can be represented by a negative resistance in series with a reactance, as shown in Figure 5.1. The negative-resistance device is represented by the amplitude- and frequency-dependent impedance

$$Z_{IN}(A, \omega) = R_{IN}(A, \omega) + jX_{IN}(A, \omega) \tag{5.1}$$

where A is the amplitude of $i(t)$ and

$$R_{IN}(A, \omega) < 0$$

The oscillator is constructed by connecting the device to a passive impedance denoted by

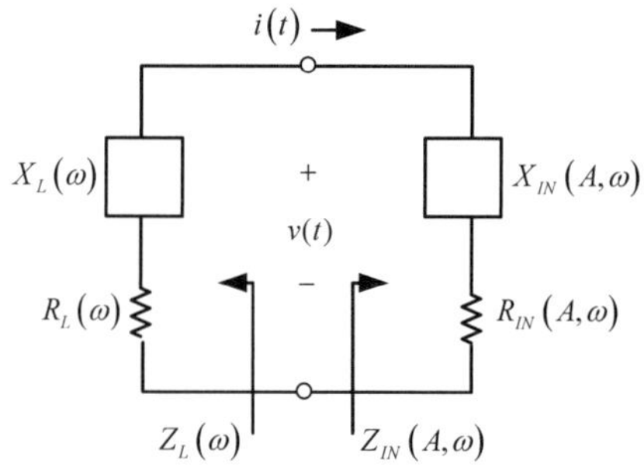

Figure 5.1 A negative-resistance model.

$$Z_L(\omega) = R_L(\omega) + jX_L(\omega) \tag{5.2}$$

If

$$R_L(\omega) + R_{IN}(A, \omega) > 0$$

the total loop resistance in Figure 5.1 is positive and the oscillation will die out. The circuit will oscillate at the frequency (ω_o) and amplitude (A_o) where

$$Z_{IN}(A_o, \omega_o) + Z_L(\omega_o) = 0 \tag{5.3}$$

Substituting (5.1) and (5.2) into (5.3), the oscillation conditions can be written as

$$R_{IN}(A_o, \omega_o) + R_L(\omega_o) = 0 \tag{5.4}$$

and

$$X_{IN}(A_o, \omega_o) + X_L(\omega_o) = 0 \tag{5.5}$$

If

$$R_L(\omega) + R_{IN}(A, \omega) < 0 \tag{5.6}$$

the oscillation is unstable and its amplitude will grow. From (5.6), the circuit in Figure 5.1 is unstable if the net loop resistance is negative; that is, when

$$\left|R_{IN}(A, \omega)\right| > R_L(\omega) \tag{5.7}$$

In a well-designed oscillator, at the start of oscillations, (5.7) is satisfied. That is, when the amplitude A is small the oscillations begin to build up. The start of oscillation condition in (5.7) is usually expressed in the form

$$\left|R_{IN}(0, \omega)\right| > R_L(\omega) \tag{5.8}$$

The oscillations will continue to build up as long as the loop resistance is negative, as required by (5.7). The amplitude of the current will reach a steady-state value (i.e., at $A = A_o$ and $\omega = \omega_o$), when the loop resistance is zero and (5.4) and (5.5) are satisfied. This is similar to satisfying the Barkhaussen criteria, which we know that can not be exactly satisfied. To satisfy the start of oscillation condition in (5.8), the build-up of oscillations in (5.7), and the oscillation conditions in (5.4) and (5.5), the impedance $Z_{IN}(A, \omega)$ must be amplitude and frequency dependent. A typical negative-resistance variation produced by the active device is such that at the start of oscillations $R_{IN}(0, \omega)$ has its maximum value and then decreases linearly as the amplitude increases. When R_{IN} reaches the value $R_{IN}(A_o, \omega)$, (5.4) and (5.5) are approximately satisfied and the circuit will oscillate with amplitude A_o at ω_o. This discussion is similar to that in Section 1.2 where a detailed analysis shows that the circuit poles move between the right-half plane and the left-half plane with (5.4) and (5.5) predicting the fundamental frequency of oscillation.

The frequency of oscillation determined by (5.4) and (5.5) might not be stable since $Z_{IN}(A, \omega)$ is amplitude and frequency dependent. Therefore, it is necessary to find another condition to guarantee a stable oscillation. If the frequency dependence of $Z_{IN}(A, \omega)$ can be neglected for small variations around ω_o, Kurokawa [2] has shown that a stable oscillation is obtained when (5.4) and (5.5) are satisfied, and the following condition is also satisfied:

$$\left.\frac{\partial R_{IN}(A)}{\partial A}\right|_{A=A_o} \left.\frac{dX_L(\omega)}{d\omega}\right|_{\omega=\omega_o} - \left.\frac{\partial X_{IN}(A)}{\partial A}\right|_{A=A_o} \left.\frac{dR_L(\omega)}{d\omega}\right|_{\omega=\omega_o} > 0 \tag{5.9}$$

The derivation of (5.9) is given in Appendix A.

In many cases

$$\frac{dR_L(\omega)}{d\omega} = 0$$

(i.e., R_L is a constant) and (5.9) simplifies accordingly. Also, when R_L is constant the term $R_L(\omega_o)$ in (5.6) is simply replaced by the constant value R_L.

For frequencies around the frequency of oscillation a reasonable approximation is to assume that $R_{IN}(A, \omega)$ and $X_{IN}(A, \omega)$ are only a function of the amplitude of $i(t)$. In some cases the variation for $R_{IN}(A, \omega)$ can be approximated by

$$R_{IN}(A, \omega) \approx R_{IN}(A) = -R_0\left(1 - \frac{A}{A_M}\right) \tag{5.10}$$

where $-R_0$ is the value of $R_{IN}(A)$ at $A = 0$, and A_M is the maximum value of A. In other words, the magnitude of the negative resistance is a linearly decreasing function of A (see Figure 5.2).

A practical way of designing R_L is to select its value for maximum oscillator power. In Figure 5.1, if R_{IN} is given by (5.10), the power delivered to R_L by R_{IN} is

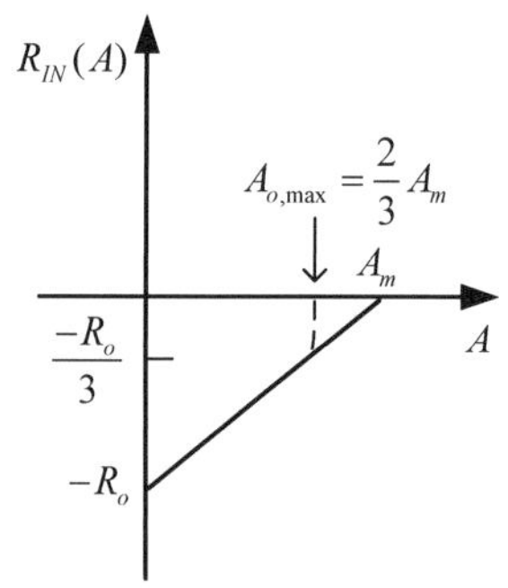

Figure 5.2 Linear variation of the negative resistance as a function of the current amplitude.

$$P = \frac{1}{2}\text{Re}(VI^*) = \frac{1}{2}|I|^2|R_{IN}(A)| = \frac{1}{2}A^2R_o\left(1 - \frac{A}{A_M}\right)$$

Hence, the value of A that maximizes the oscillation power is found from

$$\frac{dP}{dA} = \frac{1}{2}R_o\left(2A - \frac{3A^2}{A_M}\right) = 0$$

which gives the desired value of A, denoted by $A_{o,\max}$, that maximizes the power. That is,

$$A_{o,\max} = \frac{2}{3}A_M$$

At $A_{o,\max}$ the value of $R_{IN}(A_{o,\max})$ is

$$R_{IN}(A_{o,\max}) = -\frac{R_o}{3}$$

Hence, a convenient value of R_L, which maximizes the oscillator power, is

$$R_L = \frac{R_o}{3} \tag{5.11}$$

Observe that (5.11) is valid when the negative input resistance varies linearly with amplitude. In practice, the selection of R_L, according to (5.11), produces good results.

Example 5.1

(a) Determine the frequency of oscillation for the circuit in Figure 5.3, and the negative resistance R_{IN} that makes the circuit unstable. For simplicity in the analysis use $L = 1\text{H}$ and $C = 1\text{F}$.

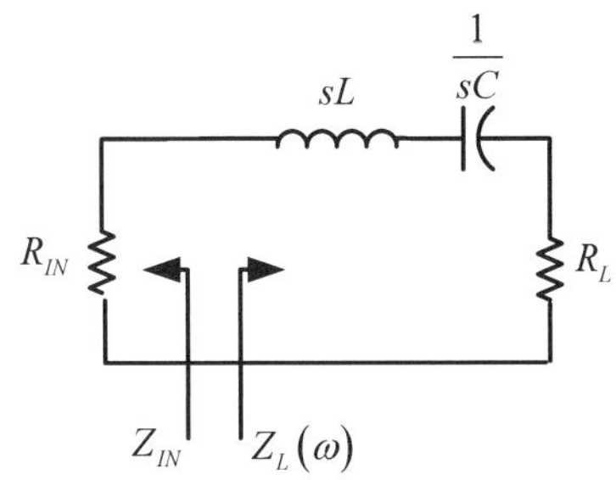

Figure 5.3 Circuit for Example 5.1.

(b) Determine if the oscillator is stable when the negative resistance is a linearly decreasing function of the amplitude.
(c) If $R_o = 45\Omega$ and $A_M = 1\text{A}$ in (5.10), determine the value of R_L for maximum oscillator power.

Solution
(a) For the circuit in Figure 5.3,

$$Z_{IN} = R_{IN}$$

and

$$Z_L(s) = s + \frac{1}{s} + R_L = \frac{s^2 + R_L s + 1}{s}$$

From (5.5), since $X_{IN} = 0$, it follows that

$$X_L(\omega_o) = 0$$

or

$$j\omega_o + \frac{1}{j\omega_o} = 0 \Rightarrow \omega_o = 1 \text{ rad/s}$$

and from (5.4), at ω_o, we obtain that

$$R_{IN} + R_L = 0$$

or

$$R_{IN} =- R_L$$

The circuit is unstable for those values of R_{IN} that make the loop resistance be negative, namely,

$$R_{IN} <- R_L$$

or the negative resistance values that satisfy

$$|R_{IN}| > R_L$$

(b) For the circuit to maintain a stable oscillation at $\omega_o = 1$ rad/s the condition (5.9) must be satisfied. With R_{IN} given by (5.10), it follows that

$$\frac{\partial R_{IN}(A)}{\partial A} = \frac{R_o}{A_M}$$

Therefore, since $R_o > 0$, we have that $\partial R_{IN}(A)/\partial A > 0$.

Also, with $X_L = (\omega - 1/\omega)$ it follows that

$$\frac{dX_L(\omega)}{d\omega} = 1 + \frac{1}{\omega^2}$$

Therefore,

$$\left.\frac{\partial R_{IN}(A)}{\partial A}\right|_{A=A_o} \left.\frac{dX_L(\omega)}{d\omega}\right|_{\omega=\omega_o} > 0$$

which shows that the oscillation is stable.

(c) From (5.10), the expression for $R_{IN}(A)$ is

$$R_{IN}(A) =- 45(1 - A)$$

Using (5.11) the value of R_L that maximizes the oscillator power is

$$R_L = \frac{R_o}{3} = \frac{45}{3} = 15\Omega$$

and the amplitude of A is

$$A_{o,\max} = \frac{2}{3}A_M = \frac{2}{3}$$

In some cases it is more convenient to represent the one-port negative-resistance oscillator using the parallel model shown in Figure 5.4. In the parallel model the voltage is the same across the two resistors. The negative-resistance device is represented by the admittance

$$Y_{IN}(A', \omega) = G_{IN}(A', \omega) + jB_{IN}(A', \omega)$$

where A' is the amplitude of $v(t)$, and $G_{IN}(A', \omega) < 0$.

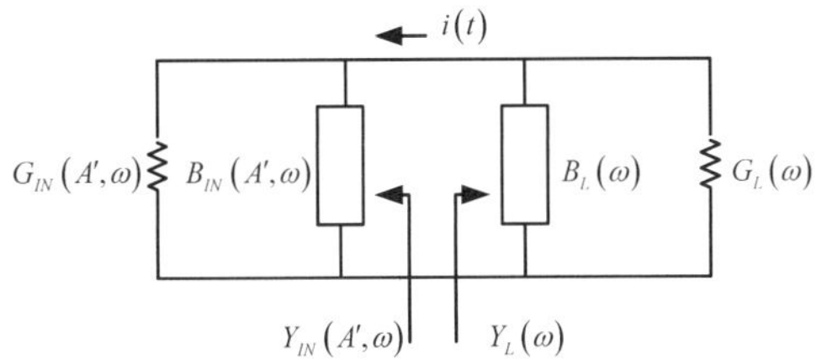

Figure 5.4 The parallel circuit model for the one-port negative-resistance oscillator.

The network in Figure 5.4 will oscillate at the amplitude A'_o and frequency $\omega = \omega_o$, when

$$Y_{IN}(A'_o, \omega_o) + Y_L(\omega_o) = 0$$

or in terms of the real and imaginary parts of the admittance we have

$$G_{IN}(A'_o, \omega_o) + G_L(\omega_o) = 0 \tag{5.12}$$

and

$$B_{IN}(A'_o, \omega_o) + B_L(\omega_o) = 0 \tag{5.13}$$

The start of oscillation condition can be expressed in the form

$$|G_{IN}(0, \omega)| > G_L(\omega)$$

For stable oscillation the Kurokawa condition must be satisfied. In terms of the circuit in Figure 5.4 the condition reads

$$\left.\frac{\partial G_{IN}(A')}{\partial A'}\right|_{A'=A_o} \left.\frac{dB_L(\omega)}{d\omega}\right|_{\omega=\omega_o} - \left.\frac{\partial B_{IN}(A')}{\partial A'}\right|_{A'=A'_o} \left.\frac{dG_L(\omega)}{d\omega}\right|_{\omega=\omega_o} > 0 \tag{5.14}$$

The selection of the impedance or admittance form for the negative resistance should not be made arbitrarily. If the impedance plot of Z_{IN} looks like a series-impedance circuit, the equivalent negative-resistance circuit should be treated using the impedance form. If the admittance plot of Y_{IN} looks likc a parallel-admittance circuit, the equivalent negative-resistance circuit should be treated using the admittance form.

Further understanding of a negative-resistance oscillator is gained by the analysis of the series RLC circuit shown in Figure 5.5. The active device is represented by the negative resistance R_{IN} (where $R_{IN} < 0$) in series with the inductor L, and the load network by R_L in series with the capacitor C. Obviously, the inductor and capacitor are interchangeable. The voltage that starts the oscillation can be modeled, for simplicity, as a step voltage with a small amplitude E. For the

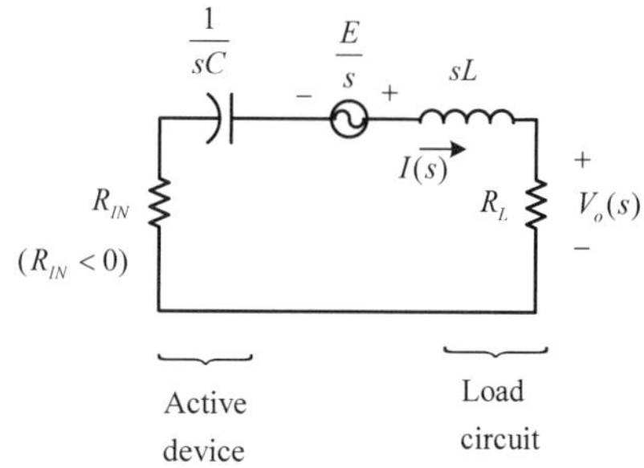

Figure 5.5 A negative-resistance circuit.

series-tuned circuit the net loop resistance is $R_L + R_{IN}$. If $|R_{IN}| > R_L$ the net loop resistance is negative. The loop equation is

$$\frac{E}{s} = I(s)(R_L + R_{IN}) + I(s)sL + \frac{I(s)}{sC}$$

or

$$I(s) = \frac{\frac{E}{L}}{s^2 + \frac{(R_L + R_{IN})}{L}s + \frac{1}{LC}}$$

The output voltage $V_o(s) = I(s)R_L$ can be conveniently expressed in the form

$$V_o(s) = \frac{\frac{ER_L}{L}}{s^2 + 2\alpha\omega_o s + \omega_o^2} \tag{5.15}$$

where

$$\omega_o = \sqrt{\frac{1}{LC}}$$

and

$$\alpha = \frac{R_L + R_{IN}}{2\omega_o L} = \frac{(R_L + R_{IN})}{2}\sqrt{\frac{C}{L}} \tag{5.16}$$

Of course, ω_o represents the resonant frequency and α the damping factor.

The time function $v_o(t)$ is obtained by taking the inverse Laplace transform of (5.15). The inverse transform depends on the location of the poles of (5.15) at

$$s = \omega_o\left(-\alpha \pm j\sqrt{1 - \alpha^2}\right) \tag{5.17}$$

There are three cases to consider in (5.17). First, (5.17) has real poles when $\alpha > 1$, and in this case the circuit does not have an oscillatory behavior, it simply behaves like a first-order circuit. Second, if $\alpha = 1$ the poles of (5.17) are equal and, consequently, the circuit is not oscillating. Third, if $\alpha < 1$ the poles in (5.17) are complex conjugates and $v_o(t)$ is given by

$$v(t) = \frac{ER_L}{L} \frac{1}{\omega_o\sqrt{1-\alpha^2}} e^{-\alpha\omega_o t} \sin\left(\omega_o\sqrt{1-\alpha^2}\, t\right)$$

which can represents either a decaying (for $0 < \alpha < 1$) or a growing oscillation (for $\alpha < 0$). The exponential factor is $e^{-\alpha\omega_o t}$ and the oscillation frequency is $\omega_o\sqrt{1-\alpha^2}$.

For an oscillation we want $\alpha = 0$, as shown in Figure 5.6. From (5.16), the ideal case of $\alpha = 0$ occurs when $R_L + R_{IN} = 0$ (or no losses in the circuit), and the oscillation frequency is

$$\omega_o = \sqrt{\frac{1}{LC}}$$

The previous analysis shows that an active component is required to implement the negative resistance. Active components are nonlinear in nature and, therefore, its impedance will be a function of the current amplitude in a series circuit. For oscillations to begin and grow we want $\alpha < 0$ (or $|R_{IN}| > R_L$). At the start of the oscillation, the term $e^{-\alpha\omega_o t}$ increases exponentially. It increases until a steady state is reached. This occurs when the nonlinearities of the active device decrease the value of R_{IN} so that $R_L + R_{IN} = 0$, or $\alpha = 0$. In the ideal case that $\alpha = 0$, the growing oscillation reaches a steady state as shown in Figure 5.6.

The concept of Q in an oscillator circuit is related to the resonator's effect (i.e., the resonator's Q) on the loop gain. The equivalent circuit of an NRO can also be drawn as shown in Figure 5.7 where the resonator losses are represented by R_R, the load resistor is R_L, and the active device negative resistance by R_{IN}. This circuit is similar to that in Figure 5.5, except that R_R was included and the resonator was viewed as composed of L, C, and R_R. The loaded Q in Figure 5.7 is infinite, since $R_R + R_L + R_{IN} = 0$. That is, in Figure 5.7, when the circuit oscillates, the negative resistance cancels $R_R + R_L$. Hence, an oscillator is basically an infinite Q_L resonator that (ideally) supports a constant amplitude sinusoidal voltage.

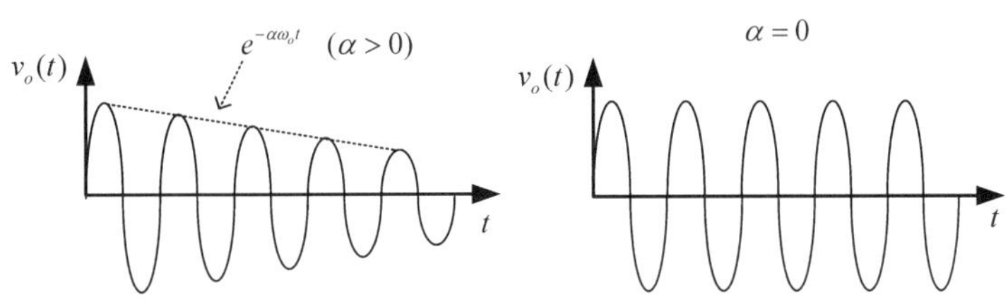

Figure 5.6 The oscillatory response of $v_o(t)$.

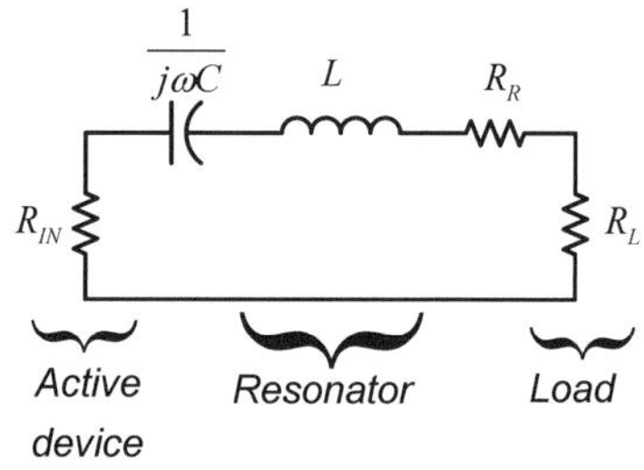

Figure 5.7 A negative-resistance circuit with R_R representing the resonator losses.

The unloaded Q of the resonator is

$$Q_U = \frac{\omega_o}{2R_R} \frac{dX}{d\omega}\bigg|_{\omega=\omega_o}$$

5.3 Oscillation Conditions—A Negative-Resistance Approach

In this section two oscillator configurations are analyzed to illustrate how the negative resistance arises. Consider the Colpitts oscillator shown in Figure 5.8(a). Its equivalent small-signal circuit, as viewed from the inductor, is shown in Figure 5.8(b).

From Figure 5.8(b), it is observed that the transistor's junction capacitance C_{be} has a negligible effect on the performance of the circuit if $C_1 \gg C_{be}$; and h_{oe} can be neglected if $|Z_{C_2}| \ll 1/h_{oe}$. Hence, from Figure 5.8(b) the loop equations are

$$v_1 = i_1(Z_1 + Z_2) - i_b(Z_1 - h_{fe}Z_2) \tag{5.18}$$

$$0 = i_1(Z_1) - i_b(Z_1 + h_{ie}) \tag{5.19}$$

Solving (5.18) and (5.19) for Z_{IN} gives

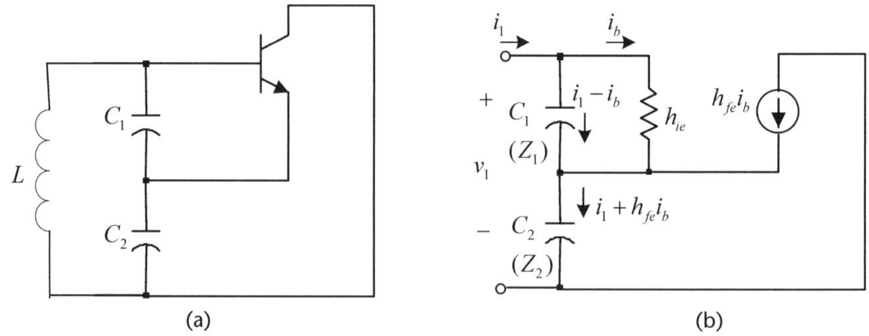

Figure 5.8 (a) A BJT Colpitts oscillator and (b) small-signal model to calculate Z_{IN}.

$$Z_{IN} = \frac{v_1}{i_1} = \frac{(1 + h_{fe})Z_1 Z_2 + h_{ie}(Z_1 + Z_2)}{Z_1 + h_{ie}} \tag{5.20}$$

In (5.20), if $|Z_1| \ll h_{ie}$ (i.e., the transistor's h_{ie} does not load the capacitor C_1), Z_{IN} can be approximated by

$$Z_{IN} \approx \frac{h_{fe}}{h_{ie}} Z_1 Z_2 + Z_1 + Z_2$$
$$= - g_m X_1 X_2 + j(X_1 + X_2)$$

or

$$Z_{IN} \approx \frac{-g_m}{\omega^2 C_1 C_2} + \frac{1}{j\omega C_T} \tag{5.21}$$

where the input capacitance C_{IN} associated with Z_{IN} is the total capacitance of the Colpitts oscillator, namely,

$$C_{IN} = C_T = \frac{C_1 C_2}{C_1 + C_2}$$

Equation (5.21) clearly illustrates that the input impedance has a negative real part (i.e., a negative resistance) in series with the total capacitance C_T. The negative resistance is given by

$$R_{IN} = \frac{-g_m}{\omega^2 C_1 C_2} \tag{5.22}$$

The equivalent negative-resistance resonant circuit is shown in Figure 5.9, where $Z_L = R_s + j\omega L$, and R_s is the inductor's loss resistance.

The gain condition for this circuit [see (5.8)] is $|R_{IN}| > R_s$ or

$$\frac{g_m}{\omega_o^2 C_1 C_2 R_s} > 1 \tag{5.23}$$

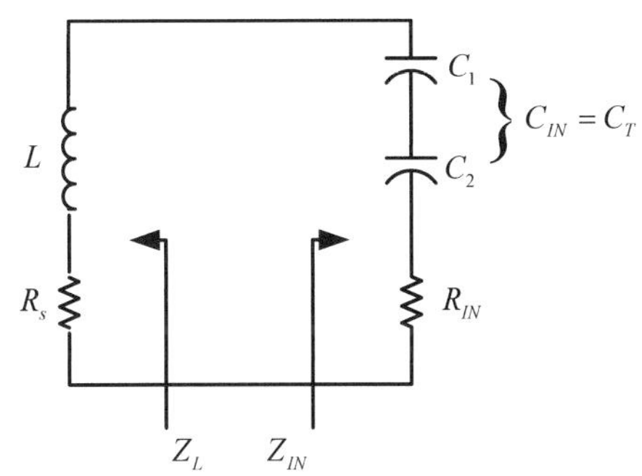

Figure 5.9 Equivalent negative-resistance circuit.

and the frequency of oscillation is

$$\omega_o = \frac{1}{\sqrt{LC_{IN}}} \tag{5.24}$$

The transistor's g_m is a function of the oscillation amplitude [i.e., $g_m(A)$]. As the amplitude builds up, the value of g_m decreases. Hence, a convenient way of writing the start of oscillation condition in (5.23) is

$$\frac{g_m(0)}{\omega_o^2 C_1 C_2 R_s} > 1 \tag{5.25}$$

where $g_m(0)$ being the small-signal value of g_m.

At the start of oscillations the loop resistance is negative. As the oscillation amplitude builds up, the value of g_m decreases, and the oscillator reaches a steady amplitude value of A_o when (5.23) is approximately equal to 1.

It is also observed that in this oscillator R_s represents the inductor's losses, which is determined by its Q_U. Hence, a design procedure based on (5.11) does not readily apply.

Equation (5.25) places a limit on the maximum value of the capacitors C_1 and C_2, since for oscillations to start, R_s must satisfy

$$R_s < \frac{g_m(0)}{\omega_o^2 C_1 C_2} \tag{5.26}$$

For the case that $C_1 = C_2$, (5.26) shows that

$$\frac{1}{\omega_o C_1} > \sqrt{\frac{R_s}{g_m(0)}}$$

or the minimum required reactance for oscillation is a function of the series loss resistance of the inductor, and of the transistor's conductance. If $C_1 \neq C_2$, (5.26) must be satisfied.

The use of a feedback resistor R, as shown in Figure 5.10, results in the following expression for Z_{IN}:

$$Z_{IN} = \frac{-X_1 X_2}{\dfrac{1}{g_m} + R} + j(X_1 + X_2) \tag{5.27}$$

Equation (5.27) shows that the negative resistance is reduced. Also, if $R \gg 1/g_m$, the dependence of the negative resistance on g_m is reduced, which might result in an improvement of the phase-noise characteristics.

The Hartley configuration shown in Figure 5.11(a) is analyzed in a similar way. Its ac model is shown in Figure 5.11(b). Neglecting the loading of the transistor's h_{ie} and C_{be}, and assuming no mutual coupling, it follows that

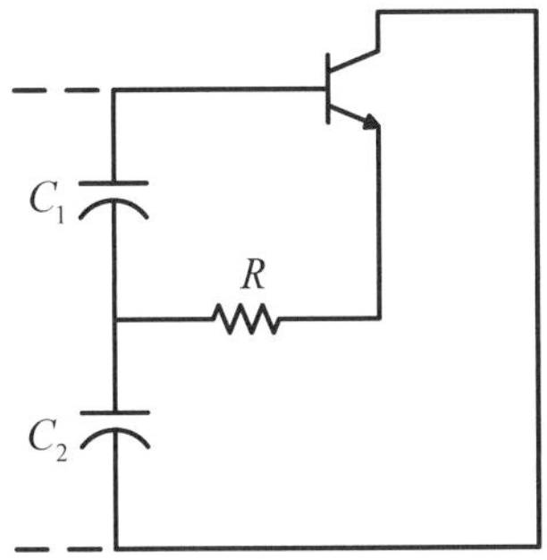

Figure 5.10 A Colpitts oscillator with a feedback resistor R.

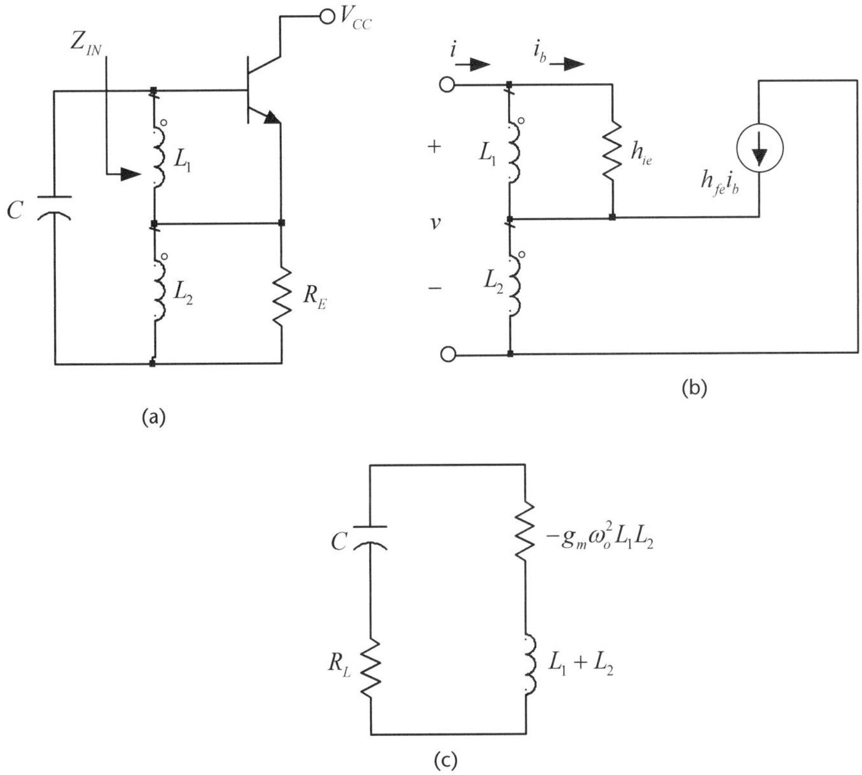

Figure 5.11 (a) A BJT Hartley oscillator, (b) its ac model, and (c) negative-resistance model.

$$Z_{IN} = \frac{v}{i} =- g_m \omega_o^2 L_1 L_2 + j\omega(L_1 + L_2) \tag{5.28}$$

The negative resistance model in Figure 5.11(c) follows from (5.28).

With the load network composed of a capacitor and a resistance, as shown in Figure 5.11(c), we obtain

$$\omega_o = \sqrt{\frac{1}{(L_1 + L_2)C}}$$

and the gain condition is

$$\frac{R_L}{L_1 L_2 \omega_o^2 g_m} < 1$$

In Figure 5.11, R_L represents either coil losses or an external load resistor.

In some configurations the selection of Z_{IN} and Z_L is not obvious. For example, consider the FET Pierce oscillator shown in Figure 5.12(a) and assume that the inductor is lossless. Its small-signal model is shown in Figure 5.12(b). Since the controlled source depends on $v_f = v_{gs}$, it is not obvious how to obtain an equivalent circuit across either C_1, C_2, or L. A particular selection of Z_L and Z_{IN} that works for this circuit is to select Z_L to be the impedance between the drain and ground; that is,

$$Z_L(j\omega) = r_d \,||\, jX_2 \,||\, (jX_3 + jX_1) \tag{5.29}$$

and Z_{IN} to be the equivalent impedance of the source $g_m v_{gs}$. Since

$$v_{gs} = v_o \frac{X_1}{X_1 + X_3}$$

then,

$$g_m v_{gs} = v_o \frac{g_m X_1}{X_1 + X_3} = \frac{v_o}{Z_{IN}}$$

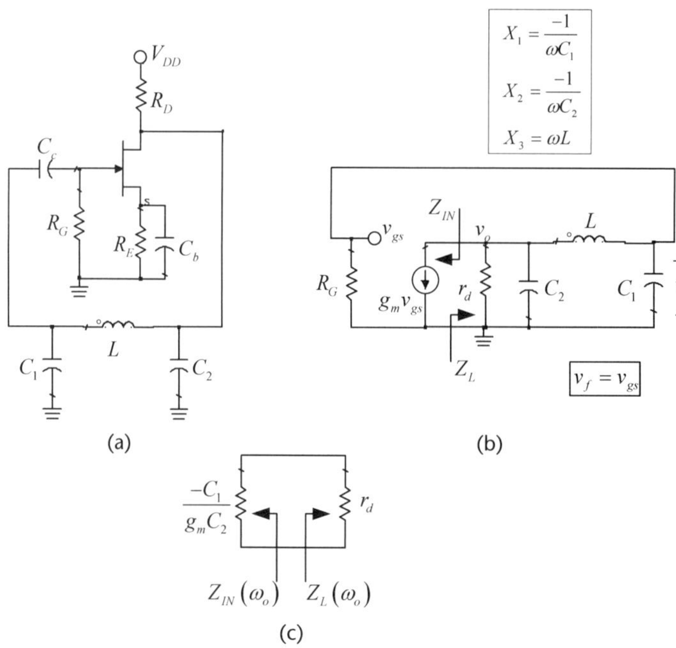

Figure 5.12 (a) An FET Pierce oscillator, (b) its small-signal model, and (c) the equivalent negative-resistance circuit at resonance.

where

$$Z_{IN} = \frac{X_1 + X_3}{g_m X_1} \tag{5.30}$$

Observe that Z_{IN} is purely real. It is a negative number if $X_1 + X_3$ is inductive, and a positive number if $X_1 + X_3$ is a capacitive. Obviously, $X_1 + X_3$ better be inductive for Z_{IN} to represent a negative resistor, and for the circuit to have a resonant frequency.

The frequency of oscillation requires that the imaginary part of $Z_{IN} + Z_L(j\omega)$ be zero. Since Z_{IN} is real, the imaginary part of $Z_L(j\omega)$ in (5.29) vanishes at the frequency ω_o where

$$X_1(j\omega_o) + X_2(j\omega_o) + X_3(j\omega_o) =- \frac{1}{\omega C_1} - \frac{1}{\omega C_2} + \omega L = 0 \tag{5.31}$$

or

$$\omega_o = \frac{1}{\sqrt{LC_T}}$$

where

$$C_T = \frac{C_1 C_2}{C_1 + C_2}$$

Furthermore, $Z_L(j\omega_o) = r_d$ is a real number.

At the frequency ω_o it follows from (5.31) and (5.30) that

$$Z_{IN} = R_{IN} = \frac{-X_2}{g_m X_1} = \frac{-C_1}{g_m C_2}$$

and the equivalent negative-resistance model at resonance is obtained, see Figure 5.12(c). From Figure 5.12(c) oscillation occurs when the loop resistance is zero. That is, when

$$\frac{-C_1}{g_m C_2} + r_d = 0$$

or

$$g_m r_d = \frac{C_1}{C_2}$$

which is the well known gain condition for the FET Pierce oscillator.

The start of oscillation condition requires that the small-signal value of g_m [i.e., $g_m(0)$] satisfy

$$g_m(0) > \frac{C_1}{C_2 r_d}$$

5.4 Traveling-Waves and Power-Waves Concepts

In this section a short review of transmission lines concepts is given. A transmission line of length $d = l$ with real characteristic impedance Z_o is shown in Figure 5.13(a). The transmission line is excited with a sinusoidal source v_s having an impedance Z_s, and terminated in the load Z_L. The phasor form of the voltage and current along a transmission line can be expressed in terms of an incident traveling-wave voltage $V^+(d)$ and a reflected traveling-wave voltage $V^-(d)$ as

$$V(d) = V^+(d) + V^-(d) = Ae^{j\beta d} + Be^{-j\beta d}$$

and

$$I(d) = I^+(d) - I^-(d) = \frac{Ae^{j\beta d}}{Z_o} - \frac{Be^{-j\beta d}}{Z_o}$$

where β is the propagation constant,

$$\beta = \frac{\omega}{v_p} = \frac{2\pi}{\lambda} \tag{5.32}$$

In (5.32), v_p is the velocity of propagation (i.e., the phase velocity) and λ is the wavelength (i.e., $\lambda = v_p/f$).

The reflection coefficient at $d = 0$, known as the load reflection coefficient Γ_o, is given by

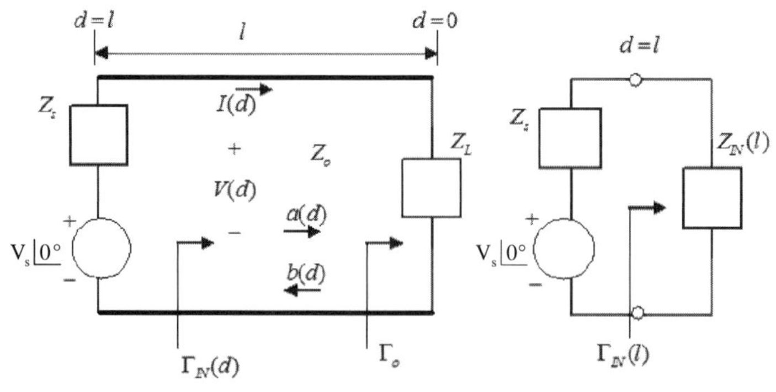

Figure 5.13 (a) A transmission line with characteristic impedance Z_o, load impedance Z_L, and length $d = l$; and (b) the equivalent circuit at $d = l$. (*From:* [1]. © 1997. Reprinted by permission of Pearson Education, Inc., NJ.)

$$\Gamma_o = \frac{Z_L - Z_o}{Z_L + Z_o} \quad (5.33)$$

The reflection coefficient at a distance d from the load is given by

$$\Gamma_{IN}(d) = \Gamma_o e^{-j2\beta d} \quad (5.34)$$

The impedance at any point along the line is given by

$$Z_{IN}(d) = \frac{V(d)}{I(d)} = Z_o \frac{1 + \Gamma_{IN}(d)}{1 - \Gamma_{IN}(d)} \quad (5.35)$$

Substituting (5.33) into (5.34) and then into (5.35) results in the following expression for $Z_{IN}(d)$:

$$Z_{IN}(d) = Z_o \frac{Z_L \cos \beta d + jZ_o \sin \beta d}{Z_o \cos \beta d + jZ_L \sin \beta d}$$

At the input of the line ($d = l$) the input reflection coefficient and input impedances are given by $\Gamma_{IN}(l)$ and $Z_{IN}(l)$, respectively. Observe, from (5.35), that

$$\Gamma_{IN}(l) = \frac{Z_{IN}(l) - Z_o}{Z_{IN}(l) + Z_o} \quad (5.36)$$

The normalized incident and reflected traveling waves, denoted by a and b, along the transmission line are defined by

$$a(d) = \frac{V^+(d)}{\sqrt{Z_o}} = \frac{1}{2\sqrt{Z_o}}[V(d) + Z_o I(d)]$$

and

$$b(d) = \frac{V^-(d)}{\sqrt{Z_o}} = \frac{1}{2\sqrt{Z_o}}[V(d) - Z_o I(d)]$$

At the input of the transmission line there is an input impedance $Z_{IN}(l)$, as shown in Figure 5.13(b). This input impedance could also be implemented with lumped elements, as shown in Figure 5.14(a). The circuit in Figure 5.14(a) has no transmission line and the input impedance is denoted by Z_{IN}. For this circuit we can still define an input reflection coefficient in terms of a real source impedance Z_s as

$$\Gamma_{IN} = \frac{Z_{IN} - Z_s}{Z_{IN} + Z_s}$$

Thus, if $Z_s = Z_o$, then

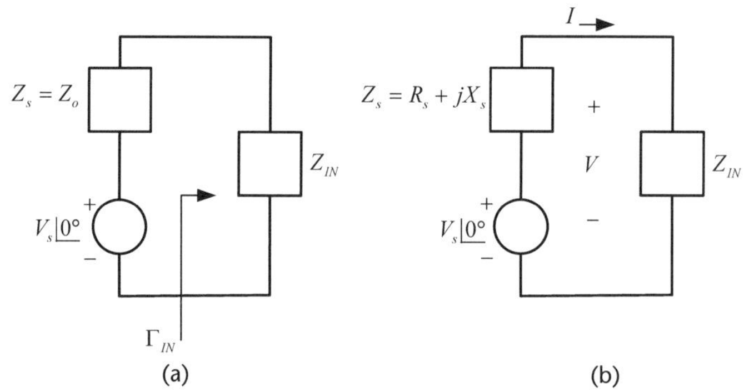

Figure 5.14 (a) Input circuit with $Z_S = Z_O$ and (b) general case where Z_S is complex.

$$\Gamma_{IN} = \frac{Z_{IN} - Z_o}{Z_{IN} + Z_o} \tag{5.37}$$

The source impedance is known as the *normalizing impedance*. If a transmission line is inserted between $Z_s = Z_o$ and a load in Figure 5.14(a), then Γ_{IN} in (5.37) represents the reflection coefficient associated with Z_{IN}.

Equation (5.36) (which applies to a transmission line) and (5.37) (which applies to a lumped network with $Z_s = Z_o$) are equivalent in form. A $(\text{VSWR})_{IN}$ can be defined in terms of Γ_{IN} for the circuit in Figure 5.14(a), namely

$$(\text{VSWR})_{IN} = \frac{1 + |\Gamma_{IN}|}{1 + |\Gamma_{IN}|} \tag{5.38}$$

Hence, there is maximum power transfer when $(\text{VSWR})_{IN} = 1$ (or $Z_{IN} = Z_s = Z_o$), and whenever $(\text{VSWR})_{IN} > 1$ there is a power mismatch. Observe that the transmission-line concepts are useful even if there is no transmission line.

A generalization of the circuit in Figure 5.14(a) is shown in Figure 5.14(b), where Z_s is complex, or $Z_s = R_s + jX_s$. Since Z_s is complex the reflection coefficient concept used in (5.37) does not apply. However, a new set of waves, known as power waves, was developed by Kurokawa to analyze such a circuit [3]. Kurokawa introduced the incident and reflected power waves (a_p and b_p, respectively) defined by

$$a_p = \frac{1}{2\sqrt{R_s}}(V + Z_s I)$$

and

$$b_p = \frac{1}{2\sqrt{R_s}}(V - Z_s^* I)$$

The reflection-coefficient Γ_p associated with the power waves, also known as the generalized reflection coefficient, is given by

$$\Gamma_p = \frac{b_p}{a_p} = \frac{Z_{IN} - Z_s^*}{Z_{IN} + Z_s^*} \tag{5.39}$$

Observe that when the normalized impedance Z_s is real and positive (say, $Z_s = Z_o$), the expressions for the power waves are identical to those of the traveling waves [see also the similarity of the power reflection coefficient in (5.39) and the input reflection coefficient in (5.37)]. A $(\text{VSWR})_{IN}$ can be defined in terms of Γ_p by replacing Γ_{IN} by Γ_p in (5.38). Equation (5.39) shows that when $Z_{IN} = Z_s^*$ there is maximum power transfer to Z_{IN}, and $(\text{VSWR})_{IN} = 1$.

5.4.1 *S* Parameters

A short review of the *S* parameters concepts is now presented. For detailed information on *S* parameter theory the reader is referred to Gonzalez [1]. For the two-port network shown in Figure 5.15, the relations between the normalized reflected waves $b_1(l_1)$ at port 1 and $b_2(l_2)$ at port 2, and the normalized incident waves $a_1(l_1)$ and $a_2(l_2)$ are

$$b_1(l_1) = S_{11}a_1(l_1) + S_{12}a_2(l_2) \tag{5.40}$$

and

$$b_2(l_2) = S_{21}a_1(l_1) + S_{22}a_2(l_2) \tag{5.41}$$

Observe that port 1 is located at $x_1 = l_1$ and port 2 is located at $x_2 = l_2$. Equations (5.40) and (5.41) can be expressed in matrix form

$$\begin{bmatrix} b_1(l_1) \\ b_2(l_1) \end{bmatrix} = \begin{bmatrix} S_{11} & S_{12} \\ S_{21} & S_{22} \end{bmatrix} \begin{bmatrix} a_1(l_1) \\ a_2(l_1) \end{bmatrix} \tag{5.42}$$

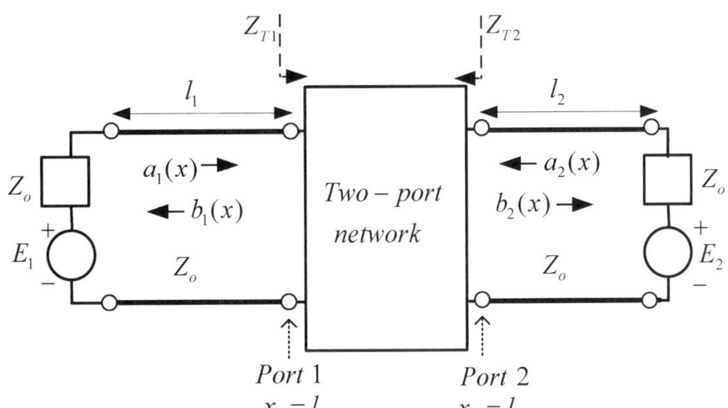

Figure 5.15 Incident and reflected waves in a two-port network. (*From:* [1]. © 1997. Reprinted by permission of Pearson Education, Inc., NJ.)

The waves $a_1(l_1)$, $a_2(l_2)$, $b_1(l_1)$, and $b_2(l_2)$ are the values of the incident and reflected waves at the specific locations denoted as port 1 and port 2 in Figure 5.15. The parameters S_{11}, S_{22}, S_{21}, and S_{22}, which represent reflection and transmission coefficients, are called the *scattering parameters* (S parameters) of the two-port network, measured at ports 1 and 2. The matrix

$$\begin{bmatrix} S_{11} & S_{12} \\ S_{21} & S_{22} \end{bmatrix}$$

is called the *scattering matrix*. The term $S_{11}a_1(l_1)$ represents the contribution to the reflected wave $b_1(l_1)$ due to the incident wave $a_1(l_1)$ at port 1. Similarly, $S_{12}a_2(l_2)$ represents the contribution to the reflected wave $b_1(l_1)$ due to the incident wave $a_2(l_2)$ at port 2, and so on.

The S parameters are seen to represent reflection or transmission coefficients. From (5.42), the S parameters measured at the specific location shown as port 1 and port 2 in Figure 5.15 are defined as follows:

$$S_{11} = \left.\frac{b_1(l_1)}{a_1(l_1)}\right|_{a_2(l_2)=0}$$

(Input reflection coefficient with output properly terminated)

$$S_{21} = \left.\frac{b_2(l_2)}{a_1(l_1)}\right|_{a_2(l_2)=0}$$

(Forward transmission coefficient with output properly terminated)

$$S_{22} = \left.\frac{b_2(l_2)}{a_2(l_2)}\right|_{a_1(l_1)=0}$$

(Output reflection coefficient with input properly terminated)

$$S_{12} = \left.\frac{b_1(l_1)}{a_2(l_2)}\right|_{a_1(l_1)=0}$$

(Reverse transmission coefficient with input properly terminated)

If the two-port network in Figure 5.15 represents a transistor, the transistor must be properly biased. Hence, the transistor S parameters are measured at a given Q point, under *small-signal conditions*. In addition, the S parameters vary with frequency; therefore, their values as the frequency is varied are usually measured.

In Figure 5.15 the two-port network is excited by the sinusoidal voltage sources represented by the phasors E_1 and E_2, and the source impedances are denoted by Z_o. The impedances Z_o are also known as the reference impedances (or the normalizing impedances). In practice, the S parameters are measured with a real

Z_o (usually $Z_o = 50\Omega$). The S parameters measured in Figure 5.15 are said to be measured in a Z_o system.

From Figure 5.15 (i.e., with $E_2 = 0$) the input reflection coefficient S_{11} at $x_1 = l_1$ is given by

$$S_{11} = \left.\frac{b_1(l_1)}{a_1(l_1)}\right|_{a_2(l_2)=0} = \frac{Z_{T1} - Z_o}{Z_{T1} - Z_o} \tag{5.43}$$

where Z_{T1} is the input impedance at port 1. Equation (5.43) shows that S_{11} is the input reflection coefficient at port 1 with port 2 terminated in its normalizing impedance Z_o (i.e., $a_2(l_2) = 0$).

Similarly, S_{22} is the reflection coefficient of port 2 with port 1 terminated in its normalizing impedance Z_o [i.e., $a_1(l_1) = 0$]. That is,

$$S_{22} = \left.\frac{b_2(l_2)}{a_2(l_2)}\right|_{a_1(l_1)=0} = \frac{Z_{T2} - Z_o}{Z_{T2} - Z_o}$$

where Z_{T2} is the input impedance at port 2.

If we consider the *quantity* $|S_{11}|^2$, it can be shown that

$$|S_{11}|^2 = \left.\left|\frac{b_1(l_1)}{a_1(l_1)}\right|^2\right|_{a_2(l_2)=0} = \frac{P_{AVS} - P_1(l_1)}{P_{AVS}}$$

or

$$P_1(l_1) = P_{AVS}\left(1 - |S_{11}|^2\right) \tag{5.44}$$

where P_{AVS} is the power available at port 1 and $P_1(l_1)$ is the input power at port 1. Equation (5.44) shows that $|S_{11}|^2$ represents the ratio of the power reflected from port 1 to the power available at port 1. If $|S_{11}| > 1$, the power reflected is larger than the power available at port 1. Therefore, in such case port 1 acts as a source of power and oscillation can occur. Similar considerations apply to port 2.

5.4.2 S_p Parameters

For the case of a two-port network with complex source impedances Z_1 and Z_2, as shown in Figure 5.16, we can define generalized scattering parameters (S_p parameters) denoted by S_{p11}, S_{p12}, S_{p21}, and S_{p22} in terms of the circuit power waves as follows:

$$b_{p1} = S_{p11}a_{p1} + S_{p12}a_{p1}$$
$$b_{p2} = S_{p21}a_{p1} + S_{p22}a_{p1}$$

where

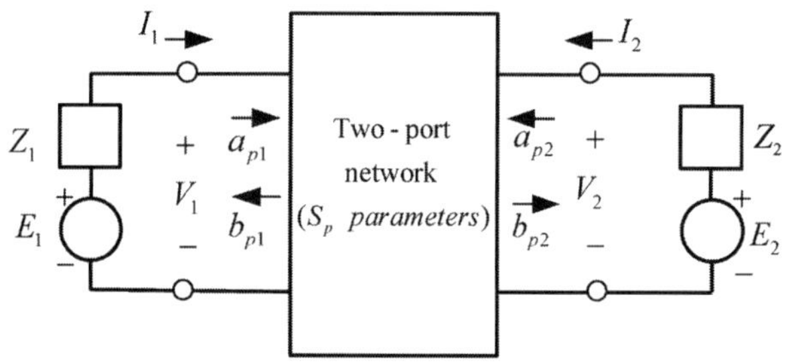

Figure 5.16 Two-port representation in terms of generalized scattering parameters.

$$a_{p1} = \frac{1}{2\sqrt{R_1}}(V_1 + Z_1 I_1)$$

$$a_{p2} = \frac{1}{2\sqrt{R_2}}(V_2 + Z_2 I_2)$$

$$b_{p2} = \frac{1}{2\sqrt{R_2}}(V_2 - Z_2^* I_2)$$

and

$$b_{p1} = \frac{1}{2\sqrt{R_1}}(V_1 - Z_1^* I_1)$$

The resistors R_1 and R_2 are $R_1 = \text{Re}[Z_1]$ and $R_2 = \text{Re}[Z_2]$. The values of the S_p parameters depend on the terminal impedances Z_1 and Z_2. These impedances are known as the references impedances.

5.5 Stability Considerations

Figure 5.17 shows a two-port network connected to a source E_s with source impedance Z_s and to a load Z_L. The two-port network is characterized by its S parameters measured in a Z_o system. The input and output reflection coefficients are given by

$$\Gamma_{IN} = S_{11} + \frac{S_{12}S_{21}\Gamma_L}{1 - S_{22}\Gamma_L} = \frac{S_{11} - \Delta\Gamma_L}{1 - S_{22}\Gamma_L} \tag{5.45}$$

and

$$\Gamma_{OUT} = S_{22} + \frac{S_{12}S_{21}\Gamma_s}{1 - S_{11}\Gamma_s} = \frac{S_{22} - \Delta\Gamma_s}{1 - S_{11}\Gamma_s} \tag{5.46}$$

where

$$\Delta = S_{11}S_{22} - S_{21}S_{12} \tag{5.47}$$

$$\Gamma_s = \frac{Z_s - Z_o}{Z_s + Z_o}$$

and

$$\Gamma_L = \frac{Z_L - Z_o}{Z_L + Z_o}$$

The source and load impedances are assumed to be passive. Hence, the real parts of Z_s and Z_L are positive, and it follows that $|\Gamma_s| < 1$ and $|\Gamma_L| < 1$.

In the two-port network in Figure 5.17 oscillations are possible when either the input or output port presents a negative resistance. This occurs when $|\Gamma_{IN}| > 1$ or $|\Gamma_{OUT}| > 1$. The two-port network is said to be potentially unstable when $|\Gamma_{IN}| > 1$ or $|\Gamma_{OUT}| > 1$. An appropriate selection of Γ_L and Γ_s will make a potentially unstable two port to oscillate.

If either $|S_{11}| > 1$ or $|S_{22}| > 1$, the two port is potentially unstable because the termination $\Gamma_L = 0$ or $\Gamma_s = 0$ will produce $|\Gamma_{IN}| > 1$ or $|\Gamma_{OUT}| > 1$.

A two-port network is unconditional stable at a given frequency if the real parts of Z_{IN} and Z_{OUT} are greater than zero for all passive load and source impedances. It can be shown that a two port is unconditionally stable if

$$K > 1 \tag{5.48}$$

and

$$|\Delta| < 1 \tag{5.49}$$

where

$$K = \frac{1 - |S_{11}|^2 - |S_{22}|^2 + |\Delta|^2}{2|S_{12}S_{21}|}$$

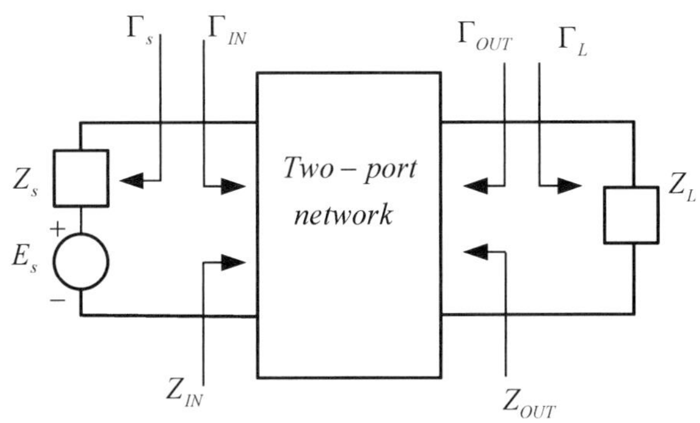

Figure 5.17 A two-port network connected to a source and a load. (*From:* [1]. © 1997. Reprinted by permission of Pearson Education, Inc., NJ.)

The unconditional stability conditions (5.48) and (5.49) can also be expressed in terms of a single condition, namely $\mu > 1$, where

$$\mu = \frac{1 - |S_{22}|^2}{|S_{11} - \Delta S_{22}^*| + |S_{12}S_{21}|} \tag{5.50}$$

That is, if $\mu > 1$ the two-port network is unconditionally stable, and if $\mu < 1$ the device is potentially unstable. Alternatively, the single $\mu > 1$ condition in (5.50) can also be expressed in terms of μ' (i.e., $\mu' > 1$), where μ' is obtained by interchanging S_{11} by S_{22} in (5.50). If $\mu > 1$, it also follows that $\mu' > 1$, and vice versa. The parameters μ represents the distance from the center of the Smith chart to the closest point on the output stability circle. Hence, if $\mu > 1$ the input stability circle is outside the Smith chart and the device is unconditionally stable. The larger the μ the more stable the device is.

When the two-port in Figure 5.17 is potentially unstable, there are values of Γ_s and Γ_L for which the real parts of Z_{IN} and Z_{OUT} are negative. These values of Γ_s and Γ_L are useful in the design of oscillators. It can be shown from (5.45) and (5.46) that the values Γ_L and Γ_s that make $|\Gamma_{IN}| = 1$ and $|\Gamma_{OUT}| = 1$ lie on circles, called *stability circles*.

The radius and center of the circle in the Γ_L plane that produces $|\Gamma_{IN}| = 1$ (i.e., the output stability circle), and in the Γ_s plane that produces $|\Gamma_{OUT}| = 1$ (i.e., the input stability circle) are as follows:

Output stability circle:

$$r_L = \left| \frac{S_{12}S_{21}}{|S_{22}|^2 - |\Delta|^2} \right| \text{ (radius)} \tag{5.51}$$

$$C_L = \frac{(S_{22} - \Delta S_{11}^*)^*}{|S_{22}|^2 - |\Delta|^2} \text{ (center)} \tag{5.52}$$

Input stability circle:

$$r_s = \left| \frac{S_{12}S_{21}}{|S_{11}|^2 - |\Delta|^2} \right| \text{ (radius)} \tag{5.53}$$

$$C_s = \frac{(S_{11} - \Delta S_{22}^*)^*}{|S_{11}|^2 - |\Delta|^2} \text{ (center)} \tag{5.54}$$

The expressions (5.51) to (5.54) can be calculated and plotted on a Smith chart for a set of S parameters, and the values of Γ_L and Γ_s that produce $|\Gamma_{IN}| = 1$ and $|\Gamma_{OUT}| = 1$ can be easily observed. Figures 5.18 and 5.19 illustrate typical constructions of the stability circles in a potentially unstable case. On one side of the stability circle boundary in the Γ_L plane, we will have $|\Gamma_{IN}| < 1$ and on the other side $|\Gamma_{IN}| > 1$. Similarly, in the Γ_s plane on one side of the stability circle boundary, we will have $|\Gamma_{OUT}| < 1$ and on the other side $|\Gamma_{OUT}| > 1$.

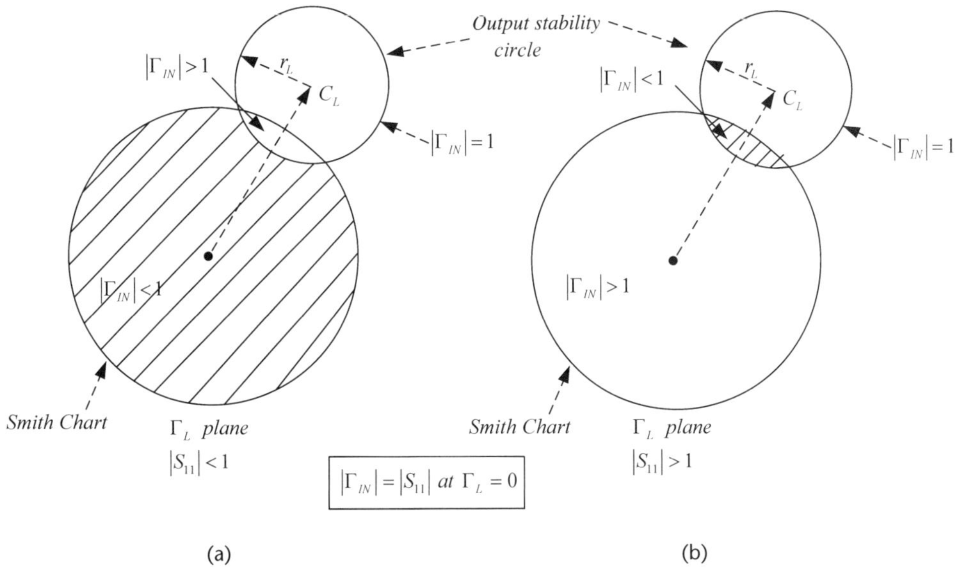

Figure 5.18 Smith chart illustrating stable and unstable regions in the Γ_L plane for (a) $|S_{11}| < 1$ and (b) $|S_{11}| > 1$. (*From:* [1]: © 1997. Reprinted by permission of Pearson Education, Inc., NJ.)

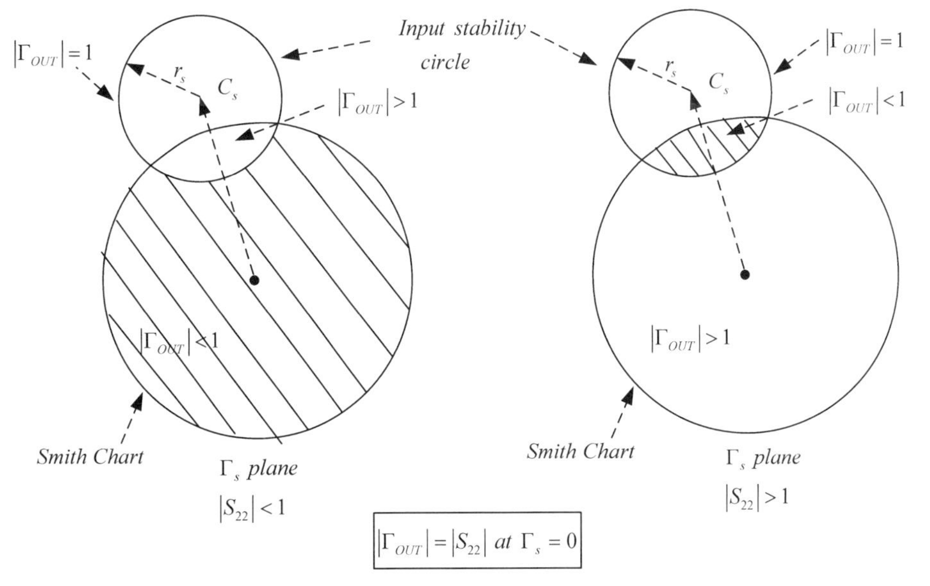

Figure 5.19 Smith chart illustrating stable and unstable regions in the Γ_s plane for (a) $|S_{22}| < 1$ and (b) $|S_{22}| > 1$. (*From:* [1]. © 1997. Reprinted by permission of Pearson Education, Inc., NJ.)

Next, we need to determine which area in the Smith chart represents the stable region and which area represents the unstable region. In other words, the stable region is where values of Γ_L $(|\Gamma_L| < 1)$ produce $|\Gamma_{IN}| < 1$ and where values of Γ_s $(|\Gamma_s| < 1)$ produce $|\Gamma_{OUT}| < 1$. The unstable region is where values of Γ_L $(|\Gamma_L| < 1)$ produce $|\Gamma_{IN}| > 1$ and where values of Γ_s $(|\Gamma_s| < 1)$ produce $|\Gamma_{OUT}| > 1$. From (5.45), for $\Gamma_L = 0$ it follows that $|\Gamma_{IN}| = |S_{11}|$. Hence, if the magnitude of S_{11} is less than 1, then $|\Gamma_{IN}| < 1$ when $\Gamma_L = 0$. That is, the center of the Smith chart in Figure 5.18(a) represents a stable operating point when $|S_{11}| < 1$. On the other hand, if $|S_{11}| > 1$ then $|\Gamma_{IN}| > 1$ when $\Gamma_L = 0$, and the center of the Smith chart represents an unstable operating point. Figure 5.18 illustrates the two cases discussed. The shaded area represents the values of Γ_L that produce a stable operation. Similarly, Figure 5.19 illustrates stable and unstable regions for Γ_s.

In an unconditional stable case (i.e., when $K > 1$ and $|\Delta| < 1$) the stability circles fall completely outside the Smith chart or completely enclosed the Smith chart. Hence, any passive terminations result in a stable condition.

5.6 Oscillation Conditions in Terms of Reflection Coefficients

In a microwave circuit a closed-loop gain function of the type in (1.4) can be developed as follows. Consider the circuit shown in Figure 5.20(a) and its flow graph in Figure 5.20(b). The coefficient $\Gamma_L(j\omega)$ represents the load reflection coefficient, and $\Gamma_{IN}(j\omega)$ the input reflection coefficient of an active device. $\Gamma_L(j\omega)$ is associated with Z_L, and $\Gamma_{IN}(j\omega)$ is associated with Z_{IN}. The input impedance Z_{IN} represents the active device impedance. The incident and reflected waves from Z_L are a_L and b_L, respectively; and from Z_{IN} the incident and reflected waves are a_{IN} and b_{IN}, respectively. The incident wave a_n represents a small noise signal generated in the circuit. From the flow graph in Figure 5.20(b) we can write

$$a_L = \frac{a_n \Gamma_{IN}(j\omega)}{1 - \Gamma_{IN}(j\omega)\Gamma_L(j\omega)} \tag{5.55}$$

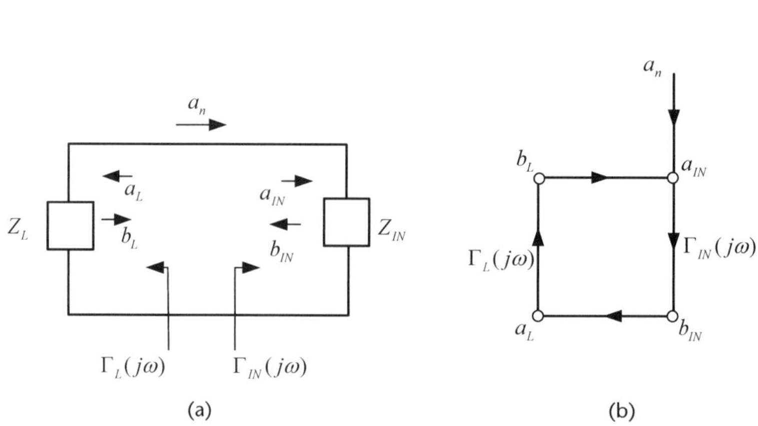

Figure 5.20 (a) A microwave circuit and (b) its flow graph.

Equation (5.55) is the closed-loop gain of the circuit in Figure 5.20(a). From (5.55) we observe that the system is unstable when some disturbance produces a growing signal a_L. This occurs when the loop gain $1 - \Gamma_{IN}(j\omega)\Gamma_L(j\omega)$ has right-half plane zeroes. The Nyquist test can be applied to analyze the function $1 - \Gamma_{IN}(j\omega)\Gamma_L(j\omega)$ by determining the encirclements of the point $\Gamma_{IN}(j\omega)\Gamma_L(j\omega) = 1$.

Equation (5.55) shows that for oscillations to occur the loop gain must be unity; that is,

$$\Gamma_{IN}(j\omega)\Gamma_L(j\omega) = 1 \tag{5.56}$$

The notation used in (5.56) is very general. A more convenient way of expressing (5.56) is to specify the frequency ω_o and the amplitude A_o at which (5.56) is satisfied; that is,

$$\Gamma_{IN}(A_o, \omega_o)\Gamma_L(\omega_o) = 1 \tag{5.57}$$

where

$$\Gamma_{IN}(A_o, \omega_o) = \frac{Z_{IN}(A_o, \omega_o) - Z_o}{Z_{IN}(A_o, \omega_o) + Z_o} \tag{5.58}$$

and

$$\Gamma_L(\omega_o) = \frac{Z_L(\omega_o) - Z_o}{Z_L(\omega_o) + Z_o} \tag{5.59}$$

where

$$Z_{IN}(A, \omega) = R_{IN}(A, \omega) + jX_{IN}(A, \omega)$$

and

$$Z_L(\omega) = R_L(\omega) + jX_L(\omega)$$

Substituting (5.58) and (5.59) into (5.57), it follows that

$$Z_{IN}(A_o, \omega_o) + Z_L(\omega_o) = 0 \tag{5.60}$$

which is, as expected, identical to (5.3). Hence, the oscillation condition (5.3) in terms of reflection coefficients is identical to the oscillation condition in terms of impedances. It also follows from (5.60) that

$$R_{IN}(A_o, \omega_o) + R_L(\omega_o) = 0$$

and

$$X_{IN}(A_o, \omega_o) + X_L(\omega_o) = 0 \tag{5.61}$$

which are the conditions for oscillation.

Example 5.2

(a) Use reflection coefficients to determine the values of R_{IN} that make the circuit in Figure 5.21(a) unstable. For simplicity in the analysis, choose $L = 1\text{H}$ and $C = 1\text{F}$.

(b) Determine the frequency of oscillation.

This example is identical to Example 5.1, except that in this example reflection coefficients are used in the analysis. In Example 5.1 we found that the circuit is unstable when $R_{IN} < -R_L$.

Solution

(a) The location of Γ_{IN} and Γ_L can be selected arbitrarily. For the circuit in Figure 5.21(a), the input and load reflection coefficients were selected at the location indicated in the figure. That is,

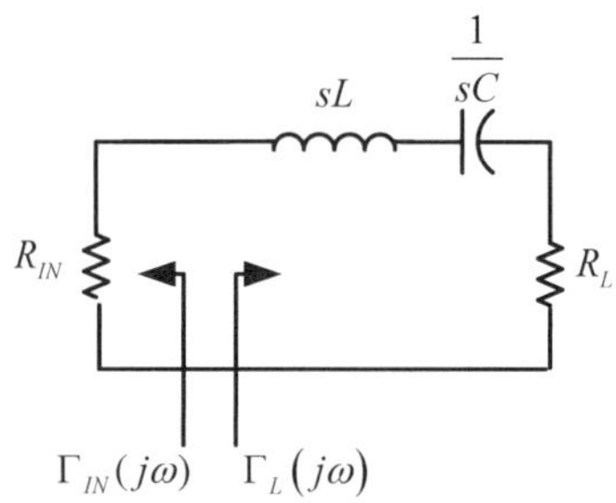

(a)

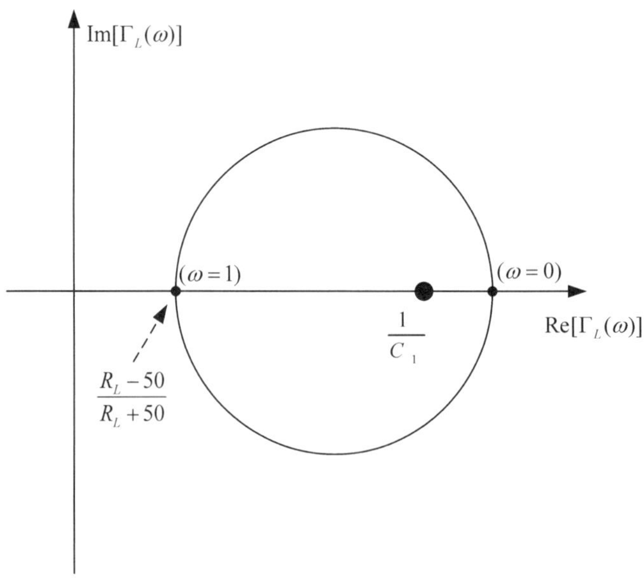

(b)

Figure 5.21 (a) Circuit for Example 5.2 and (b) the Nyquist plot of $\Gamma_{IN}(j\omega)\Gamma_L(j\omega)$. (*From:* [1]. © 1997. Reprinted by permission of Pearson Education, Inc., NJ.)

$$Z_{IN} = R_{IN}$$

and

$$Z_L(s) = s + \frac{1}{s} + R_L = \frac{s^2 + R_L s + 1}{s}$$

where $s = j\omega$. Then, in a 50-Ω system

$$\Gamma_{IN}(s) = \frac{R_{IN} - 50}{R_{IN} + 50} = C_1$$

and

$$\Gamma_L(s) = \frac{Z_L(s) - 50}{Z_L(s) + 50} = \frac{s^2 + s(R_L - 50) + 1}{s^2 + s(R_L + 50) + 1}$$

The selection of $Z_{IN} = R_{IN}$ made $\Gamma_{IN}(s)$ a constant, which is denoted by C_1.

In order to apply the Nyquist test, we form

$$\Gamma_{IN}(j\omega)\Gamma_L(j\omega) = 1$$

or

$$\frac{(1 - \omega^2) + j\omega(R_L - 50)}{(1 - \omega^2) + j\omega(R_L + 50)} = \frac{1}{C_1} \tag{5.62}$$

The Nyquist plot of (5.62) is shown in Figure 5.21(b) for $C_1 > 0$ (the plot for $C_1 < 0$ is quite similar). The circuit is unstable when the number of encirclements of the point $1/C_1$ is nonzero. Figure 5.21(b) shows that the circuit is unstable when

$$\frac{R_L - 50}{R_L + 50} < \frac{1}{C_1} < 1$$

Since $1/C_1 < 1$ for any value of C_1, it follows that the circuit is unstable when

$$\frac{R_L - 50}{R_L + 50} < \frac{R_{IN} + 50}{R_{IN} - 50}$$

Since $R_{IN} < 0$, we can express the inequality in the form

$$(R_L - 50)(R_{IN} - 50) > (R_{IN} + 50)(R_L + 50)$$

which reduces to

$$R_L + R_{IN} < 0$$

or simply

$$R_{IN} <- R_L$$

which is the expected result for the circuit in Figure 5.21(a) to be unstable.
(b) From (5.61), with $X_{IN} = 0$, the frequency of oscillation is given by

$$X_L(j\omega_o) = j\omega_o + \frac{1}{j\omega_o} = j\left(\omega_o - \frac{1}{\omega_o}\right) = 0$$

or

$$\omega_o = 1 \text{ rad/s}$$

5.7 Two-Port Negative-Resistance Oscillators

The schematic diagrams for one-port negative-resistance oscillators are shown in Figures 5.1 and 5.4. The general block diagrams for two-ports negative-resistance oscillators are shown in Figure 5.22(a, b). The transistor network is characterized by its S parameters, Z_T is the terminating network impedance, and Z_L is the load impedance. Observe the notation used in Figure 5.22(a, b), which shows that in an oscillator either port of the transistor can be used as the terminating port. Once the terminating port is selected, the other port is referred to as the input port. The load-matching network is connected to the input port, in agreement with the one-port notation.

When the two-port is potentially unstable, an appropriate Z_T permits the two-port to be represented as a one-port negative-resistance device with input impedance Z_{IN}, as shown in Figure 5.1, or with input admittance Y_{IN}, as shown in Figure 5.4.

When the input port is made to oscillate, the terminating port also oscillates. The fact that both ports are oscillating can be proven as follows. The input port is oscillating when

$$\Gamma_{IN}\Gamma_L = 1 \tag{5.63}$$

and from (5.45) and (5.63) (using the nomenclature in Figure 5.2.2), it follows that

$$\Gamma_L = \frac{1}{\Gamma_{IN}} = \frac{1 - S_{22}\Gamma_T}{S_{11} - \Delta\Gamma_T}$$

or

$$\Gamma_T = \frac{1 - S_{11}\Gamma_L}{S_{22} - \Delta\Gamma_L} \tag{5.64}$$

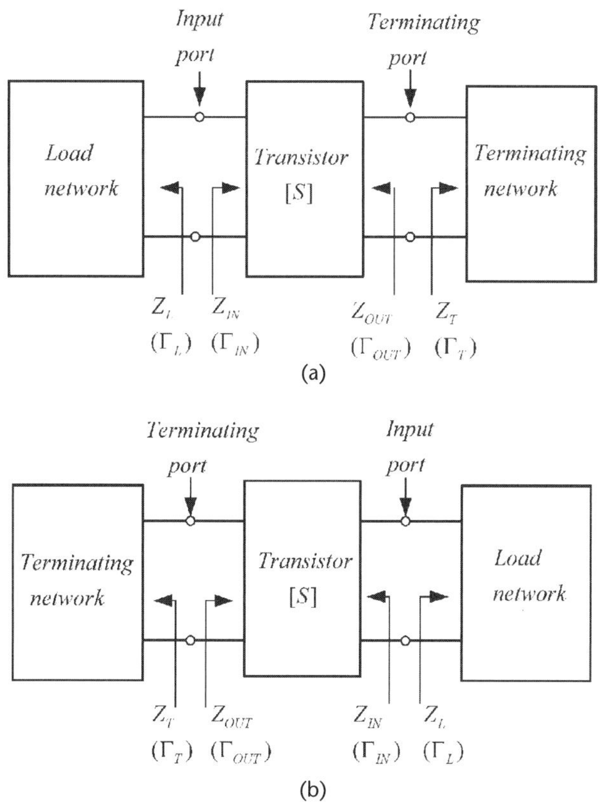

Figure 5.22 (a) Two-port oscillator models and (b) alternate representation. (*From:* [1]. © 1997. Reprinted by permission of Pearson Education, Inc., NJ.)

Also, from (5.46)

$$\Gamma_{OUT} = \frac{S_{22} - \Delta\Gamma_L}{1 - S_{11}\Gamma_L} \tag{5.65}$$

Hence, from (5.64) and (5.65) it follows that

$$\Gamma_{OUT}\Gamma_T = 1$$

which shows that the terminating port is also oscillating.

A basic design procedure, based on the small-signal S parameters, for a two-port oscillator can be as follows:

1. Use a potentially unstable transistor at the desired frequency of oscillation ω_O.
2. Design the terminating network to make $|\Gamma_{IN}| > 1$. Series or shunt feedback can be used to increase $|\Gamma_{IN}|$.
3. Design the load network to resonate Z_{IN}, and to satisfy the start of oscillation condition in (5.8). That is, let

$$X_L(\omega_o) = -X_{IN}(\omega_o) \tag{5.66}$$

and

$$R_L = \frac{|R_{IN}(0, \omega)|}{3} \tag{5.67}$$

This design procedure is popular due to its high rate of success. However, it is based on the small-signal S parameters and, therefore, cannot be used to properly characterize the oscillator performance. For example, the frequency of oscillation will shift somewhat from its designed value at ω_o. This occurs because the oscillation power increases until the negative resistance is equal to the load resistance and X_{IN} varies as a function of A (i.e., as a function of the oscillation power). Also, the procedure cannot be used to predict the oscillator power, or to calculate the harmonics.

The following example illustrates the previous design procedure.

Example 5.3

Design the common-gate (CG) microwave oscillator shown in Figure 5.23(a) to oscillate at 9 GHz. The S parameters of the FET at 9 GHz in a CG configuration, at a given Q point, are

$$S_{11} = 0.5 \angle -95^\circ$$

$$S_{21} = 1.1 \angle 44^\circ$$

$$S_{12} = 0.35 \angle -31^\circ$$

$$S_{22} = 0.8 \angle 46^\circ$$

Solution

This example only considers the ac part of the design. From (5.48) and (5.49) it follows that at 9 GHz: $K = 0.355$ and $|\Delta| = 0.405$. Therefore, since $K < 1$ the transistor is potentially unstable. Either port can be used as the terminating port. For this design, port 2 is selected as the terminating port, as shown in Figure 5.23(b). Using (5.51) and (5.52), the stability circle radius and center at port 2 are: $r_T = 0.808$ and $C_T = 1.493 \angle -59.8^\circ$. The terminating-port stability circle is drawn in Figure 5.23(c). Any Γ_T in the shaded region produces a negative resistance (i.e., $|\Gamma_{IN}| > 1$). The reflection coefficient Γ_T is selected at point A in Figure 5.23(c), or $\Gamma_T = 1 \angle -67.38^\circ$ (i.e., $Z_T = -j75\Omega$). This capacitive reactance at 9 GHz can be implemented with an open-circuited 50-Ω microstrip line of length 0.092λ.

The value of Γ_{IN} follows from (5.45), namely $\Gamma_{IN} = 1.49 \angle -100.43^\circ$. The associated input impedance is $Z_{IN} = -16.25 - j38.95\Omega$. The load-matching network is designed using (5.66) and (5.67), namely,

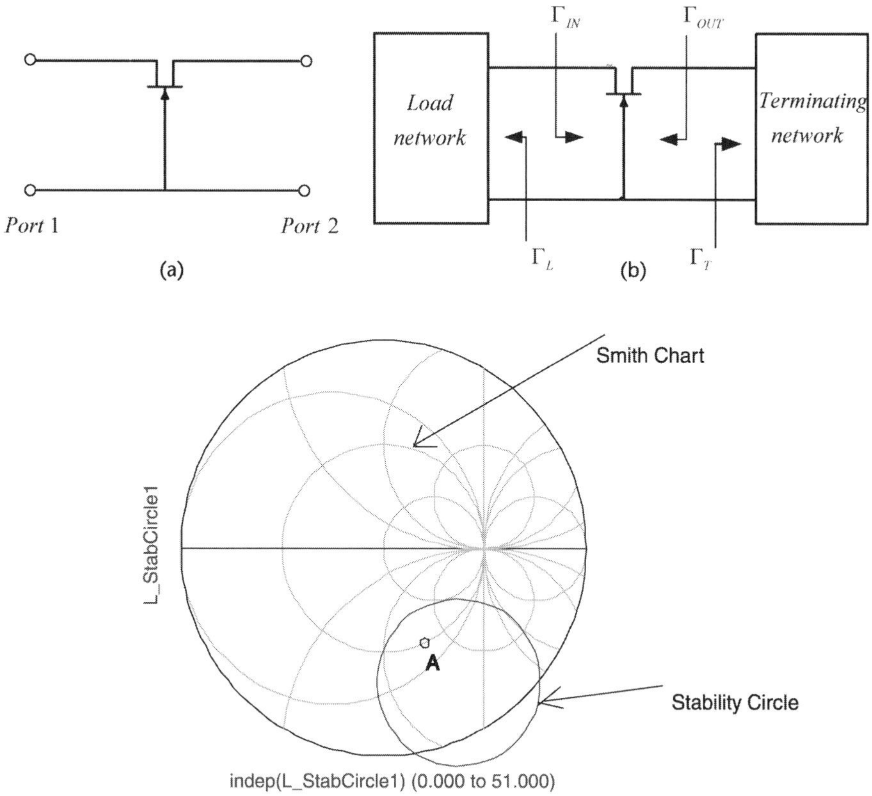

freq	K	delta
9.000 GHz	0.355	0.405 / –106.179

(c)

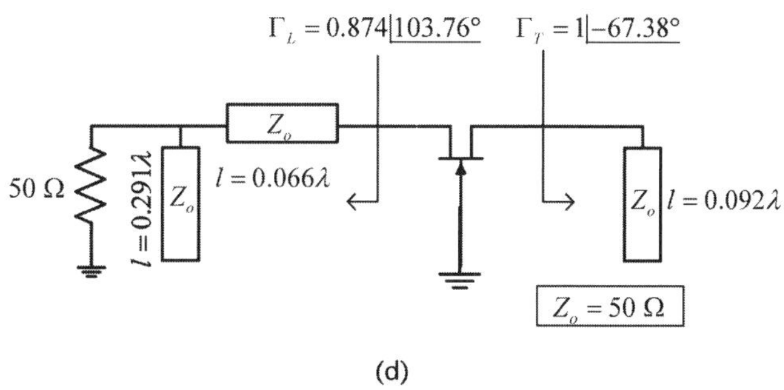

Figure 5.23 (a) The FET in a CG configuration, (b) a CG oscillator circuit, (c) the terminating port stability circle, and (d) implementation of the oscillator.

$$X_L = -X_{IN} = 38.95\Omega$$

and

$$R_L = \frac{|R_{IN}|}{3} = \frac{6.25}{3} = 5.42\Omega$$

Hence, $Z_L = 5.42 + j38.95\Omega$, or $\Gamma_L = 0.874\,\underline{|103.76°}$. The microstrip implementation of Z_L, as well as the complete ac implementation of the oscillator, is shown in Figure 5.23(d). Further analysis and simulation of this oscillator requires the large-signal model of the FET.

For negative-resistance oscillators at microwave frequencies using BJTs the common-base (CB) configuration is normally used. For FETs, the CG configuration is commonly used. As we will see in these configurations the transistors are usually potentially unstable, and this is a very desirable condition in oscillator design.

In a negative-resistance oscillator, the capacitances of the transistor provide some or all of the feedback needed for oscillation. However, a properly designed series-feedback network can significantly increase the negative-resistance presented by the two-port network. In other words, the unstable region of the two-port network can be enhanced by the use of series feedback. In general, for BJTs a series-feedback inductor of a few nanohenries, as shown in Figure 5.24(a), enhances the unstable regions. For FETs, see Figure 5.24(b), either a series-feedback inductor or a series-feedback capacitor of a few picofarads enhances the unstable regions. The purpose of the series feedback element is to provide positive feedback, making the configuration more unstable. Analytical and mapping methods that can be used to design the series-feedback network are presented in Appendix B.

It is not always necessary to design the feedback network to maximize $|\Gamma_{IN}|$ or $|\Gamma_{OUT}|$. In fact, the next step in many designs involves the selection of the terminating network and the observation of its effect on the input impedance. In a potentially unstable two port, Γ_T must be selected to be in the unstable region so that $|\Gamma_{IN}| > 1$. Series feedback can easily enhance the unstable region so that (in some cases), even with $\Gamma_T = 0$ (i.e., a 50-Ω termination), $|\Gamma_{IN}|$ is greater than 1.

The next example illustrates in a simple manner the use of series feedback and the selection of the terminating network.

Example 5.4

Design a 2.75-GHz oscillator using a BJT in a CB configuration. The S parameters at 2.75 GHz, at a given Q point, are

$$S_{11} = 0.9\,\underline{|150°}$$

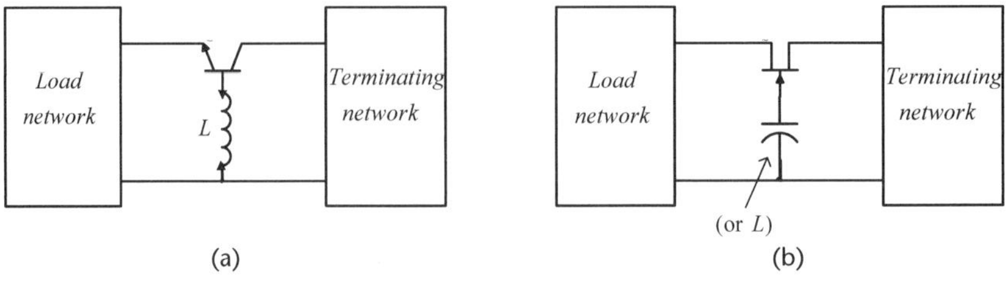

Figure 5.24 Typical series feedback for (a) a BJT oscillator and (b) an FET oscillator.

$$S_{21} = 1.7 \angle{-80^\circ}$$

$$S_{12} = 0.07 \angle{120^\circ}$$

$$S_{22} = 1.08 \angle{-56^\circ}$$

Solution

The transistor is potentially unstable at 2.75 GHz (since $K =- 0.64$). The region of instability in the Smith chart can be increased using series feedback, as shown in Figure 5.24(a). Using the methods discussed in Appendix B, or using the tuning capabilities of ADS, it follows that the largest magnitudes of S_{11} and S_{22} were obtained using a series feedback inductor of value $L = 1.45$ nH. With $L = 1.45$ nH, the resulting S parameters for the network in Figure 5.25(a) are

$$S_{11} = 1.72 \angle{100^\circ}$$

$$S_{21} = 2.08 \angle{-136^\circ}$$

$$S_{12} = 0.712 \angle{94^\circ}$$

$$S_{22} = 1.16 \angle{-102^\circ}$$

Either port can be used as the terminating port. In this design the emitter-to-ground port was selected for the load matching network, and the collector-to-ground port for the terminating network. The terminating port stability circle is shown in Figure 5.25(b). Observe the large region of instability and note that the center of the Smith chart is unstable. Hence, a 50-Ω resistor at the terminating port will ensure that $|\Gamma_{IN}| > 1$. However, a 50-Ω termination might not be the most suitable since for the associated value of Γ_{IN} the required value of Γ_L for the oscillator might be difficult to implement. In addition, some tuning capabilities are needed in the matching network to attain the desired frequency of oscillation.

The values of Γ_T and Γ_{IN} are related by

$$\Gamma_{IN} = S_{11} + \frac{S_{12} S_{21} \Gamma_T}{1 - S_{22} \Gamma_T}$$

One way that the value of Γ_T can be selected is by varying Γ_T in the unstable region and calculating the associated value of Γ_{IN}. The calculations are shown in Table 5.1, where the magnitude of Γ_T was set at 0.25, 0.5, 0.75, and 1, and the angle of Γ_T was varied from 90° to 270° in increments of 36°. From Table 5.1 it is seen that there are many convenient values of Γ_T that can be selected for this design. The value selected was $\Gamma_T = 0.5 \angle{162^\circ}$ and it follows that $\Gamma_{IN} = 2.31 \angle{117.6^\circ}$ (or $Z_{IN} =- 25.6 + j24\Omega$). From (5.66) and (5.67) the impedance of the load-matching network can be selected as $Z_L = 8.5 - j24\Omega$.

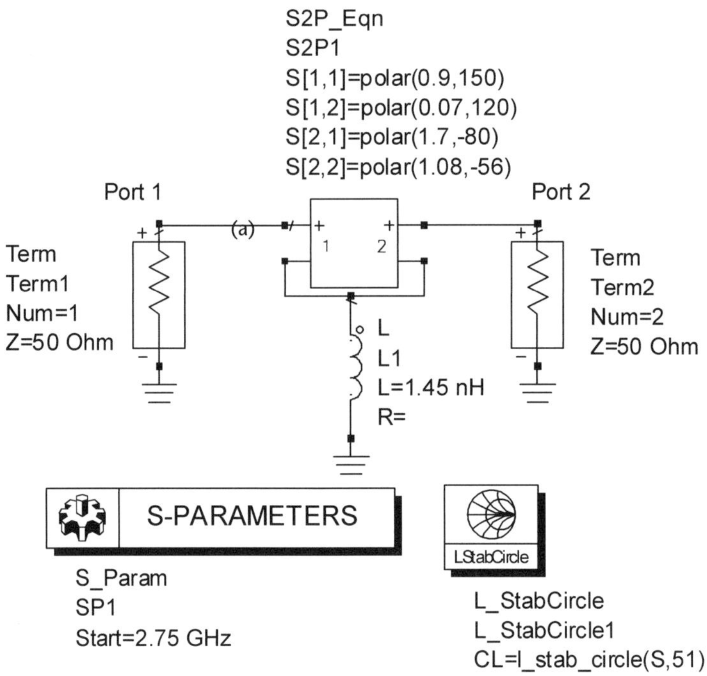

(a)

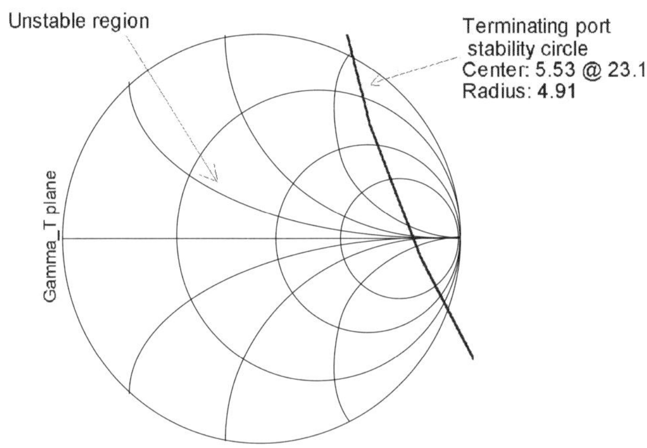

(b)

Figure 5.25 (a) BJT with series inductive feedback, (b) terminating port stability circle, and (c) the oscillator circuit.

A design for the terminating and load networks is illustrated in Figure 5.25(c). Power is delivered to the resistors in the load and terminating networks, the 50-Ω resistor can be used to extract the oscillator signal.

In what follows, design procedures for negative resistance oscillators will be discussed and design examples given to illustrate the procedures. The following

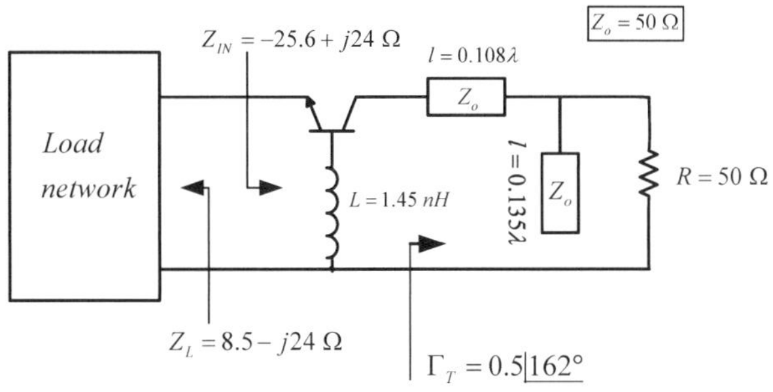

(c)

Figure 5.25 (Continued.)

Table 5.1 Γ_{IN} Versus Γ_T

Γ_T	Γ_{IN}	Γ_T	Γ_{IN}
$1\angle 90°$	$3.67\angle -67.1°$	$0.5\angle 90°$	$2.80\angle 67°$
$1\angle 126°$	$3.79\angle 154.5°$	$0.5\angle 126°$	$3.11\angle 104.8°$
$1\angle 162°$	$3.67\angle 137.3°$	$0.5\angle 162°$	$2.31\angle 117.6°$
$1\angle 198°$	$1.78\angle 130.5°$	$0.5\angle 198°$	$1.88\angle 118.9°$
$1\angle 234°$	$1.52\angle 125.9°$	$0.5\angle 234°$	$1.62\angle 117.2°$
$1\angle 270°$	$1.33\angle 121.7°$	$0.5\angle 270°$	$1.44\angle 113.9°$
$0.75\angle 90°$	$4.68\angle 18.5°$	$0.25\angle 90°$	$2.05\angle 87.8°$
$0.75\angle 126°$	$4.13\angle 127.1°$	$0.25\angle 126°$	$2.21\angle 98.5°$
$0.75\angle 162°$	$2.35\angle 128.7°$	$0.25\angle 162°$	$2.07\angle 106.8°$
$0.75\angle 198°$	$1.84\angle 125.6°$	$0.25\angle 198°$	$1.86\angle 110.2°$
$0.75\angle 234°$	$1.57\angle 122.2°$	$0.25\angle 234°$	$1.68\angle 110.2°$
$0.75\angle 270°$	$1.38\angle 118.3°$	$0.25\angle 270°$	$1.55\angle 108.1°$

examples deal with the specific design of a 1-GHz oscillator using the BJT AT41411 in a CB configuration. The Agilent AT41411 is a BJT (npn type) for use in low-noise and wideband amplifiers, mixers, and oscillators at RF and microwave frequencies.

Example 5.5

(a) Analyze the stability of the BJT AT41411 in a CB configuration at 1 GHz.
(b) Analyze the effects of a series-feedback inductor [as in Figure 5.24(a)] on the stability of the BJT AT41411.

Solution

(a) The typical S parameters of the AT41411 in a CE configuration at $V_{CE} = 8$V and $I_C = 25$ mA in a 50-Ω system are shown in Table 5.2.

From the CE S-parameters, the CB S-parameters are calculated (see Table 5.3). The simulation shown in Figure 5.26(a) also gives the CB S-parameters.

Table 5.2 CE *S*-Parameters of the AT41411

		S		
Frequency	*S*(1, 1)	*S*(1, 2)	*S*(2, 1)	*S*(2, 2)
500.0 MHz	0.460/–137.000	0.025/56.000	13.210/100.000	0.570/–26.000
1.000 GHz	0.430/–175.000	0.038/58.000	6.850/80.000	0.520/–29.000
1.500 GHz	0.440/163.000	0.048/61.000	4.630/67.000	0.510/–32.000
2.000 GHz	0.470/148.000	0.062/61.000	3.470/56.000	0.500/–37.000
2.500 GHz	0.500/140.000	0.071/60.000	2.820/50.000	0.470/–39.000
3.000 GHz	0.530/132.000	0.092/61.000	2.360/40.000	0.460/–48.000
3.500 GHz	0.550/122.000	0.105/57.000	2.020/30.000	0.450/–60.000
4.000 GHz	0.560/112.000	0.122/53.000	1.740/19.000	0.450/–73.000

Table 5.3 CB *S*-Parameters of the AT41411

		S		
Frequency	*S*(1, 1)	*S*(1, 2)	*S*(2, 1)	*S*(2, 2)
500.0 MHz	0.929/175.209	0.009/127.126	1.925/–8.209	1.015/–4.481
1.000 GHz	0.982/170.132	0.037/152.131	2.002/–17.026	1.068/–9.239
1.500 GHz	1.084/164.744	0.083/157.542	2.148/–26.523	1.150/–14.274
2.000 GHz	1.252/158.536	0.162/160.089	2.396/–37.870	1.289/–20.885
2.500 GHz	1.410/151.586	0.245/152.997	2.590/–50.032	1.392/–29.495
3.000 GHz	1.778/140.811	0.450/148.793	3.194/–67.976	1.740/–42.549
3.500 GHz	2.467/117.608	0.934/128.975	4.303/–99.046	2.406/–68.392
4.000 GHz	2.784/66.432	1.550/81.511	4.846/–160.030	2.757/–122.137

The stability of the transistor in the CB configuration can be analyzed using the ADS simulation shown in Figure 5.26(a) by varying the frequency from 500 MHz to 4 GHz. That is, setting in the S_Param component: Start = 500 MHz and Stop = 4 GHz. The AT41411 in a CB configuration is potentially unstable at all frequencies. Specifically, in Figure 5.26(a, b) the *S* parameters for the device in the CB configuration at 1 GHz are listed and the stability circle of port 1 is drawn.
(b) There are several methods that we can use to analyze the effects of a series inductor on the stability of the circuit. The most complete way is to map the feedback plane onto the Γ_{IN} plane. Another way is to find the value of L that maximizes $|\Gamma_{IN}|$. Both of these methods are discussed in Appendix B. A simple way to analyze the effects of the feedback inductor is to calculate its effect on the μ factor; that is, to plot the μ factor versus the series inductance value. One could also analyze the effects of the series inductance on the K factor, since this is the factor that usually becomes less than 1 (in the K and $|\Delta|$ test) in BJTs. The previous method uses a 50-Ω termination. Later, the effect of a different termination is analyzed.

Figure 5.27 shows the ADS simulation used and the resulting plot of μ versus the series inductance at 1 GHz. It is seen that a value of 2 nH makes the configuration more unstable, since μ decreases to about –0.9. However, for this circuit the series inductor is not needed since the configuration is already unstable at 1 GHz and the improvement was minimal. In fact, a plot of the Port 1 stability circle for the circuit in Figure 5.27 is almost similar to that in Figure 5.26(b).

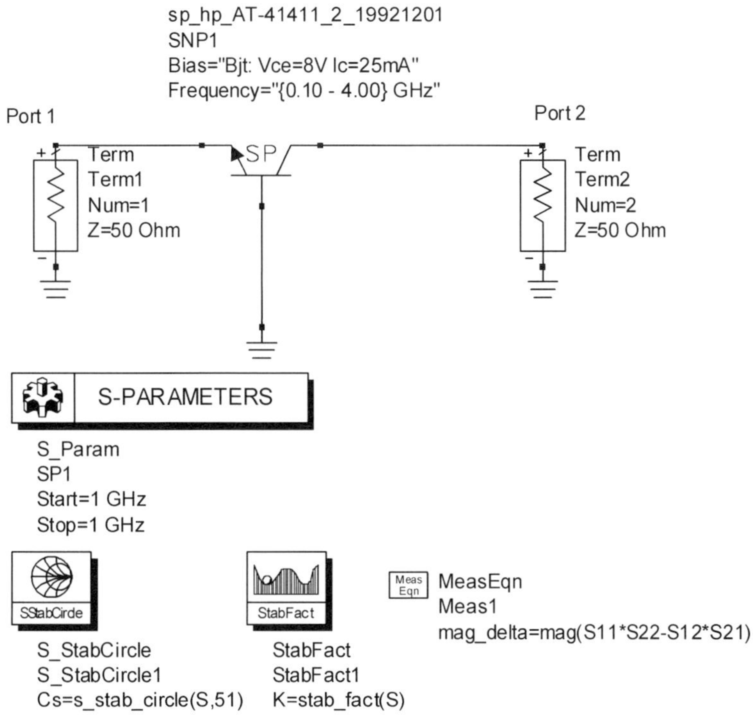

(a)

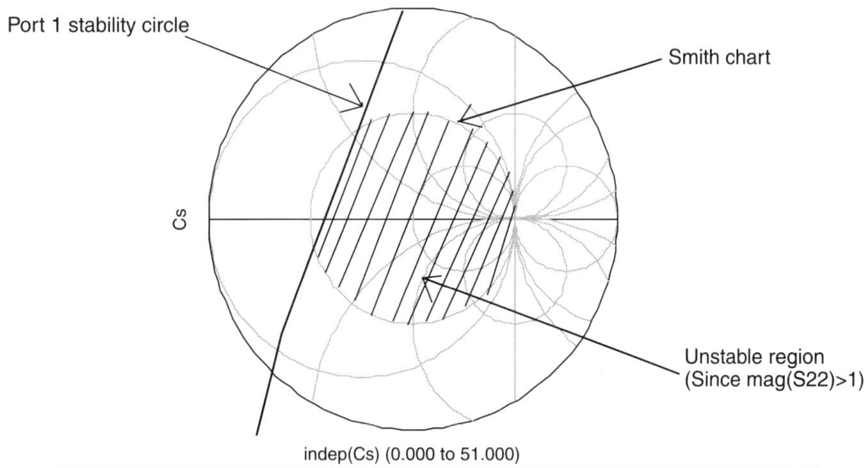

freq	S			
	S(1,1)	S(1,2)	S(2,1)	S(2,2)
1.000GHz	0.982 / 170.132	0.037 / 152.131	2.002 / -17.026	1.068 / -9.239

freq	K	mag_delta
1.000GHz	-0.941	0.981

(b)

Figure 5.26 (a) Stability simulation for the AT-41411 and (b) port 1 stability circle at 1 GHz.

5.8 The Terminating Network

The general block diagrams for two-port negative resistance oscillators are shown in Figure 5.22. The input reflection coefficient Γ_{IN} and the terminating network coefficient Γ_T are related by

$$\Gamma_{IN} = \frac{S_{11} - \Delta\Gamma_T}{1 - S_{22}\Gamma_T} \tag{5.68}$$

Equation (5.68) describes the mapping of the Γ_T plane onto the Γ_{IN} plane. The mapping of $|\Gamma_T| = 1$ onto the Γ_{IN} plane gives information about the passive impedances at the terminating port that will make $|\Gamma_{IN}| > 1$. The $|\Gamma_T| = 1$ circle maps onto a circle in the Γ_{IN} plane with center at

$$C_{IN} = \frac{S_{11} - \Delta S_{22}^*}{1 - |S_{22}|^2}$$

and radius of

$$r_{IN} = \frac{|S_{12}S_{21}|}{\left|1 - |S_{22}|^2\right|}$$

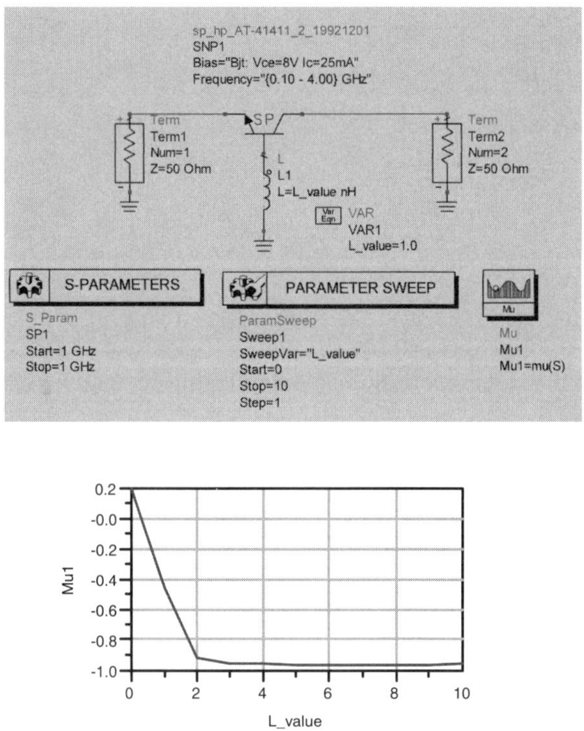

Figure 5.27 Stability simulation for the AT-41411 with a series inductor.

Figure 5.28 shows a typical mapping of the Γ_T onto the Γ_{IN} plane, where $|\Gamma_{IN}|_{max}$ is the largest magnitude that Γ_{IN} can attain. $|\Gamma_{IN}|_{max}$ is obtained when the value of Γ_{IN} on the $|\Gamma_T| = 1$ circle has the same phase angle as C_{IN}, or

$$|\Gamma_{IN}|_{max} = \left(|C_{IN}| + r_{IN}\right) \underline{\lfloor C_{IN}}$$

Of course, the selection of Γ_T does not have to produce the maximum value of $|\Gamma_{IN}|_{max}$. The main point is that the selection of Γ_T must produce sufficient negative resistance in Z_{IN}.

The mapping in Figure 5.28 can also be generated with ADS. Figure 5.29 shows the simulation used to map the Γ_T plane onto the Γ_{IN} plane. The terminating network is represented by the S1P_Eqn component which is a one-port reflection coefficient defined by $S(1, 1)$, where $S(1, 1) = \Gamma_T$. The values of Γ_T are generated by the program, as described in the VAR component. The ParamSweep component sweeps the resistance (r) values by fixing the reactance (x), and vise versa. By fixing x at 0 and ±1 and sweeping r, the simulation generates the mapping of the constant normalized reactances. When r is fixed at 0 and 1 and x is swept, the mapping of the constant normalized resistances is generated. The selected values of r and x are usually sufficient to properly view the mapping. To generate a smooth mapping the number of points calculated for both variables were selected as described in Figure 5.29.

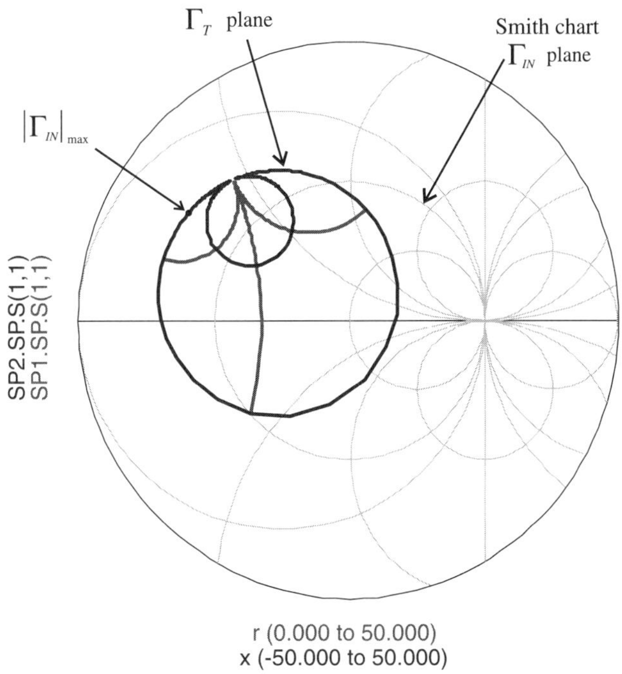

Figure 5.28 Typical mapping of Γ_T onto Γ_{IN}.

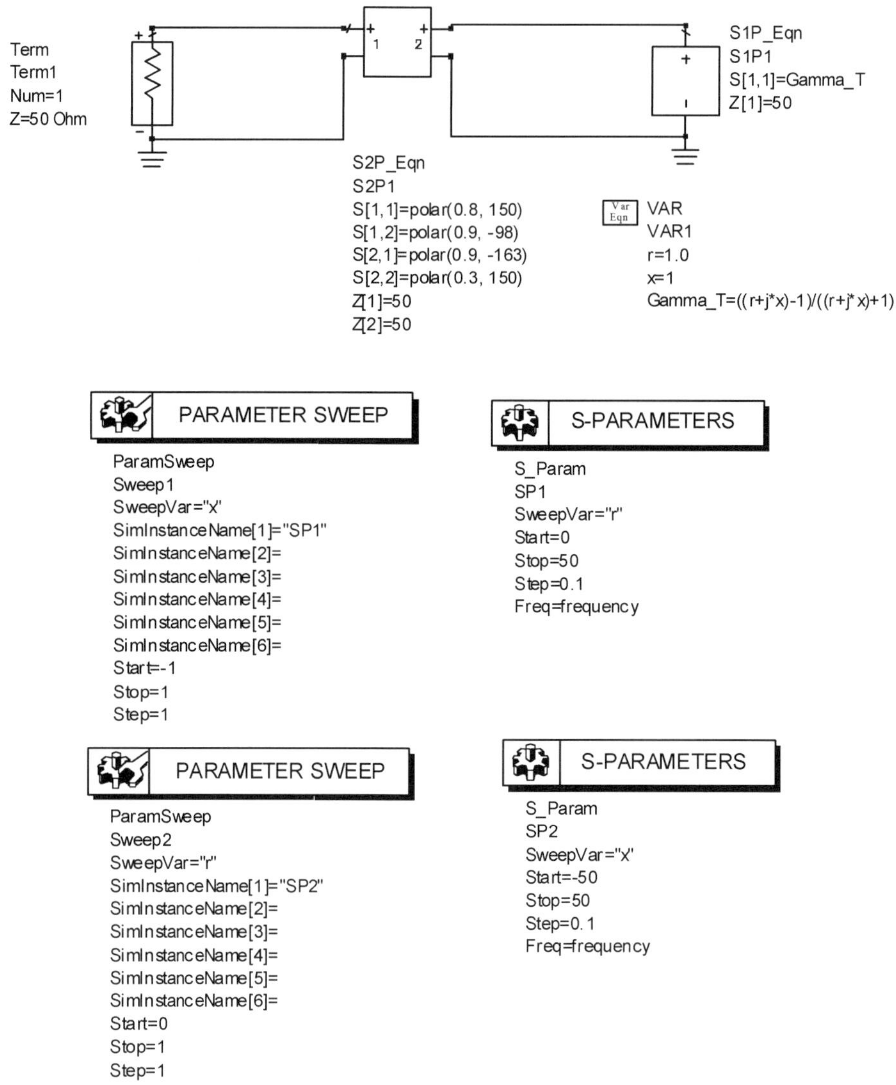

Figure 5.29 ADS simulation used to generate the mapping of the Γ_T plane onto the Γ_{IN} plane.

While the mapping can be generated as described in Figure 5.29 (i.e., using the transistor's S parameter) it is more accurate to generate the mapping using the nonlinear model for the transistor.

Example 5.6

(a) Show the mapping of the Γ_T plane onto the Γ_{IN} plane for the AT41411 in a CB configuration at 1 GHz.

(b) Select an appropriate value of Γ_T (and the associated Γ_{IN}) for part (a).

Solution

(a) Figure 5.30 shows the ADS simulation used and the resulting mapping of Γ_T onto the Γ_{IN} plane for the AT41411 at 1 GHz. The nonlinear model of the transistor was used, and the Q point is at $V_{CE} = 8\text{V}$ and $I_C = 25$ mA. The mapping shows an inversion since the $r = 1$ circle maps outside the $r = 0$ circle.

(b) From Figure 5.30, Γ_T and its resulting Γ_{IN} can be selected so that $|\Gamma_{IN}| > 1$. A value of $|\Gamma_{IN}| > 1$ can be selected with Γ_T on the $r = 1$ circle at the point m1. The value of Γ_T associated with Γ_{IN} is obtained from

$$\Gamma_T = \frac{\Gamma_{IN} - S_{11}}{\Gamma_{IN} S_{22} - \Delta} \tag{5.69}$$

that is, at m1:

$$\Gamma_T = 0.961 \underline{|15.93^\circ}$$

and

$$\Gamma_{IN} = 1.545 \underline{|154.23^\circ}$$

These values are displayed in Figure 5.30.

5.9 Oscillation-Conditions Simulations

ADS uses the components OscTest and OscPort to analyze and simulate the oscillator. The purposes of these components are as follows.

5.9.1 OscTest

OscTest is an ADS probe component that is used to evaluate the loop-gain associated with a circuit. The OscTest probe component and its three-port S parameter values are shown in Figure 5.31. OscTest is basically a three-port circulator with port 1 excited by the incident wave a_1, and with ports 2 and 3 defined by the reflection coefficients Γ_2 and Γ_3, respectively. The S parameters and the reflection coefficients are defined in a Z_o system (usually 50Ω). That is,

$$\Gamma_2 = \frac{a_2}{b_2} = \frac{Z_2 - Z_o}{Z_2 + Z_o}$$

and

$$\Gamma_3 = \frac{a_3}{b_3} = \frac{Z_3 - Z_o}{Z_3 + Z_o}$$

Since $S_{21} = S_{32} = S_{13} = 1$ and all other S parameters are zero, it follows that $b_2 = a_1$, $b_3 = a_2$, and $b_1 = a_3$. Then,

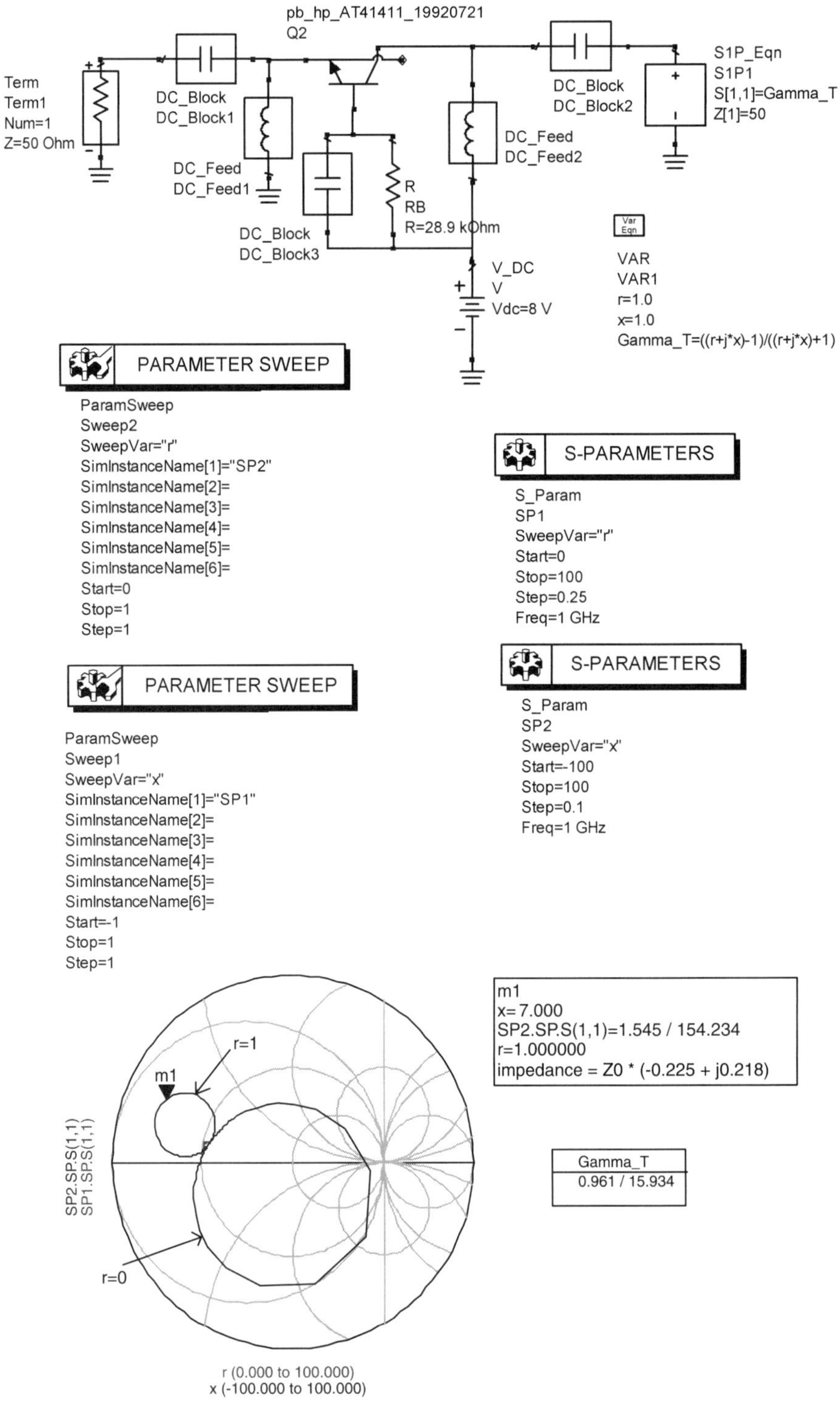

Figure 5.30 Mapping of the Γ_T plane for the A-41411.

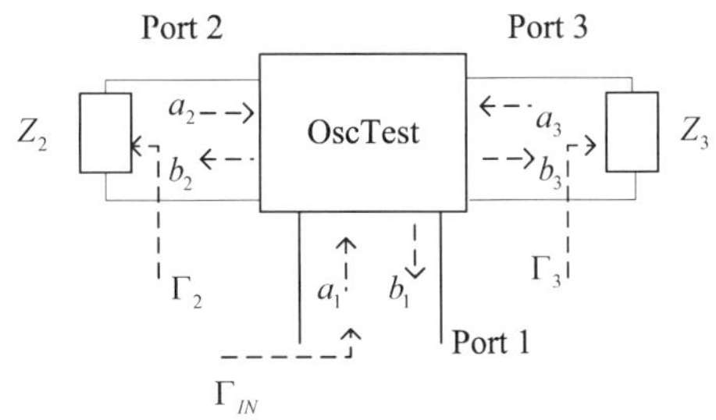

$$[S] = \begin{bmatrix} S_{11} = 0 & S_{12} = 0 & S_{13} = 1 \\ S_{21} = 1 & S_{22} = 0 & S_{23} = 0 \\ S_{31} = 0 & S_{32} = 1 & S_{33} = 0 \end{bmatrix}$$

Figure 5.31 The OscTest component.

$$b_1 = a_3 = \Gamma_3 b_3 = \Gamma_3 a_2 = \Gamma_3 \Gamma_2 b_2 = \Gamma_3 \Gamma_2 a_1$$

Defining the ratio of b_1 to a_1 as $\Gamma_{1,IN}$ gives

$$\Gamma_{1,IN} = \frac{b_1}{a_1} = \Gamma_2 \Gamma_3$$

In other words, $\Gamma_{1,IN}$ is the loop gain.

OscTest calculates $\Gamma_{1,IN}$, and it is denoted in ADS simply by $S(1, 1)$ [or $S(1, 1) = \Gamma_2\Gamma_3$].

In an NRO, Γ_2 and Γ_3 represent the input and the load reflection coefficients, respectively. The use of OscTest in an NRO is illustrated in Figure 5.32. It is recommended that OscTest be placed between the negative resistance and the load network, pointing in the direction of the load network.

In the case of a tuned-circuit oscillator the OscTest should be placed between the amplifier and the tuned circuit, as shown in Figure 5.33. The tuned circuit provides the positive feedback. Figure 5.33 shows that OscTest is placed in the direction of gain injection at the point where the feedback loop is normally broken for the calculation of the open-loop gain.

A typical result of the loop gain from OscTest is shown in Figure 5.34(a). It is a magnitude-phase plot of $S(1, 1)$, which is equal to the loop gain [i.e., $S(1,1) = \Gamma_2\Gamma_3$ in Figure 5.32]. This particular plot shows that the circuit has the potential to oscillate at 800 MHz, where $|S(1, 1)| > 1$, and $\angle S(1, 1) = 0°$.

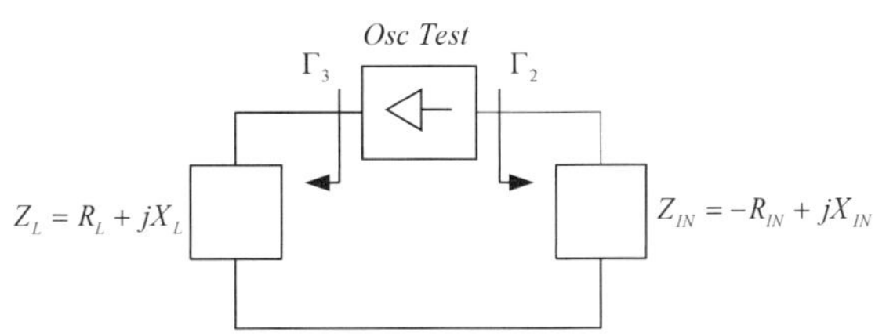

Figure 5.32 OscTest in a negative-resistance oscillator.

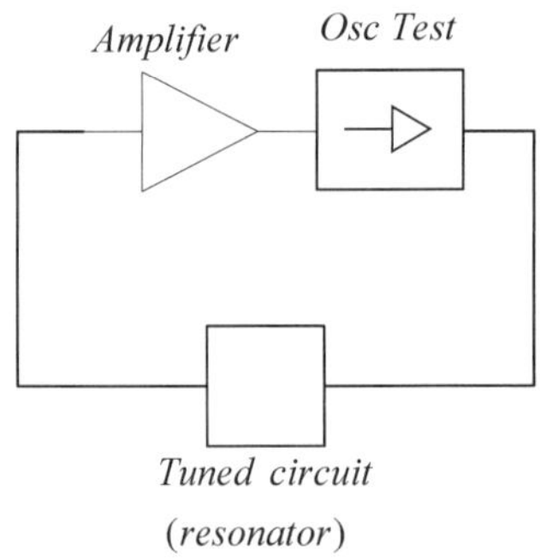

Figure 5.33 OscTest in a tuned-circuit oscillator.

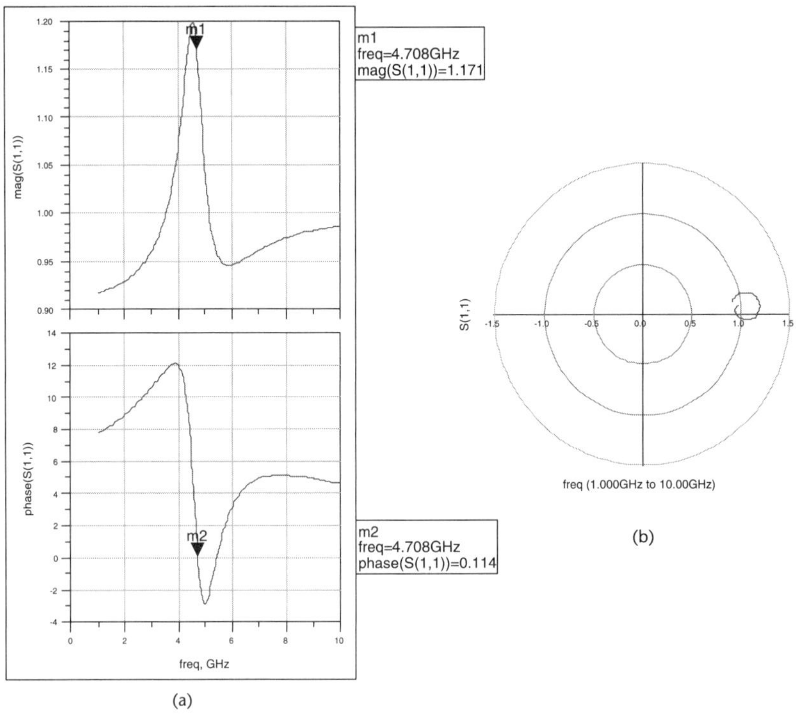

Figure 5.34 (a) Typical loop-gain results from OscTest and (b) a typical Nyquist plot from OscTest.

While most microwave circuits that have the potential to oscillate using the loop-gain test will properly oscillate, this test is not sufficient to determine if the circuit will oscillate. Hence, the next step is to generate the Nyquist plot and to apply the Nyquist test. The Nyquist plot is also generated by OscTest. It is a polar plot of $S(1, 1)$ as shown in Figure 5.34(b).

5.9.2 Nyquist Test

The stability of an active circuit can be determined from the Nyquist plot of its loop gain. The Nyquist plot is simply the frequency response of $S(1, 1)$ (obtained from OscTest) in a polar graph. For the type of oscillators encountered at RF and microwave frequencies it usually follows that the number of clockwise encirclements of the $1 + j0$ point in the Nyquist plot indicates the number of right-half plane poles of the circuit. Thus, if the Nyquist plot encloses the $1 + j0$ point clockwise it is reasonable to conclude that the system is unstable and oscillations will occur.

5.9.3 OscPort

OscPort is an ADS probe component used to calculate the oscillator waveform using a harmonic-balance simulation. It calculates the large-signal steady state form of the oscillatory signal. The OscPort component should be inserted at the point where OscTest was used and a harmonic-balance simulator must be inserted in the program. Figure 5.35 illustrates the use of the OscPort component.

5.10 Large-Signal Analysis for NROs

The design method discussed in Section 5.7 has limitations since it is based on small-signal S parameters. In this section a detailed method based on large-signal measurements is developed for the design of oscillators. Basically, the method consists of designing the terminating network so that the two-port presents an

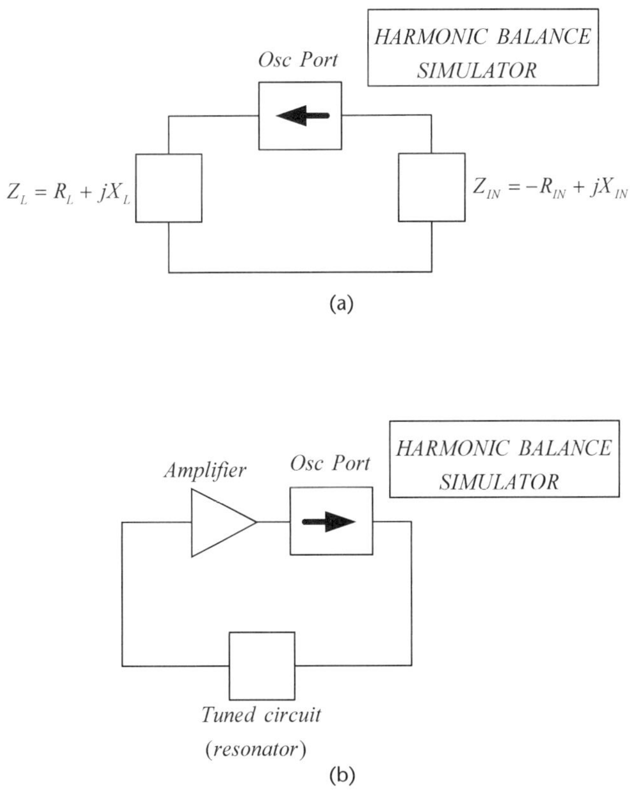

Figure 5.35 (a) OscPort in an NRO and (b) OscPort in a tuned-circuit oscillator.

appropriate negative resistance at the load port. Then, the large signal properties of the resulting one port are measured to properly characterize the oscillator. Large-signal design information can only be obtained using nonlinear techniques.

Small signal considerations can still be used to find the terminating impedance that results in a negative resistance at the input port. However, a large-signal simulation is necessary to provide the characterization of the oscillator performance, such as oscillator power and harmonic information. In large-signal analysis a nonlinear transistor model is used at the appropriate bias point.

The resulting one-port negative-resistance circuit at the load port can be characterized by determining the input impedance as a function of the input power at the oscillation frequency ω_O. This is a large-signal characterization of the one-port network, which is also called a *device-line characterization*.

The large-signal characterization is achieved by implementing the circuit shown in Figure 5.36(a) and calculating the input impedance $Z_{IN}(A, \omega_o)$ as a function of the input power at the frequency of oscillation ω_o, and calculating the added power. The added power P_{ADD} is defined as the reflected minus the available power from the source. It is given by

$$P_{ADD} = P_{AVS}\left(|\Gamma_{IN}|^2 - 1\right) \tag{5.70}$$

where

$$P_{AVS} = \frac{V_s^2}{8R_s} \tag{5.71}$$

and

$$\Gamma_{IN} = \frac{Z_{IN} - R_s}{Z_{IN} + R_s} \tag{5.72}$$

Substituting (5.71) and (5.72) into (5.70) we obtain

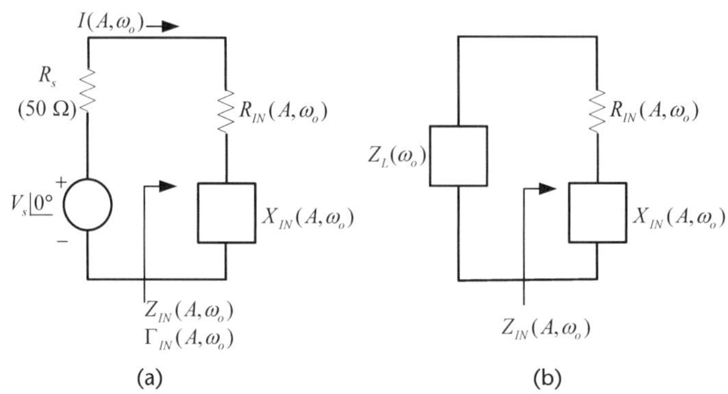

Figure 5.36 (a) Large-signal measuring circuit and (b) power delivered to Z_L.

$$P_{ADD} = \frac{V_s^2 |R_{IN}(A, \omega_o)|}{2\{[R_{IN}(A, \omega_o) + R_s]^2 + X_{IN}^2\}} = \frac{1}{2}|I(A, \omega_o)|^2 |R_{IN}(A, \omega_o)| \tag{5.73}$$

where $I(A, \omega_o)$ is the current in the circuit.

Equation (5.73) shows that the added power is the power that the negative resistance of the one-port will deliver when the current in the circuit is $I(A, \omega_o)$. If the one-port is now terminated in the load impedance $Z_L(\omega_o)$, as shown in Figure 5.36(b), where

$$Z_L(\omega_o) =- Z_{IN}(A, \omega_o) \tag{5.74}$$

the power delivered to Z_L when the current in the circuit is $I(A, \omega_o)$ is P_{ADD}. The calculation of P_{ADD} as a function of $Z_{IN}(A, \omega)$ and the use of (5.74) predicts fairly well the frequency of oscillation and the resulting oscillator power.

While in a simulation the source impedance R_s can be made arbitrary, measurement equipment has a typical source resistance of 50Ω. Hence, for the measurement of P_{ADD} to be stable the negative resistance must satisfy $|R_{IN}| < 50\Omega$.

There are several ways of implementing the load impedance $Z_L(\omega_o)$ and, of course, not all of them will lead to a stable oscillation. For stable oscillation we have to check that $Z_L(\omega_o)$ satisfies the Kurokawa condition [see (5.9) and (5.14)]. The verification of the Kurokawa condition is easily simulated in ADS.

Example 5.7

Calculate P_{ADD} and Z_{IN} for the oscillator design in Example 5.6 (see Figure 5.30).

Solution

The ADS simulation shown in Figure 5.37 is used to calculate P_{ADD} and Z_{IN} for the oscillator using the AT41411. The data is generated by sweeping the input power from –40 to 10 dBm. Table 5.4 illustrates the resulting P_{ADD} and Γ_{IN}. Also, the value of Z_{IN} is listed. The negative resistance in this design varies from a small signal value of –11Ω to a large-signal value of –6Ω, and the associated P_{ADD} values vary from about –30 to –12 dBm. The behavior of the negative resistance as a function of amplitude can be deduced from the data in Table 5.4. It is also observed that the negative resistance equivalent circuit should be treated using the impedance form.

From the results in Table 5.4, the oscillator can be designed with a power of –13.57 dBm with $Z_L =- Z_{IN} = 9.85 - j13.09\Omega$. The ac schematic of the oscillator is shown in Figure 5.37(b) where the reactance $-j13.09\Omega$ at 1 GHz is implemented with $C = 12.16$ pF.

The oscillator design in Figure 5.37(b) is now analyzed with OscTest and OscPort. OscTest is used to obtain the Nyquist plot in order to check the oscillation conditions. Also, the Kurokawa condition for stable oscillation must be checked. In Appendix B the Kurokawa condition is expressed in terms of reflection coefficients,

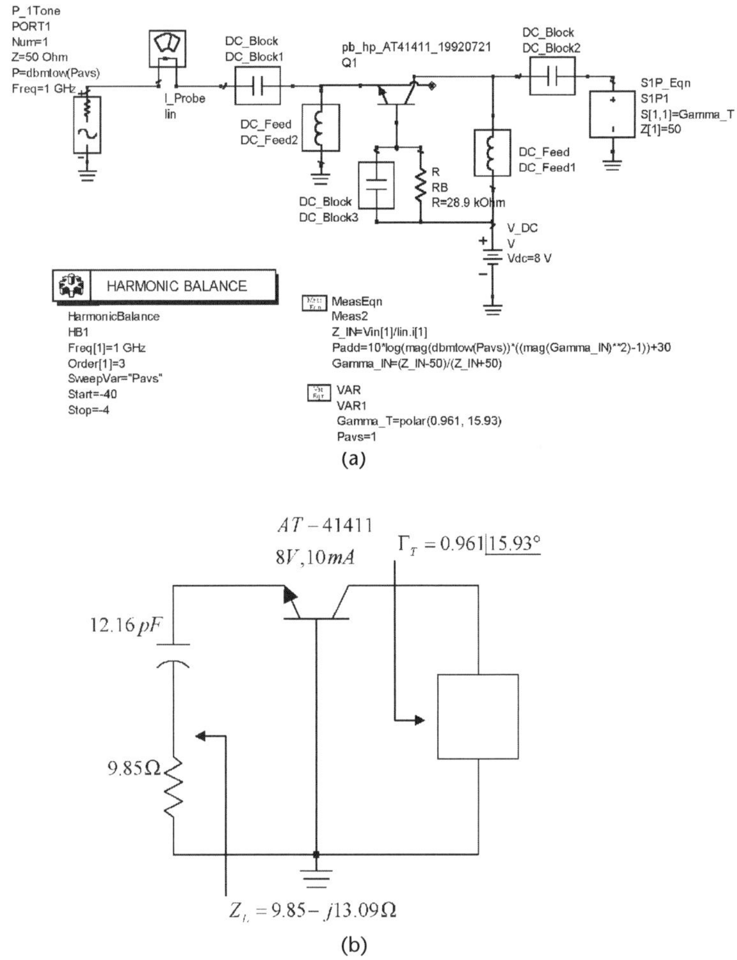

Figure 5.37 (a) ADS schematic for the calculation of P_{ADD} and Z_{IN} and (b) an ac schematic of the oscillator for P_{ADD} =– 13.45 dBm at 1 GHz.

which are a convenient form for simulation. OscPort is used to obtain the resulting waveform, the frequency of oscillation, and the oscillation power.

Example 5.8

Verify the oscillation condition and the stability of the oscillation for the 1-GHz oscillator design in Example 5.7 (see Figure 5.37).

Solution

(a) The oscillator schematic in Figure 5.38(a) is the oscillator in Figure 5.37 with $\Gamma_T = 0.961\,\underline{|15.93^\circ}$ and $Z_L = 9.85 - j13.09\Omega$ (or $R = 9.85\Omega$ and $C = 12.16$ pF at 1 GHz). The probe component OscTest is inserted between the input and load network. The value for Z in OscTest is chosen or varied until a readable plot is obtained. The loop gain and the resulting Nyquist plot are shown in Figure 5.38(b).

Table 5.4 Data of P_{ADD} Versus Z_{IN} for the Oscillator

P_{AVS}	Z_{IN}	Γ_{IN}	P_{ADD}
−40.000	−11.028 + *j*10.845	1.532/154.374	38.704/180.000
−38.000	−11.027 + *j*10.848	1.532/154.366	36.705/180.000
−36.000	−11.025 + *j*10.853	1.532/154.354	34.706/180.000
−34.000	−11.022 + *j*10.862	1.532/154.335	32.708/180.000
−32.000	−11.018 + *j*10.876	1.531/154.305	32.712/180.000
−30.000	−11.010 + *j*10.897	1.531/154.258	28.717/180.000
−28.000	−10.999 + *j*10.931	1.530/154.183	26.726/180.000
−26.000	−10.980 + *j*10.986	1.529/154.064	24.741/180.000
−24.000	−10.949 + *j*11.072	1.526/153.876	22.765/180.000
−22.000	−10.897 + *j*11.208	1.522/153.577	20.804/180.000
−20.000	−10.809 + *j*11.425	1.516/153.107	18.870/180.000
−18.000	−10.656 + *j*11.767	1.505/152.371	16.984/180.000
−16.000	−10.375 + *j*12.298	1.485/151.244	15.188/180.000
−14.000	−9.851 + *j*13.091	1.451/149.604	13.567/180.000
−12.000	−8.887 + *j*14.149	1.393/147.498	12.268/180.000
−10.000	−6.274 + *j*14.889	1.260/146.375	12.305/180.000
−8.000	4.210 + *j*16.679	0.859/142.885	19.421/135.370
−6.000	17.056 + *j*13.828	0.522/145.579	15.512/118.411
−4.000	30.883 + *j*5.083	0.244/161.515	14.295/107.366

The figure shows that the $1 + j0$ point is encircled in a clockwise direction. Therefore, the circuit is unstable and oscillations are possible.

The ADS simulation in Figure 5.38(c) is used to check the Kurokawa condition, since the simulation generates the data necessary to plot $1/\Gamma_{IN}(A, \omega_o)$ at 1 GHz by sweeping the power level. The data required to plot $\Gamma_L(\omega)$ by sweeping the frequency is generated by the load circuit shown in the schematic. The results from the simulation are displayed in the Smith chart. Observe that the direction of $\Gamma_L(\omega)$ is upward (from m1 to m2), and the direction of $1/\Gamma_{IN}(A, \omega_o)$ is downward (from m3 to m4). The plots intersect at a point where the angle going counterclockwise from $\Gamma_L(\omega)$ to $1/\Gamma_{IN}(A, \omega_o)$ is between 0° and 180°, which satisfies the Kurokawa condition.

The OscPort probe can now be used to calculate the frequency spectrum of the oscillator voltage, as well as the steady-state time-domain waveform. The number of harmonics used in the simulation is defined in the harmonic-balance controller. Usually, seven harmonics are more than enough for an oscillator simulation.

Example 5.9

Determine the fundamental frequency of oscillation, the harmonics, and the oscillation power for the 1-GHz oscillator design in Example 5.8 [see Figure 5.38(a)].

Solution

(a) The ADS simulation using OscPort is shown in Figure 5.39(a) with $Z_L = 9.85 - j13.09\Omega$. For clarity, a simplified dc bias network was used in Figure 5.39(a). A practical implementation requires an appropriate dc bias network. The oscillator

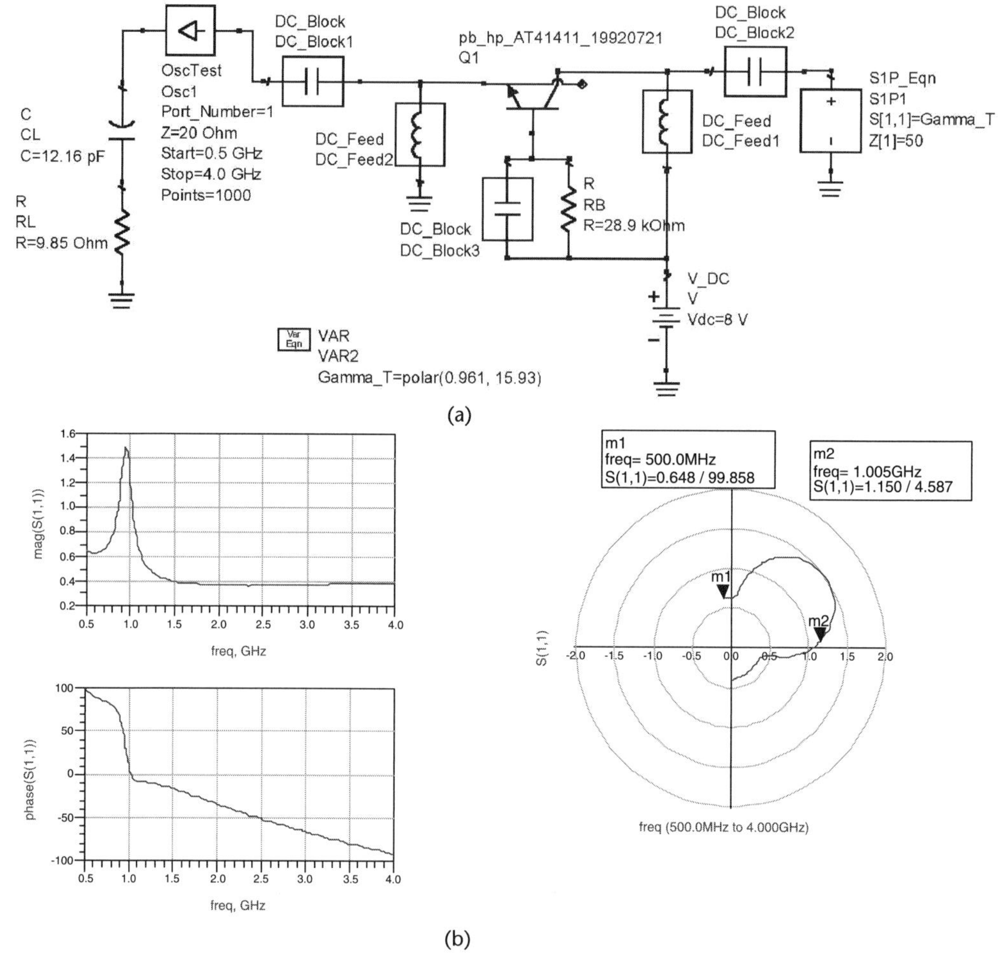

Figure 5.38 (a) Oscillator design with the Osctest component, (b) the loop gain and the Nyquist plot, and (c) the Kurokawa condition.

voltage is observed at the load (i.e., at VL). The resulting frequency spectrum is shown in Figure 5.39(b), as well as the oscillator waveform at VL. The fundamental frequency of oscillation is, as expected, at 1 GHz, and the second harmonic is –20 dB below the fundamental. The power dissipated in the 9.85-Ω resistor at the fundamental frequency is –13.5 dBm, which agrees closely with the predicted P_{ADD}.

The previous design can be used as the starting point for a VCO centered at 1 GHz by replacing the capacitor with an appropriate varactor diode.

5.11 Design of Feedback Oscillators Using the Negative-Resistance Method

In this section, some of the classical feedback oscillators are designed using the negative-resistance method. In the following examples the negative-resistance technique is applied to the design of a Colpitts oscillator and a GB oscillator.

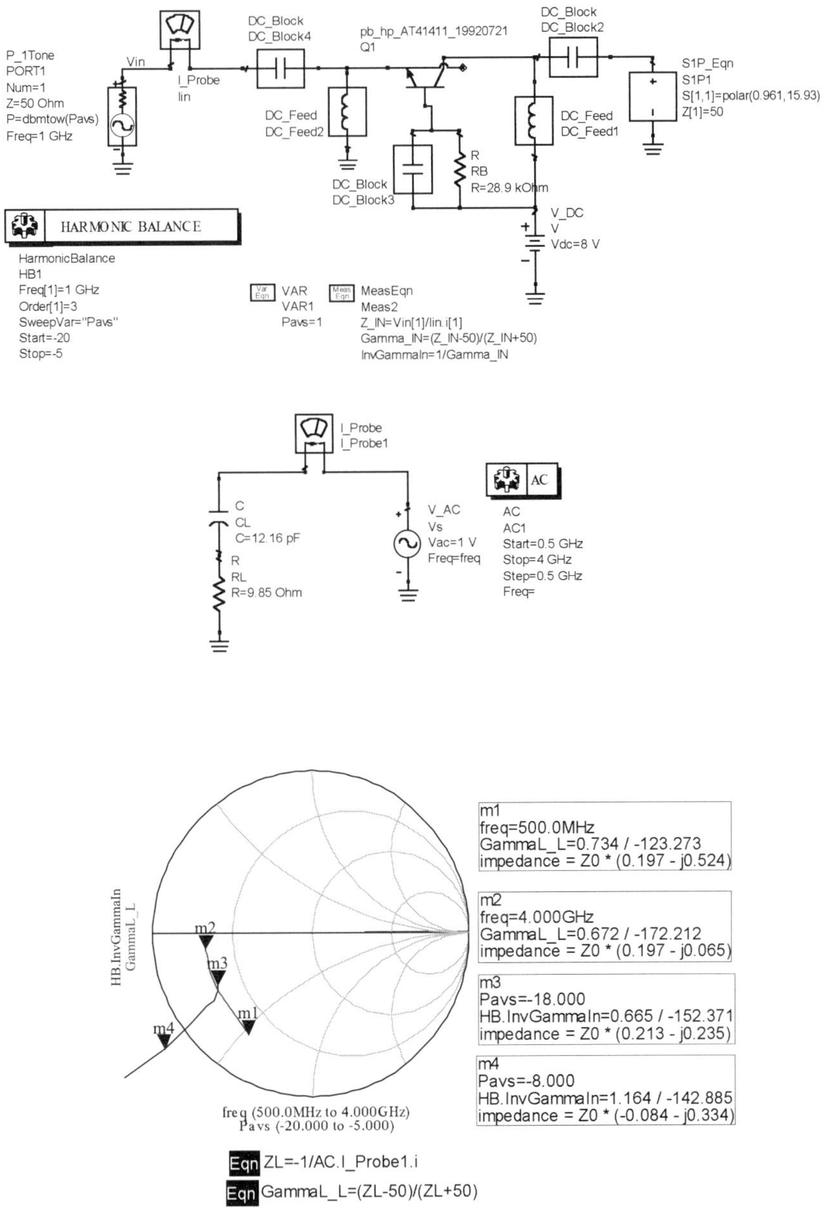

(c)

Figure 5.38 (Continued.)

Example 5.10

Design the Colpitts oscillator in Example 3.6 using the negative-resistance method.

Solution

In Example 3.6 a 1-MHz Colpitts oscillator was designed using a 2N2222 BJT biased at 8V, 10 mA. The oscillator circuit is shown in Figure 5.40(a) with the

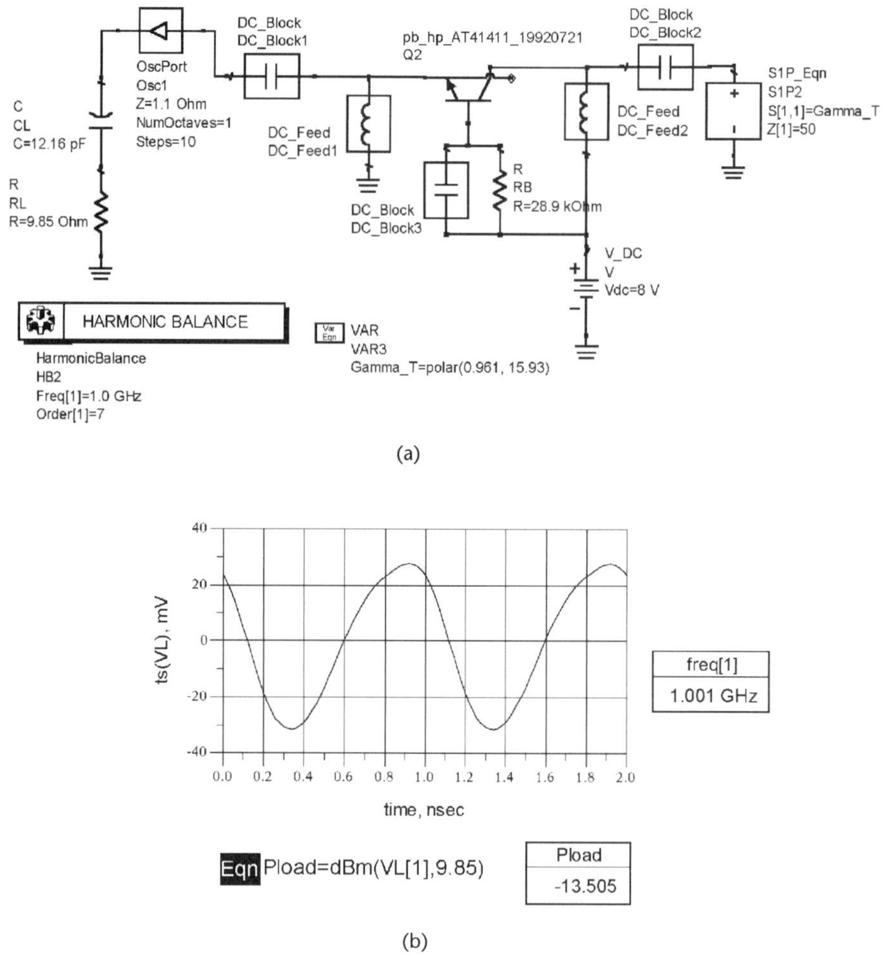

Figure 5.39 (a) Simulation of the 1-GHz oscillator design using OscPort and (b) waveform plot.

appropriate source component (i.e., P_1TONE) to measure its large-signal characteristics at 1 MHz, namely P_{ADD} and $Z_{IN}(A, \omega_o)$. The results of the simulation are shown in Figure 5.40(b).

From the data in Figure 5.40(b) the oscillator can be designed for an added power of about 16 dBm with $Z_L \approx 0.64 + j32\Omega$ (or $R = 0.64\Omega$ and $L = 5.07\ \mu$H). The use of a series circuit for Z_L agrees with the negative resistance behavior [see Figure 5.40(b)]. These are the values used in Example 3.6 for Z_L. Figure 5.40(c) shows an implementation of Z_L and the simulation for the Nyquist plot using the OscTest probe. Observe that the Nyquist plot, shown in Figure 5.40(c), shows that the circuit is indeed unstable.

Figure 5.40(d) shows the simulation using OscPort, and the resulting data. The oscillation voltage across R shows that the power is 15.8 dBm in close agreement with P_{ADD}. The oscillation voltage across R_e is also shown. The power in R_e is 13.4 dBm. The fundamental frequency of oscillation is at 998.1 kHz.

The GB oscillator was discussed in Section 3.3 using a feedback approach. In the following example the negative-resistance method is used to design the oscillator. In

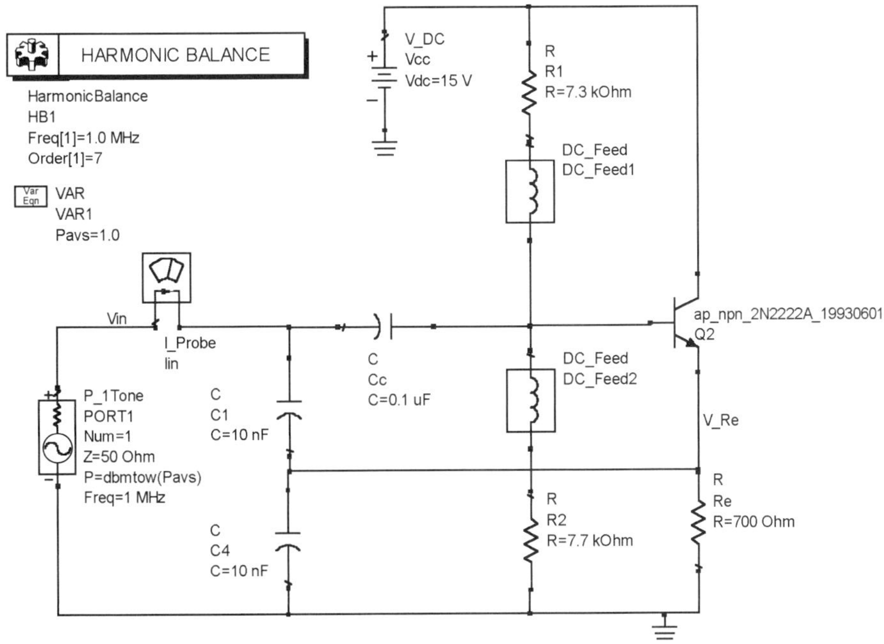

(a)

Pavs	Zin	Gamma_in	Padd
-20.000	-66.356 - j34.747	3.162 / -48.165	10.458 / 180.000
-18.000	-64.349 - j34.905	3.168 / -50.678	8.440 / 180.000
-16.000	-61.148 - j35.095	3.165 / -54.853	6.448 / 180.000
-14.000	-56.357 - j35.253	3.128 / -61.440	4.563 / 180.000
-12.000	-50.230 - j35.261	3.013 / -70.245	2.926 / 180.000
-10.000	-43.820 - j35.069	2.813 / -79.500	1.604 / 180.000
-8.000	-37.889 - j34.742	2.569 / -87.650	0.520 / 180.000
-6.000	-32.607 - j34.355	2.323 / -94.270	0.433 / 0.000
-4.000	-27.933 - j33.955	2.099 / -99.477	1.323 / 0.000
-2.000	-23.802 - j33.579	1.904 / -103.496	2.190 / 0.000
0.000	-20.158 - j33.235	1.738 / -106.574	3.055 / 0.000
2.000	-16.969 - j32.924	1.600 / -108.912	3.932 / 0.000
4.000	-14.202 - j32.681	1.486 / -110.628	4.824 / 0.000
6.000	-11.813 - j32.507	1.393 / -111.854	5.729 / 0.000
8.000	-9.771 - j32.356	1.317 / -112.762	6.652 / 0.000
10.000	-8.052 - j32.200	1.255 / -113.473	7.604 / 0.000
12.000	-6.622 - j32.052	1.206 / -114.026	8.583 / 0.000
14.000	-5.437 - j31.943	1.167 / -114.415	9.583 / 0.000
16.000	-4.454 - j31.891	1.135 / -114.646	10.596 / 0.000
18.000	-3.643 - j31.874	1.109 / -114.770	11.621 / 0.000
20.000	-2.981 - j31.867	1.088 / -114.847	12.667 / 0.000
22.000	-2.445 - j31.857	1.072 / -114.906	13.740 / 0.000
24.000	-2.013 - j31.845	1.059 / -114.954	14.842 / 0.000
26.000	-1.666 - j31.832	1.049 / -114.995	15.978 / 0.000
28.000	-1.187 - j31.857	1.034 / -114.974	16.442 / 0.000
30.000	-0.656 - j31.900	1.019 / -114.918	15.798 / 0.000
32.000	-0.226 - j31.925	1.006 / -114.882	13.113 / 0.000
34.000	0.117 - j31.866	0.997 / -114.980	18.320 / 48.136
36.000	0.391 - j31.799	0.989 / -115.086	23.739 / 35.081

Eqn Zin=Vin[1]/Iin.i[1]

Eqn Gamma_in=(Zin-50)/(Zin+50)

Eqn Padd=10*log(mag(dbmtow(Pavs))*((mag(Gamma_in)**2)-1))+30

(b)

Figure 5.40 (a) Simulation of the Colpitts oscillator in Example 5.10 for P_{ADD} and Z_{IN}, (b) simulation data, (c) simulation using OscTest and the Nyquist plot, and (d) simulation of the oscillator using OscPort and the oscillator waveforms.

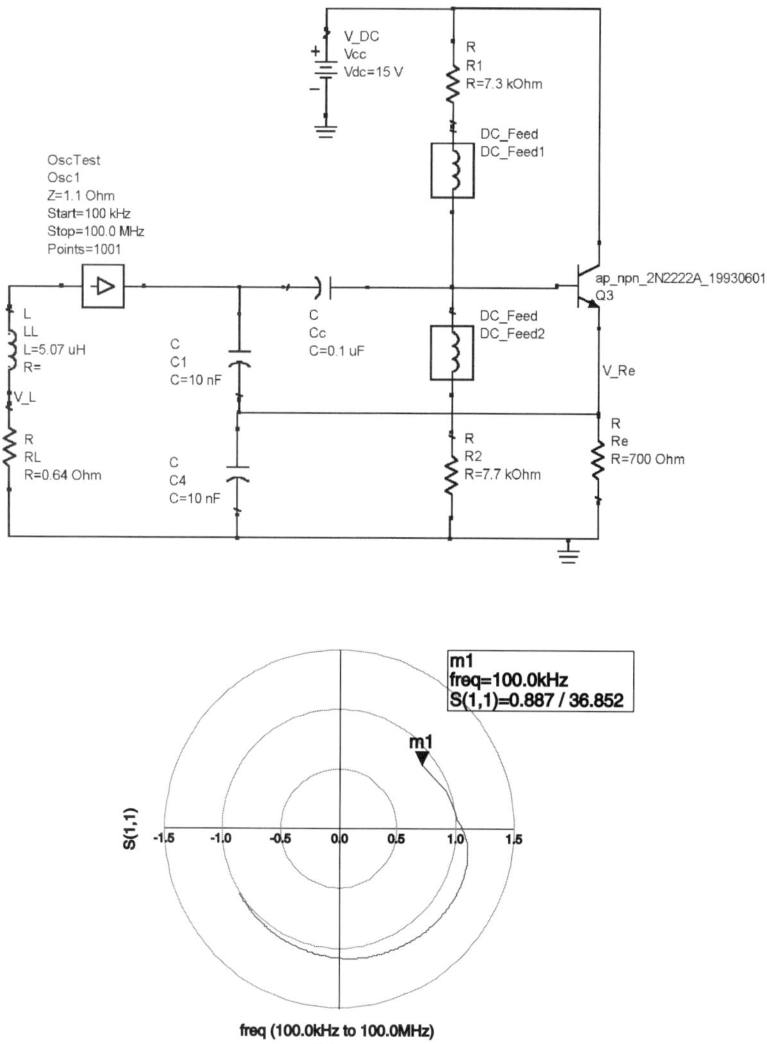

(c)

Figure 5.40 (Continued.)

fact, the negative-resistance method is a convenient method to use in the design of the GB oscillator.

Example 5.11

Design a GB oscillator to oscillate at 1.25 GHz using the negative-resistance method.

Solution

The GB oscillator configuration is illustrated in Figure 5.41(a). The BJT used is the Phillips BRF520, which has an f_T of 8 GHz. The bias point selected is at $V_{CE} = 5\text{V}$ and $I_C = 5$ mA.

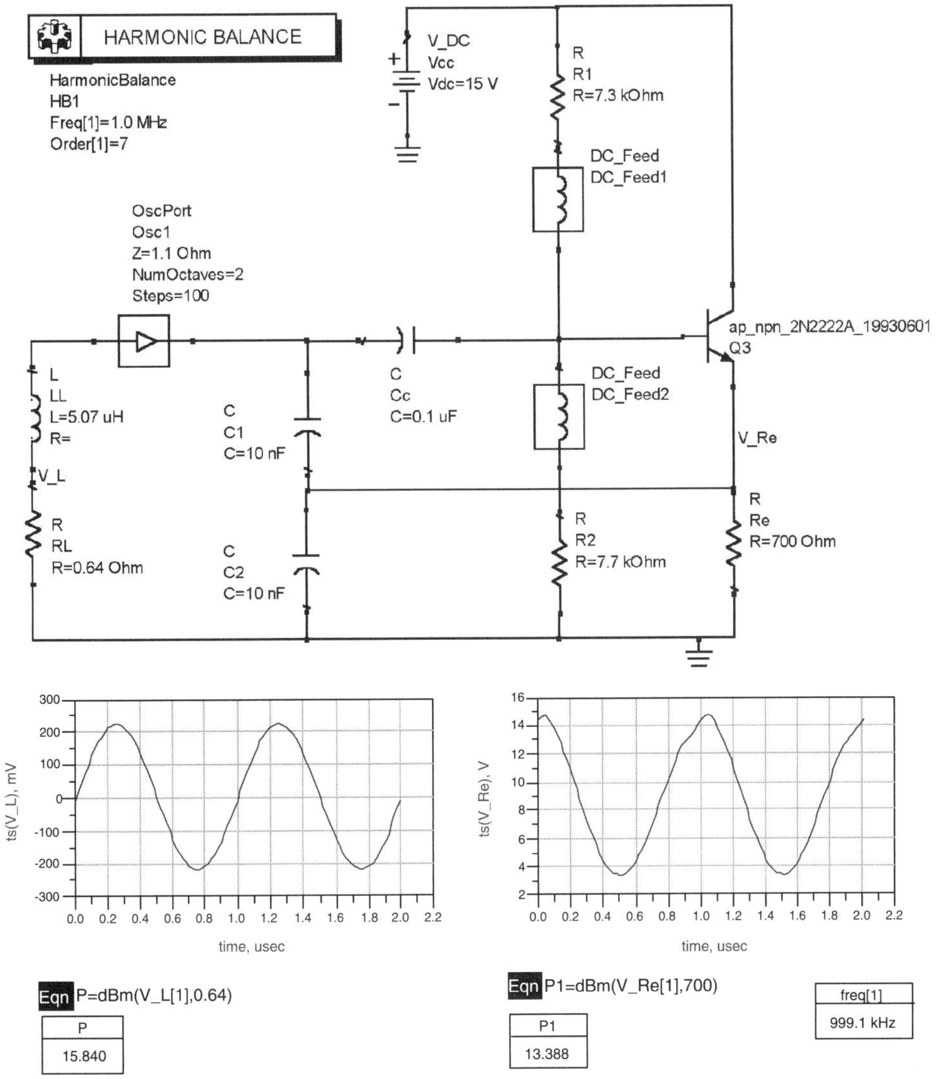

(d)

Figure 5.40 (Continued.)

The ADS simulation in Figure 5.41(a) is used to generate the mapping of Γ_T onto Γ_{IN}. The resulting data is shown in Figure 5.41(b). With Γ_T selected at point m1 (i.e., at $r = 0$ and $x =- 0.26$), the termination can be implemented with a 9.8 pF capacitor. The resulting, small-signal Γ_{IN} is $2.32\underline{|-60.4^\circ}$ (or $Z_{IN} =- 53.5 - j49.2\Omega$).

With a 9.8 pf at the terminating port, the added power versus Z_{IN} is obtained using the simulation shown in Figure 5.41(c). From the data obtained (see Figure 5.41(c)), it is seen that with $Z_L =- Z_{IN} == 43 + j102\Omega$ (or $R = 43\Omega$ and $L = 13$ nH) an added power of 6.15 dBm is obtained. The OscTest probe was used to obtain the Nyquist and Kurokawa plots (not shown). As expected, the plots showed that a stable oscillation will occur around 1.25 GHz.

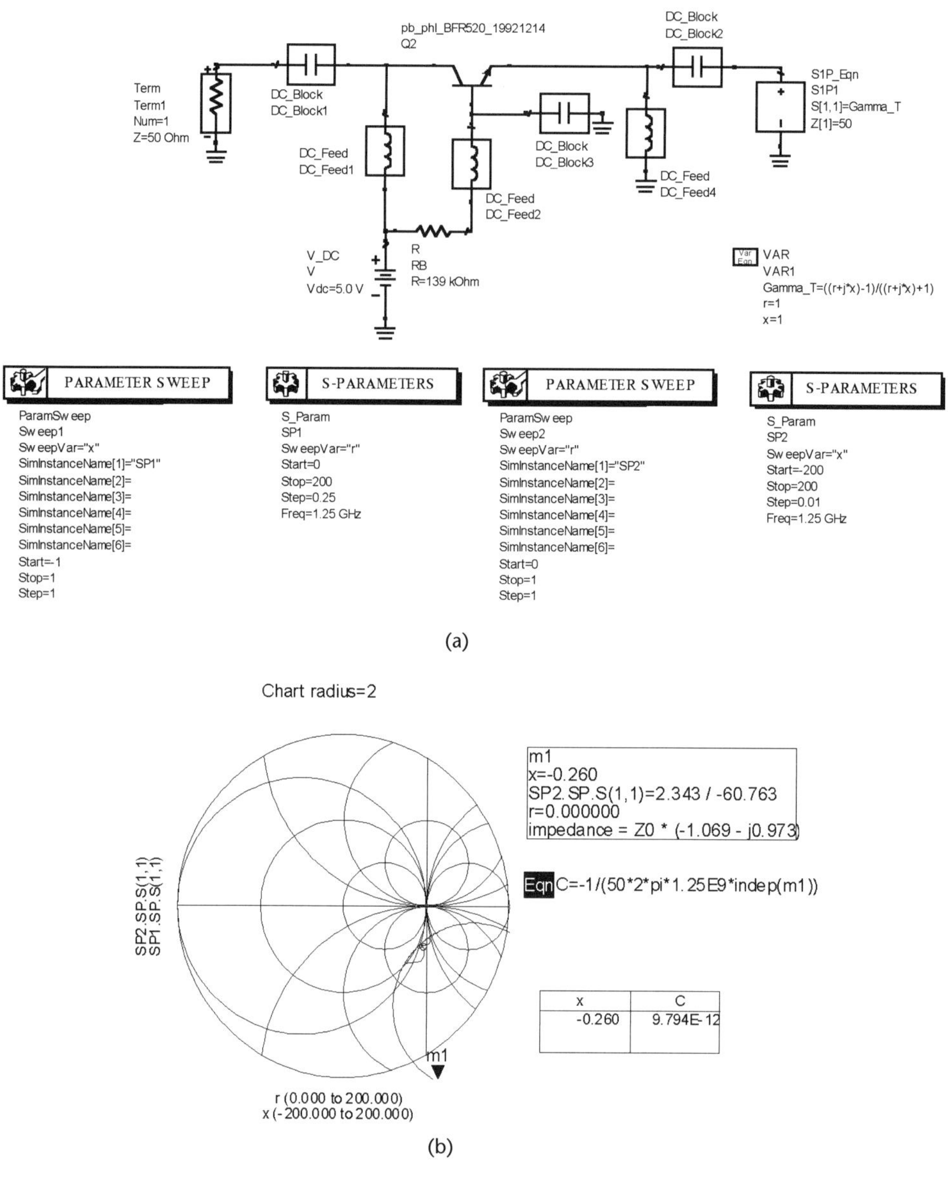

Figure 5.41 (a) The GB oscillator configuration for Example 5.11, (b) mapping of the Γ_T plane onto the Γ_{IN} plane, (c) simulation for P_{ADD} and Z_{IN} and the resulting data, and (d) harmonic-balance simulation and the oscillator signal.

The harmonic-balance simulation with OscPort and the resulting data are shown in Figure 5.41(d). The oscillator waveform shows that the fundamental frequency of oscillation is 1.27 GHz and the power delivered to the 43-Ω resistor is 6.8 dBm. The output signal can be coupled using a buffer amplifier; or since at 1.25 GHz the impedance of $C = 9.8$ pF has a magnitude close to one, a resistive load much larger than the magnitude of this impedance can be connected in parallel with C to couple the oscillator signal.

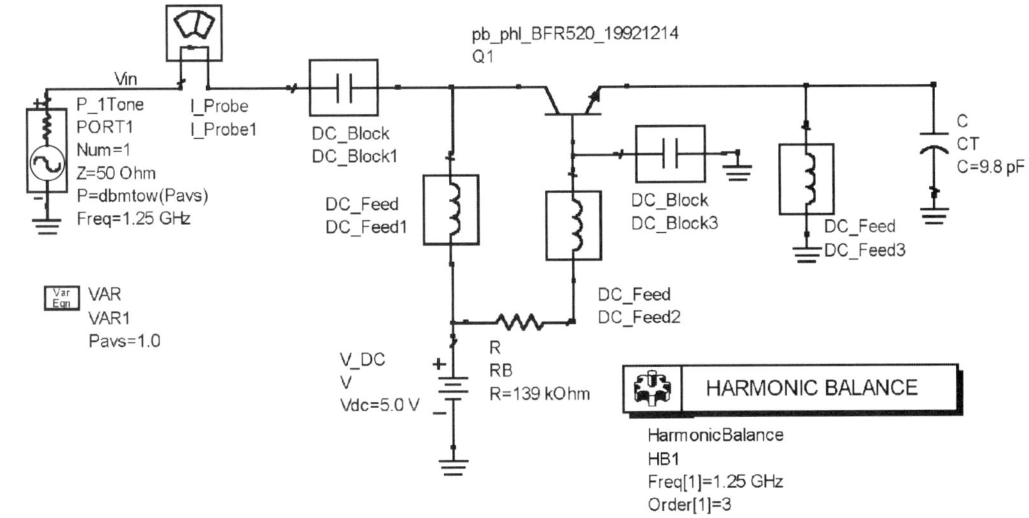

Pavs	Zin	Gamma_In	Reflection	Padd
-15.000	-53.949 - j50.262	2.290 / -59.702	4.245	-8.721
-14.000	-54.092 - j50.642	2.278 / -59.437	4.191	-7.777
-13.000	-54.263 - j51.111	2.264 / -59.117	4.126	-6.845
-12.000	-54.468 - j51.689	2.247 / -58.734	4.047	-5.928
-11.000	-54.709 - j52.395	2.226 / -58.281	3.954	-5.030
10.000	54.988 j53.256	2.201 / 57.752	3.844	4.152
9.000	55.305 j54.296	2.172 / 57.144	3.717	3.299
-8.000	-55.655 - j55.542	2.138 / -56.456	3.571	-2.472
-7.000	-56.029 - j57.023	2.100 / -55.693	3.408	-1.675
-6.000	-56.411 - j58.765	2.056 / -54.865	3.229	-0.910
-5.000	-56.775 - j60.790	2.009 / -53.987	3.035	-0.178
-4.000	-57.087 - j63.117	1.957 / -53.078	2.830	0.518
-3.000	-57.300 - j65.758	1.902 / -52.163	2.618	1.180
-2.000	-57.360 - j68.711	1.845 / -51.267	2.402	1.806
-1.000	-57.201 - j71.965	1.785 / -50.412	2.187	2.399
0.000	-56.759 - j75.491	1.725 / -49.619	1.976	2.958
1.000	55.972 j79.240	1.665 / 48.903	1.773	3.487
2.000	-54.791 - j83.147	1.606 / -48.272	1.580	3.986
3.000	-53.189 - j87.129	1.549 / -47.727	1.399	4.459
4.000	-51.165 - j91.093	1.494 / -47.266	1.233	4.910
5.000	-48.743 - j94.941	1.443 / -46.883	1.081	5.340
6.000	-45.974 - j98.582	1.394 / -46.571	0.945	5.752
7.000	-42.928 - j101.937	1.350 / -46.322	0.822	6.150
8.000	-39.681 - j104.945	1.309 / -46.131	0.714	6.535
9.000	-36.312 - j107.561	1.272 / -45.997	0.618	6.908
10.000	-32.889 - j109.760	1.238 / -45.920	0.533	7.268
11.000	-29.471 - j111.525	1.208 / -45.903	0.458	7.612
12.000	-26.099 - j112.850	1.180 / -45.951	0.392	7.936
13.000	-22.806 - j113.734	1.155 / -46.072	0.334	8.232
14.000	19.609 j114.181	1.132 / 46.273	0.281	8.486
15.000	16.517 j114.209	1.111 / 46.557	0.233	8.678
16.000	-13.534 - j113.893	1.091 / -46.908	0.189	8.771
17.000	-10.665 - j113.506	1.071 / -47.237	0.148	8.697
18.000	-7.936 - j113.716	1.053 / -47.298	0.108	8.333
19.000	-0.710 - j118.295	1.004 / -45.824	0.009	-1.634
20.000	6.790 - j118.808	0.960 / -45.534	-0.078	8.938
21.000	7.674 - j121.025	0.956 / -44.756	-0.085	10.314

Eqn Zin=Vin[1]/I_Probe1.i[1]

Eqn Gamma_In=(Zin-50)/(Zin+50)

Eqn Reflection=(mag(Gamma_In)**2)-1

Eqn Padd1=dbmtow(Pavs)*mag(Reflection)

Eqn Padd=10*log10(Padd1/0.001)

(c)

Figure 5.41 (continued.)

5.12 Dielectric-Resonator Oscillators

Dielectric resonators (DRs) for oscillator applications in the RF and microwave frequencies are of two types: those that operate in the TEM mode, and those that operate in a TE mode.

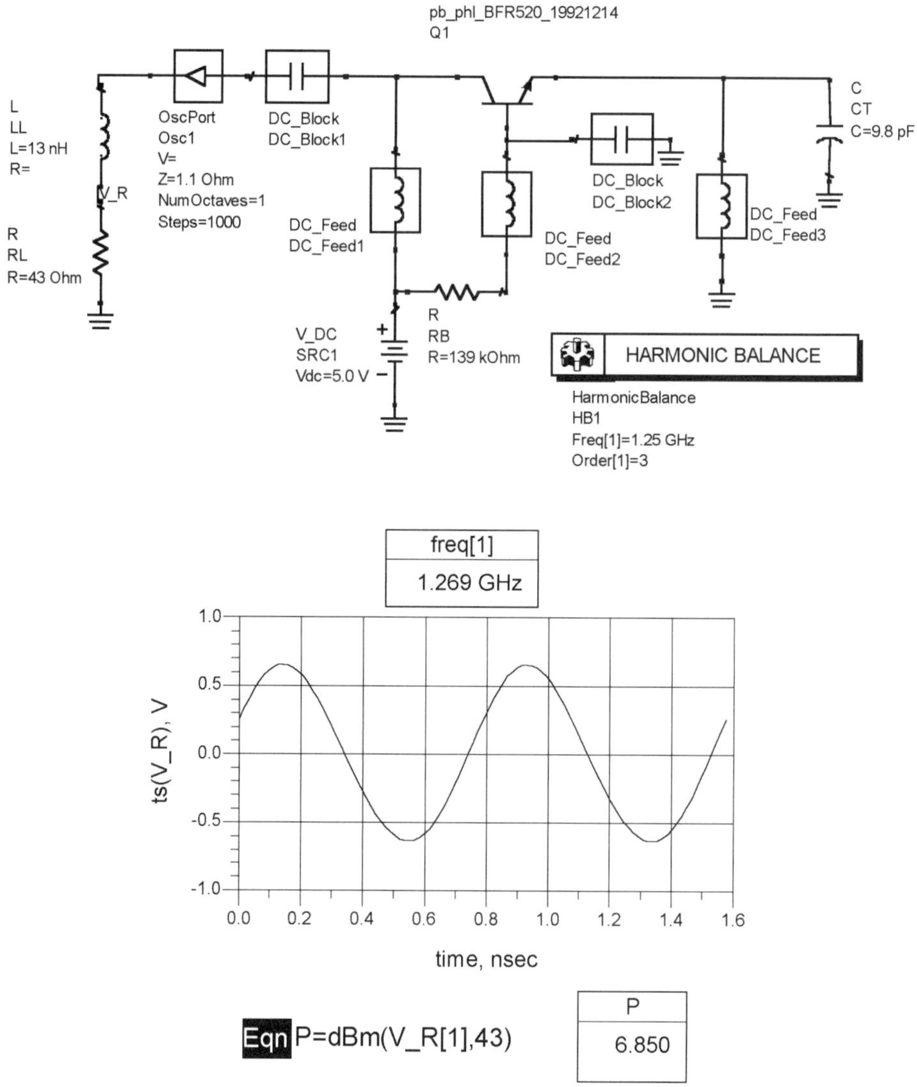

(d)

Figure 5.41 (continued.)

5.12.1 TEM-Mode DRs

A TEM-mode DR is similar to a coaxial cable. The main use of DR resonators is to implement high-Q inductors in oscillators to resonate with a capacitor or varactor.

DR resonators are available as quarter-wave resonators with one end shorted, or as half-wave resonators with both ends open. For example, at resonance a short-circuited quarter-wave transmission line appears as an open circuit, and at frequencies below resonance it behaves like an inductor. These resonators are usually made with ceramic dielectric material and, therefore, are also called TEM-mode ceramic resonators (TEM-mode CRs). These particular resonators have found applications in many oscillators used in the wireless communication field. For example, they can be used in VCOs in the range of 300 MHz to about 5 GHz.

Good performance is obtained by the use of a coaxial resonator, where the resonator is shaped in the form of a cube of length d with a coaxial bore and shorted at one end, as shown in Figure 5.42. The inner and outer surfaces are plated with copper or with silver. The conductivity of the plating affects the Q of the resonator. Those plated with copper are cheaper than those plated with silver. However, when high-Q values are required, the silver-plated resonator is best.

The resonators are fabricated with a high dielectric constant in order to make the component small and, therefore, to be useful as an inductive component in microwave oscillators. The high value of the relative dielectric constant ϵ_r is required for circuit miniaturization since the size of the component is inversely proportional to $\sqrt{\epsilon_r}$. This is a consequence of the fact that the wavelength in the dielectric material λ_d is given by

$$\lambda_d = \frac{\lambda_o}{\sqrt{\epsilon_r}}$$

where λ_o is the free space wavelength.

Typical values of ϵ_r are from about 10 to 100 (e.g., $\epsilon_r = 21$, 38, 88, and 92), with typical lengths of 3 to 14 mm. The high values, $\epsilon_r = 88$ and 92 are found in DRs for the 300-MHz to about 1-GHz range, the value $\epsilon_r = 38$ are found in DRs operating in the 800-MHz to 2.5-GHz range, and $\epsilon_r = 21$ in the 1- to 4-GHz range. Typical values of the characteristic impedance are from 4Ω to 20Ω.

The Q of the DR describes the losses of the component, namely dielectric losses and conduction losses. The temperature coefficient TC (in ppm/°C) indicates the thermal stability of the DR. It describes how much the resonant frequency drifts as a function of temperature.

The Murata, copper plated, DRR030 is a $\lambda/4$ resonators that can be fabricated to order in the 900-MHz to 1.6-GHz range with $Q > 230$ and characteristic impedance of $Z_o = 7.4\Omega$. They are also available in the 1.9- to 3-GHz range with $Q > 380$ and $Z_o = 15.4\Omega$. The length d of the resonator is given by

$$d = \frac{3 \times 10^{11}}{n\sqrt{\epsilon_r} f_o} \text{ mm}$$

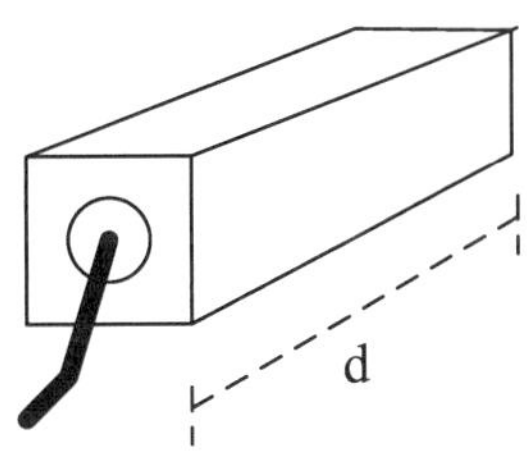

Figure 5.42 Shape of a TEM-mode DR.

where $n = 4$ for a $\lambda/4$ resonator and $n = 2$ for a $\lambda/2$ resonator. With $n = 4$ and $\epsilon_r = 92$ we obtain

$$d = \frac{3 \times 10^{11}}{4\sqrt{92} f_o} = \frac{7.82 \times 10^9}{f_o} \text{ mm}$$

At 1 GHz the length is $d = 7.83$ mm.

For a short-circuited coaxial-cable resonator [see Figure 5.43(a)] its capacitance per unit length is given by

$$C = \frac{2\pi\epsilon_o \epsilon_r}{\ln\left(\frac{D_1}{D_2}\right)}$$

and its inductance per unit length is given by

$$L = \frac{\mu_o}{2\pi} \ln\left(\frac{D_1}{D_2}\right)$$

Hence, the characteristic impedance of the resonator is

$$Z_o = \sqrt{\frac{L}{C}} = \frac{60}{\sqrt{\epsilon_r}} \ln\left(\frac{D_1}{D_2}\right)$$

For the geometry of the resonator in Figure 5.42, a correction to the D_1/D_2 ratio is required. The correction can be obtained from the manufacturers.

The DR is usually designed with a $\lambda/4$ length at a given frequency f_o. The impedance of a lossless short-circuited resonator is

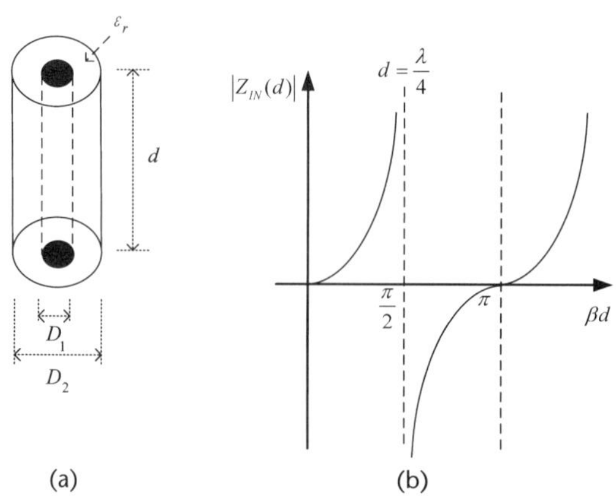

Figure 5.43 (a) Coaxial line geometry and (b) plot of $|Z_{IN}(d)|$ for a short-circuited resonator.

$$Z_{IN}(d) = jZ_o \tan \beta d$$

A plot of $Z_{IN}(d)|$ is shown in Figure 5.43(b). The plot shows that for $0 < \beta d < \pi/2$ (or $0 < d < \lambda/4$) the shorted line is inductive, while for $\pi/2 < \beta d < \pi$ the shorted line is capacitive. Hence, at a frequency lower than the resonant frequency of the resonator, the line is inductive.

Quarter-wave resonators behave like parallel-resonant circuits. To obtain an equivalent input parallel resonant circuit the input admittance of a $\lambda/4$ short-circuited lossy transmission line is considered, namely,

$$\begin{aligned} Y_{IN}(d) &= Y_o \coth(\alpha + j\beta) \\ &= Y_o \frac{1 + j\tan \beta d \tanh \alpha d}{\tanh \alpha d + j\tan \beta d} \end{aligned} \tag{5.75}$$

For low losses (i.e., for $\alpha d \ll 1$) we have

$$\tanh \alpha d \approx \alpha d = \frac{\alpha\lambda}{4} \tag{5.76}$$

Since

$$\beta d = \frac{2\pi d}{\lambda} = \frac{\omega d}{v_p}$$

and with $d = \lambda/4$ at $\omega = \omega_o$, and $\omega = \omega_o + \Delta\omega$, we can write

$$\beta d = \frac{\omega_o d}{v_p} + \frac{\Delta\omega d}{v_p} = \frac{\pi}{2} + \frac{\pi\Delta\omega}{2\omega_o}$$

Therefore,

$$\tan \beta d = \tan\left(\frac{\pi}{2} + \frac{\pi\Delta\omega}{2\omega_o}\right) = \frac{-1}{\tan\dfrac{\pi\Delta\omega}{2\omega_o}} \approx \frac{-2\omega_o}{\pi\Delta\omega} \tag{5.77}$$

Substituting (5.76) and (5.77) into (5.75) gives

$$Y_{IN} \approx Y_o \frac{1 - j\dfrac{2\omega_o}{\pi\Delta\omega}\dfrac{\alpha\lambda}{4}}{\dfrac{\alpha\lambda}{4} - j\dfrac{2\omega_o}{\pi\Delta\omega}} = Y_o \frac{j\dfrac{\pi\Delta\omega}{2\omega_o} + \dfrac{\alpha\lambda}{4}}{j\dfrac{\alpha\lambda}{4}\dfrac{\pi\Delta\omega}{2\omega_o} + 1} \tag{5.78}$$

Since $\dfrac{\alpha\lambda}{4}\dfrac{\pi\Delta\omega}{2\omega_o} \ll 1$ we can further approximate (5.78) as

$$Y_{IN} \approx Y_o\left(\frac{\alpha\lambda}{4} + j\frac{\pi\Delta\omega}{2\omega_o}\right) \tag{5.79}$$

In order to develop an equivalent circuit from (5.79), we consider the input admittance of the parallel-tuned circuit shown in Figure 5.44. That is,

$$Y_{IN} = G + \frac{1}{j\omega L} + j\omega C \tag{5.80}$$

With $\omega = \omega_o + \Delta\omega$ and

$$\frac{1}{\omega_o + \Delta\omega_o} \approx \frac{1 - \dfrac{\Delta\omega_o}{\omega_o}}{\omega_o}$$

(5.80) is rewritten as

$$Y_{IN} = G + \frac{1 - \dfrac{\Delta\omega}{\omega_o}}{j\omega_o L} + j\omega_o C + j\Delta\omega C \tag{5.81}$$

Observing that

$$\frac{1}{j\omega_o L} + j\omega_o C = 0 \text{ or } \omega_o = \sqrt{\frac{1}{LC}}$$

(5.81) reduces to

$$\begin{aligned} Y_{IN} &= G + j\frac{\Delta\omega}{\omega_o^2 L} + j\Delta\omega C \\ &\approx G + j2\Delta\omega C \end{aligned} \tag{5.82}$$

Comparing (5.79) and (5.82), it follows that

$$R = \frac{4Z_o}{\alpha\lambda} \tag{5.83}$$

$$C = \frac{\pi}{4\omega_o Z_o} \tag{5.84}$$

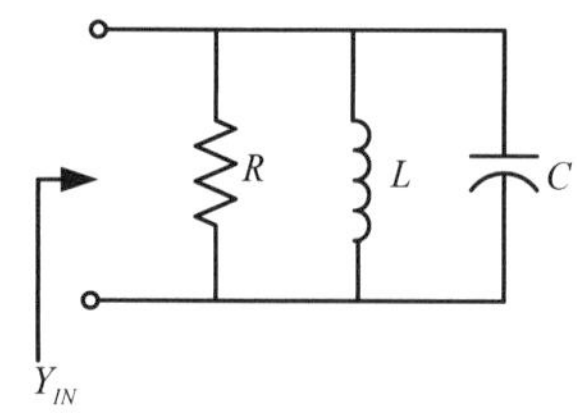

Figure 5.44 Equivalent parallel circuit of a resonator.

and

$$L = \frac{1}{\omega_o^2 C} \tag{5.85}$$

At resonance the input impedance is

$$Z_{IN}(\omega_o) = R$$

which, of course, is very high since α is small. The Q_U of the resonator is

$$Q_U = \omega_o RC = \frac{\pi}{\alpha\lambda} = \frac{\beta}{2\alpha}$$

In terms of Q_U, (5.83) can be expressed as

$$R = \frac{Q_U}{\omega_o C} \tag{5.86}$$

For a given resonator the parameters Z_o, ω_o, and its Q_U are known, then (5.84), (5.85), and (5.86) are used to find the equivalent parallel circuit. Hence, the $\lambda/4$ resonator can be modeled by the parallel-resonant circuit in Figure 5.45 around the resonant frequency ω_o.

An application of a DR to a tuned varactor diode circuit is shown in Figure 5.45. For example, oscillator circuits in the 900-MHz region require very low values of inductors (say, 5 nH). These inductor values can be conveniently implemented using DR components.

5.12.2 TE-Mode DRs

A dielectric resonator oscillator (DRO) is a high-Q, temperature-stable oscillator that is used in many practical applications at microwave frequencies. BJTs are used in DROs with oscillation frequencies up to about 15 GHz; with GaAs FETs the frequency of oscillation of the DROs can be extended to about 35 GHz. With transistor DROs power levels of 10 to 15 dBm are possible.

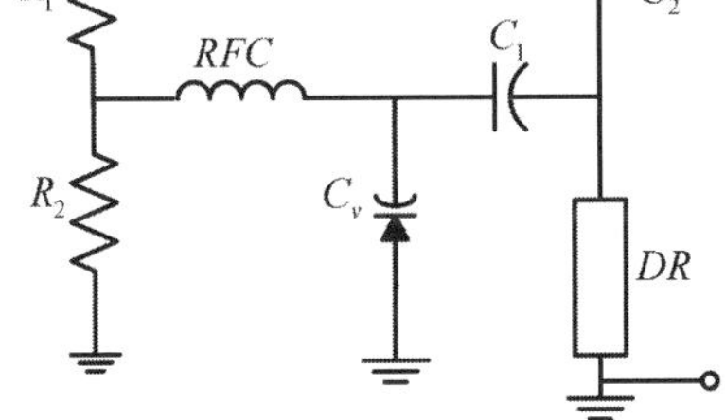

Figure 5.45 A varactor circuit tuned with a DR inductor.

Several compounds with dielectric constants between 20 and 80 are used in the construction of dielectric resonators (DR). For microwave transistor oscillators, dielectric resonators having a solid cylindrical shape can cover the frequency range from about 1 to 40 GHz. In fact, dielectric resonators can operate up to 100 GHz. The problem with dielectric resonators at the lower frequencies is the large dimensions of the resonator. Dielectric resonators having a coaxial tubular shape (i.e., a hollow center) have been made for practical applications down to 500 MHz.

A good reference on DRs and applications is found in the text *Dielectric Resonators* by Kajfez and Guillon [4]. A dielectric resonator will resonate in several modes. The most commonly used mode in cylindrical resonators is a *TE* mode (specifically the $TE_{01\delta}$ mode). The $TE_{01\delta}$ mode is the lowest-order mode of resonance, so operating in this mode avoids the chances of the oscillator operating at the frequency of resonance of a higher order mode. In order to understand the coupling of a dielectric resonator to a microstrip line, consider Figure 5.46, which shows the field distribution of the $TE_{01\delta}$ mode. The electric field lines are concentric circles around the z-axis, and there is no z component of the electric field. The magnetic field lines are illustrated in Figure 5.46.

Figure 5.47 shows the dielectric resonator coupled to the microstrip line. The dielectric resonator is placed on top of the substrate at a distance d from the microstrip line. The distance d and the DR characteristics determine the coupling.

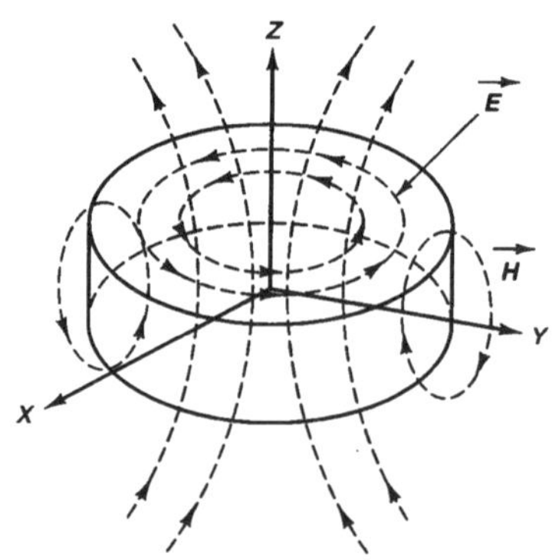

Figure 5.46 Field distribution of the $TE_{01\delta}$ mode in a cylindrical resonator.

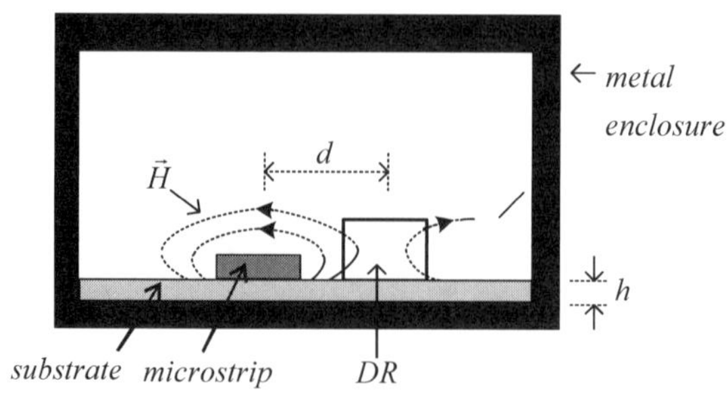

Figure 5.47 Coupling of a dielectric resonator to a microstrip line.

Observe the magnetic coupling between the DR and the microstrip line. The metallic enclosure is used to minimize the radiation losses and, therefore, to increase the resonator Q. The $TE_{01\delta}$ is excited in the DR by the electromagnetic field produced by the microstrip line. In turn, the DR reflects RF energy at its resonant frequency, resulting in a high-Q resonator.

Figure 5.48(a) shows the DR coupled to a microstrip line with characteristic impedance Z_o and terminated in Z_o impedances (usually $Z_o = 50\Omega$). Due to the magnetic-field coupling, the resonator appears in series with transmission line. The DR model is shown in Figure 5.48(b) where the magnetic coupling is modeled by the transformer. The equivalent circuit of the DR coupled to the microstrip line is illustrated in Figure 5.48(c). It consists of a parallel tuned circuit placed in series, at the position XX', with the transmission lines. The values of R, L, and C in the equivalent circuit of the DR depend on the DR characteristics and the distance d. The impedance Z is given by

$$Z = \frac{1}{C}\frac{s}{s^2 + \frac{s}{RC} + \frac{1}{LC}} = \frac{1}{C}\frac{s}{s^2 + 2\alpha s + \omega_o^2} \tag{5.87}$$

where the bandwidth and resonant frequency are

$$BW = 2\alpha = \frac{1}{RC}$$

and

$$\omega_o = \frac{1}{\sqrt{LC}}$$

Letting $s = j\omega$, (5.87) can be expressed as

$$Z = \frac{R}{1 + jQ_U\frac{\left(\omega^2 - \omega_o^2\right)}{\omega\omega_o}} \tag{5.88}$$

where

$$Q_U = \frac{\omega_o}{2\alpha} = \omega_o RC = \frac{R}{\omega_o L} \tag{5.89}$$

Since the frequency of operation is very close to ω_o (i.e., $\omega + \omega_o \approx 2\omega_o$), we can approximate (5.88) by

$$Z = \frac{R}{1 + j2Q_U\delta} \tag{5.90}$$

where

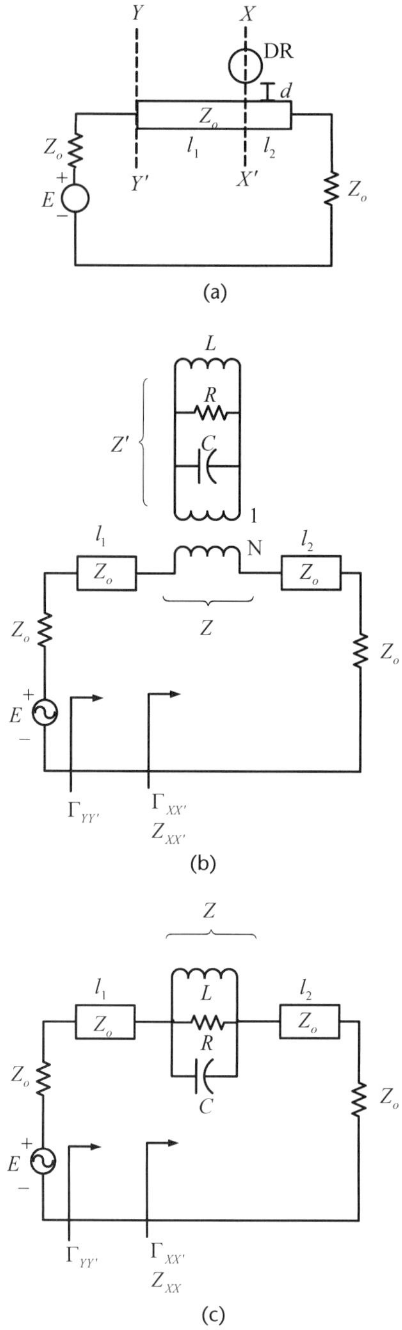

Figure 5.48 (a) Dielectric resonator coupled to a microstrip line, (b) magnetic coupling model for the DR, and (c) equivalent circuit at the *XX′* plane.

$$\delta = \frac{\omega - \omega_O}{\omega_O}$$

At the reference plane XX' the input impedance is given by

$$Z_{XX'} = Z + Z_o$$

or

$$z_{XX'} = \frac{Z_{XX'}}{Z_o} = \frac{\frac{R}{Z_o}}{1 + j2Q_U\delta} + 1 \tag{5.91}$$

For convenience, a coupling coefficient β is defined to measure the coupling between the resonator and the microstrip transmission line. The coefficient β is defined as the ratio Q_U to the external Q, where

$$Q_E = \frac{2Z_o}{\omega L}$$

The term $2Z_o$ represents the total external resistance in Figure 5.48(c). Hence,

$$\beta = \frac{Q_U}{Q_E} = \frac{R}{2Z_o} \tag{5.92}$$

In cases that the resonator is terminated in a short circuit (i.e., $Z_{XX} = Z$), then the coupling coefficient is given by $\beta = R/Z_o$. A short-circuit termination is easily implemented with an open circuited transmission line of length $\lambda/4$.

Since Q_U, Q_E, and the loaded Q (i.e., Q_L) are related by

$$\frac{1}{Q_L} = \frac{1}{Q_U} + \frac{1}{Q_E}$$

it follows that

$$Q_L = \frac{Q_U}{\beta + 1}$$

This relation shows how β affects Q_L.

In terms of β, we can write (5.91) as

$$z_{XX'} = \frac{2\beta}{1 + j2Q_U\delta} + 1$$

Observe that at $\omega = \omega_o$ (or $\delta = 0$) it follows that $z_{XX'} = 2\beta + 1$. Then, the reflection coefficient at $\omega = \omega_o$, at the XX' plane, is

$$\Gamma_{XX'}(\omega_o) = \frac{z_{XX'} - 1}{z_{XX'} + 1} = \frac{\beta}{\beta + 1} \tag{5.93}$$

In general, the reflection coefficient at the XX' plane is given by

$$\Gamma_{XX'} = \frac{z_{XX'} - 1}{z_{XX'} + 1} = \frac{\beta}{\beta + 1 + j2Q_U\delta}$$

Hence, the reflection coefficient seen at the input of the line, denoted by $\Gamma_{YY'}$ is

$$\Gamma_{YY'} = \frac{\beta}{\sqrt{(\beta+1)^2 + (2Q_U\delta)^2}} e^{-j2\left(\theta + \tan^{-1}\frac{2Q_U\delta}{\beta+1}\right)} \tag{5.94}$$

At the resonant frequency (i.e., at $\omega = \omega_o$ or $\delta = 0$), (5.94) reduces to

$$\Gamma_{YY'}(\omega_o) = \Gamma_{XX'}(\omega_o)e^{-j2\theta} = \frac{\beta}{\beta+1}e^{-j2\theta} \tag{5.95}$$

This relation shows that if β is a constant and the length of the line is varied from $\theta = 0°$ to $180°$, the values of $\Gamma_{YY'}$ lie on a circle in the Smith chart. Thus, with the appropriate selection of β, the reflection coefficient $\Gamma_{YY'}$ can implement any passive impedance. Equation (5.95) is used to select the desired coupling and electrical length of the transmission line for a particular input impedance.

The previous discussion shows that the parameters β, ω_o, and Q_U describe the operation of the DR. These parameters are either measured or given by the manufacturer of the DR. The coupling coefficient is calculated by measuring $\Gamma_{XX'}$ at the resonant frequency. The values of R, L, and C in Figure 5.48(c) can be calculated in terms of β, ω_o, and Q_u. The value of R follows from (5.92), and the values of L and C from (5.89).

From Figure 5.48(c), at the resonant frequency ω_o, the S parameters of the tuned circuit (i.e., the DR coupled to the transmission line) are

$$[S(\omega_o)] = \begin{bmatrix} \frac{\beta}{\beta+1} & \frac{1}{\beta+1} \\ \frac{1}{\beta+1} & \frac{\beta}{\beta+1} \end{bmatrix} \tag{5.96}$$

From (5.96) it follows that β can be expressed in terms of the $S_{11}(\omega_o)$ and $S_{21}(\omega_o)$ parameters; that is,

$$\beta = \frac{S_{11}(\omega_o)}{1 - S_{11}(\omega_o)} = \frac{1 - S_{21}(\omega_o)}{S_{21}(\omega_o)}$$

For a closely coupled resonator, typical values of β are from 2 to 20.

The value of Q_U can be found by measuring S_{21}. A typical plot of S_{21} in decibels is shown in Figure 5.49. The insertion loss L_o is given by

$$L_o\ (\text{dB}) = -20 \log |S_{21}(\omega_o)|$$

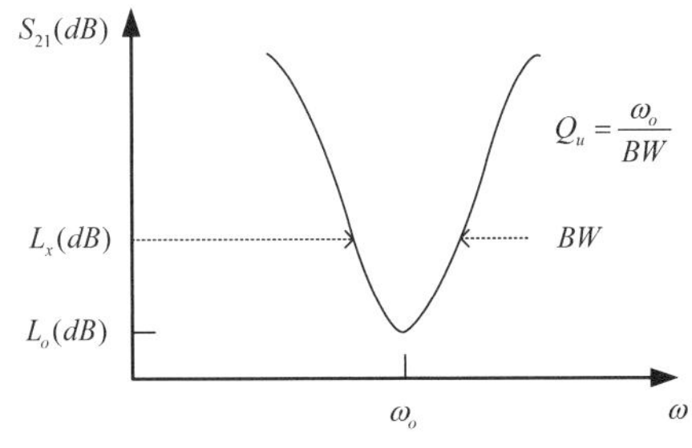

Figure 5.49 A typical plot of S_{21} (dB).

and the insertion loss value L_x is given by

$$L_x \text{ (dB)} = L_o - 3 + 10 \log(1 + 10^{-0.1L_o})$$

Hence, Q_U is found by dividing ω_o by the bandwidth where the insertion loss is L_x.

Some practical circuit configurations for DROs are shown in Figure 5.50. In the configurations shown in Figure 5.50(a, b) the DR acts as a series-feedback element. The DR is connected to the terminating port, and the DR and transistor produce a negative resistance at the load port.

In the configurations shown in Figure 5.50(c) and Figure 5.50(d) the DR acts as a parallel feedback element. The forward gain of the transistor compensates for the insertion loss of the DR. The DR is a high-Q tuned circuit, and therefore, oscillation will occur when the oscillation conditions are satisfied. An advantage of the parallel-feedback configuration is its wide tuning range.

The series-feedback configurations are simple to construct. The parallel configuration is somewhat more difficult to construct because of the coupling of the two lines.

The ADS program has a component called DR, shown in Figure 5.51(a), that models the typical housing used in the coupling of a DR to a microstrip line. The housing is shown in Figure 5.51(b). The DR manufacturer specifies the radius (*Rad*) and height (H) of the resonator for a particular resonant frequency along with Qdr. E_{rL} and H_L are the dielectric constant and height of the substrate on which the DR is placed. These are the same parameters that affect the characteristic impedance of the microstrip line to which the DR is coupled. *Mode* is another important parameter, which controls the resonant mode of the DR. When Mode is set to 0, the resonator will resonate in the $TE_{01\delta}$ mode.

DROs can be tuned over a narrow range of frequency using a metallic shield with a tuning screw, as shown in Figure 5.52. Basically, the screw affects the height H_U and, therefore, the resonant frequency. The height h should be greater than 0.5, in order not to decrease the Q of the resonator. Tuning bandwidths of the order of 0.1% to 1% of the resonant frequency are possible.

The DROs can also be tuned electrically. A varactor tuned DRO is shown in Figure 5.53. The varactor is coupled to the DR, resulting in two coupled circuits. The varactor's capacitance can be changed with an appropriate dc bias voltage,

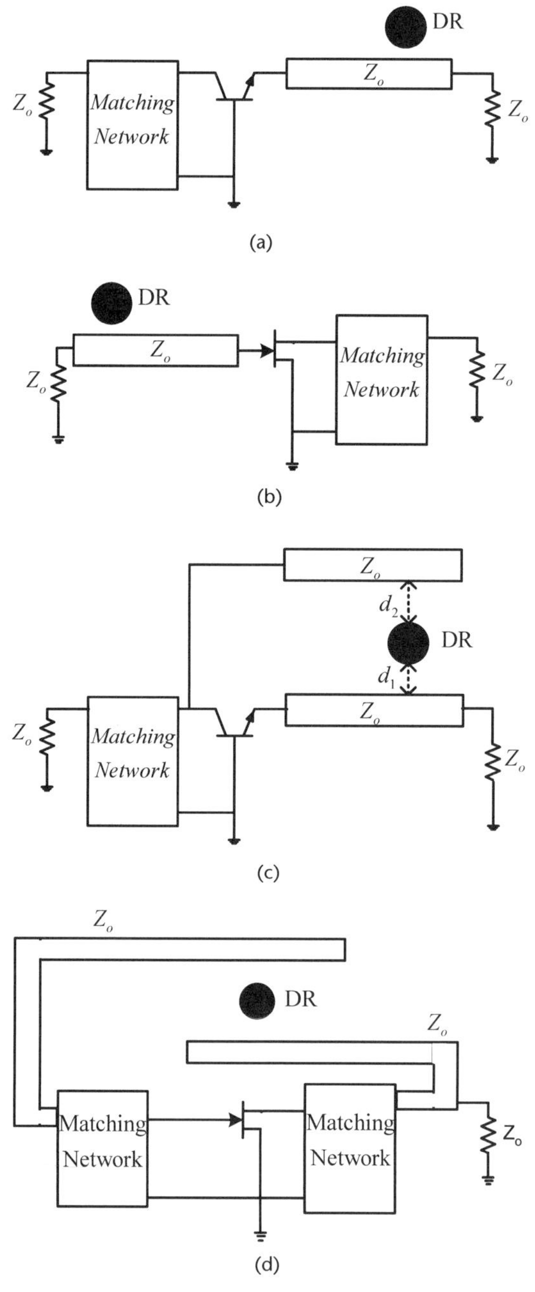

Figure 5.50 (a) A series-feedback DRO using a BJT, (b) a series-feedback DRO using a GaAs FET, (c) a parallel-feedback DRO using a BJT, and (d) a parallel-feedback DRO using a GaAs FET.

and this in turn changes the resonant frequency of the DR. The varactor can provide a resonant frequency tuning range of 1%. There are other ways of electrically tuning a DR.

The temperature stability of a DR is expressed in parts per million per degree centigrade (ppm/°C). DRs with temperature coefficients ranging from −10 to

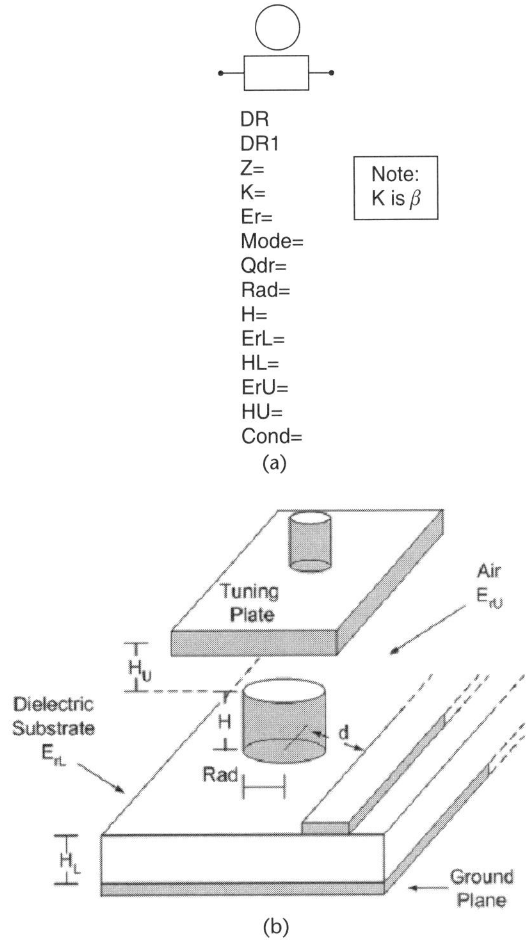

Figure 5.51 (a) ADS model for a DR and (b) DR housing.

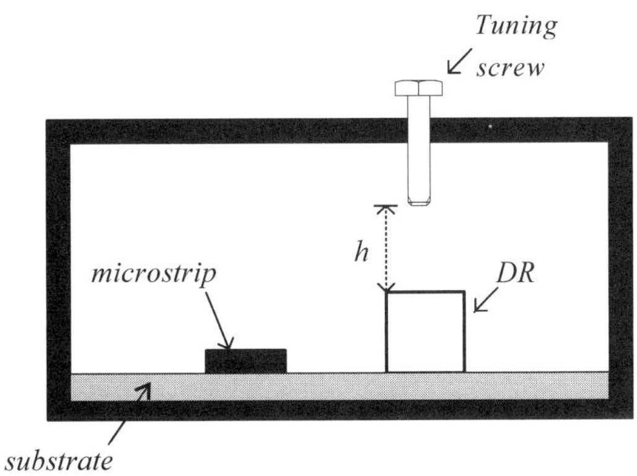

Figure 5.52 A mechanical tuning arrangement for DROs.

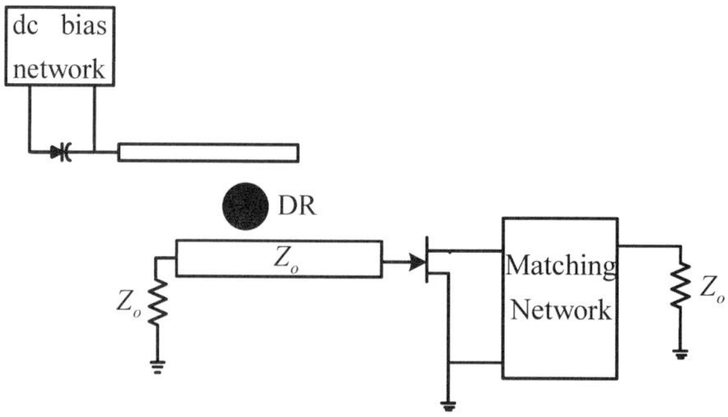

Figure 5.53 A varactor tuned DRO.

10 ppm/°C are available. For example, an 8-GHz DRO can be specified to have a frequency stability of 500 ppm over a temperature range of –50°C to 80°C. This requirement will keep the frequency drift to 4 MHz.

Example 5.12

Design a 10-GHz DRO using the ATF-13786 GaAs FET whose *S* parameters (at 10 GHz) in a CS configuration are:

$$S_{11} = 0.63\angle 130^\circ$$

$$S_{12} = 0.15\angle 6^\circ$$

$$S_{21} = 2.04\angle 4^\circ$$

$$S_{22} = 0.19\angle 134^\circ$$

The transistor power at the 1-dB compression point is $P_{1\,dB} = 15$ dBm.

Solution

The ATF-13786 GaAs FET is biased at 3V, 40 mA. It is unconditional stable at 10 GHz since $K = 1.14$ and $\Delta = 0.358\angle 151.3^\circ$. Using a series-feedback capacitor having a series reactance of $Z = -j120\Omega$ [implemented as an open-circuited shunt stub, with $l = 0.064\lambda$ (or 22.68°), in Figure 5.54(a)] results in a potentially unstable configuration in a 50-Ω system with the following *S* parameters:

$$S_{11} = 3.68\angle -175.1^\circ$$

$$S_{12} = 3.86\angle -38.28^\circ$$

$$S_{21} = 4\angle 30.05^\circ$$

$$S_{22} = 2.77\angle 176.15^\circ$$

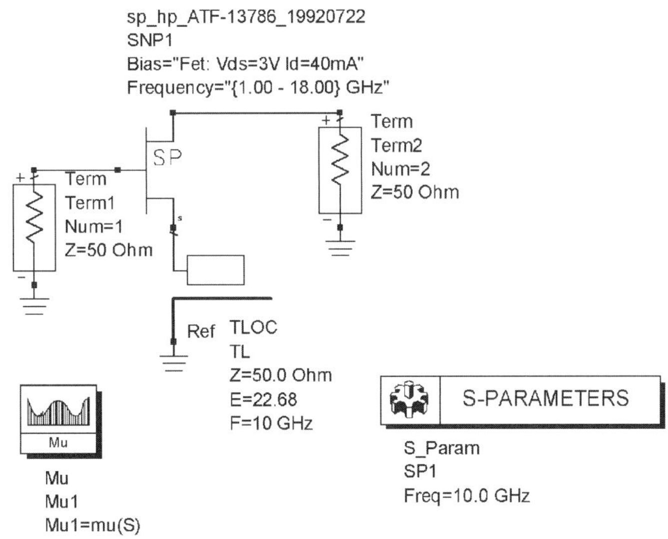

freq	S(1,1)	S(1,2)	S(2,1)	S(2,2)	Mu1
10.00 GHz	3.676 / -175.139	3.864 / -38.280	4.001 / 30.054	2.767 / 176.154	-0.323

(a)

Figure 5.54 (a) GaAs FET with series feedback, (b) stability circle at the terminating port, (c) simulation of the DR oscillator, and (d) simulation using the DR component.

The previous calculations were made using the average small-signal S parameters of the transistor. In what follows the nonlinear model of the device is used. The gate-to-ground port was selected as the terminating port. The oscillator configuration is shown in Figure 5.54(b), as well as the stability simulation. The stability circle at the terminating port is drawn in the Smith chart shown in Figure 5.54(b). Designing for Γ_T at point A (i.e., $\Gamma_T = 0.666$) and with $l_1 = \lambda/2$ (i.e., $\theta = \pi$), it follows from (5.93) that

$$\Gamma_T = \Gamma_{XX'} = \frac{\beta}{\beta + 1} = 0.666 \Rightarrow \beta = 2$$

The DRO can be represented by a parallel tuned circuit, as shown in Figure 5.54(c). With a typical $Q_U = 1{,}000$, it follows that

$$R = \beta(2Z_o) = 2(2)50 = 200\Omega$$

$$C = \frac{Q_U}{\omega_o R} = \frac{1{,}000}{2\pi \times 10^{10}(200)} = 79.6 \text{ pF}$$

and

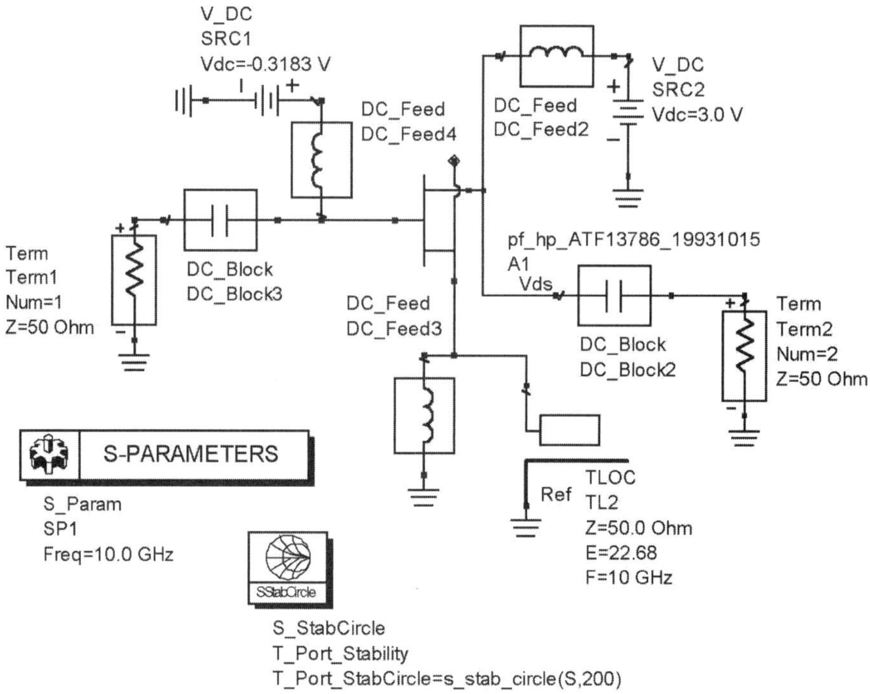

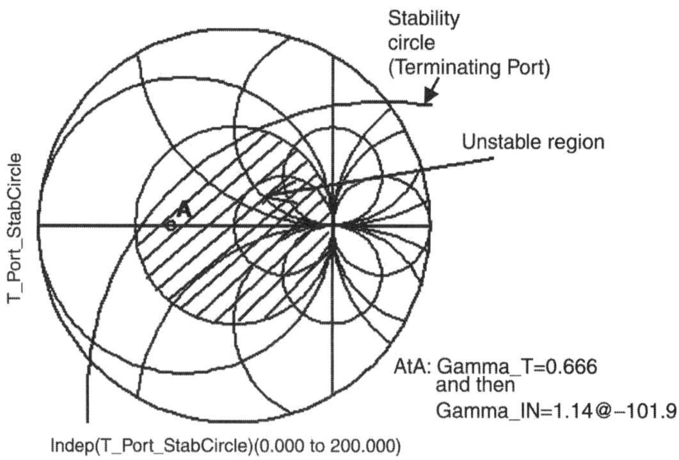

(b)

Figure 5.54 (continued.)

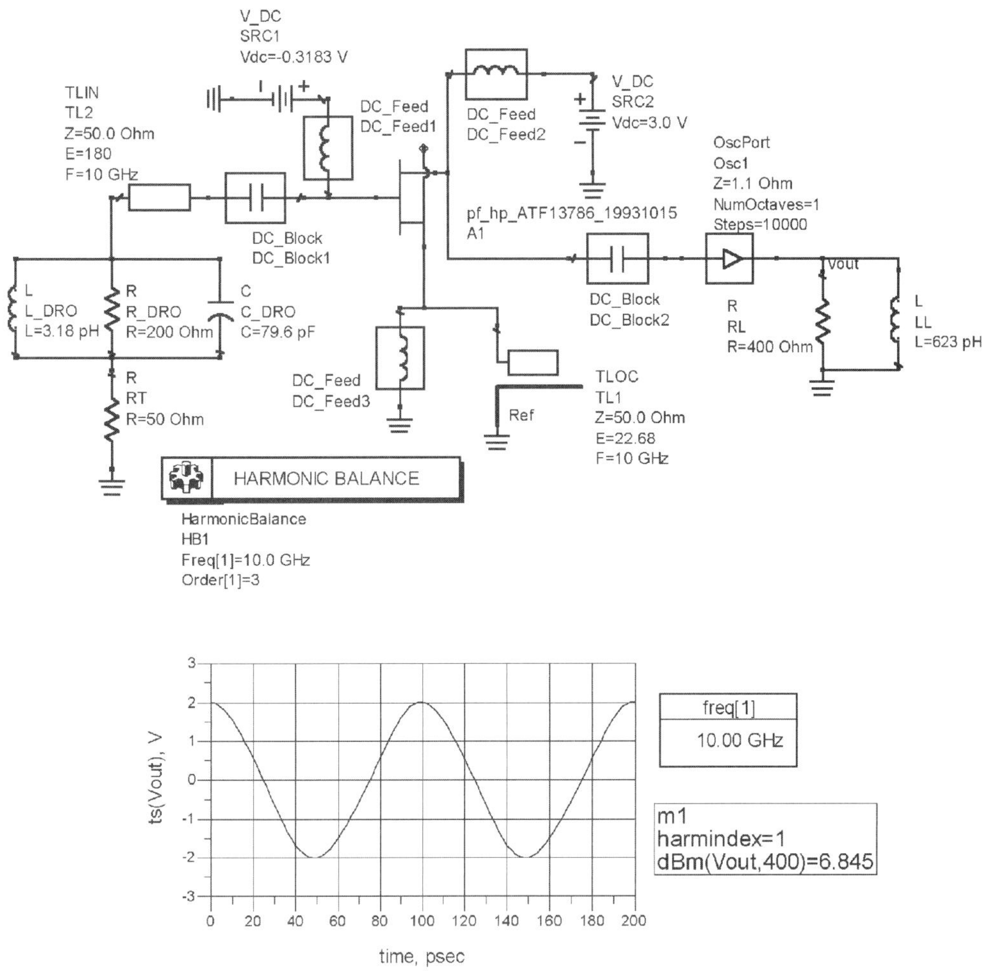

(c)

Figure 5.54 (continued.)

$$L = \frac{1}{\omega_o^2 C} = \frac{1}{(2\pi \times 10^{10})79.6 \times 10^{-12}} = 3.183 \text{ pH}$$

The DRO implementation is shown in Figure 5.54(c). For $\Gamma_T = 0.666$ the small-signal value of Γ_{IN} is $1.14 \underline{|-102^\circ}$ (or $Z_{IN} =- 5.4 - j40.3\Omega$). The simulation to obtain P_{ADD} and Z_{IN} at the load port of the oscillator was performed. The results showed that the load network $Z_L \approx 3.7 + j39\Omega$, implemented in parallel form, produces an added power of 6.9 dBm. The implementation of the load network is shown in Figure 5.54(c). The component OscPort was first changed to OscTEst to verify the loop-gain condition for oscillation using the Nyquist plot, as well as the Kurokawa condition.

Because of the many variables involved in the implementation of the DRO, the previous design should be taken as a preliminary design for further CAD

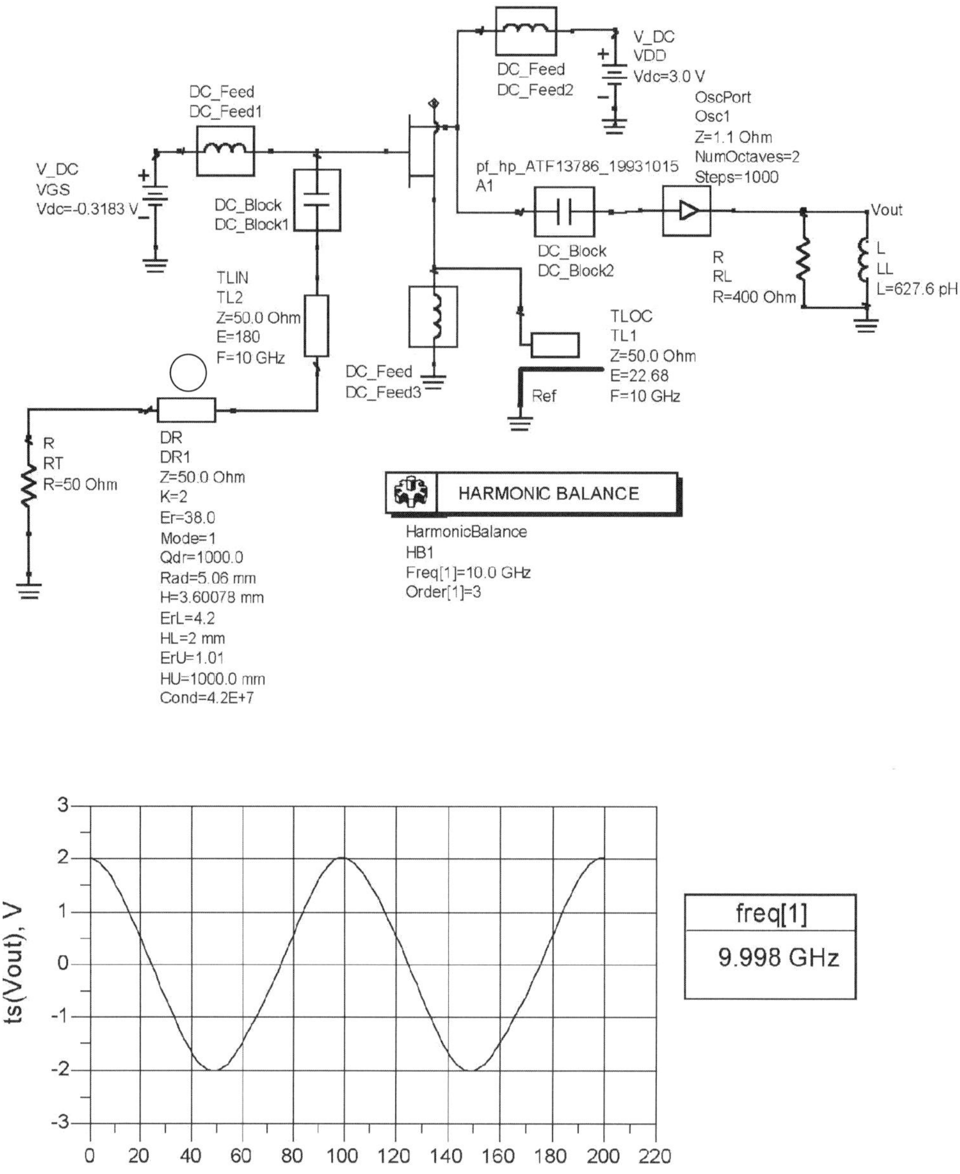

(d)

Figure 5.54 (continued.)

analysis and optimization. The output waveform in Figure 5.54(c) shows that the fundamental oscillation is at 10 GHz with an output power of 6.845 dBm.

The simulation using the DR component in Figure 5.51 is shown in Figure 5.54(d). As expected, the results agree with those in Figure 5.54(c).

The following DRO example and the parallel-coupled implementation were adapted from the Master's thesis of J. Vasiliadis [5].

Example 5.13

Design a 2-GHz DRO using the Infineon BFR183 BJT.

Solution

This transistor lists f_T = 8 GHz. The ADS library has the SPICE model for the BFR183 in package form. However, for this design the development of the transistor model for the BFR183 in package form is illustrated. The SPICE data for the BFR183 (given by Infineon) is shown in Figure 5.55(a). The transistor chip is represented by the component *BJT_model*. There are also package parasitics that should be included in the model. The package model is shown in Figure 5.55(a) and the packaged component is denoted by *Ch5_Ex12pt3_DCmodel*.

The BFR183 BJT is biased at V_{CE} = 8.2V and I_C = 15 mA, as shown in Figure 5.55(b). The resulting biased BJT component is denoted by *Ch5_Ex_12pt3_BJTmodel_DCbias*.

The transistor stability is checked using the ADS simulation shown in Figure 5.55(c). The data shows that $|S_{11}| > 1$ and $|S_{22}| > 1$ at 2 GHz (with $K = -0.937$); therefore, the transistor is potentially unstable. The load port (i.e., the collector-to-base port) and the terminating port (i.e., the emitter-to-base port) stability circles, shown in Figure 5.55(c), show that most of the Smith chart is unstable.

The ADS simulation shown in Figure 5.55(d) maps the Γ_T plane onto the Γ_{IN} plane, where the terminating port (i.e., Γ_T plane) was selected to be the emitter to base (or ground) port. The point m1 corresponds to $\Gamma_{IN} = 2.425\,\underline{|-6.79^\circ}$ and the corresponding value of Γ_T is $\Gamma_T = 0.724\,\underline{|-126.3^\circ}$. The associated negative resistance is $Z_{IN} = -118.2 - j13.9\Omega$.

From (5.95), $\Gamma_T = 0.724\,\underline{|-126.3^\circ}$ is realized with $\beta = 2.623$ and a line length of 0.1754λ. Using an alumina substrate with $\epsilon_r = 9.6$ and height of 30 mils, a line width of 29.7 mils produces a Z_o of 50Ω, and the effective dielectric constant is $\epsilon_{ff} = 6.46$. Hence, the line length of 0.1754λ at 2 GHz is 10.35 mm long. A Murata DR for the 2-GHz range was used (see the Murata RESOMICS catalog O95E-8). The DR component used is described in Figure 5.56. As a check on the design of Γ_T the ADS simulation shown in Figure 5.56 was used. First, the phase goal was deactivated and K and HU were optimized for the magnitude goal. This simulation produced the following values: HU = 2.885 mm and K = 2.63. Then, the ADS simulation shown in Figure 5.56 with both goals activated was used. The data in Figure 5.56 shows that the final values produced by the optimization, namely $\Gamma_T = 0.725\,\underline{|-126.3^\circ}$ with $\beta = 2.605$ and $L = 10.38$ mm. Also, the equivalent RLC circuit for the DR is given.

Next, the added power simulation is performed as shown in Figure 5.57. The data shows that the small-signal resistance is $-118 - j13.9\Omega$. The large-signal data shows that an added power of 12.5 dBm can be obtained with $Z_L = -Z_{IN} = 109.2 + j64.7\Omega$. The load circuit design is shown in Figure 5.58(a). It consists of a 6.55-mm short-circuited stub followed by a 7.78-mm transmission line that produces the desired load impedance Z_L.

The design simulation, using OscPort, is shown in Figure 5.58(a). First, OscPort was changed to OscTest to verify that the oscillation conditions are satisfied. As expected, the circuit should have a stable oscillation at 2 GHz.

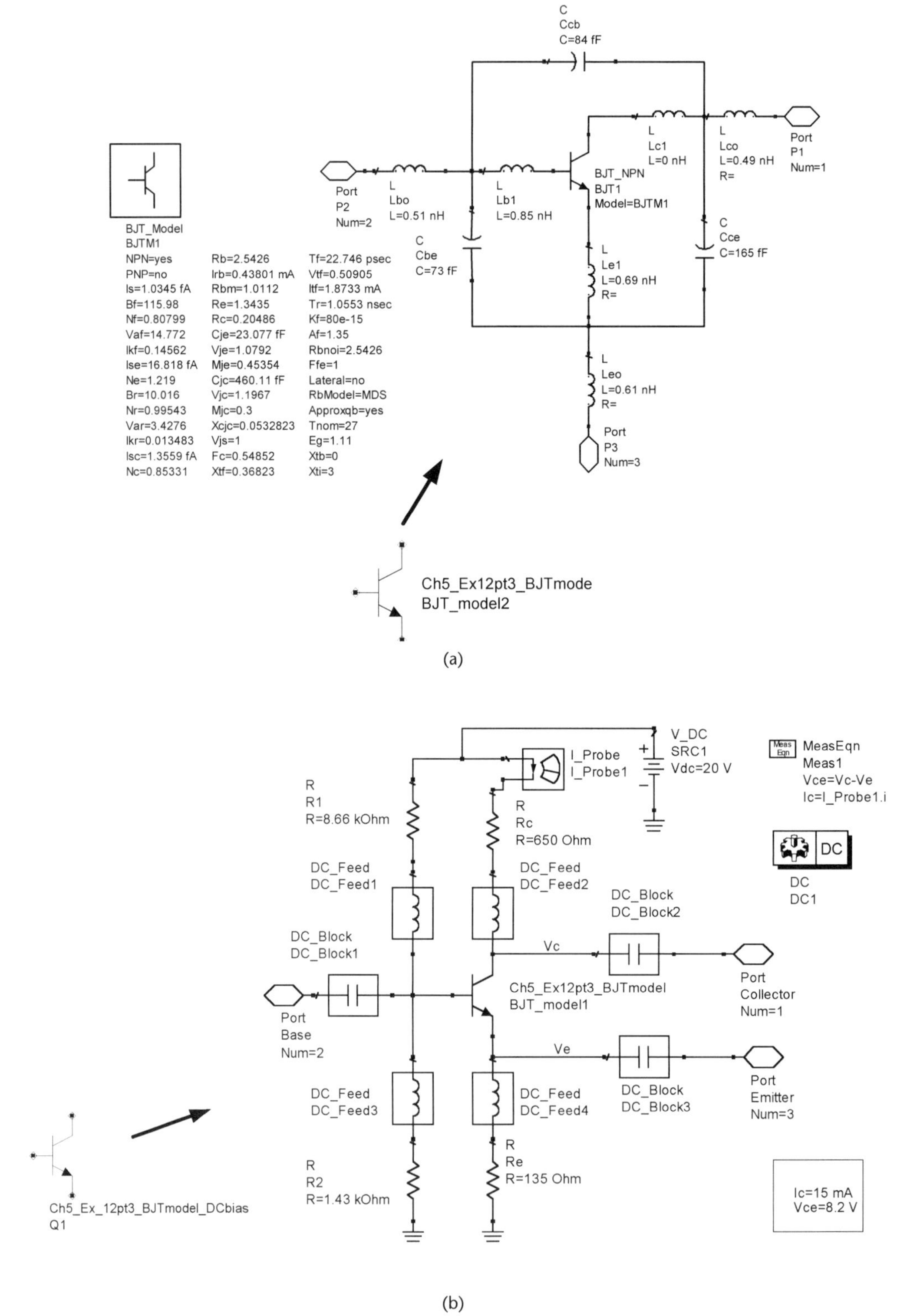

Figure 5.55 (a) SPICE model for the BFR183 transistor and its package model, (b) the DC bias network used for the packaged transistor, (c) stability simulation, and (d) mapping of the Γ_T plane onto the Γ_{IN} plane.

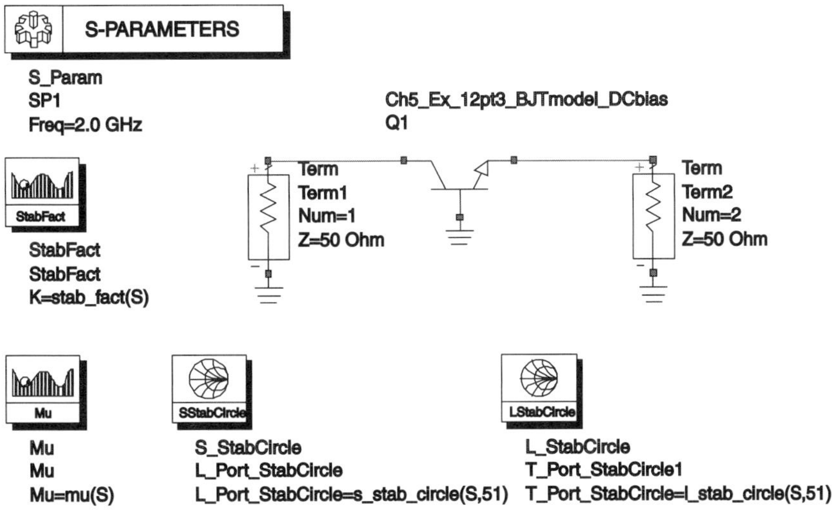

freq	S			
	S(1,1)	S(1,2)	S(2,1)	S(2,2)
2.000 GHz	1.215 / -32.375	2.067 / -55.217	0.187 / 144.637	1.232 / 136.217

freq	K	Mu
2.000 GHz	-0.937	-0.846

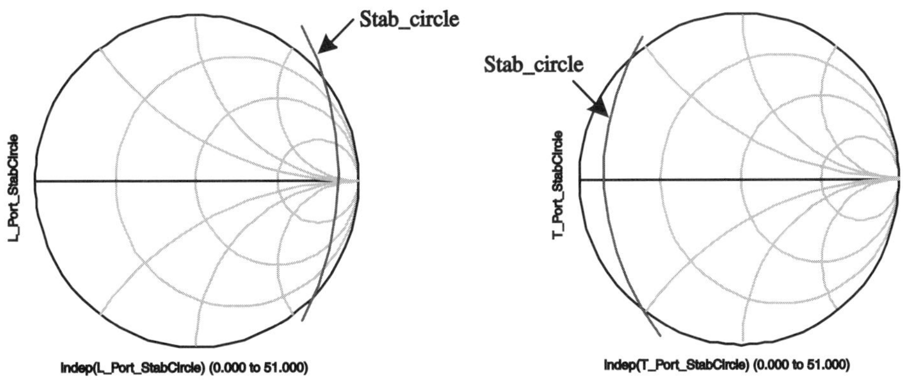

(c)

Figure 5.55 (continued.)

The harmonic balance simulation in Figure 5.58(a) was performed, and the resulting waveform of the oscillator is shown in Figure 5.58(b). The fundamental frequency of oscillation is at 2 GHz and the oscillator power is 12.2 dBm in good agreement with the predicted values.

The phase-noise performance of the oscillator is now analyzed. The phase-noise simulation is shown in Figure 5.59 where the phase noise is calculated from 1 Hz to 10 MHz. The SPICE flicker-noise parameters were not listed by the

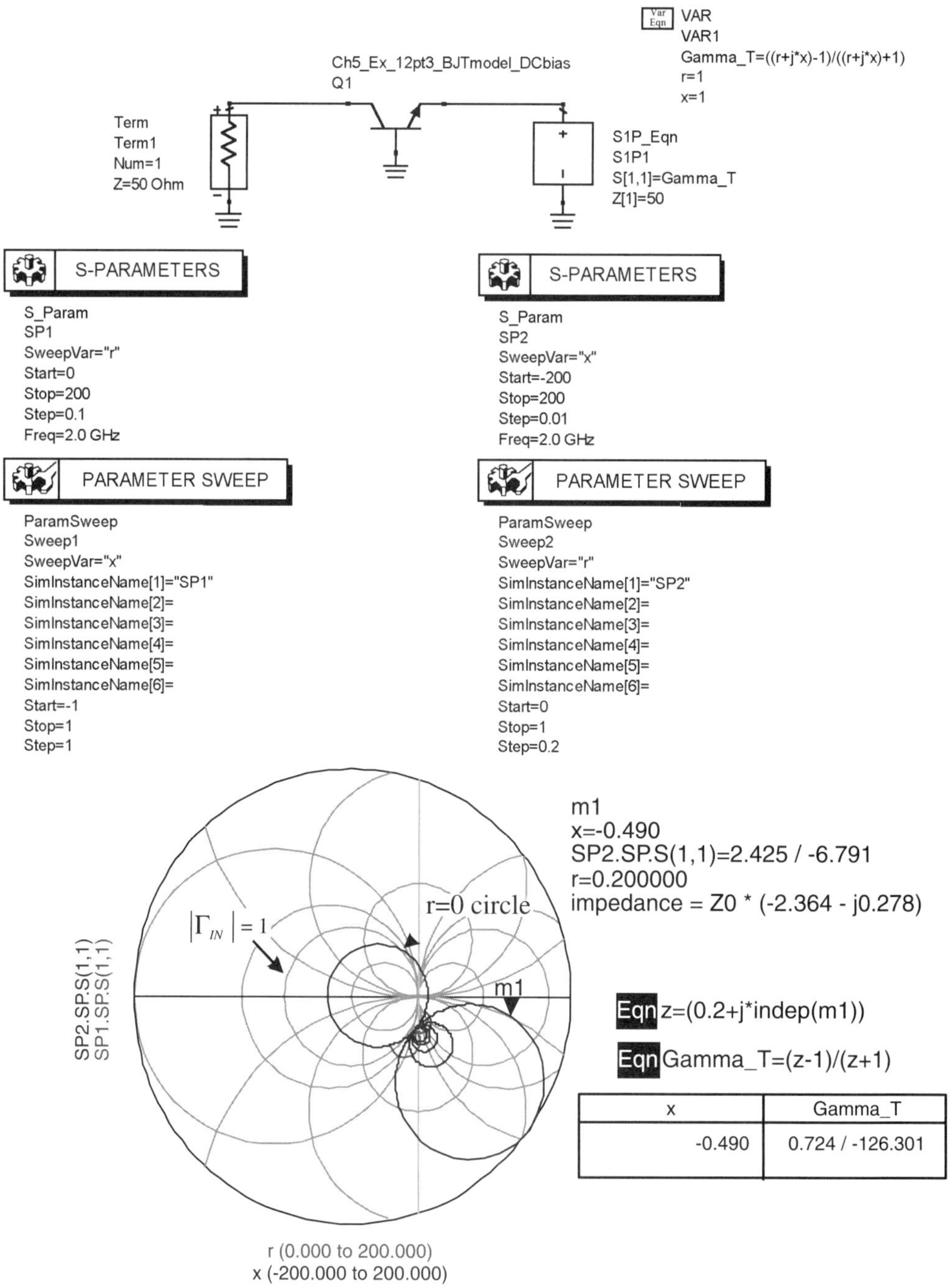

(d)

Figure 5.55 (continued.)

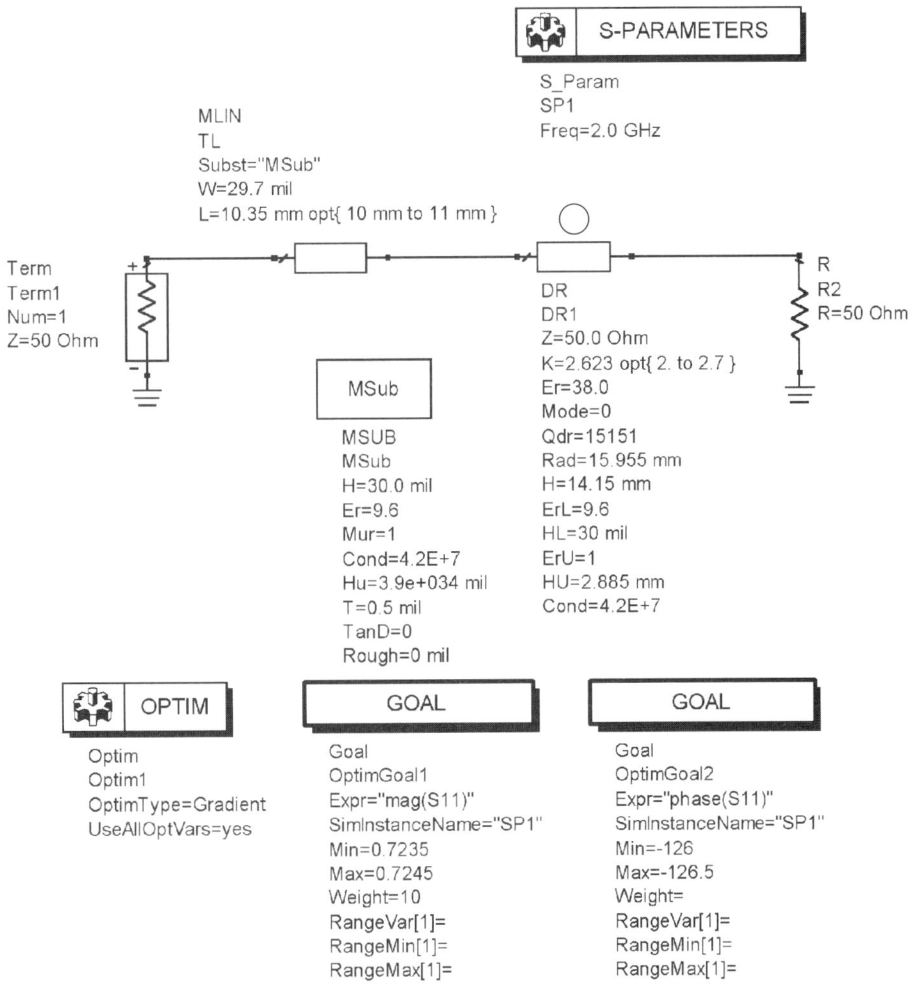

optIter	K	L
10	2.605	0.010

freq	S(1,1) optIter=10
2.000 GHz	0.725 / -126.250

DR1: Q due to the conductor plates = 9646.371266
DR1: Resonant Freq = 2.000006 GHz, and Unloaded Q =5893.857437
DR1: (Equivalent Parallel RLC: R=260.529069 Ohm, L=0.003518 nH, and C=1800.248014 pF)

Figure 5.56 The design of Γ_T using the DR model in ADS.

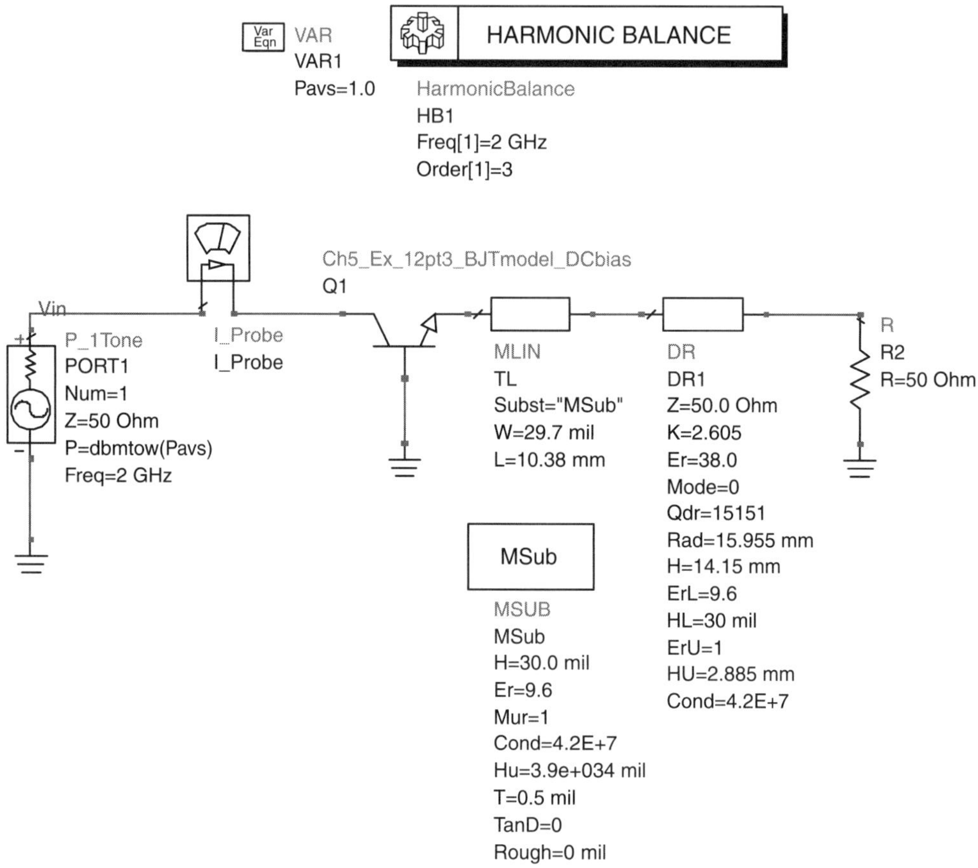

Eqn Zin=Vin[1]/I_Probe.i[1]

Eqn Gamma_IN=(Zin-50)/(Zin+50)

Eqn Padd=P_reflected-P_available

Eqn P_available=dbmtow(Pavs)

Eqn P_reflected=P_available*mag(Gamma_IN)**2

Eqn Padd_dBm=10*log10(mag(Padd)/0.001)

Pavs	Zin	Gamma_IN	Padd_dBm
-20.000	-118.668 - j13.555	2.418 / -6.572	-13.147
-18.000	-118.671 - j13.570	2.417 / -6.578	-11.148
-16.000	-118.677 - j13.594	2.417 / -6.589	-9.149
-14.000	-118.685 - j13.632	2.417 / -6.606	-7.151
-12.000	-118.699 - j13.694	2.416 / -6.632	-5.153
-10.000	-118.721 - j13.792	2.415 / -6.675	-3.158
-8.000	-118.757 - j13.952	2.414 / -6.744	-1.165
-6.000	-118.815 - j14.213	2.411 / -6.857	0.824
-4.000	-118.910 - j14.651	2.407 / -7.045	2.805
-2.000	-119.071 - j15.408	2.399 / -7.368	4.772
0.000	-119.353 - j16.801	2.385 / -7.952	6.710
2.000	-119.890 - j19.664	2.356 / -9.112	8.579
4.000	-120.957 - j26.807	2.281 / -11.784	10.237
6.000	-120.046 - j44.671	2.116 / -17.808	11.414
8.000	-109.233 - j64.700	1.959 / -25.413	12.532
10.000	-94.937 - j76.932	1.842 / -31.751	13.788
12.000	-85.435 - j87.943	1.703 / -35.057	14.789
14.000	-76.372 - j99.174	1.565 / -36.985	15.615
16.000	-65.223 - j107.440	1.452 / -38.937	16.445
18.000	-52.358 - j112.376	1.352 / -41.127	17.185
20.000	-37.174 - j114.394	1.249 / -43.706	17.490
22.000	-26.526 - j117.109	1.171 / -44.497	17.704
24.000	-22.580 - j119.215	1.141 / -44.287	18.797
26.000	-14.906 - j129.961	1.079 / -41.650	18.162
28.000	-13.061 - j166.323	1.044 / -33.286	17.542
30.000	-1.898 - j168.266	1.006 / -33.095	10.933
32.000	7.558 - j167.603	0.976 / -33.164	18.825
34.000	14.738 - j172.101	0.955 / -32.193	23.404
36.000	19.404 - j176.683	0.945 / -31.270	26.322
38.000	23.738 - j181.203	0.936 / -30.390	28.936
40.000	27.790 - j185.638	0.929 / -29.558	31.373

Figure 5.57 Added power simulation and the load circuit design.

manufacturer. For the simulation we used $A_F = 1.35$ and two values of K_f: $K_f = 80\text{e}{-13}$ and $K_f = 80\text{e}{-15}$. The resulting plot of $\mathcal{L}(f_m)$ shows a slope of –30 dB/dec until f_c with $K_f = 80\text{e}{-13}$, and essentially –20 dB/dec with $K_f = 80\text{e}{-15}$. It is also observed that the $\mathcal{L}(f_m)$ has a value of –137.58 dBc/Hz at a 10-kHz offset from the carrier, and the noise floor is at –170 dBc/Hz.

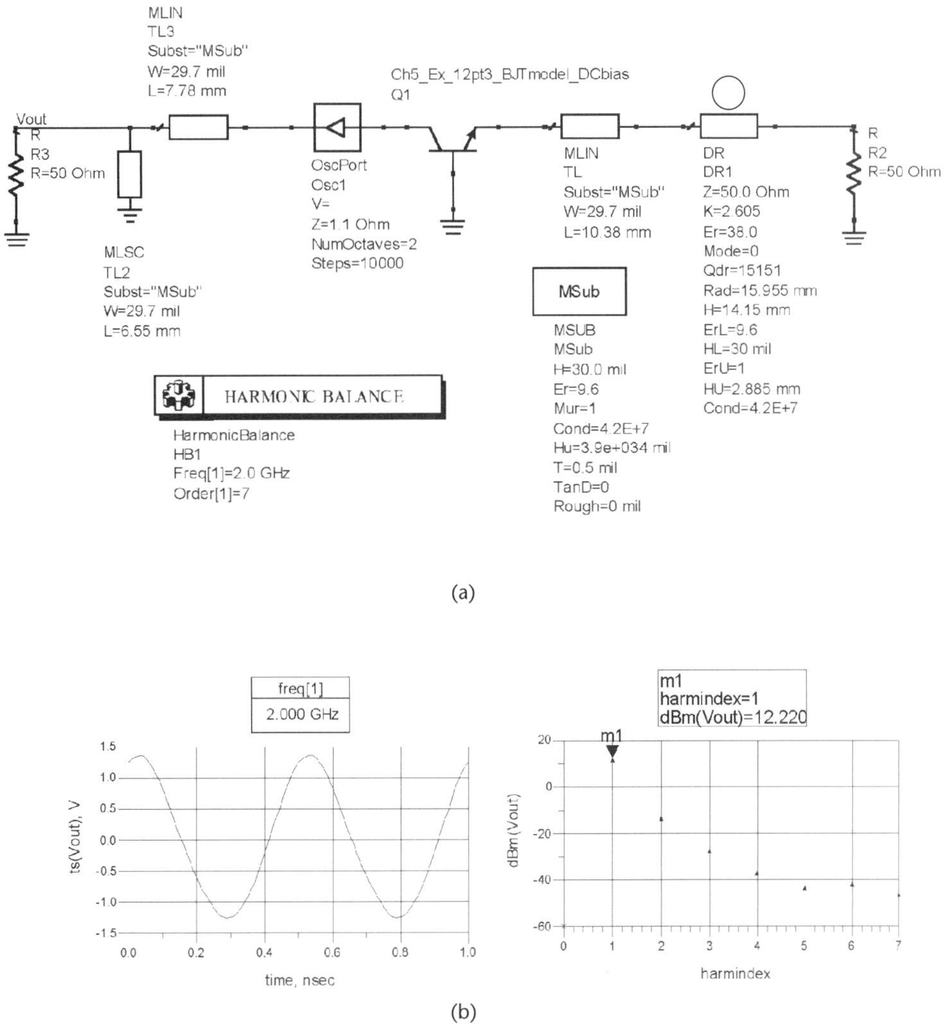

Figure 5.58 (a) The simulation of the oscillator and (b) the oscillator waveform.

A yield analysis is performed next. Three yield specifications were considered: the resonant frequency, the output power, and the SSB phase noise at 10 kHz from the carrier. The circuit is said to "pass" the design specifications if the resonant frequency is within 0.5% of 2 GHz, the output power is above 10 dBm, and the SSB phase noise is less than –133 dBc/Hz at 10 kHz from the carrier.

The yield-analysis simulation is shown in Figure 5.60(a), and the simulation results (i.e., the histograms) are shown in Figure 5.60(b). Typical process parameters variations were entered for all components in the design.

ADS sets the oscillation frequency to 0 Hz when the oscillator fails to oscillate. The size of the variable "indices" in Figure 5.60(b) shows the number of simulations that did not oscillate. Only 8 of the 100 simulations did not meet the oscillation conditions. The overall yield of the circuit is 15%, which is certainly unsatisfactory

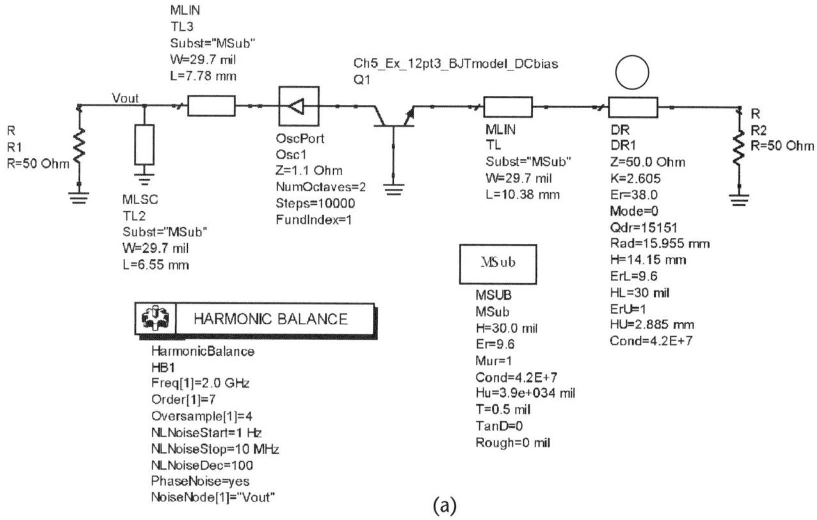

(a)

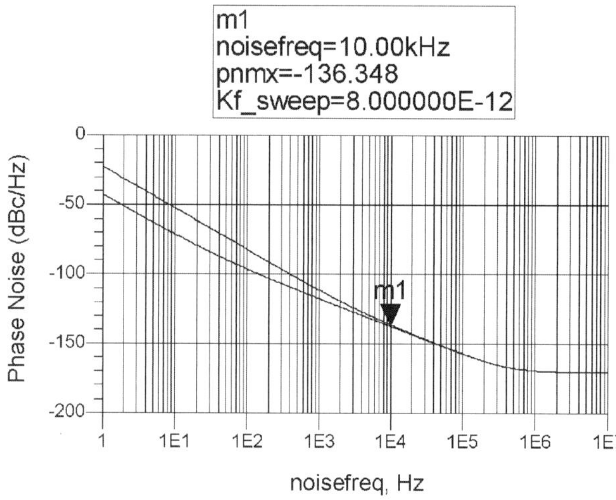

(b)

Figure 5.59 (a) Phase noise simulation and (b) the SSB phase noise.

for any fabrication process. It is also observed from the histograms that most designs failed due to the frequency specifications.

Components sensitivity histograms are shown in Figure 5.60(c). These results show that the design is already centered.

In order to improve the overall yield, the process parameters in dielectric resonator section were analyzed as shown in Figure 5.61(a). The variations in the radius and height of the DR affect significantly the resonant frequency. The simulation controller Spec1 is set up so that the trial fails if the magnitude of Γ_T is greater than 0.728 within the frequency range of 1.99 to 2.01 GHz, which corresponds to the frequency tolerance. The simulation data for $|S_{11}|$, shown in Figure 5.61(b), shows that all values $|S_{11}|$ are less than 0.728, so the Spec1 yield is 100%. However, the data also shows that very few trials satisfy the resonant frequency specification

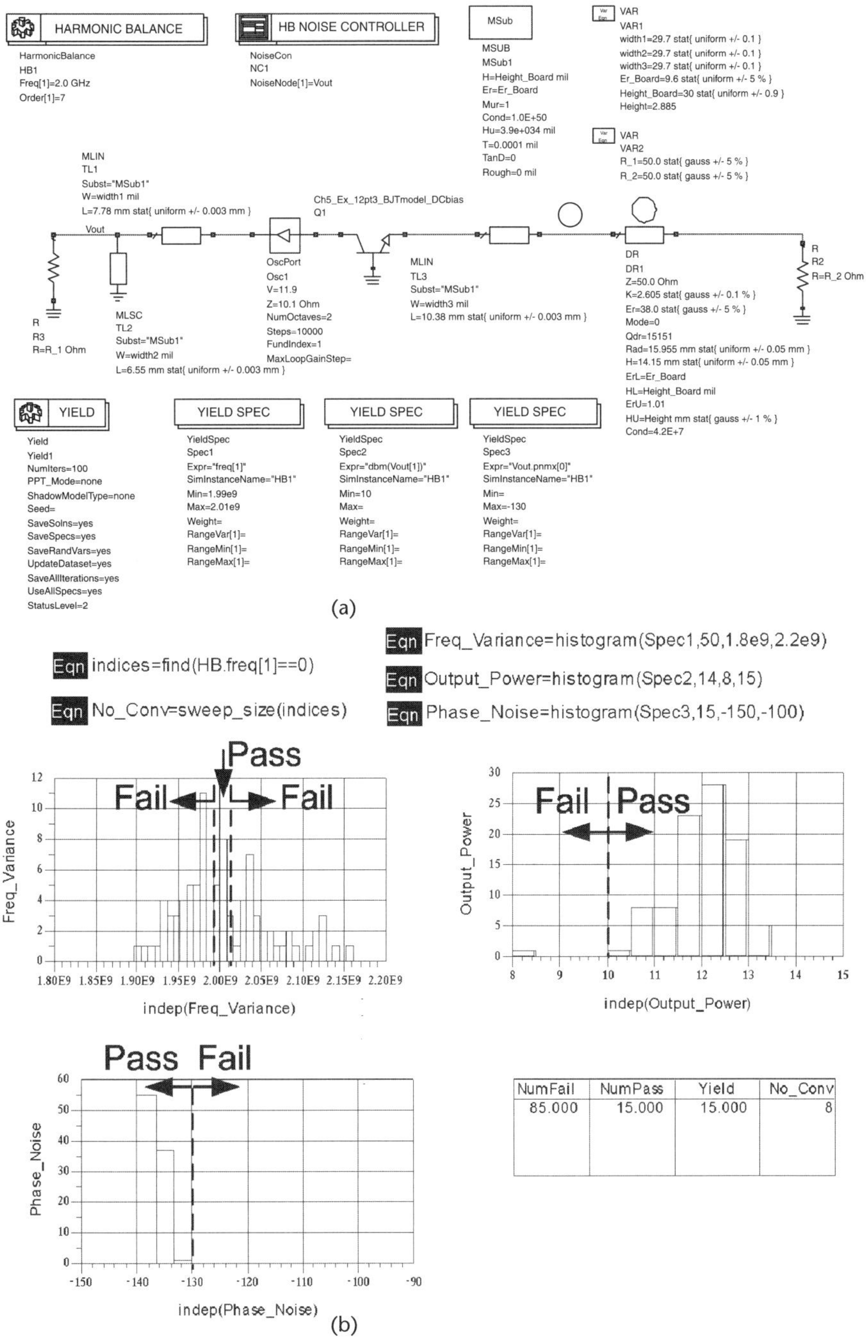

Figure 5.60 (a) The yield-analysis simulation, (b) the simulation results, and (c) components sensitivity histograms.

of 1.99 GHz < f_o < 1.01 GHz. Hence, the variations in the resonant frequency of the DR are determining the overall yield.

For a high yield the resonator frequency must be tuned using external means, such as using a tuning screw to vary *Hu*. The ADS simulation in Figure 5.62 is

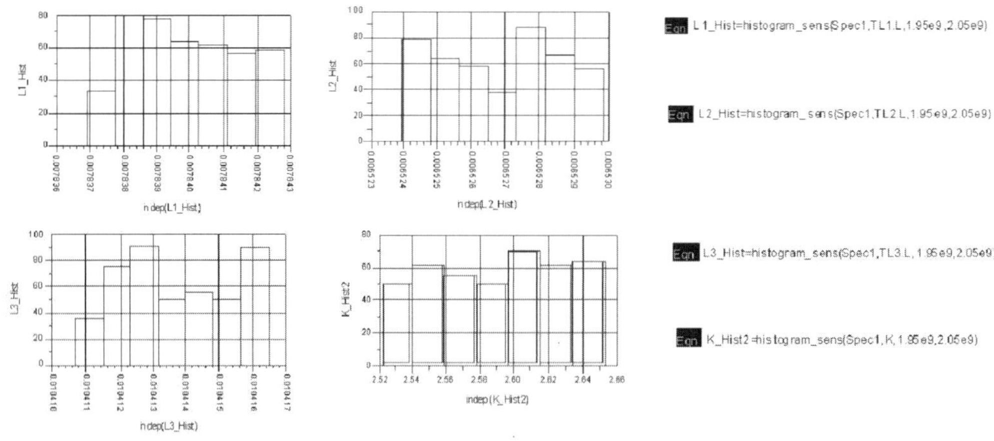

(c)

Figure 5.60 (continued.)

used to tune the height of the tuning plate to make the resonant frequency of the DR fall within the desired frequency tolerance. The height of the tuning plate is allowed to vary from 1 to 10 mm. The yield-analysis controller is set up to convert failures and therefore adjusts the value of *Hu* until the circuit passes the design specifications. The postproduction tuning results, shown in Table 5.5, show that using the tuning screw in the DR housing to adjust the frequency of oscillation can bring the yield to about 87%.

In some designs the DR is coupled to the microstrip line as shown in Figure 5.63. The $\lambda/4$ open-circuited transmission line presents a short circuit at the position of the DR. Hence, the magnetic coupling is increased since the current is a maximum at the DR location. From Figure 5.63, with $\Gamma_L = -1$, we obtain at ω_o that

$$\Gamma_{XX'} = S_{22} + \frac{S_{12}S_{21}(-1)}{1 - S_{22}(-1)} = \frac{\beta - 1}{\beta + 1}$$

and

$$\Gamma_T = \frac{\beta - 1}{\beta + 1} e^{-j2\theta}$$

Example 5.14

Design the terminating network for a 10-GHz oscillator using the DRO coupling shown in Figure 5.63.

Solution

At the resonator location the open-circuited $\lambda/4$ line provides the short circuit. Hence,

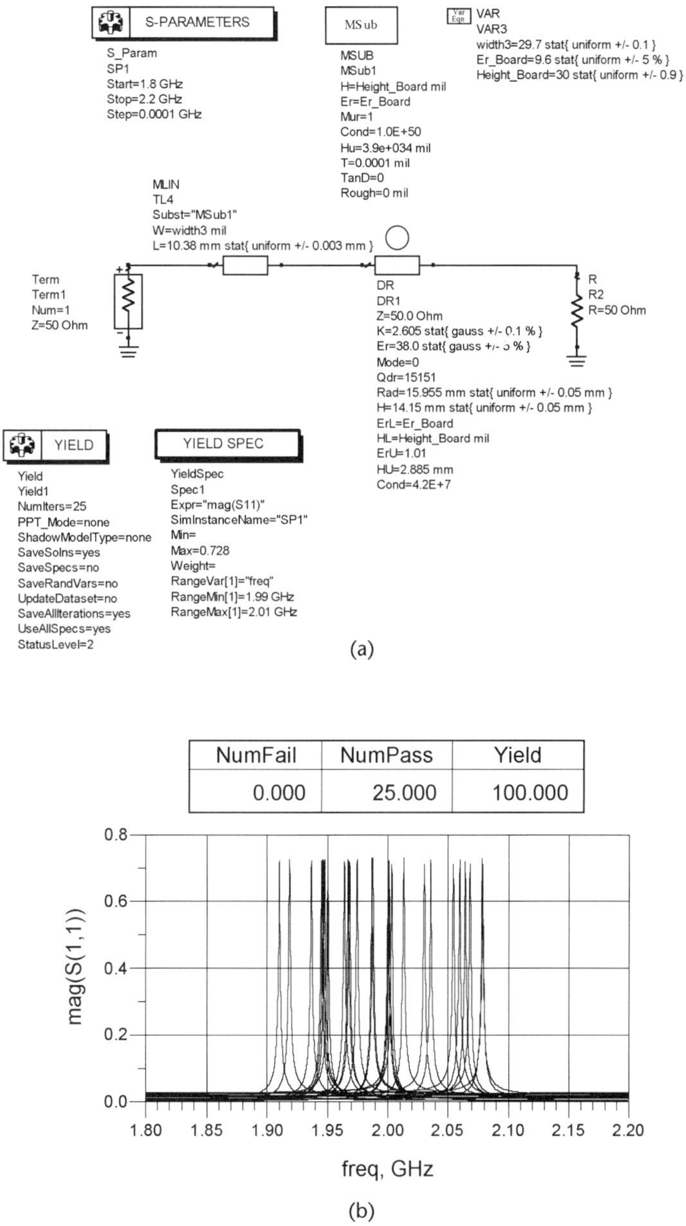

NumFail	NumPass	Yield
0.000	25.000	100.000

Figure 5.61 (a) The yield of the terminating network and (b) the simulation data.

$$\Gamma_{XX'} = \frac{\beta - 1}{\beta + 1} = 0.666 \Rightarrow \beta = 0.047$$

Then, with $l_2 = \lambda/2$ it follows that $\Gamma_T = 0.666$. Of course, in such a case $\beta = 0.047$, which corresponds to an under-coupled case. The rest of the design is similar to the previous example.

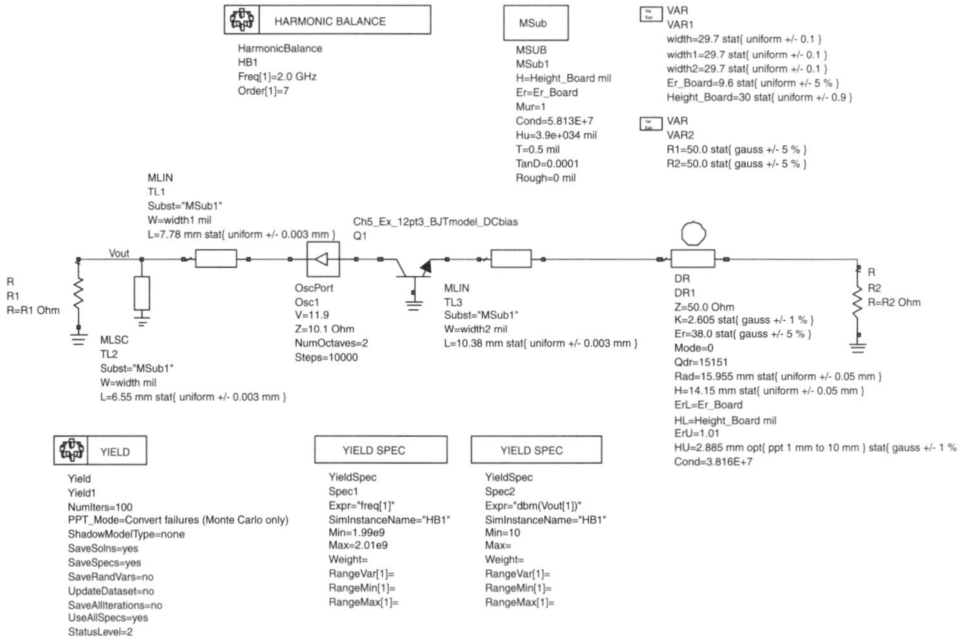

Figure 5.62 Postproduction tuning simulation.

Table 5.5 Simulation Results

PreTuneYield	*PPT_Yield*	*PPT_Converts*
21.000	87.000	66.000

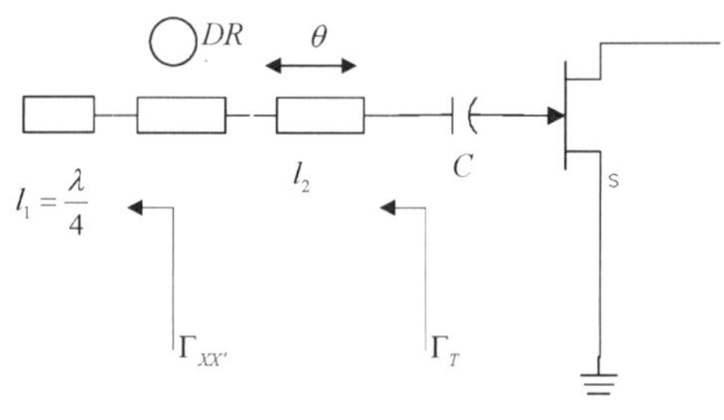

Figure 5.63 A DRO coupling with $l_1 = \lambda/4$.

5.12.3 Parallel-Coupled DRO

The circuit in Figure 5.64 shows a parallel-coupled dielectric resonator oscillator. The DR in this configuration acts as a parallel-feedback element, which, along with the transistor, presents a negative resistance at the load port.

The DR and coupling circuits in Figure 5.65 can model the circuit in Figure 5.64 by appropriately choosing the R, L, and C values of the parallel resonant circuit (which models the DR) as well as the transformer turn ratios N_1 and N_2 (which are determined by the distance between the DR and the microstrip lines).

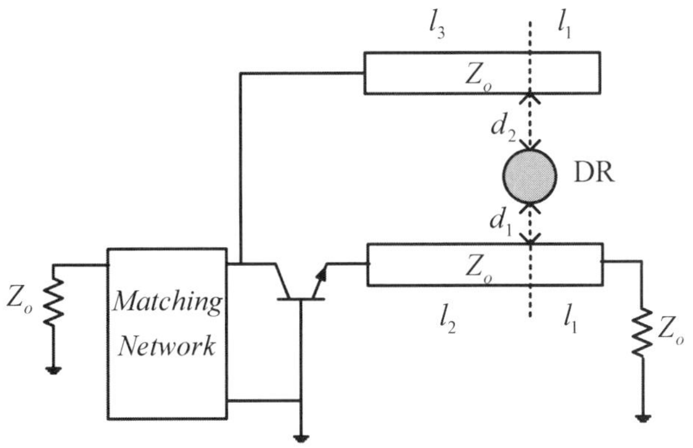

Figure 5.64 Parallel-coupled dielectric resonator oscillator.

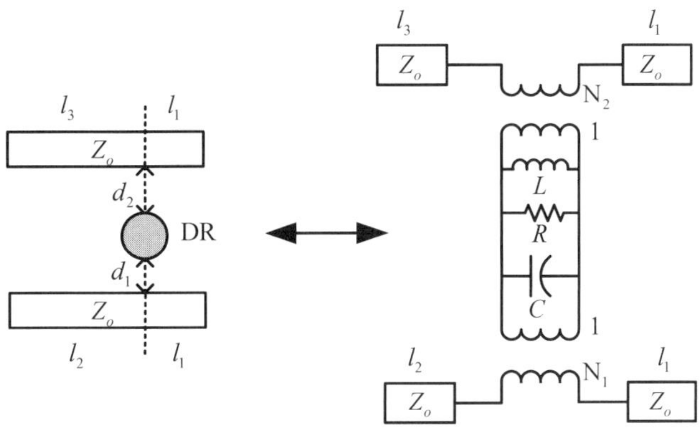

Figure 5.65 Parallel-coupled DR model.

The feedback through the microstrip line from emitter to collector impedes the isolation of one port from the other making the whole design cumbersome.

A design procedure for a parallel-coupled DR oscillator at 2 GHz is now illustrated. The complexity presented by the feedback can be analyzed using a combination of iterative simulations. The simulations will measure the effect of varying the above-mentioned parameters so as to obtain an adequate negative resistance at the load port.

The circuit in Figure 5.66 shows the ADS simulation used to perform the first of these iterative simulations. As a starting point, the DR is replaced by its equivalent parallel-resonant circuit at 2 GHz, which was obtained in the simulation shown in Figure 5.56. The length of microstrip lines "TL1" and "TL2" are set to a quarter-wave length (i.e., to $l = 14.75$ mm), and "TL3" is set to 360°. Both transformer turns-ratio coefficients are initially set to 1 and will be optimized in a subsequent simulation. *Pavs* is set to 1, and the resulting input reflection coefficients as a function of the line length TL4 are calculated. The simulation data in Table 5.6 shows that a line length for TL4 of 12 mm produces a negative resistance at the load port (i.e., $|\Gamma_{IN}| = 1.481\,\underline{|41.46^\circ}$).

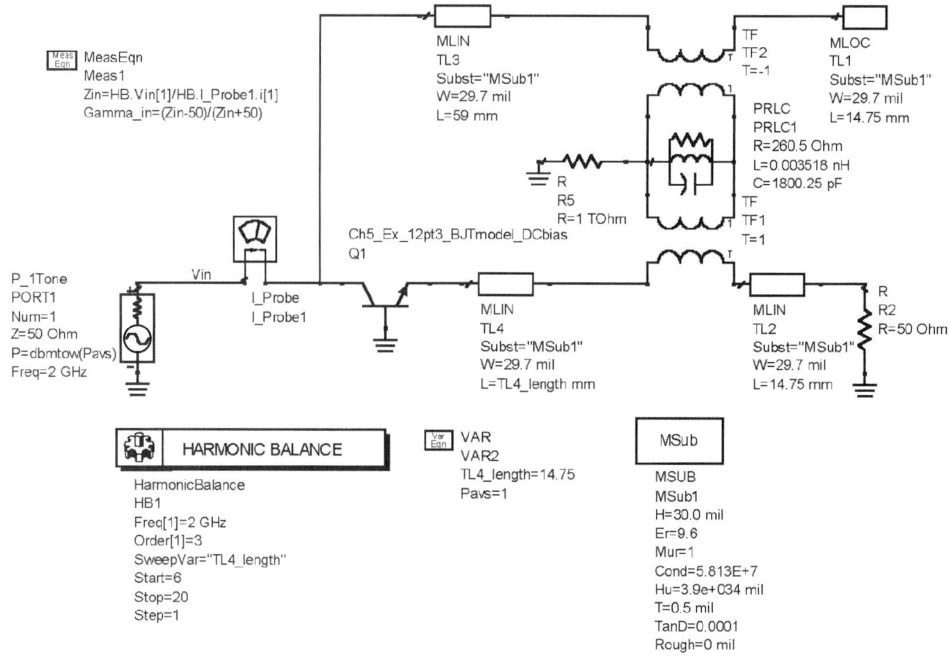

Figure 5.66 Parallel-coupled DR line length optimization.

Table 5.6 Load-Port Reflection Coefficient Versus Line Length TL4

TL4_length	Γ_{IN}
6.000	0.548/99.759
7.000	0.640/91.707
8.000	0.756/83.714
9.000	0.902/75.396
10.000	1.082/66.203
11.000	1.290/55.268
12.000	1.481/41.459
13.000	1.508/27.821
14.000	1.477/10.077
15.000	1.593/–9.080
16.000	1.579/–16.555
17.000	1.494/–32.002
18.000	1.487/–48.741
19.000	1.345/–57.745
20.000	1.194/–63.741

Next, the effects of the transformers' turns ratios are analyzed. The simulation is shown in Figure 5.67. The simulation data (Table 5.7) shows that there are many combinations for the two transformer ratios, which result in a negative resistance at the load port. Figure 5.68 shows the simulation with the turns ratios set equal to unity. The resulting frequency of oscillation is at 2 GHz. It was also observed that small values of the coupling coefficients resulted in lower phase noise but lower output power.

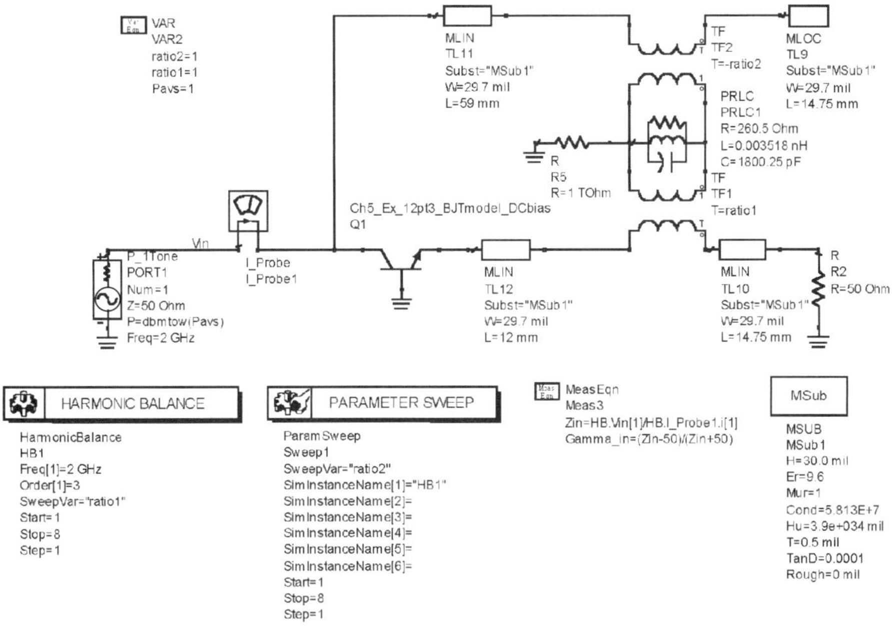

Figure 5.67 Turns-ratio effect on the load-port negative resistance.

5.13 YIG Oscillators

Yittrium Iron Garnet (YIG) is a ferrimagnetic material. A YIG resonator uses the resonant properties of a YIG sphere under the influence of a dc and ac magnetic field. The YIG sphere can be modeled by a parallel RLC resonant circuit. The value of the elements depends on the magnetization, coupling, and resonance line width of the YIG sphere and on the applied dc magnetic field. The uniform dc magnetic field is applied with an electromagnet with a single gap. The gap design is important since a nonuniform dc magnetic field results in a tuning hysteresis and spurious responses. A common-gate GaAs FET oscillator using a YIG resonator is shown in Figure 5.69. A dc magnetic field is applied in the plane of the coupling loop and an RF current is used to generate the ac field at a right angle to the dc magnetic field. The design of the electromagnet system is cumbersome.

The YIG sphere is strongly coupled to the transmission line that connects the active device. Assuming that the YIG sphere is magnetically saturated and that the sphere diameter is much smaller than $\lambda/4$, the YIG device can be modeled by a parallel resonant circuit, as shown in Figure 5.70. The element values are given by

$$G_o = \frac{d^2}{\mu_o V \omega_m Q_U}$$

$$L_o = \frac{\mu_o V \omega_m}{\omega_o d^2}$$

Table 5.7 Simulation Data

	Γ_{IN}							
Ratio 1	*Ratio 2 = 1.000*	*Ratio 2 = 2.000*	*Ratio 2 = 3.000*	*Ratio 2 = 4.000*	*Ratio 2 = 5.000*	*Ratio 2 = 6.000*	*Ratio 2 = 7.000*	*Ratio 2 = 8.000*
1.000	1.481/41.4	1.567/11.6	1.470/–1.8	1.407/–8.8	1.366/–13.1	1.338/–16.0	1.317/–18.1	1.302/–19.6
2.000	1.044/96.2	1.662/50.2	1.804/29.4	1.684/14.7	1.588/5.54	1.520/–0.58	1.471/–4.9	1.434/–8.2
3.000	0.930/118	1.287/80.0	1.690/52.0	1.877/37.7	1.796/24.5	1.707/15.2	1.633/8.58	1.575/3.5
4.000	0.843/131	1.201/100	1.392/71.9	1.701/52.7	1.895/42.1	1.856/30.9	1.783/22.2	1.715/15.4
5.000	0.817/141	1.051/108	1.306/91.6	1.456/67.4	1.705/53.0	1.891/44.6	1.887/35.4	1.833/27.3
6.000	0.813/148	0.949/116	1.217/100	1.346/83.5	1.501/64.7	1.708/53.2	1.878/46.2	1.903/38.5
7.000	0.818/153	0.886/124	1.106/105	1.307/95.0	1.383/77.7	1.532/62.8	1.710/53.4	1.862/47.2
8.000	0.826/157	0.849/130	1.018/110	1.223/99.9	1.339/89.4	1.417/73.6	1.557/61.5	1.712/53.4

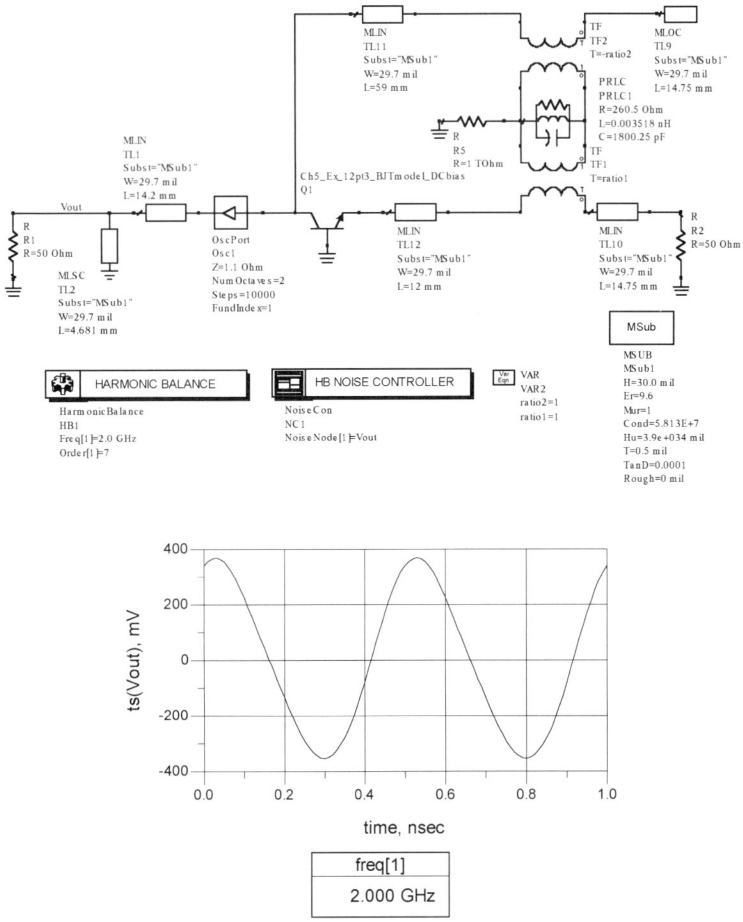

Figure 5.68 Simulation of the oscillator.

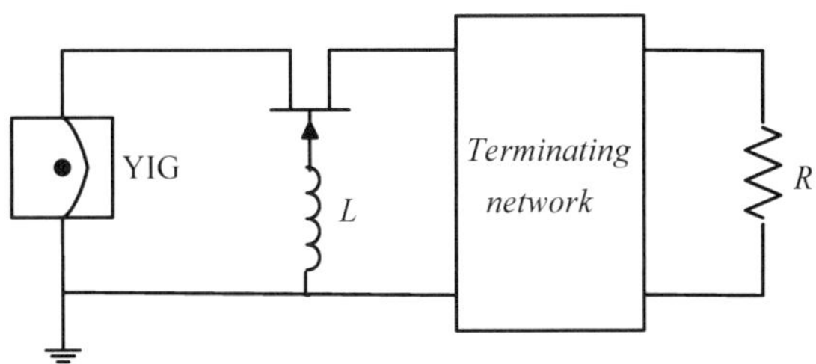

Figure 5.69 A YIG-tuned oscillator.

$$C_o = \frac{1}{\omega_o^2 L_o}$$

where

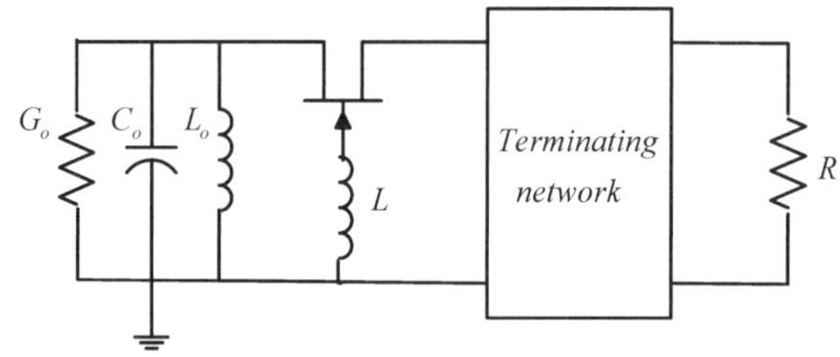

Figure 5.70 Equivalent RLC network for the YIG sphere.

$$\omega_m = \gamma 8\pi^2 M_s$$

$$\omega_o = 2\pi\gamma H_o$$

$$Q_U = \frac{H_o - \dfrac{4\pi M_s}{3}}{\Delta H}$$

The quantity $4\pi M_s$ is the saturation magnetization of the sphere, $\mu_o = 4\pi(10^{-7})$ H/m, V is the volume of the sphere, d is the coupling loop diameter, γ is the gyromagnetic ratio (2.8 MHz/Oe), H_o is the applied dc magnetic field, Q_U is the unloaded Q, ΔH is the resonance line width (approximately 0.2 Oe), ω_o is the center frequency of resonance, and H_o is the dc magnetic field.

The frequency of resonance can be controlled accurately with the dc field from an electromagnet. As the dc is reduced, the losses increase and at some point the YIG sphere stops acting as a resonator.

5.14 Other Negative-Resistance Devices

Several two-terminal negative-resistance devices have been developed. Two that have found practical application as oscillators in the microwave and millimeter range are the Gunn diodes and the Impatt diodes. A brief description of oscillators using these diodes is given.

5.14.1 Gunn Diodes

The Gunn diode symbol and characteristics are shown in Figure 5.71(a). The device exhibits a negative resistance region. The Gunn diode is called a diode because of its two terminals. However, it has no pn junction, nor does it resemble a Schottky diode. In fact, Gunn diodes are bulk effect devices where two of the bulk semiconductor material used in their fabrication are n-type GaAs and InP (Indium Phosphide). The frequency of operation is from about 10 to 180 GHz. In CW operation, oscillation power levels of about 500 mW can be obtained at the lower frequencies, and about 50 mW at the higher frequencies.

Figure 5.71(b) shows that at the threshold voltage (V_{th}) the current reaches a maximum value (i.e., I_{th}). As the bias is further increased the current decreases

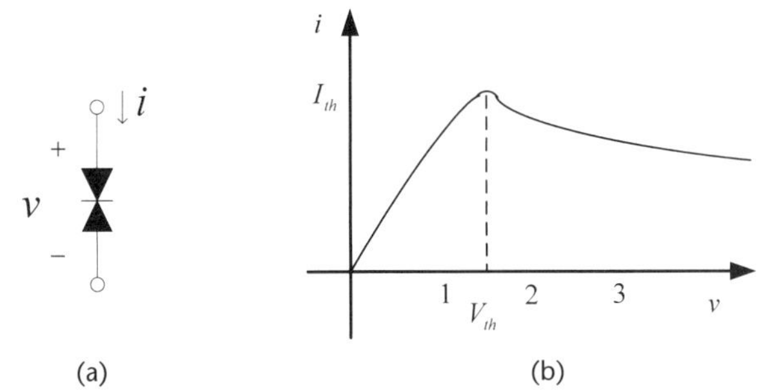

Figure 5.71 (a) Gunn diode symbol and (b) characteristics.

and the device exhibit a negative resistance. The current will continue to decrease until a breakdown voltage is reached were the diode will burn.

Different modes of operation are possible with Gunn diodes. The operation is controlled either by the design of the diode or by the external circuit. The physics of the device operation is quite involved. However, for oscillator design purposes what is needed is an equivalent model of the device in the negative-resistance region. In the negative-resistance region the diode can be represented by the model is shown in Figure 5.72(a), where R_s is the diode series resistance, L_p and C_p are the package inductance and capacitance, C_G is the diode capacitance, and $-R_G$ is the negative resistance of the diode. Typical values for a 10-GHz Gunn diode are: $L_p = 1$ nH, $C_p = 0.5$ pF, $R_s = 0.5\Omega$, and $-R_G$ -20Ω. In general, the elements L_p and R_s can be neglected and the model reduces to a simple parallel $R_G C_G$ circuit with package capacitance C_p, as shown in Figure 5.72(b).

A basic Gunn diode oscillator is shown in Figure 5.73(a), where C_b is a bypass capacitor. The ac model of the oscillator is shown in Figure 5.73(b). The load network, represented by a parallel RLC circuit, is designed to satisfy the oscillation conditions.

The resulting model in Figure 5.73(b) is that of a parallel negative-resistance device. Hence, oscillations build up when $|G_G| > G_L$. The fundamental frequency of oscillation is given by

$$\omega_o = \frac{1}{\sqrt{L(C_G + C_p + C)}}$$

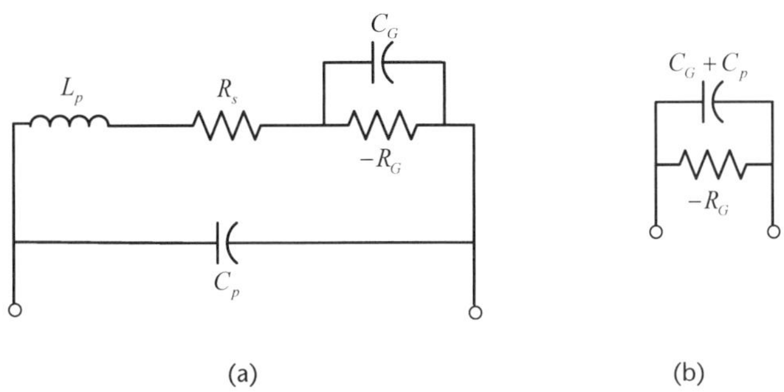

Figure 5.72 (a) Gunn diode model and (b) simplified model.

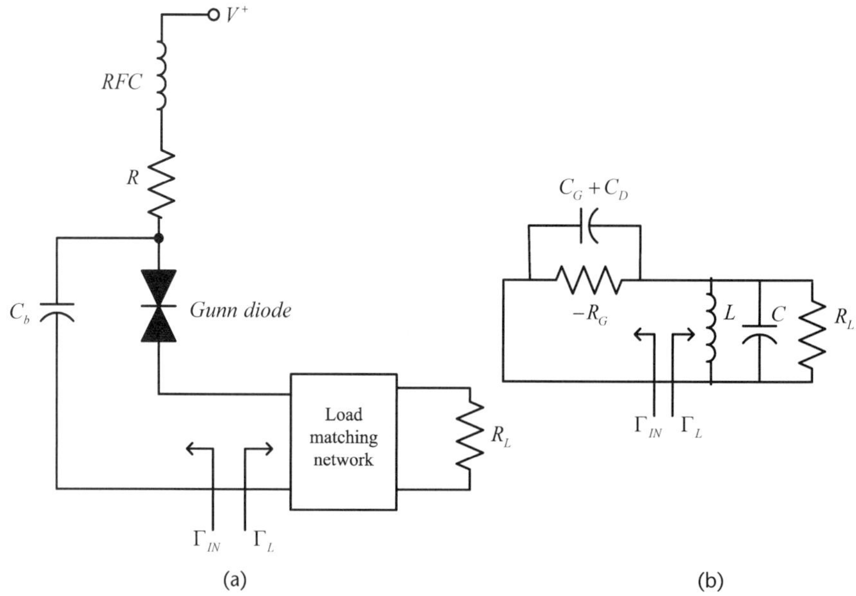

Figure 5.73 (a) Basic circuit for a Gunn-diode oscillator and (b) its ac model.

In practice, Gunn-diode oscillators are implemented using either a coaxial or rectangular waveguide cavity resonator. A rectangular waveguide circuit is shown in Figure 5.74. In this resonator the dc is applied through a metal post to the Gunn diode. The current in the post excite the fields inside the waveguide (usually a *TE* mode), which is coupled to the load through the opening (i.e., iris). A tuning screw is used to adjust the cavity resonant frequency. The shorted waveguide is designed to provide the needed inductance to resonate the Gunn's diode capacitance.

The design of the waveguide structure with a post and the diode is involved. Numerical techniques have been used to calculate the impedance seen by the diode and to completely describe the structure.

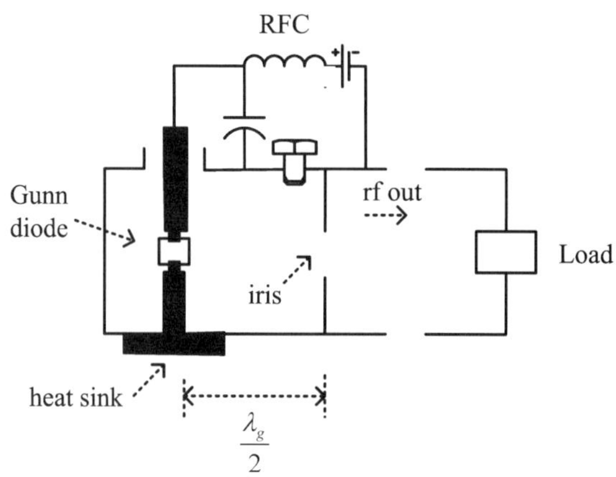

Figure 5.74 Waveguide circuit for the Gunn-diode oscillator.

Gunn-diode oscillators generate power with low efficiency (2% to 10%). Hence, a good heat sink is necessary.

Gunn-diode VCOs with a varactor tuner are available with fairly constant output power over a wide frequency range. Pulse operation is also used with resulting larger powers than in CW operation.

5.14.2 Impatt Diodes

Impatt diodes (or Impact Ionization Avalanche Transit Time diodes) were developed using the physics of avalanche and saturation velocity of carriers in a semiconductor. The device is useful at millimeter frequencies where it can generate a large amount of power. Impatt diodes have higher efficiency than Gunn diodes (about 20% efficiency). However, they are more noisy because of the avalanche mechanism.

Impatt diodes are fabricated using Si and GaAs. The package is similar to that of the Gunn diodes. The Impatt-diode characteristics and equivalent circuit are shown in Figure 5.75. The device operation is in the breakdown region.

Because of the breakdown operation the applied voltage is high. For example, in the X-band region, the required voltages are from 60V to 90V, while at millimeter frequencies (say 100 GHz) the required voltage is around 10V. Pulse operation is possible with significant increase of the output power.

The equivalent circuit of an Impatt diode is a series RC circuit as shown in Figure 5.75(b). Typical negative-resistance values of -0.5Ω to -4Ω are typical for RF currents of 50 mA to 1A. L_p and C_p are the package inductance and capacitance.

The matching of the low negative resistance associated with Impatt diodes requires some ingenuity in the coaxial and rectangular waveguide structures. A typical waveguide circuit for an Impatt-diode oscillator is shown in Figure 5.76. The configuration shows that a graded transformer was used to attain the match.

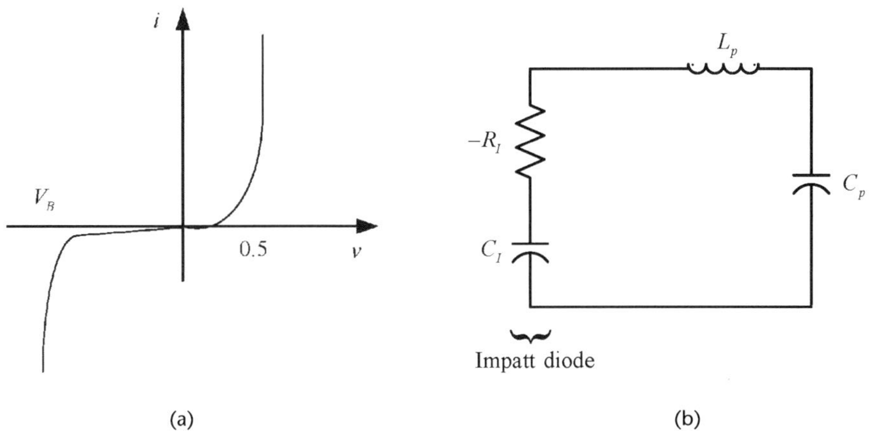

Figure 5.75 (a) Impatt diode characteristics and (b) equivalent circuit.

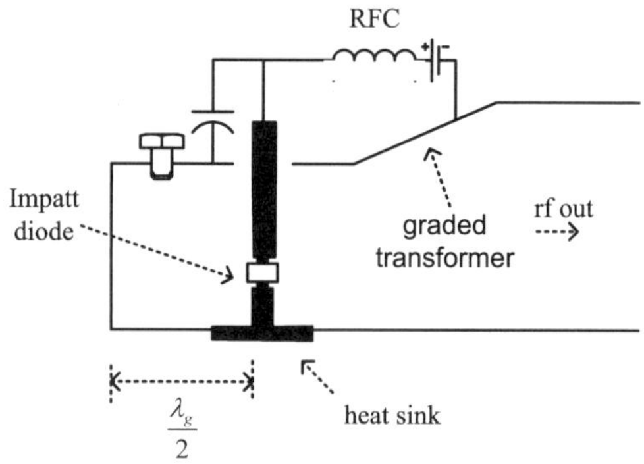

Figure 5.76 Rectangular waveguide circuit for an Impatt-diode oscillator.

References

[1] Gonzalez, G., *Microwave Transistor Amplifiers: Analysis and Design,* Second Edition, Upper Saddle River, NJ: Prentice-Hall, 1997.

[2] Kurokawa, K, "Some Basic Characteristics of Broadband Negative Resistance Oscillator Circuit," *Bell System Technical Journal,* Vol. 48, July 1969, pp. 1937–1955.

[3] Kurokawa, K, "Power Waves and the Scattering Matrix," *IEEE Transactions on MTT,* Vol. 13, March 1965, pp. 194–202.

[4] Kajfez, D., and P. Guillon, (Eds.), *Dielectric Resonators,* Norwood, MA: Artech House, 1986.

[5] Vasiliadis, J., "Design and Statistical Analysis of a DRO Using CAD Techniques," M.S. Thesis, University of Miami, August 2004.

CHAPTER 6

Nonsinusoidal Oscillators

6.1 Introduction

In this chapter a variety of nonsinusoidal oscillators are discussed. These oscillators are usually referred to as relaxation oscillators. A relaxation oscillator is a circuit that switches between two states, producing a periodic signal whose period depends on the charging and discharging of a capacitor.

Relaxation oscillators are used to produce square waves, pulses, and triangular waves. Square waves (also referred as a clock oscillator) are one of the most important signals in digital circuits. A relaxation oscillator is basically an astable multivibrator. That is, an electronic circuit that operates between two quasi-stable states (i.e., on and off) at a certain frequency.

6.2 Various Relaxation Oscillators

6.2.1 Relaxation Oscillators Using Operational Amplifiers

A relaxation oscillator using an op amp is shown Figure 6.1. The operational amplifier is connected in a Schmitt-trigger configuration with positive feedback through R_1 and R_2.

In a Schmitt trigger the output voltage is at one of its saturation values, denoted by V_s^+ and V_s^-. The saturation values are about 1V below V^+ and 1V above V^-; that is,

$$V_s^+ \approx V^+ - 1$$

and

$$V_s^- \approx V^- + 1$$

When $v_o = V_s^+$ it follows that

$$v_+ = \frac{V_s^+ R_1}{R_1 + R_2}$$

and with $R_1 = R_2$ then

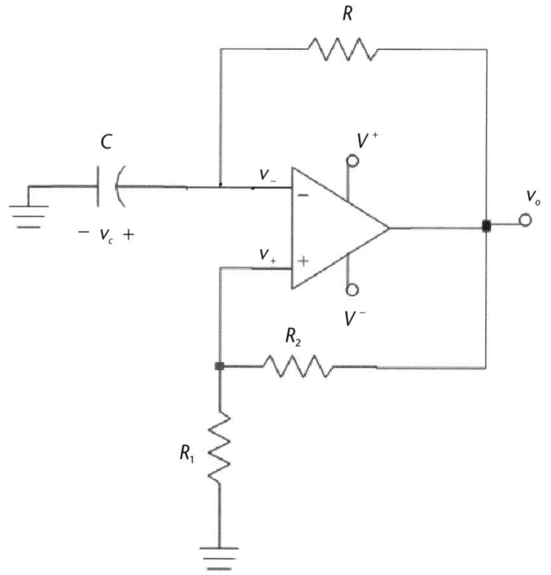

Figure 6.1 A relaxation oscillator.

$$v_+ = \frac{V_s^+}{2}$$

When $v_o = V_s^+$, capacitor C charges towards V_s^+. When the capacitor voltage reaches $V_s^+/2$, then $v_- > v_+$ and the Schmitt trigger output changes from V_s^+ to V_s^-. With $v_o = V_s^-$, it follows that

$$v_+ = \frac{V_s^- R_1}{R_1 + R_2} = \frac{V_s^-}{2}$$

and the capacitor voltage decreases toward V_s^-. When the capacitor voltage reaches $V_s^-/2$, then $v_- < v_+$ and the Schmitt trigger output changes from V_s^- to V_s^+. The output switching between V_s^+ and V_s^- continues and its frequency is controlled by the RC time constant. The resulting waveforms are shown in Figure 6.2.

To calculate the period of the oscillation, consider the charging of the capacitor shown in Figure 6.3 where at time $t = t_a$ (see Figure 6.2) the output voltage has just changed to V_s^+ and the capacitor voltage is at $V_s^-/2$. For $t > t_a$ (see Figure 6.3), the capacitor voltage is given by

$$v_c(t) = A + Be^{-(t-t_a)/RC} \text{ for } t_a \le t \le t_b \tag{6.1}$$

Since $v_c(t_a) = V_s^-/2$ and $v_c(\infty) = V_s^+$, it follows from (6.1) that

$$\frac{V_s^-}{2} = A + B$$

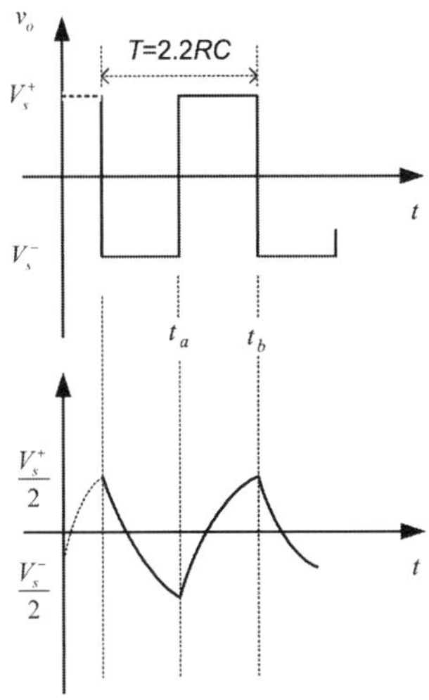

Figure 6.2 The Schmitt trigger output voltage and the capacitor voltage.

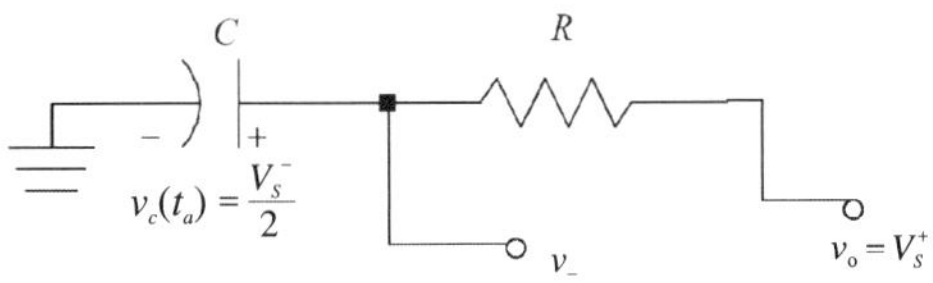

Figure 6.3 Charging of the capacitor.

and

$$V_s^+ = A$$

Hence, since $V_s^+ =- V_s^-$ we obtain

$$B = \frac{V_s^-}{2} - A =- \frac{V_s^+}{2} - V_s^+ =- \frac{3}{2} V_s^+$$

and from (6.1)

$$v_c(t) = V_s^+ - \frac{3}{2} V_s^+ e^{-(t-t_a)/RC} \tag{6.2}$$

From Figure 6.2 at $t = t_b$ the capacitor voltage reaches $V_s^+/2$. Substituting $t = t_b$ in (6.2) gives

$$v_c(t_b) = \frac{V_s^+}{2} = V_s^+ - \frac{3}{2} V_s^+ e^{-(t_b-t_a)/RC}$$

Solving for $t_b - t_a$ gives

$$t_b - t_a = RC \ln 3 = 1.0986RC$$

From symmetry (see Figure 6.2) the time that it takes the capacitor to discharge from $V_s^+/2$ to $V_s^-/2$ is also given by $1.0986RC$. Hence, the period of the oscillation is

$$T = 2(t_b - t_a) = 2.197RC \approx 2.2RC \tag{6.3}$$

Since the output voltage of the op amp in Figure 6.1 changes from V_s^+ to V_s^-, the slew rate of the op amp limits the frequency of operation.

For the case that $R_1 \neq R_2$, the period is given by

$$T = 2RC \ln\left(1 + \frac{2R_1}{R_2}\right) \tag{6.4}$$

Observe that if $R_1 = R_2$, $T \approx 2.2RC$ in (6.4), which is identical to (6.3).

Example 6.1

Design the relaxation oscillator in Figure 6.1 to oscillate at 8 kHz.

Solution

Let $R_1 = R_2 = 10$ kΩ and $C = 10$ nF. Then, from (6.3),

$$R = \frac{1}{2.2fC} = \frac{1}{2.2(8 \times 10^3)(10 \times 10^{-9})} = 5.68 \text{ k}\Omega$$

The simulation of this oscillator is shown in Figure 6.4. The frequency of oscillation is 8.13 kHz, which can be easily adjusted with a variable R.

Another op amp relaxation oscillator is shown in Figure 6.5. In this oscillator the output voltage amplitude varies between $V_z + 0.7$V to $-V_z - 0.7$V. The resistor R_A limits the current for proper operation of the Zener diodes.

The waveform for v_o is shown in Figure 6.5(b). The period of the oscillator in Figure 6.5(a) is similar to that in Figure 6.1 [compare Figure 6.2 with Figure 6.5(b)], and it is given by either (6.3) or (6.4).

6.2.2 Relaxation Oscillators with Digital Gates

Relaxation oscillators can also be constructed using digital gates. A typical oscillator using NAND gates is shown in Figure 6.6. The NAND gates in Figure 6.6 are connected as inverters, such that when the digital inputs are 0 the output is 1, and vice versa. The closing of the switch starts the oscillation.

Let us assume that the threshold voltage (V_T) of the CMOS NAND gates in Figure 6.6 is $V_T = V^+/2$; that is, when the input voltage to the gate goes above or

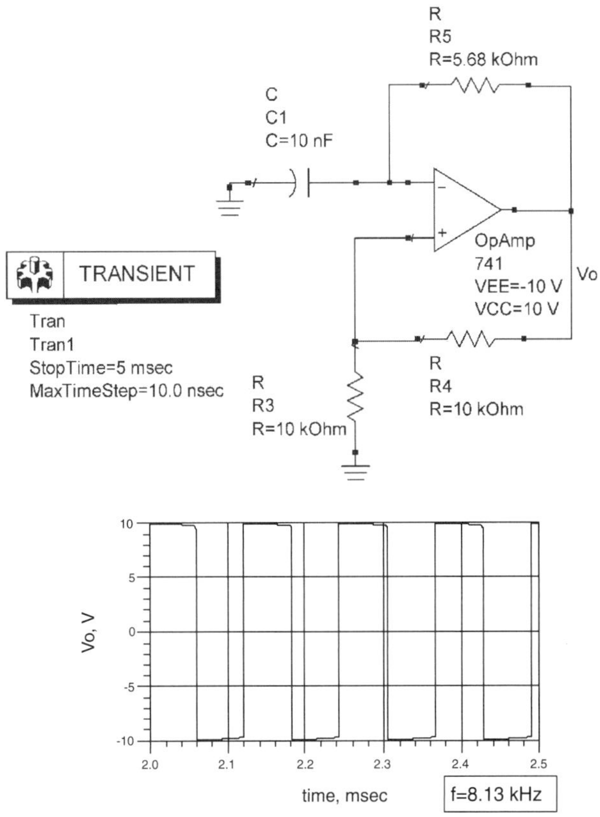

Figure 6.4 Simulation of the relaxation oscillator in Example 6.1.

below V_T, the gates change states accordingly. To analyze this oscillator assume that $v_{o1} = V^+$, $v_{o2} = 0$, and the capacitor is uncharged. This situation is illustrated in Figure 6.7(a). The capacitor in Figure 6.7(a) will charge toward $-V^+$, with an RC time constant. Therefore,

$$v_x(t) =- v_c(t) = V^+\left(1 - e^{-t/RC}\right)$$

When $v_x(t)$ reaches $V_T = V_s^+/2$, say at $t = t_a$, the gates change states. The waveforms are shown in Figure 6.7(b) where at $t = t_a^+$ we have that $v_{o1} = 0$, $v_{o2} = V^+$, and $v_c(t_a) =- V_T$. The circuit during the time interval $t_a^+ \le t \le t_b^-$ where $v_{o1} = 0$ and $v_{o2} = V^+$ is shown in Figure 6.7(c). Observe that $v_x(t_a^-) = V_T$, and $v_x(t_a^+) = 1.5V_T$ since the capacitor voltage can not change instantaneously, or

$$v_c(t_a^-) = v_c(t_a^+) =- V_T$$

For $t_a^+ \le t \le t_b^-$, the capacitor voltage is given by

$$v_c(t) = A + Be^{-(t-t_a)/RC} \quad t_a \le t \le t_b \tag{6.5}$$

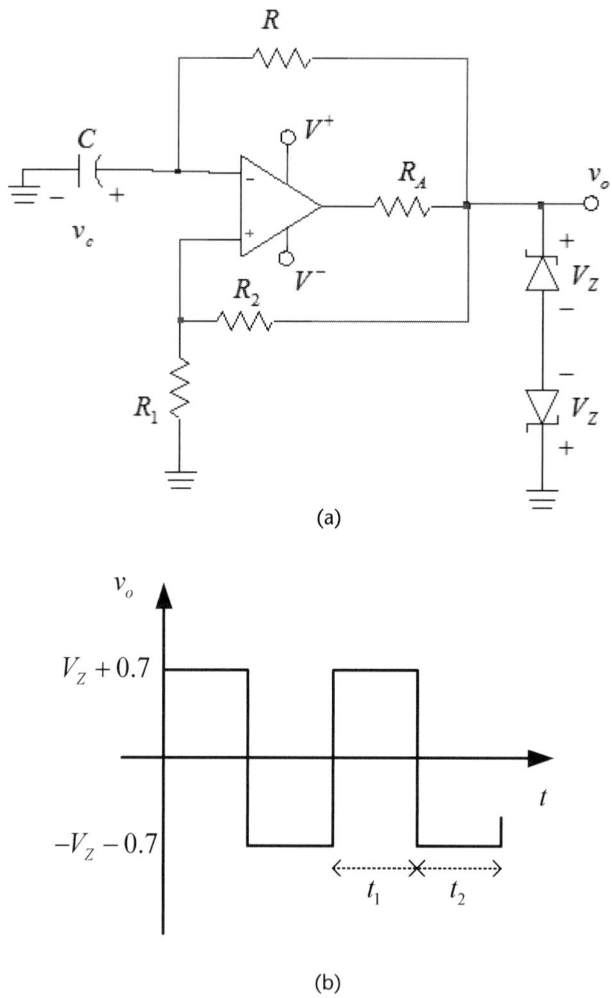

Figure 6.5 (a) An op amp relaxation oscillator with a controlled output voltage and (b) the oscillator waveform.

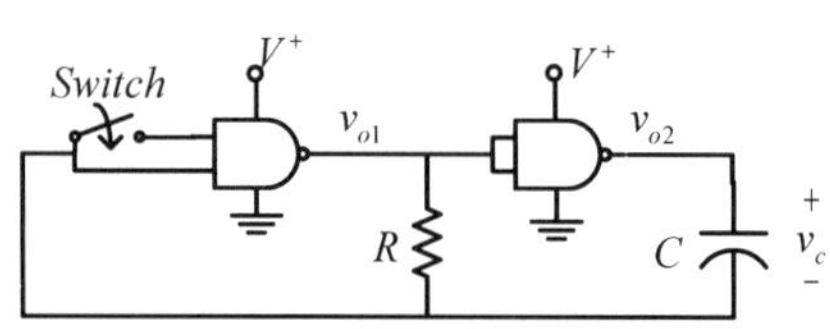

Figure 6.6 A relaxation oscillator using NAND gates.

Since $v_c(t_a) =- V_T$ and $v_c(\infty) = V^+$, it follows that $A = V^+$ and $B =- 1.5V_T$. Hence, (6.5) is expressed in the form

$$v_c(t) = V^+ - 1.5V^+e^{-(t-t_a)/RC} \quad t_a \leq t \leq t_b \tag{6.6}$$

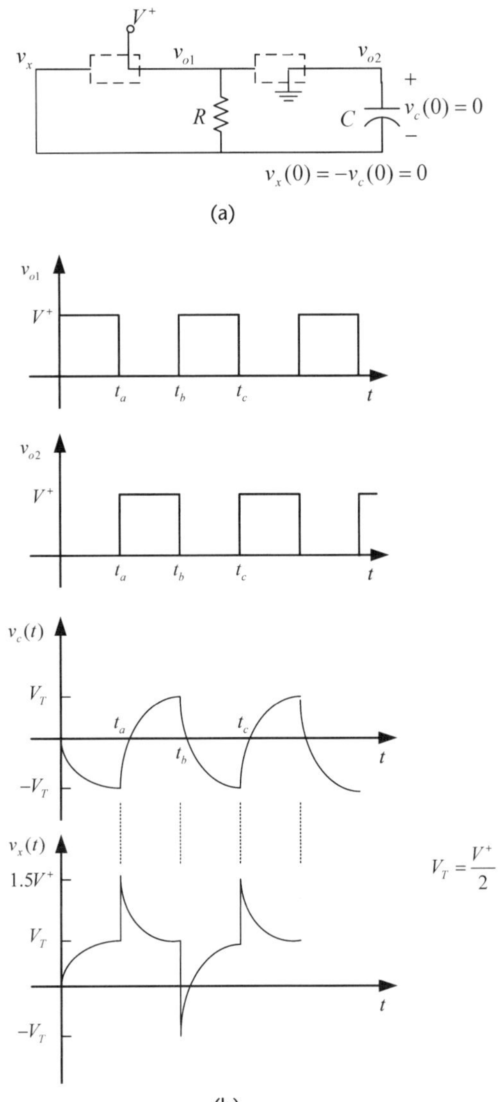

Figure 6.7 (a) Oscillator circuit for $0 < t < t_a^-$ (at $t = t_a^-$ the capacitor voltage is $v_c(t_a^-) = v_c(t_a^+) = -V_T = -V^+/2$), (b) oscillator waveforms, (c) oscillator circuit for $t_a^+ \le t \le t_b^-$, and (d) oscillator circuit for $t_b^+ \le t \le t_c^-$.

Also, for $t_a^+ \le t \le t_b^-$, since $v_x(t) = V^+ - v_c(t)$ it follows that

$$v_x(t) = 1.5V^+ e^{-(t-t_a)/RC} \quad t_a^+ \le t \le t_b^-$$

The waveforms for $t_a \le t \le t_b$ are shown in Figure 6.7(b).

At $t = t_b^-$ the voltage v_x decreases to V_T, and $v_c(t_b^-) = V_T$. Hence, the gates switch states again. That is, $v_{o1} = V^+$ and $v_{o2} = 0$ at $t = t_b^+$, as shown in Figure 6.7(b), and the circuit during the time interval $t_b^+ \le t \le t_c^-$ is shown in Figure 6.7(d).

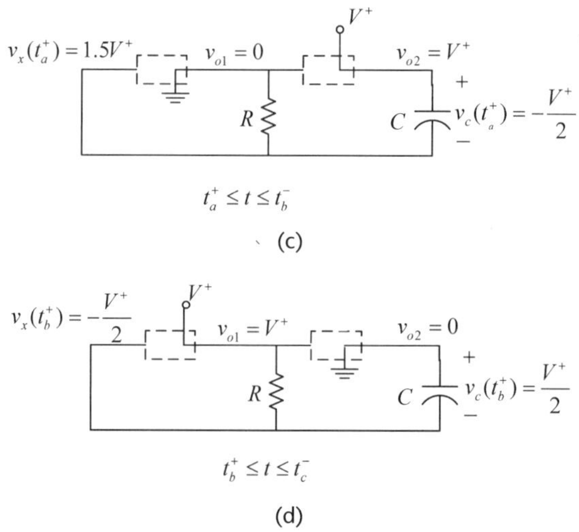

Figure 6.7 (continued.)

The time $t_b - t_a$ can be calculated using (6.6); that is,

$$v_c(t_b) = \frac{V^+}{2} = V^+ - 1.5V^+ e^{-(t_b - t_a)/RC}$$

or

$$t_c - t_b = RC \ln 3 = 1.09RC$$

At $t = t_b^+$, the voltage across the capacitor remains at $V^+/2$. However, from Figure 6.7(b) it is seen that $v_x(t_b^+) =- V_T$. In the interval $t_b^+ \le t \le t_c^-$ the capacitor charges toward $-V^+$, and $v_c(t)$ is given by

$$v_c(t) =- V^+ + 1.5V^+ e^{-(t - t_b)/RC}$$

In the relaxation oscillator the above process repeats itself; that is, as shown in Figure 6.7(b), $v_c(t)$ varies between V_T and $-V_T$, and $v_x(t)$ varies between $1.5V_T$ and $-V_T$.

From symmetry, the period and frequency are given by

$$T = \frac{1}{f} = 2(t_b - t_a) \approx 2.2RC \tag{6.7}$$

While the inverters in Figure 6.6 were implemented in using CMOS NAND gates, there are other ways of implementing inverters—for example, a CMOS NOR gate with the two inputs connected together functions as an inverter.

Example 6.2

Design the relaxation oscillator shown in Figure 6.6 to oscillate at 10 kHz.

Solution

Let C = 1 nF, then from (6.7),

$$R = \frac{1}{2.2(10 \times 10^3)(10^9)} = 45.4 \text{ k}\Omega$$

The internal operation of the gates used to build the oscillator in Figure 6.6 affects the waveforms associated with the circuit and, therefore, the oscillator frequency. For example, CMOS NAND gates have protection diodes, which clamp the input voltage between zero and V^+. Hence, the switching action that makes the voltage v_x go negative or above V^+ is prevented by the protection diodes of the first gate. A modification to the relaxation oscillator in Figure 6.6 that improves the frequency stability is shown in Figure 6.8. The resistor R_a lets the voltage v_x go negative and above V^+, as described in Figure 6.7(b), by providing the needed voltage drop for v_x' to be at the clamp values of the gate. The resistor R_a is usually designed to be 10 times larger than R_1.

A modification that produces a relaxation oscillator with an asymmetrical square wave is shown in Figure 6.9. The diode D makes C to be charged by R_1 in parallel with R_2 when $v_{o1} = 0$ and $v_{o2} = V^+$, and to be charged by R_2 only when $v_{o1} = V^+$ and $v_{o2} = 0$. Hence, the resistor R_2 affects the time interval when $v_{o2} = V^+$.

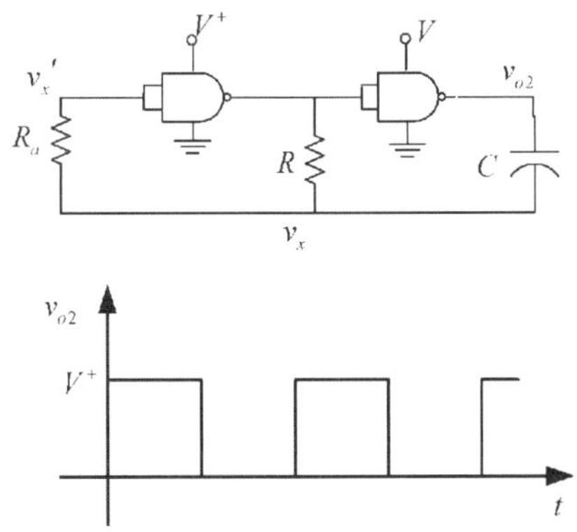

Figure 6.8 A relaxation oscillator with improved frequency stability.

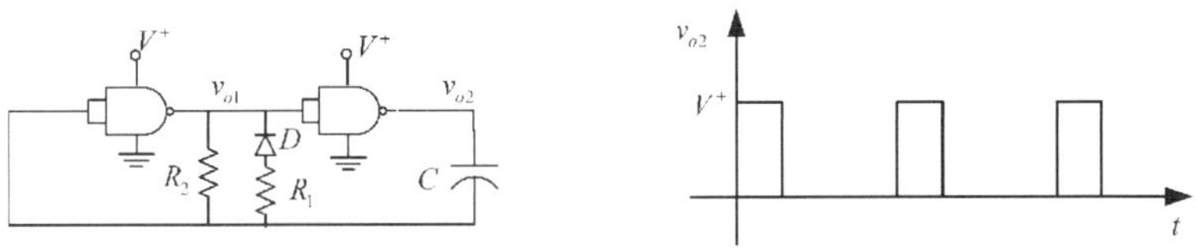

Figure 6.9 A relaxation oscillator that produces an asymmetrical square wave.

A resistor R_a could be added to the input of the first gate (as in Figure 6.8) to improve the circuit performance. Also, the addition of a second diode in series with R_2 could be used to control the interval when $v_{o2} = 0$.

Gated operation of the previous relaxation oscillator is obtained by using an electronic switch to turn the oscillator on and off. For CMOS NAND gates, the switch can be implemented as shown in Figure 6.10(a), and for CMOS NOR gates as shown in Figure 6.10(b). The circuit in Figure 6.10(a) is turned on by a logic 1 input at v_x, and that in Figure 6.10(b) by a logic 0 at the input v_x.

If BJT inverters are used (such as the 7404 inverters) instead of CMOS inverters, the resulting waveforms would be different since the voltages at which the gates switch are different. The transfer characteristic of a 7404 inverter is shown in Figure 6.11. The high (H) and low (L) values of the input and output voltages are indicated, as well as the listed input current.

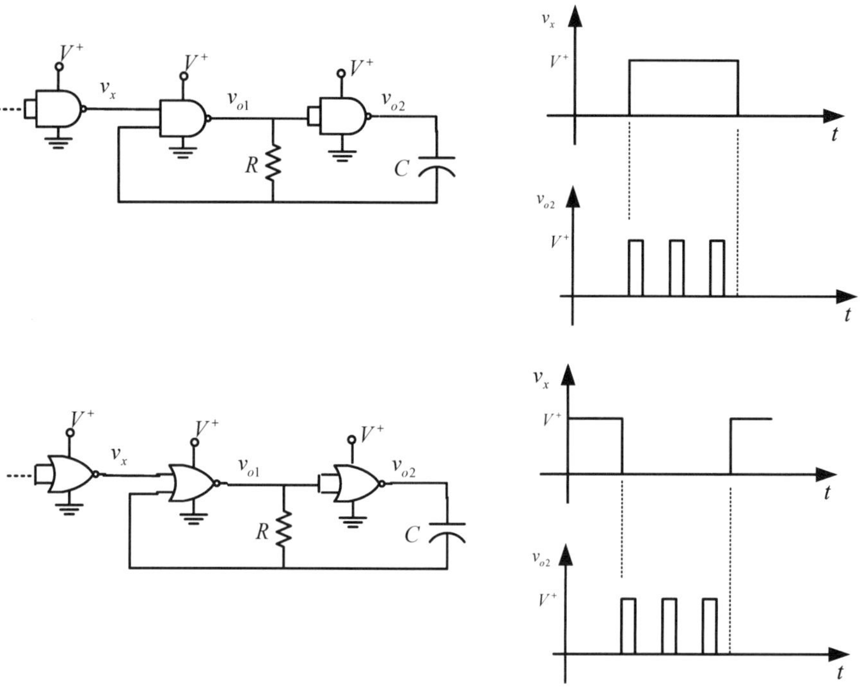

Figure 6.10 Relaxation oscillator with gated operation.

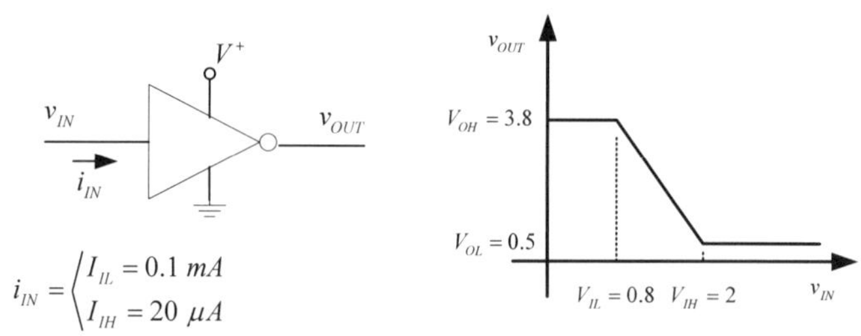

Figure 6.11 Characteristics of the 7404 inverter.

When such inverters are used in the relaxation oscillator shown in Figure 6.6, the period and frequency of oscillation of v_{o2} can be shown to be given by

$$T = \frac{1}{f} = T_1 + T_2 = RC\left(\ln \frac{V_{OH} + V_{IL} - V_{OL}}{V_{IH} - V_{OL}} + \ln \frac{V_{IH} - 2V_{OH}}{V_{IL} - V_{OH} - I_{IL}R}\right) \tag{6.8}$$

where T_1 is the time that v_{o1} is high at V_{OH}, and T_2 is the time that is low at V_{OL}.

A check of (6.8) can be done by assuming that the inverters are CMOS gates. Then, $V_{OL} = 0$, $V_{OH} = V^+$, $V_{IH} = V_{IL} = V_T$, and $I_{IL} = 0$. Substituting into (6.8) we obtain

$$T = \frac{1}{f} = RC(\ln 3 + \ln 3) \approx 2.2RC$$

which is identical to (6.3).

A relaxation oscillator implemented with BJTs is shown in Figure 6.12. This oscillator is also known as a BJT astable multivibrator. The resistors are selected so that with no capacitors the BJTs are saturated (i.e., let $R_{B1} < \beta R_{C1}$ and $R_{B2} < \beta R_{C2}$). The operation of the oscillator can be described as follows. If $T1$ and $T2$ are originally in the active region, any variation in i_{C1} or i_{C2} will drive one of the transistors into saturation and the other into cut off. For example, if i_{C1} increases then v_{BE2} decreases. This produces a decrease in i_{C2} and an increase in v_{CE2}. The increase in v_{CE2} produces an increase in v_{BE1} and eventually $T1$ saturates and $T2$ is cut off. At this point the capacitor C_1 charges through R_{B2}, and this causes v_{BE2} to increase towards approximately 0.7V, at which point $T2$ saturates and $T1$ is cut off. With $T2$ saturated the capacitor C_2 charges through R_{B1} until $T1$ saturates and $T2$ is cut off. The process repeat itself at a frequency determined by the values of the base resistor and the capacitors.

In order to analyze the circuit in Figure 6.12 we begin by assuming that $T1$ is saturated and $T2$ is cut off, as shown in Figure 6.13(a). For simplicity we assume

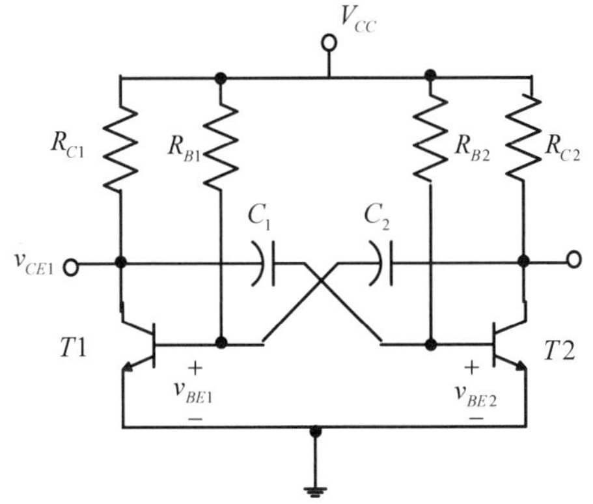

Figure 6.12 A BJT astable multivibrator.

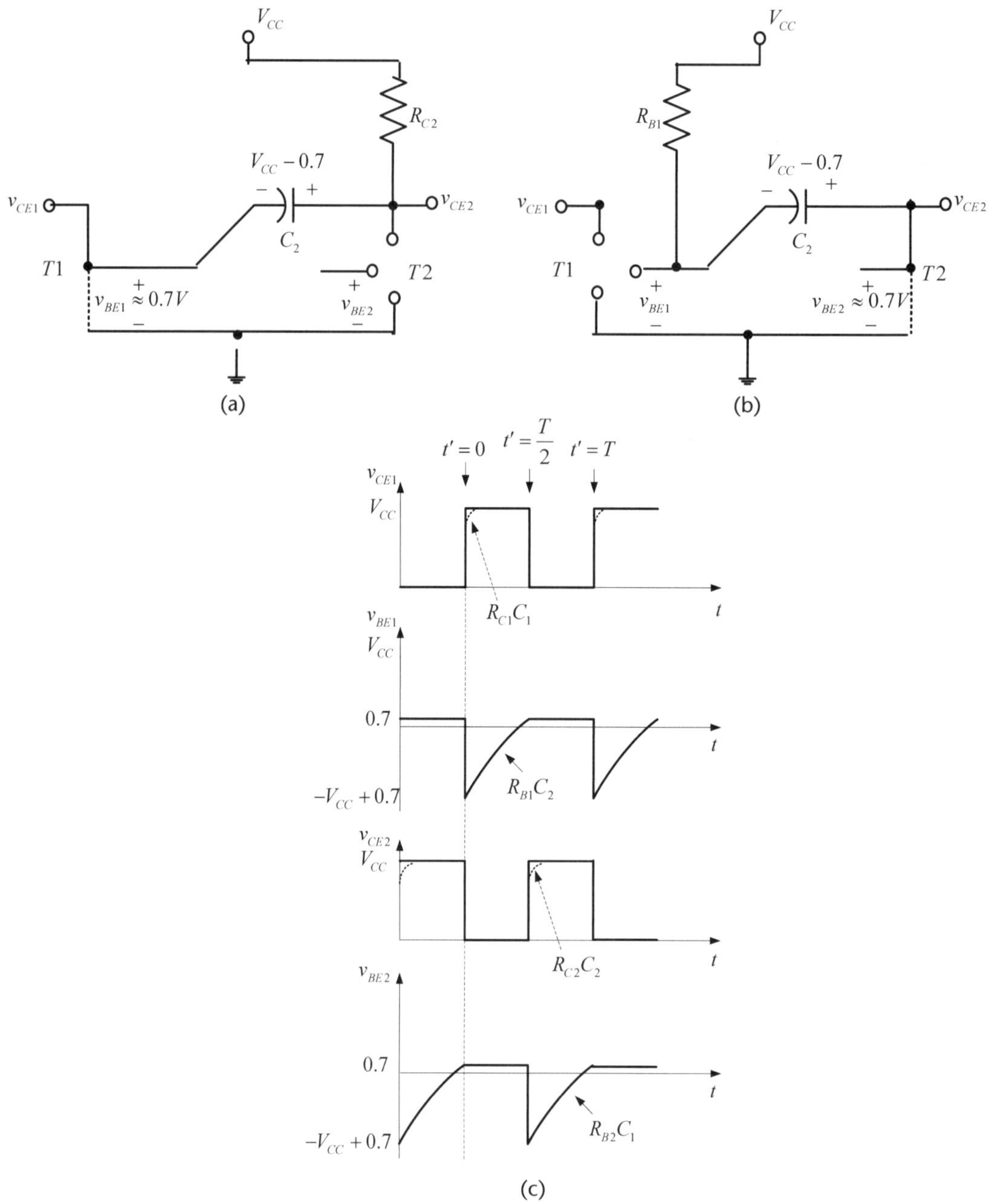

Figure 6.13 (a) Portion of the circuit showing *T1* saturated and *T2* in cut off, (b) portion showing *T1* in cut off and *T2* saturated, and (c) the waveforms of the oscillator.

that when a transistor is saturated the base to emitter voltage is 0.7 and its collector to emitter voltage is zero. In Figure 6.13(a), the time constant $R_{C2}C_2$ is such that the capacitor charges fast to $V_{CC} - 0.7$. Next, consider the time $t' = 0$ when the transistors switch states [see Figure 6.13(c)]. That is, at $t' = 0$, $T1$ becomes cut off and $T2$ is saturated. At $t' = 0$ capacitor C_2 is charged to $V_{CC} - 0.7$. Figure 6.13(b) shows the oscillator circuit at $t' = 0^+$ with the capacitor C_2 voltage at $V_{CC} - 0.7$.

At this time v_{BE1} begins to rise toward 0.7 since C_2 charges towards $-V_{CC}$ with a time constant of $R_{B1}C_2$. During this charging time $T2$ is saturated and v_{CE2} remains at 0V, as shown in Figure 6.13(b). From Figure 6.13(b), the expression for v_{BE1} is

$$v_{BE1} = V_{CC} - (2V_{CC} - 0.7)e^{-t'/R_{B1}C_2} \tag{6.9}$$

which shows that at time $t' = 0$ the voltage is $v_{BE1} =- V_{CC} + 0.7$ and at $t' = \infty$ the voltage is $v_{BE1} = V_{CC}$. At $t' = T/2$ the process repeats itself with $R_{B2}C_1$ determining the pulse width from $t' = T/2$ to $t' = T$. Figure 6.13(c) shows the waveforms associated with the oscillator.

When $t' = T/2$, the voltage v_{BE1} rises to 0.7V. From (6.9)

$$0.7 = V_{CC} - (2V_{CC} - 0.7)e^{-T/2R_{B1}C_2}$$

or

$$\frac{T}{2} =- R_{B1}C_2 \ln \frac{V_{CC} - 0.7}{2V_{CC} - 0.7} \approx R_{B1}C_2 \ln 2$$

where the approximation $V_{CC} \gg 0.7$V was used. If $R_{B1} = R_{B2}$ and $C_1 = C_2$, the frequency of oscillation is

$$f = \frac{1}{T} = \frac{1}{2R_{B1}C_2 \ln 2}$$

If $R_{B1} \neq R_{B2}$ and $C_1 \neq C_2$, the frequency of oscillation is

$$f = \frac{1}{T} = \frac{1}{2(R_{B1}C_2 + R_{B2}C_1) \ln 2}$$

Detail analysis of the waveshapes, which includes the effect of the cut-in voltage, the finite saturation voltage, and the switching time of the transistors, can be performed easily in SPICE. The actual results will be in close agreement with the waveshapes shown in Figure 6.13(c).

6.2.3 The Ring Oscillator

An odd number of inverters will oscillate in the ring configuration shown in Figure 6.14. This configuration is best analyzed in terms of the time delay (or propagation delay) that the signal experiences as it goes through the ring inverters. If the propagation delay of the inverter is t_p, then the frequency of oscillation is given by

$$f_o = \frac{1}{2nt_p}$$

where n is the number of gates.

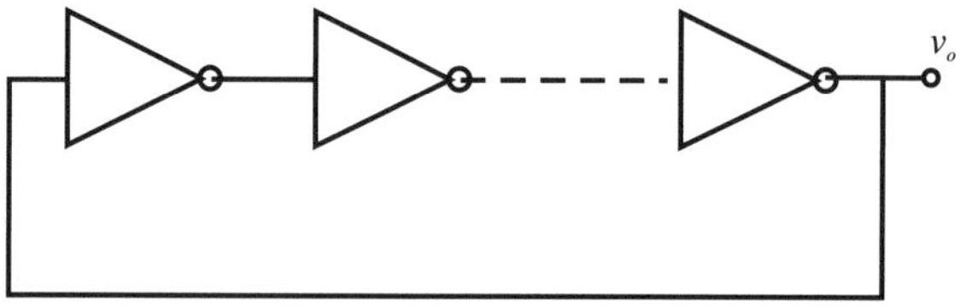

Figure 6.14 Ring-oscillator configuration.

Manufacturers of inverters provide data regarding the propagation delay as a function or the device load capacitance. For example, if n = 3 and the output capacitance of the inverter has a nominal value of 10 pF producing a nominal propagation delay of 20 ns, the frequency of oscillation is

$$f_o = \frac{1}{2(3)20 \times 10^{-9}} = 8.33 \text{ MHz}$$

However, if such an oscillator is constructed in the laboratory, the resultant frequency of oscillation can vary considerable from the predicted 8.33 MHz.

Since the frequency of oscillation is controlled by the propagation delay, it is difficult to properly control it. Too many factors affect the propagation delay, including the variation in output capacitance from unit to unit. To better control the frequency of oscillation, external elements are added.

While the two-stage relaxation oscillator is a general-purpose square-wave oscillator, it might not be suitable for many timing circuits, especially in high-speed digital circuits. A better square-wave oscillator is shown in Figure 6.15. This type of configuration is known as a *ring of three*.

The frequency of oscillation for the circuit in Figure 6.15 is given by

$$f = \frac{1}{2R_1 C\left(\dfrac{0.405R_2}{R_1 + R_2} + 0.693\right)} \tag{6.10}$$

In the case that $R_2 = 0$, (6. 10) gives

$$f = \frac{1.731}{R_1 C}$$

and with $R_1 = R_2 = R$, it reduces to

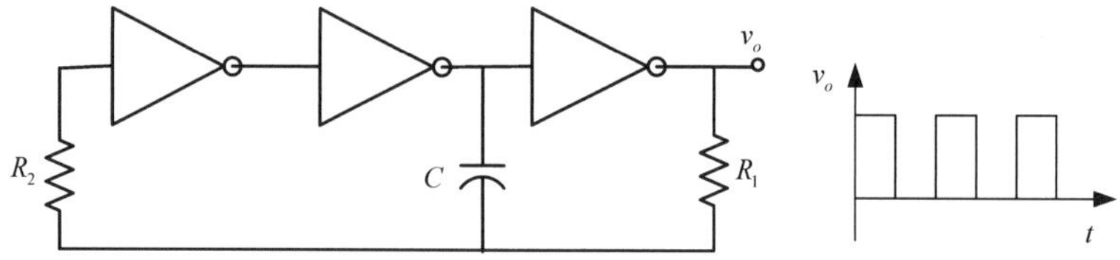

Figure 6.15 Ring-of-three oscillator.

$$f = \frac{0.6}{RC}$$

The ring-of-three oscillator can also be gated, as shown in Figure 6.16, for a NAND CMOS implementation. The logic gate control voltage is v_G. The oscillator in Figure 6.16(a) has a normally low output, while the oscillator in Figure 6.16(b) has a normally high output. The ring-of-three oscillator can also be constructed with NOR gates.

6.3 Triangular-Wave Oscillators

In an RC circuit the capacitor waveform is exponential. However, if a capacitor is charged from a constant current source the waveshape is linear (i.e., It/C). Hence, a circuit that changes the direction of the charging current source in a periodical manner can be used to generate a triangular-wave oscillator. Such a circuit can be implemented using a Schmitt trigger (i.e., a comparator with hysteresis). Two Schmitt trigger circuits are shown in Figure 6.17 with the associated characteristics.

The characteristics of the Schmitt trigger in Figure 6.17(a) are derived by observing that v_x can only be at either V_s^+ or V_s^-, and the voltage at the positive input (v_+) is

$$v_+ = v_x \frac{R_1}{R_1 + R_2} + v_I \frac{R_2}{R_1 + R_2} \tag{6.11}$$

When $v_x = V_s^+$, the required input voltage to change the output from V_s^+ to V_s^- is denoted by $v_I = V_{LT}$. From (6.11), setting $v_+ = 0$, $v_x = V_s^+$, and $v_I = V_{LT}$, gives

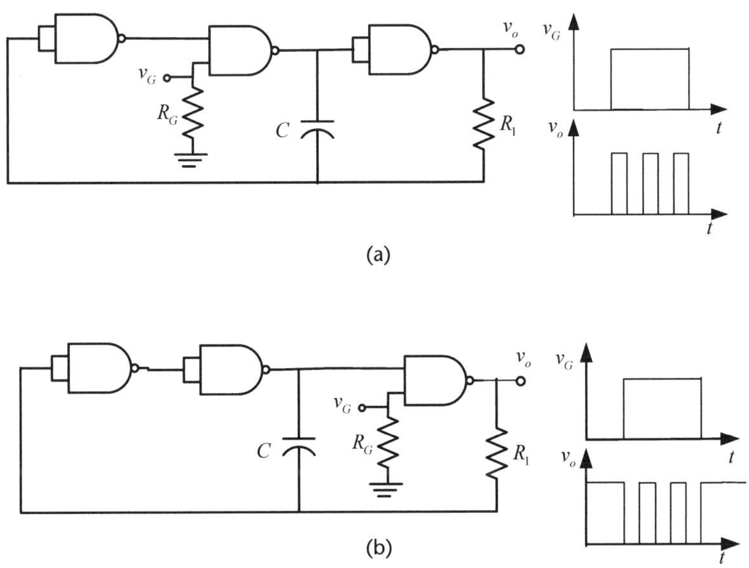

Figure 6.16 Gated ring-of-three with (a) normally low output and (b) normally high output.

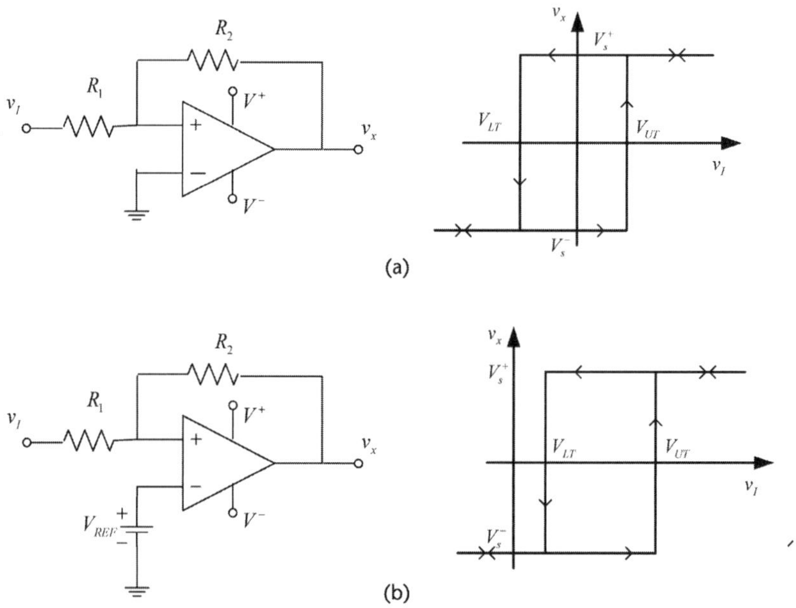

Figure 6.17 (a) A Schmitt trigger circuit and its characteristics and (b) a Schmitt trigger circuit with a reference voltage and its characteristics.

$$V_{LT} = -\frac{R_1}{R_2} V_s^+$$

V_{LT} is known as the lower threshold voltage of the Schmitt trigger.

When $v_x = V_s^-$, the required input voltage to change the output from V_s^- to V_s^+ is denoted by $v_I = V_{UT}$. From (6.11), setting $v_+ = 0$, $v_x = V_s^-$, and $v_I = V_{UT}$, gives

$$V_{UT} = -\frac{R_1}{R_2} V_s^-$$

V_{UT} is known as the upper threshold voltage of the Schmitt trigger.

If a reference voltage V_{REF} is applied to the negative input [as shown in Figure 6.17(b)], the resulting characteristics are shifted and V_{LT} can be made positive. For this circuit it follows that

$$V_{LT} = -\frac{R_1}{R_2} V_s^+ + V_{REF}\left(1 + \frac{R_1}{R_2}\right)$$

and

$$V_{UT} = -\frac{R_1}{R_2} V_s^- + V_{REF}\left(1 + \frac{R_1}{R_2}\right)$$

A triangular-wave oscillator using the Schmitt trigger in Figure 6.17(a) is shown in Figure 6.18. The oscillator in Figure 6.18 can be analyzed as follows. Assume

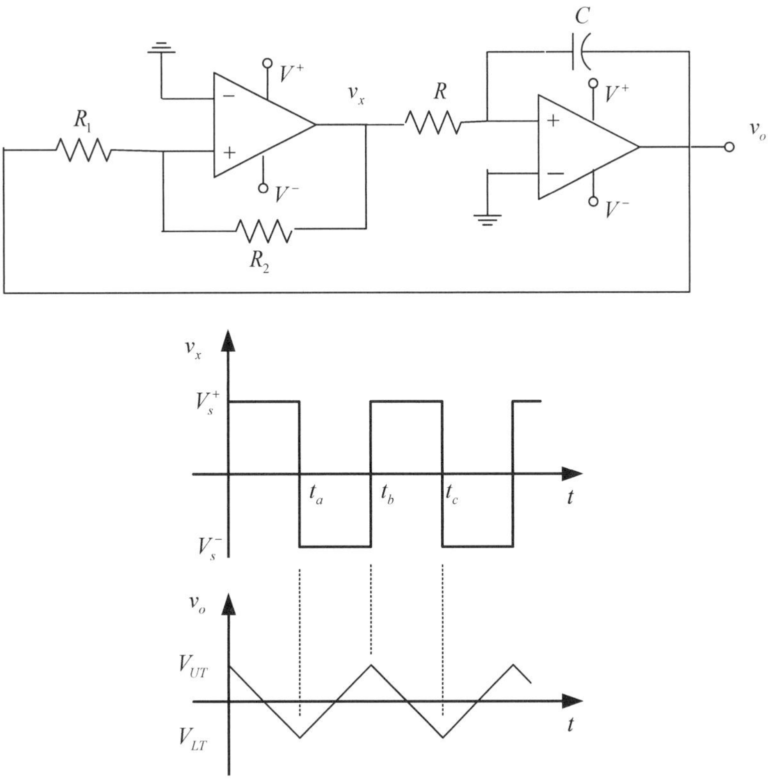

Figure 6.18 A triangular-wave oscillator and the circuit waveform.

that at $t = 0$, v_x is at the saturation value V_s^+ and $v_o = 0$. Hence, the current into the integrator is

$$I = \frac{V_s^+}{R}$$

and the output voltage for $0 \le t \le t_a$ is given by

$$v_o(t) =- \frac{1}{C}\int_0^t I\,dt =- \frac{1}{RC}\int_0^t V_s^+\,dt =- \frac{V_s^+ t}{RC} \tag{6.12}$$

Equation (6.12) shows that the voltage decreases linearly. At $t = t_a$, the output voltage reaches the lower threshold value V_{LT}. At $t = t_a^+$, v_x changes from V_s^+ to V_s^- and the current into the integrator reverses direction. Its value is given by

$$I = \frac{V_s^-}{R} \tag{6.13}$$

The current I, given by (6.13), has a negative value since V_s^- is negative.

The output voltage for $t_a \le t \le t_b$ is

$$v_o(t) = v_o(t_a) - \frac{I}{C}(t - t_a) \tag{6.14}$$

$$= V_{LT} - \frac{V_s^-}{RC}(t - t_a)$$

Hence, the voltage increases linearly from its value of V_{LT} at $t = t_a$. At $t = t_b$ we have $v_o(t_b) = V_{UT}$, and from (6.14),

$$V_{UT} = V_{LT} - \frac{V_s^-}{RC}(t_b - t_a)$$

or

$$t_b - t_a = \frac{(V_{LT} - V_{UT})RC}{V_s^-} = 2\frac{R_1}{R_2}RC \tag{6.15}$$

For $t_b \le t \le t_c$, the current into the integrator is $I = V_s^+/R$; therefore, $v_o(t)$ is given by

$$v_o(t) = v_o(t_b) - \frac{I}{C}(t - t_b) \tag{6.16}$$

$$= V_{UT} - \frac{V_s^+}{RC}(t - t_b)$$

which represents the linearly decreasing voltage for $t_b \le t \le t_c$.

From symmetry, $t_c - t_b = t_b - t_a$; therefore, the frequency of oscillation is

$$f = \frac{1}{T} = \frac{1}{2(t_b - t_a)} = \frac{R_2}{4R_1 RC} \tag{6.17}$$

Since v_x changes from V_s^+ to V_s^-, the slew rate of the comparator op amp limits the maximum frequency of oscillation. Also, V_{UT} and V_{LT} depend on V_s^+ to V_s^-, which cannot be controlled exactly.

The output waveform of the oscillator in Figure 6.18(a) can be displaced with respect to ground using a reference voltage in the comparator, as shown in Figure 6.17(b). Appropriate values of V_{REF} can make $V_{LT} > 0$, producing the triangular waveform shown in Figure 6.19.

An oscillator that produces a triangular waveform with $V_{LT} = 0$ is shown in Figure 6.20(a). The comparator uses a diode in series with R_2 such that when v_x goes to V_s^- the diode conducts and the resulting v_o decreases linearly. When v_x goes to V_s^+, D is off and therefore $V_{LT} = 0$. The comparator characteristics are shown in Figure 6.20(b), and the output waveform of the oscillator in Figure 6.20(c).

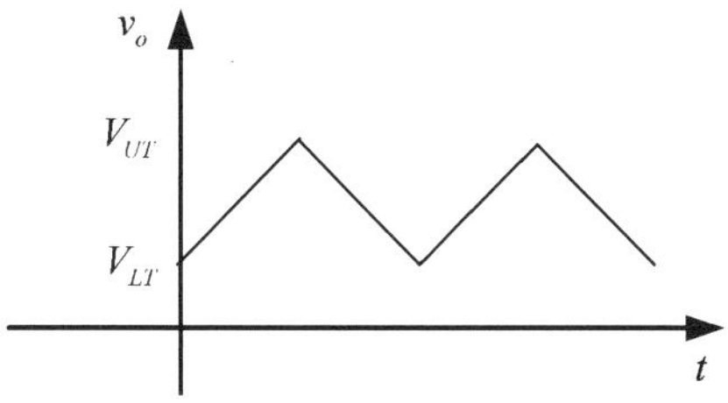

Figure 6.19 Triangular waveform displaced above ground.

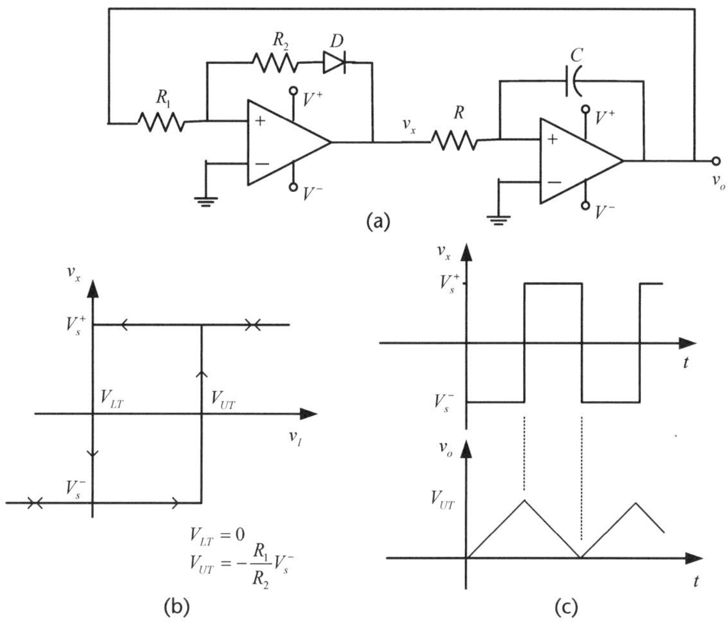

Figure 6.20 (a) A triangular wave oscillator (with $V_{LT} = 0$), (b) the comparator characteristics, and (c) the oscillator waveform.

If the integrator circuit in Figure 6.18 is replaced by the integrator circuit shown in Figure 6.21(a), the resulting waveform is shown in Figure 6.21(b) where $T_1 \neq T_2$. For this circuit, D1 conducts (ideally) when $v_x > 0$, charging the capacitor through R_{11}. When $v_x < 0$, D2 conducts and the capacitor gets charged through R_{12}. Hence, the resulting waveform (with $R_{11} \neq R_{12}$) exhibit different T_1 and T_2, where $T = T_1 + T_2$.

A circuit that can be used to control the values of V_{UT} and V_{LT} in a triangular-wave oscillator is shown in Figure 6.22(a), and the resulting waveforms in Figure 6.22(b). The resistor R_A is used to provide the appropriate current to the Zener diodes. When the comparator saturates, the resulting values of v_x are $v_x = V_z + 0.7$ and $v_x =- V_z - 0.7$. Hence, the upper and lower threshold values in this oscillator are given by

$$V_{UT} = \frac{R_1}{R_2}(V_Z + 0.7) \tag{6.18}$$

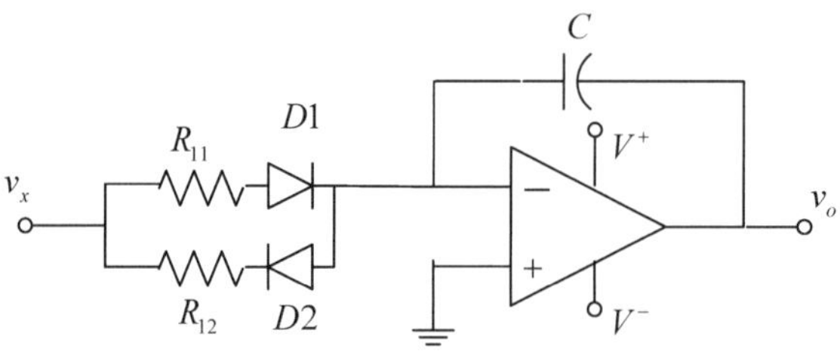

(a)

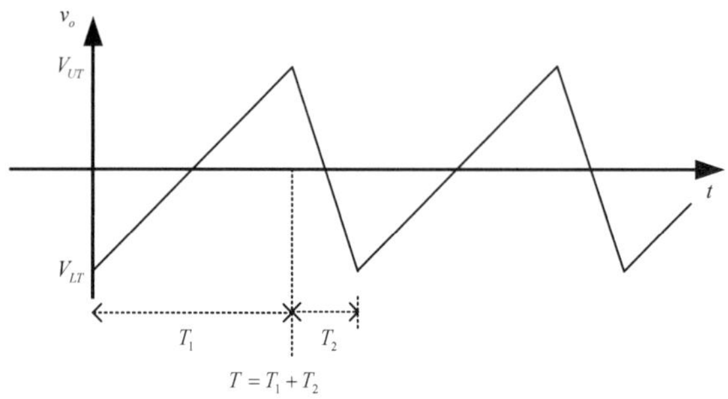

(b)

Figure 6.21 (a) An integrator circuit for the generation of an asymmetrical triangular wave and (b) the output waveform.

and

$$V_{LT} =- \frac{R_1}{R_2}(V_Z + 0.7) \tag{6.19}$$

The analysis of the oscillator in Figure 6.22(a) is similar to that in Figure 6.18(a). Let us begin the analysis at $t = t_a$ when v_x is at the saturation value $v_x =- V_z - 0.7$ and $v_o = V_{LT}$. Then, the current into the integrator is

$$I = \frac{-V_z - 0.7}{R}$$

and the output voltage for $t_a \leq t \leq t_b$ is given by (6.14) with V_{LT} given by (6.19) and V_s^- replaced by $-V_z - 0.7$.

At $t = t_b$, the voltage reaches the upper threshold value V_{UT}. At this time, v_x changes from $v_x =- V_z - 0.7$ to $v_x = V_z + 0.7$, and the current into the integrator reverses direction. Its value is given by

$$I = \frac{V_z + 0.7}{R}$$

Hence, the output voltage for $t_b \leq t \leq t_c$ is given by (6.16) with V_{UT} given by (6.18) and V_s^+ replaced by $V_z + 0.7$.

The frequency of oscillation is given by (6.17), namely,

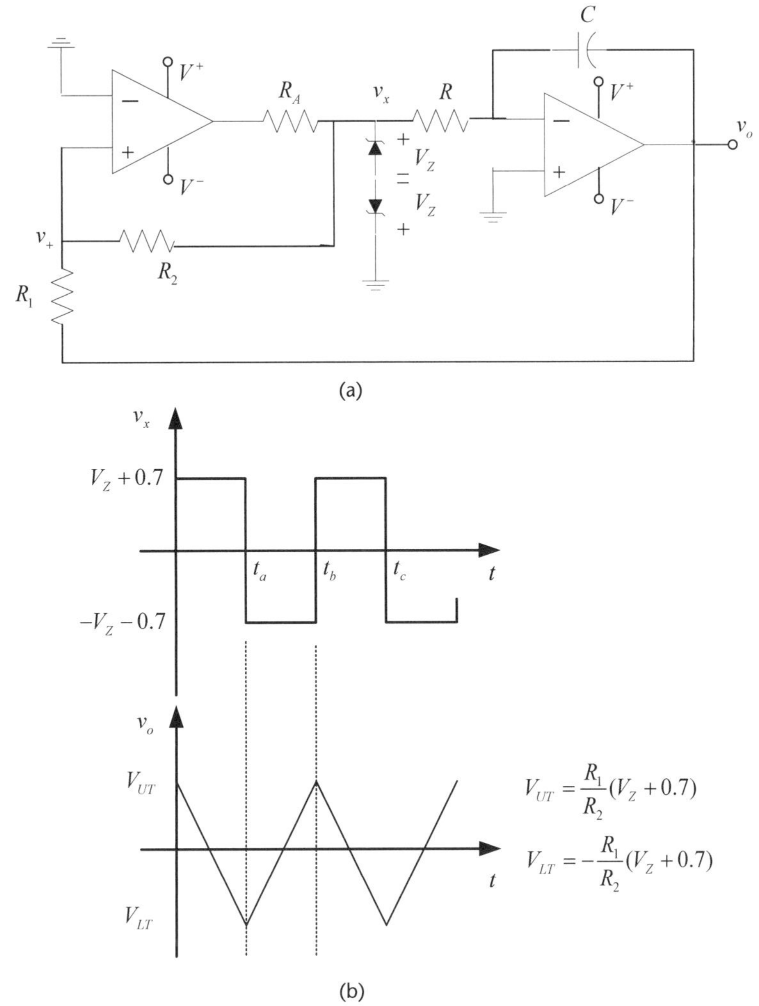

Figure 6.22 (a) A triangular-wave oscillator using Zener diodes to control V_{UT} and V_{LT} and (b) the oscillator waveforms.

$$f=\frac{R_2}{4R_1RC} \tag{6.20}$$

In the design of the triangular-wave oscillator in Figure 6.22, it is common to make R_1 and R_2 variable resistors. Also, the peak of the triangular waveform must be larger than the output offset voltage of the operational amplifier.

Example 6.3

The triangular waveform oscillator in Figure 6.22 is designed with $R_1 = R_2 =$ 10 kΩ, $R = 5$ kΩ, and $C = 0.1$ μF. The Zener diodes are 1N4734 diodes, with $V_z = 5.6$V. Two 741 op amps are used. Calculate the frequency of oscillation and show the resulting waveform.

Solution

Using (6.20), the frequency of oscillation is

$$f = \frac{10 \times 10^3}{4(10 \times 10^3)(5 \times 10^3)(0.1 \times 10^{-6})} = 500 \text{ Hz}$$

The simulation of the oscillator and the resulting output waveform are shown Figure 6.23.

A sawtooth waveform can be obtained with the oscillator in Figure 6.22 by making the time $t_c - t_b$ smaller than $t_b - t_a$. Also, removing one of the Zener

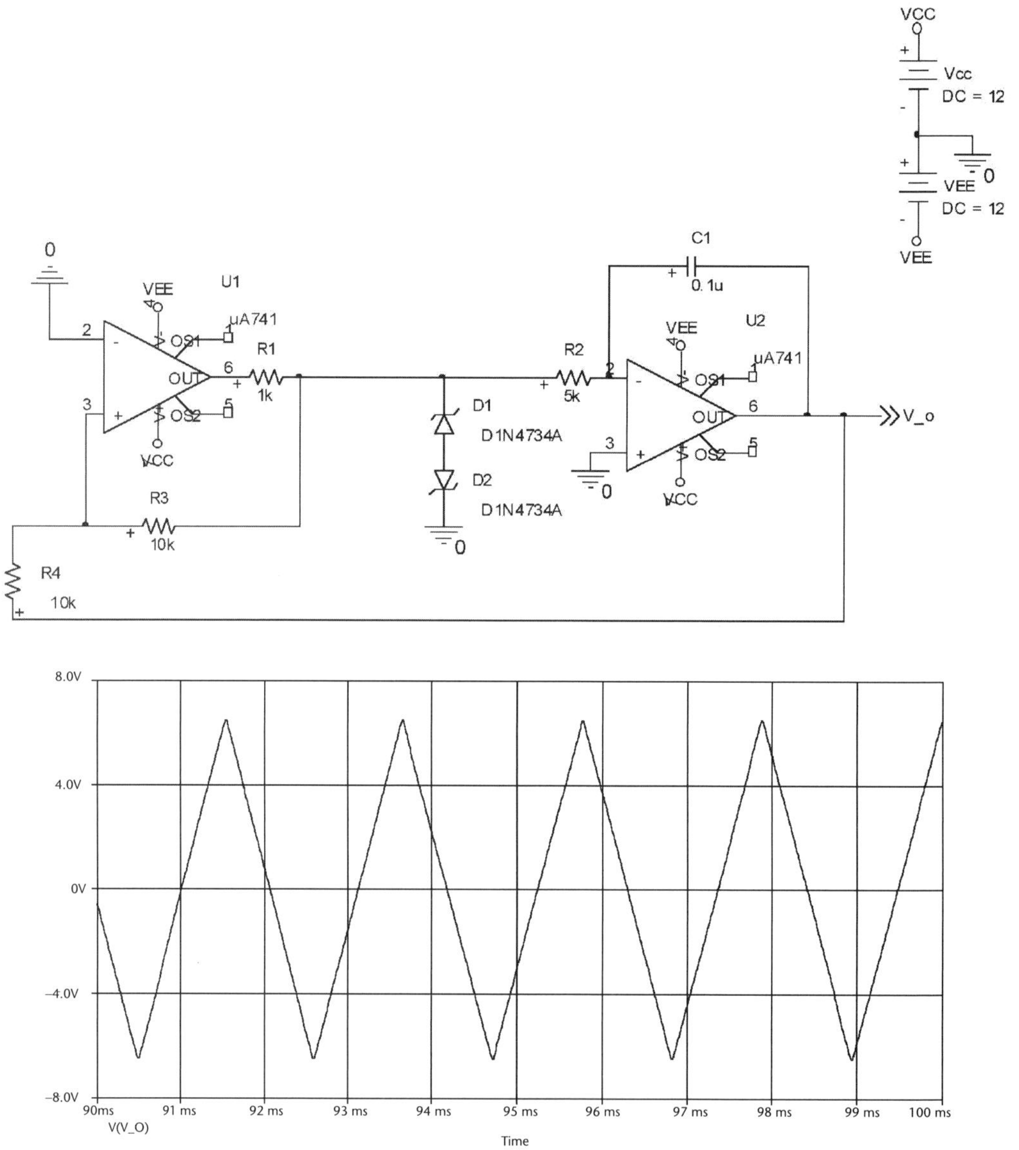

Figure 6.23 PSPICE simulation of the oscillator in Example 6.3.

diodes, as shown in Figure 6.24, produces a sawtooth waveform with v_x changing between –0.7V and V_Z, and v_o changing between

$$V_{UT} = \frac{R_1}{R_2}(0.7) \tag{6.21}$$

and

$$V_{LT} =- \frac{R_1}{R_2} V_Z$$

For this oscillator the time intervals are given by

$$t_b - t_a = \frac{(V_{UT} - V_{LT})RC}{0.7} \tag{6.22}$$

and

$$t_c - t_b = \frac{(V_{UT} - V_{LT})RC}{V_Z} \tag{6.23}$$

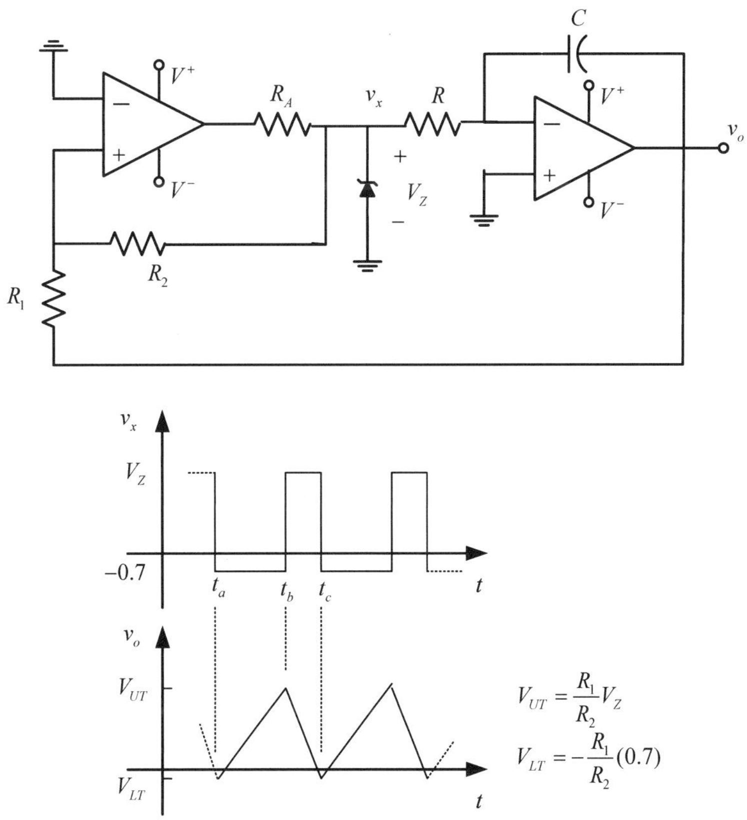

Figure 6.24 A sawtooth generator.

where

$$V_{UT} - V_{LT} = \frac{R_1}{R_2}(V_Z + 0.7)$$

Hence, the period is given by

$$T = t_c - t_a = \frac{R_1}{R_2} RC(V_Z + 0.7)\left(\frac{1}{0.7} + \frac{1}{V_Z}\right) \quad (6.24)$$

Example 6.4

The sawtooth oscillator in Figure 6.24 is to be designed with a period of 15 ms and a rise time of 14 ms.

Solution

With $T = 15$ ms and $t_b - t_a = 14$ ms, then $t_c - t_b = 1$ ms. From (6.22) and (6.23) we have

$$\frac{t_b - t_a}{t_c - t_b} = \frac{14(10^{-3})}{1(10^{-3})} = \frac{V_z}{0.7}$$

Hence, the required Zener voltage is $V_z = 9.8$V.

From (6.21), with $R_1 = R_2$ (say, 10 kΩ) it follows that $V_{UT} = 0.7$V and $V_{LT} =- 9.8$V.

Finally, from (6.24)

$$15(10^{-3}) = RC(9.8 + 0.7)\left(\frac{1}{0.7} + \frac{1}{9.8}\right) \Rightarrow RC = 0.933 \times 10^{-3}$$

This relation can be satisfied with $C = 50$ nF and $R = 18.67$ kΩ. The output voltage for this oscillator is shown in Figure 6.25, as well as the voltage v_x.

A more sophisticated triangular-wave oscillator is shown in Figure 6.26. The first op amp operates as a comparator. A 311 comparator can be used. The 311 has an open-collector output that changes from V^+ to ground (as suggested by the simplified diagram of the comparator in Figure 6.26. The second op amp circuit is an integrator. The MOSFET switch has a threshold value such that it closes when $v_x = V^+$, and opens when $v_x = 0$. The voltage V_{IN} is a dc voltage generated from V^+ (i.e., $V_{IN} < V^+$), which will determine the oscillation frequency.

When $v_x = V^+$, the switch closes and, using superposition, the voltage v_+ at the comparator (denoted by $v_{+(1)}$) is

$$v_{+(1)} = 2\left[\frac{V^+\left(\frac{R_A}{2}\right)}{\frac{R_A}{2} + R_A}\right] = \frac{2}{3}V^+$$

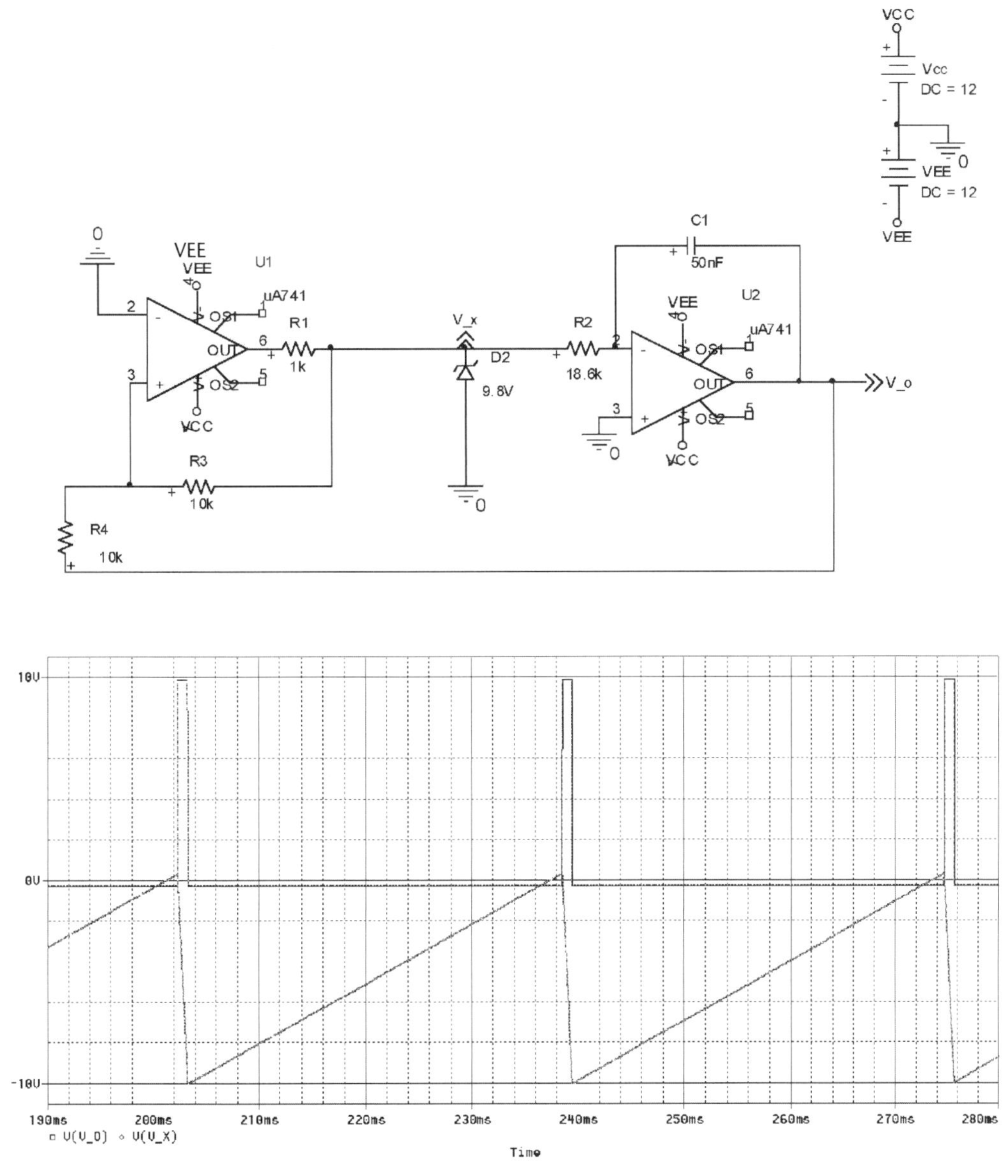

Figure 6.25 Sawtooth signal for the design in Example 6.4.

With the switch closed, the capacitor charges and the voltage v_o increases linearly until it reaches $2V^+/3$. When v_o reaches $2V^+/3$, the output of the comparator changes to $v_x = 0$ and the switch opens. With $v_x = 0$, the $v_{+(1)}$) becomes

$$v_{+(1)} = \frac{V^+\left(\frac{R_A}{2}\right)}{\frac{R_A}{2} + R_A} = \frac{V^+}{3}$$

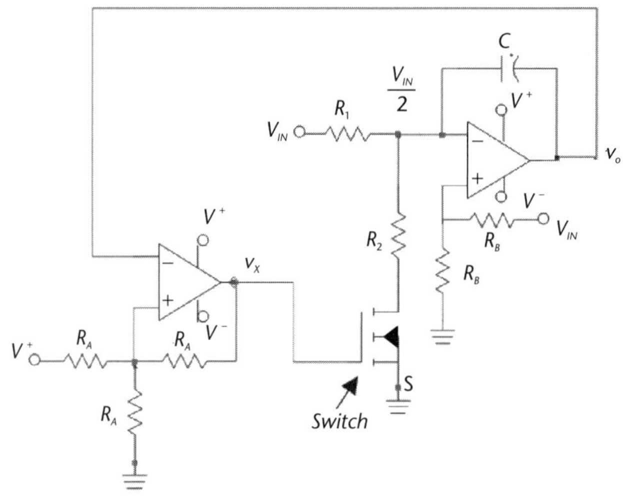

Figure 6.26 A triangular wave generator.

When the switch opens the output voltage decreases linearly from $2V^+/3$ to $V^+/3$. Hence, the output voltage changes between $V^+/3$ to $2V^+/3$ as v_X switches between 0 and V^+.

The equivalent integrator circuit with the switch closed is shown in Figure 6.27(a). The voltage v_+ of the integrator (denoted by $v_{+(2)}$) is

$$v_{+(2)} = \frac{V_{IN}R_B}{R_B + R_B} = \frac{V_{IN}}{2}$$

That is, the integrator has a reference voltage of $V_{IN}/2$ at its positive input. This is necessary for the switch to operate properly (i.e., $v_{-(2)}) = v_{+(2)}) = V_{IN}/2$).

At the $v_{-(2)}$ node we can write

$$\frac{V_{IN} - \frac{V_{IN}}{2}}{R_1} - \frac{\frac{V_{IN}}{2}}{R_2} =- C\frac{dv_o}{dt}$$

or

$$v_o(t) = v_o(t_a) + \frac{1}{R_{eq}C}\int_{t_a}^{t} V_{IN}\,dt = \frac{V^+}{3} + \frac{V_{IN}}{R_{eq}C}(t - t_a) \tag{6.25}$$

where

$$R_{eq} = \frac{2R_1R_2}{R_1 - R_2} \tag{6.26}$$

The output voltage for $t_a \le t \le t_b$, as described by (6.25), is shown in Figure 6.27(c).

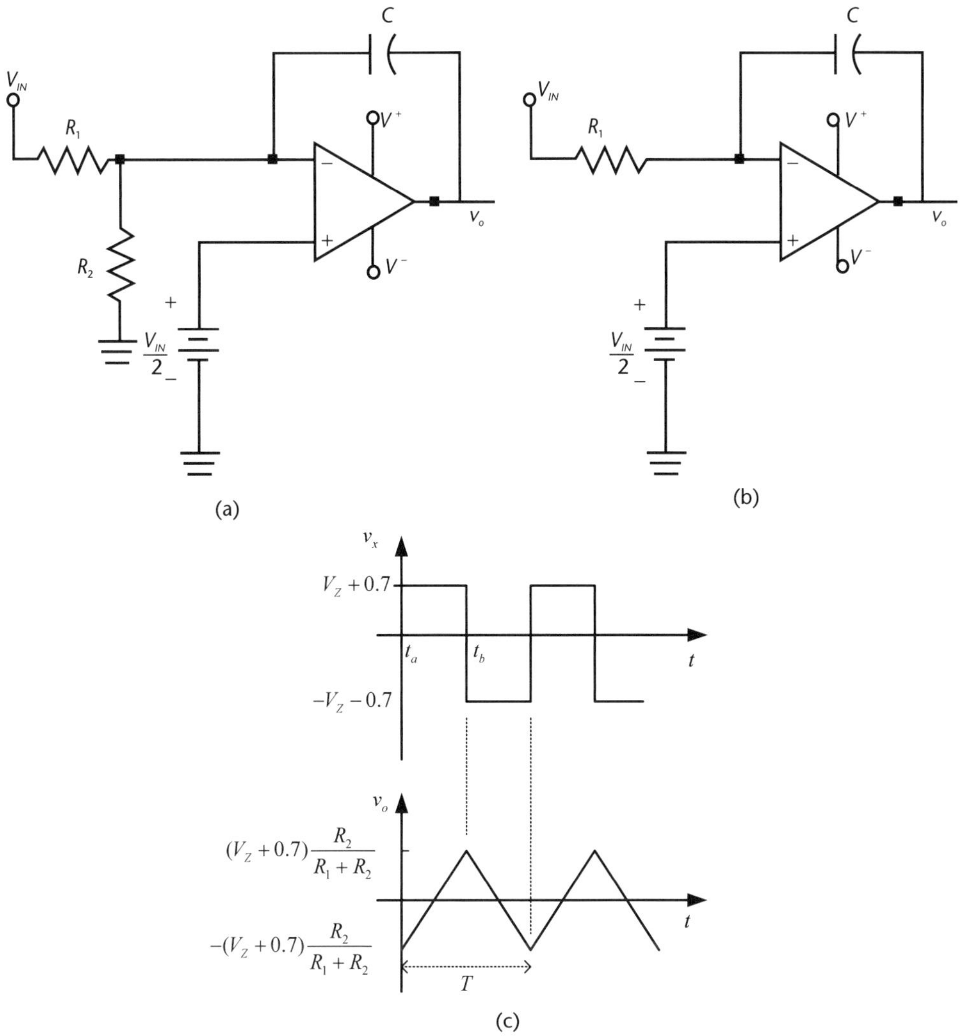

Figure 6.27 (a) Integrator circuit with the switch closed (i.e., with $v_X = V^+$), (b) integrator circuit with the switch open (i.e., with $v_X = 0$), and (c) the output waveform and v_X.

The equivalent integrator circuit with the switch open is shown in Figure 6.27(b), and at the $v_{-(2)}$ node of the integrator we write

$$\frac{V_{IN} - \dfrac{V_{IN}}{2}}{R_1} =- C\frac{dv_o}{dt}$$

or

$$v_o(t) = v_o(t_b) - \frac{1}{2R_1C}\int_{t_b}^{t} V_{IN}\,dt = \frac{2V^+}{3} - \frac{V_{IN}}{2R_1C}(t - t_b) \qquad (6.27)$$

The plot of (6.27) for $t_b \le t \le t_c$ is shown in Figure 6.27(c). When v_o reaches $V^+/3$, the output of the comparator changes to $v_x = V^+$ and the process repeats.

Observe that (6.25) and (6.27) show that if $R_{eq} = 2R_1$ (i.e., when $R_1 = 2R_2$) the circuit provides the same current to the capacitor whether the switch is closed or open.

From Figure 6.27(c), at $t = t_a$ the output voltage is $V^+/3$, and at $t = t_b$ it is $2V^+/3$. Hence, from (6.25) it follows that

$$v_o(t_b) = \frac{2}{3}V^+ = \frac{V^+}{3} + \frac{V_{IN}}{CR_{eq}}(t_b - t_a)$$

or

$$t_b - t_a = C\frac{R_{eq}}{3}\frac{V^+}{V_{IN}}$$

From symmetry (i.e., with $R_1 = 2R_2$) it follows that

$$t_c - t_b = C\frac{R_{eq}}{3}\frac{V^+}{V_{IN}}$$

Hence,

$$f = \frac{1}{T} = \frac{1}{t_c - t_a} = \frac{1}{2(t_b - t_a)} = \frac{3V_{IN}}{2CR_{eq}V^+} \tag{6.28}$$

Equation (6.28) shows that the frequency of oscillation is a linear function of V_{IN}. Furthermore, this oscillator can be used as a VCO if V_{IN} is replaced by a modulating signal $v_m(t)$.

Example 6.5

The triangular wave oscillator in Figure 6.26 is constructed with: $R_A = 50\ \text{k}\Omega$, $R_B = 25\ \text{k}\Omega$, $R_1 = 50\ \text{k}\Omega$, $R_2 = 25\ \text{k}\Omega$ (i.e., $R_1 = 2R_2$), and $C = 0.01\ \mu\text{F}$.
(a) Determine the frequency of oscillation.
(b) Determine V_{IN} for an oscillation frequency of 500 Hz.

Solution

(a) From (6.26),

$$R_{eq} = \frac{2(50 \times 10^3)25 \times 10^3}{50 \times 10^3 - 25 \times 10^3} = 100\ \text{k}\Omega$$

From (6.28), with $C = 0.01\ \mu\text{F}$, we obtain

$$f = \frac{3V_{IN}}{2(0.01 \times 10^{-6})100 \times 10^3 V^+} = \frac{1{,}500 V_{IN}}{V^+} \tag{6.29}$$

Hence, for $0 < V_{IN} < V^+$ the frequency range of the oscillator is $0 < f < 1{,}500$ Hz. (b) From (6.29), for a 500-Hz triangular wave we obtain

$$V_{IN} = \frac{V^+}{3}$$

A resistive divider from V^+ can be designed to satisfy this relation.

Another triangular wave generator that can be used as a VCO is shown in Figure 6.28(a). The first op amp circuit operates as a Schmitt trigger. The CMOS operates as a single-pole double-throw switch. The v_m and $-v_m$ signals control the switch. The second op amp operates as a unity gain buffer stage, and the third op amp circuit is an integrator.

When $v_x = V_Z + 0.7$, $T1$ is off and $T2$ is on. Hence, $-v_m$ is applied to the integrator producing a linearly increasing output waveform [i.e., $v_o(t) = v_a(t) + v_m/RC$]. When the comparator output changes to $v_x = -V_Z - 0.7$, $T1$ is on and $T2$ is off, resulting in v_m being applied to the integrator and the output waveform decreasing linearly [or $v_o(t) = v_o(t_b) - v_m/RC$]. Hence, it follows that the frequency of oscillation is given by

$$f = \frac{R_1 + R_2}{4R_2CR} \frac{v_m}{(V_Z + 0.7)}$$

The oscillator waveform is shown in Figure 6.28(b).

6.4 Sawtooth Oscillators

A sawtooth oscillator is shown in Figure 6.24. There are other ways of generating a sawtooth wave. An integrator produces a ramp waveform. Hence, an integrator that is electronically controlled to discharge its capacitor (i.e., to place a switch across the capacitor) can be used to produce a sawtooth wave. One such circuit is shown in Figure 6.29(a).

With $v_x = V_s^-$, the input voltage V_{IN} to the integrator in Figure 6.29(a) produces an output voltage of the form

$$v_o(t) = \frac{1}{RC} \int_0^t V_{IN}\, dt = \frac{V_{IN}}{RC} t \tag{6.30}$$

The resulting ramp waveform is shown in Figure 6.29(b).

When v_o reaches V_{REF}, the comparator output v_x changes from V_s^- to V_s^+. When $v_x = V_s^+$, D is forward biased and the BJT conducts (i.e., saturates). The resistor R_{BB} is needed to limit the base current of the BJT. The BJT acts like a

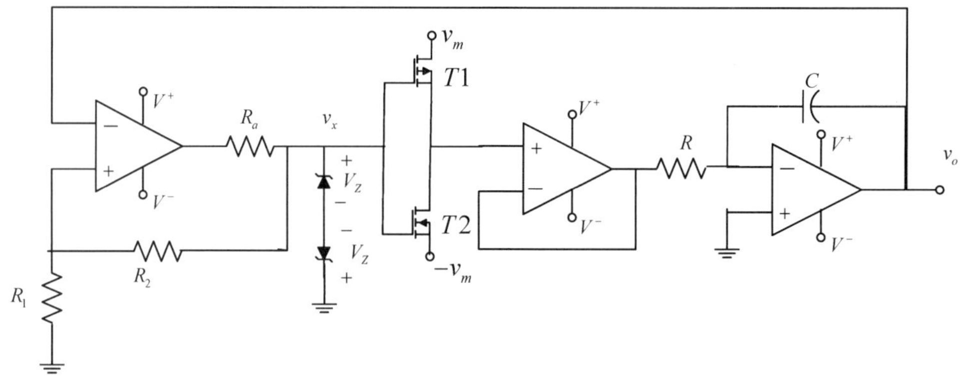

(a)

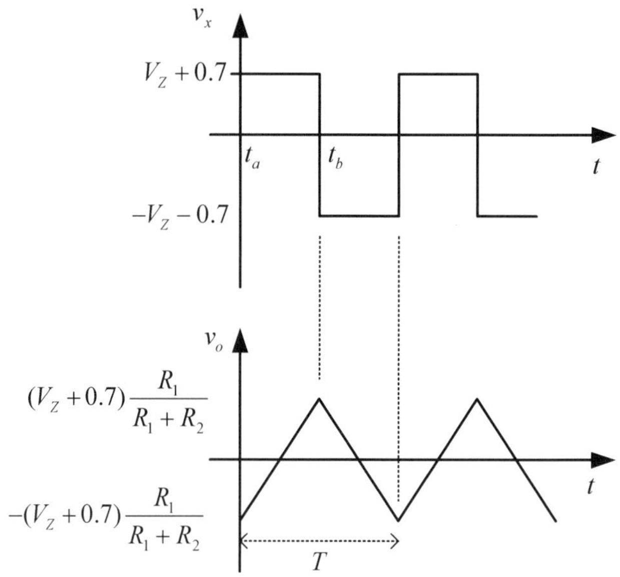

(b)

Figure 6.28 (a) A triangular VCO, where the frequency of oscillation is a linear function of the modulating voltage v_m; and (b) the circuit waveforms.

switch, discharging the capacitor and making v_O go to zero and $v_X = V_s^-$. Then, the cycle repeats itself, as shown in Figure 6.29(b).

The frequency of oscillation is approximately given by (6.30) with $v_O(T) = V_{REF}$. That is,

$$f = \frac{1}{RC}\left(\frac{V_{IN}}{V_{REF}}\right)$$

6.5 Oscillators Using the 555 Timer

This versatile IC is used extensively in a variety of square-waves oscillators and pulse-generator applications. The 8-pin DIP schematic and block diagram of the

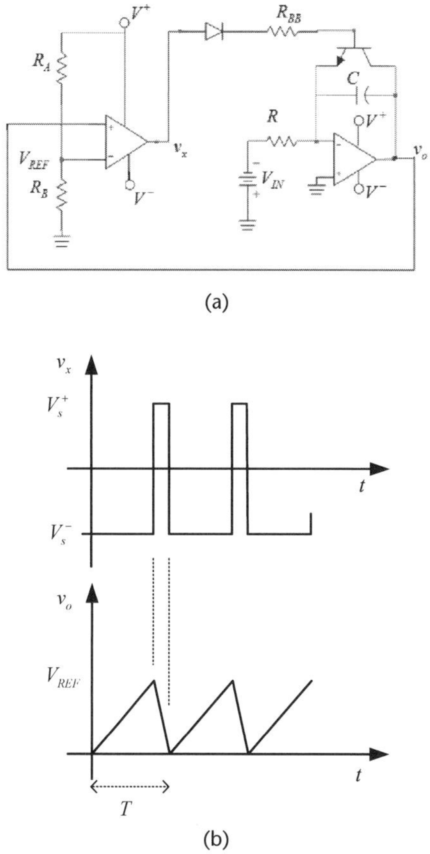

Figure 6.29 (a) A sawtooth oscillator and (b) the oscillator waveforms.

555 timer are shown in Figure 6.30. It consists of two comparators that control the output of a flip-flop. The flip-flop output is coupled to the output pin (pin 3) through an output driver stage (not shown). Also, the flip-flop output controls the on and off state of transistor *T*. The transistor is also referred to as the discharge transistor since it acts as a switch and controls the charging or discharging of an external capacitor that is connected from pin 7 to ground. The three internal resistors (usually $R = 5$ kΩ) produce a voltage divider that sets the positive input terminal of comparator *I* at $V^+/3$, and the negative input terminal of comparator *II* at $2V^+/3$. These reference voltages determine the operation of the comparators.

An overall view of the pins operation is as follows:

- *Pin 1* (*Ground*): The ground pin.
- *Pin 2* (*Trigger*): When the input voltage at pin 2 falls below $V^+/3$, the output of comparator *I* changes state (goes from low to high), causing the flip-flop to set ($S = 1$) and the *Q* output of the flip-flop to go high. This will produce a high at the output pin 3.
- *Pin 3* (*Output*): The output is normally low, and goes high when the trigger signal drops bellow $V^+/3$. The output driver (or power) stage can source or sink a maximum current of 200 mA.

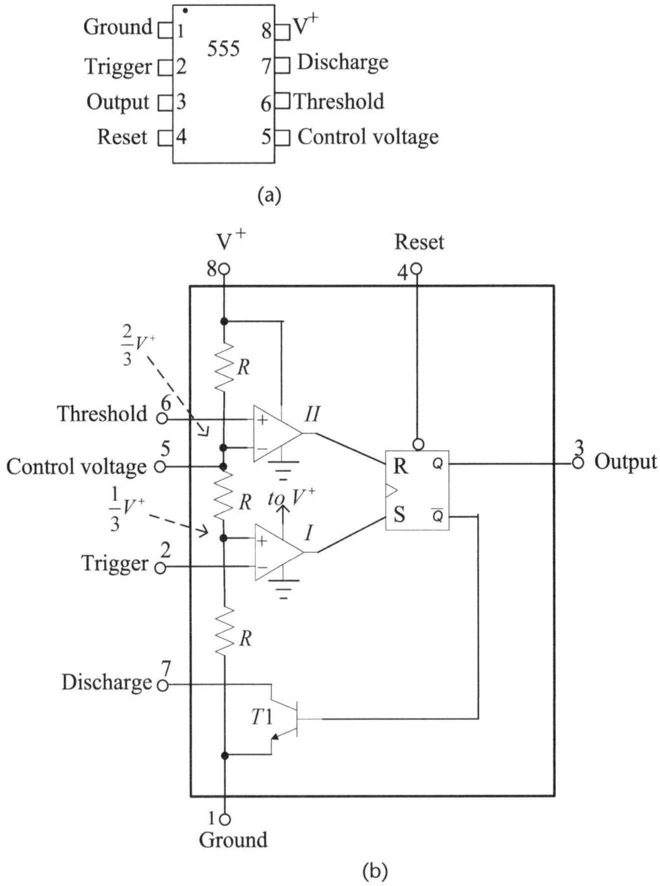

Figure 6.30 (a) Eight-pin DIP schematic of the 555 timer and (b) block diagram of the 555 timer.

- *Pin 4* (*Reset*): The timer can be triggered only when the reset pin voltage is above 1V. If pin 4 is grounded, the trigger cycle will be interrupted. Under normal operation pin 4 is usually connected to V^+.
- *Pin 5* (*Control Voltage*): An external voltage connected to pin 5 can be used to change the comparators reference voltage. Under normal operation, the $2V^+/3$ reference voltage is used and pin 5 is left open, or a capacitor of 0.01 μF is connected from pin 5 to ground to filter any noise that can cause false triggering.
- *Pin 6* (*Threshold*): When the input voltage at pin 6 is greater than $2V^+/3$, the output of comparator *II* changes state (goes from low to high). The flip-flop is reset (i.e., $R = 1$) causing Q to go low, and the voltage at the output pin 3 goes low. Since $\overline{Q} = 1$, the transistor T is on, connecting pin 7 to ground.
- *Pin 7* (*Discharge*): The discharge pin is connected to the collector of transistor T. The transistor is off when $Q = 1$ (i.e., output is high) and $\overline{Q} = 0$, and the transistor is on when $Q = 0$ (i.e., output is low) and $\overline{Q} = 1$. This pin is

normally used to control the charging or discharging of a capacitor, which is connected between pin 7 and ground.

- *Pin 8 (Supply Voltage)*: The supply voltage can range from about 4.5V to 18V. The 555 was originally designed for 5V operation to be compatible with TTL devices.

The way that the 555 generates a pulse can be described by assuming a quiescent state where the threshold voltage at pin 6 is low (i.e., approximately 0V) and the trigger signal at pin 2 is high (i.e., at V^+). Thus, the output of comparator I is low (i.e., ≈ 0V), making $S = 0$, $Q = 0$, and $\overline{Q} = 1$ (i.e., high). The high at $\overline{Q}$ saturates transistor $T1$ (i.e., turns-on $T1$), causing pin 7 to be connected to the ground pin 1. Also, with a low at Q the output pin 3 is low (i.e., ≈ 0V). A negative trigger signal at pin 2, lower than $V^+/3$, will produce a high at the output of comparator I, making $S = 1$, $Q = 1$ and $\overline{Q} = 0$. Hence, the output pin 3 will go high, the transistor T is off, and pin 7 is open.

The flip-flop is reset when the threshold voltage (i.e., the voltage at pin 6) exceeds $2V^+/3$ (so that $R = 1$) and the trigger pulse is not present (i.e., the voltage at pin 2 greater than $V^+/3$ so that $S = 0$). Then, the Q output goes low, $\overline{Q} = 1$, and the transistor $T1$ saturates, causing pin 7 to be connected to ground.

In some 555 timers, when operated with $V^+ = 5$V the low state has a maximum value of 0.25V, and the high state has a typical value of 3.3V with a guaranteed maximum value of 2.75V.

A Pulse Generator

The 555 timer connected as a pulse generator is shown in Figure 6.31(a). This circuit is used to provide a timing pulse (i.e., a one shot) whose duration is controlled by the external capacitor C.

Referring to Figures 6.31(a) and 6.30(b), it follows that with the trigger pin at V^+, the Q output of the internal flip-flop is low and, therefore, the output at pin 3 is low (i.e., $v_o = 0$). The internal transistor T is saturated since $\overline{Q} = 1$ (i.e., $T1$ is on). Hence, the external capacitor C is shorted to ground because pin 7 is shorted to pin 1. When a negative trigger pulse is applied to pin 2, [as shown in Figure 6.31(a)], the output of the internal comparator I goes high, so $S = 1$, $Q = 1$ and the output voltage at pin 3 goes high. The transistor goes off, thus removing the short across C. The voltage across C [i.e., $v_6(t)$] rises exponentially towards V^+, according to

$$v_6(t) = V^+\left(1 - e^{-t'/RC}\right) \tag{6.31}$$

where the charging time constant is controlled by the external RC components.

The capacitor is also connected to pin 6. Hence, when the voltage $v_6(t)$ reaches $2V^+/3$, comparator II will reset the flip-flop, and the transistor will turn on, grounding pin 7. Hence, the capacitor will quickly discharge through the transistor, and the output will go low.

The waveforms generated by the trigger signal are shown in Figure 6.31(a). The voltage v_o is high during the time interval that it takes for $v_6(t)$ to reach $2V^+/3$. This interval, denoted by t_x, can be calculated from (6.31) as follows:

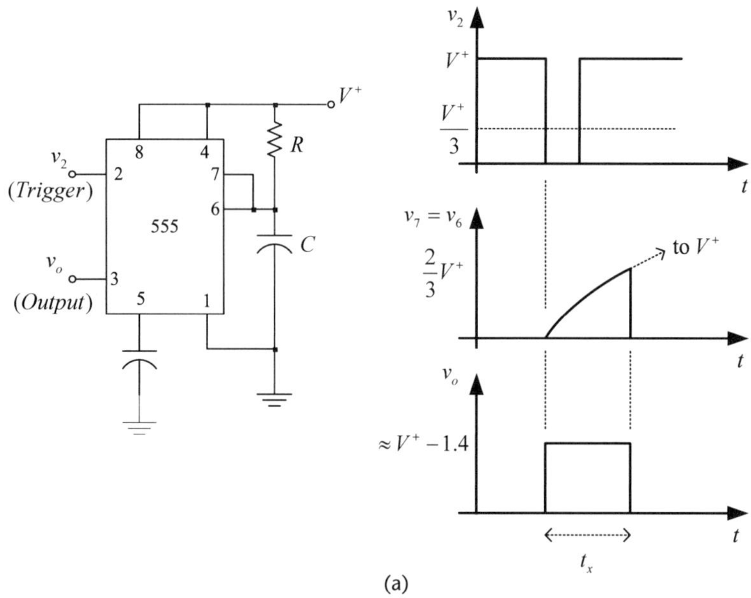

(a)

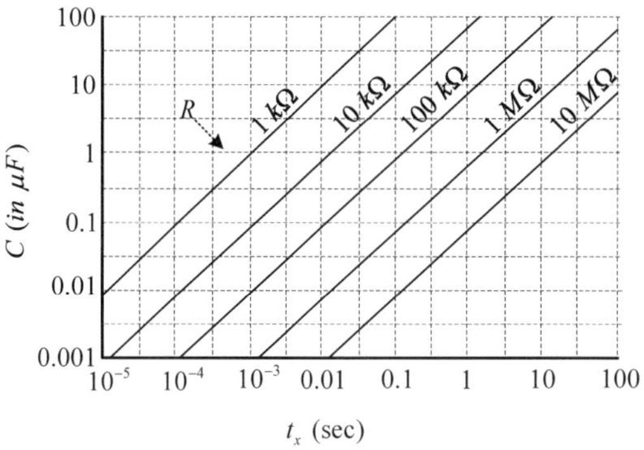

(b)

Figure 6.31 (a) A monostable multivibrator using the 555 timer and (b) values of *R* and *C* for a given pulse width t_x.

$$\frac{2}{3}V^+ = V^+\left(1 - e^{-t_x/RC}\right)$$

which gives

$$t_x = 1.1RC \tag{6.32}$$

The relation (6.32) is plotted in Figure 6.31(b) to show the various combinations of *R* and *C* that produces a given pulse width t_x.

The triggering of the 555 occurs when the input trigger pulse falls below $V^+/3$. For proper operation the trigger signal voltage level should not remain below $V^+/3$ for a time longer than t_x. Typical triggering signals are usually those associated with the high-to-low output transition of various digital circuits. Hence, certain periodic transitions can be used to trigger the 555, resulting in pulses whose width is controlled by the time t_x.

A Square-Wave Oscillator

The 555 timer connected as a square-wave oscillator is shown in Figure 6.32. In Figure 6.32(a) the capacitor is connected to pins 2 and 6. The capacitor charges towards V^+ through R_A and R_B. When the capacitor voltage (i.e., pin 6 voltage) reaches $2V^+/3$, comparator *II* resets the flip-flop, the transistor *T* turns-on shorting pin 7 to pin 1 (ground), and the capacitor discharges towards 0V through R_B. However, when the capacitor voltage drops to $V^+/3$, the decaying voltage at pin 2 will set the internal flip flop (i.e., $Q = 1$ and $\overline{Q} = 0$), making the transistor *T* turn off. Hence, the capacitor begins to charge again towards V^+. In summary,

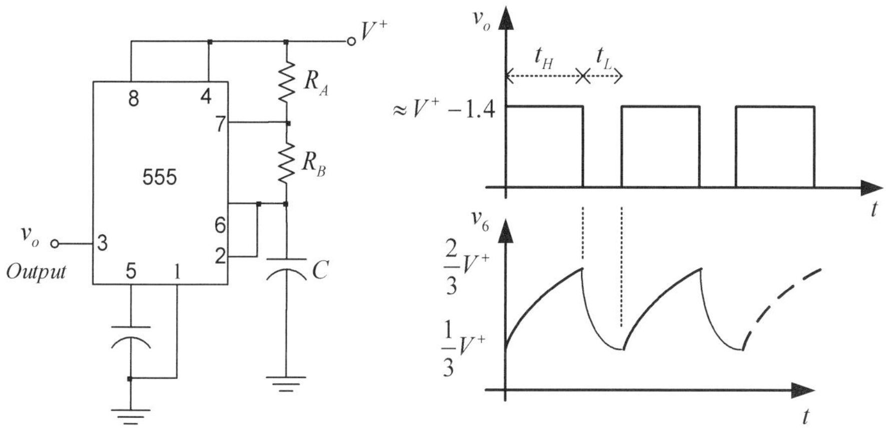

(a)

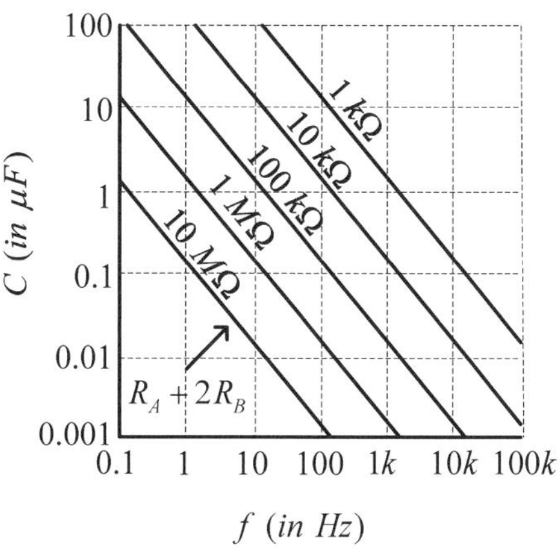

(b)

Figure 6.32 (a) A square-wave oscillator using the 555 timer and (b) a design monograph for the oscillator.

the capacitor charges and discharges between $V^+/3$ and $2V^+/3$, as shown in Figure 6.32(a).

The charge of the capacitor from an initial value of $V^+/3$ towards V^+ is described by

$$v_6(t) = \frac{V^+}{3} + \frac{2V^+}{3}\left(1 - e^{-t/(R_A+R_B)C}\right) \quad 0 \le t \le t_H \tag{6.33}$$

Hence, the time that the output pulse is high (t_H) is obtained from (6.33) by setting $v_6(t_H) = 2V^+/3$, namely,

$$t_H = 0.693(R_A + R_B)C \tag{6.34}$$

Similarly, the discharge time of the capacitor from an initial value of $2V^+/3$ towards 0 is described by

$$v_6(t) = \frac{2V^+}{3} e^{-(t-t_H)/R_B C} \quad 0 \le t \le t_L \tag{6.35}$$

Hence, the time that the output pulse is low (t_L), is obtained from (6.35) by setting $v_6(t_H + t_L) = V^+/3$. Thus, we obtain

$$t_L = 0.693 R_B C \tag{6.36}$$

The frequency of the square wave is

$$f = \frac{1}{T} = \frac{1}{t_H + t_L} = \frac{1.44}{(R_A + 2R_B)C} \tag{6.37}$$

In general, the 555 timer can oscillate at frequencies up to 500 kHz. Some versions of the 555 using CMOS technology can oscillate at frequencies up to 2 MHz. A monograph for (6.37) is shown in Figure 6.32(b).

The duty cycle D can be defined as the ratio of the time where the output is high (t_H) to the total cycle time, or

$$D = \frac{t_H}{T} = \frac{R_A + R_B}{R_A + 2R_B} \tag{6.38}$$

Sometimes the duty cycle is defined as the ratio of t_L to T.

Example 6.6

Design the square-wave oscillator in Figure 6.32 to oscillate at a frequency of 10 kHz with a duty cycle of 70%, and an output voltage of 5V.

Solution

From (6.38), the duty cycle is given by

$$D = \frac{t_H}{T} = \frac{R_A + R_B}{R_A + 2R_B} = 0.7 \tag{6.39}$$

Hence, $T = 1/f = 100\ \mu s$, $t_H = 70\ \mu s$, and $t_L = 30\ \mu s$.

From (6.37), selecting $C = 0.01\ \mu F$, it follows that

$$R_A + 2R_B = \frac{1.44}{fC} = \frac{1.44}{(10 \times 10^3)(0.01 \times 10^{-6})} = 14.4\ k\Omega \tag{6.40}$$

The simultaneous solution of (6.39) and (6.40) gives R_A = 5.79 kΩ and R_B = 4.34 kΩ.

Observe that (6.38) shows that a symmetrical square wave (i.e., D = 50%) can be approximated by making R_B much larger than R_A. An oscillator that better approximates a symmetrical square wave (D = 50%) is shown in Figure 6.33. In this oscillator the capacitor charges through R_A and the diode, and discharges through R_B. Hence, the period of oscillation is given approximately given by

$$T = 0.693(R_A + R_B)C$$

If $R_A = R_B$ the duty cycle is 50%.

6.6 ICs Function Generators

Some monolithic ICs capable of producing sine, square, triangle, sawtooth, and pulse waveforms are the Exar XR-2206, the Intersil ICL8038, the Maxim MAX038, and the NTE864.

The block diagram of the XR-2206 is shown in Figure 6.34(a), and the pin configuration in Figure 6.34(b). The XR-2206 consists of a VCO, an analog

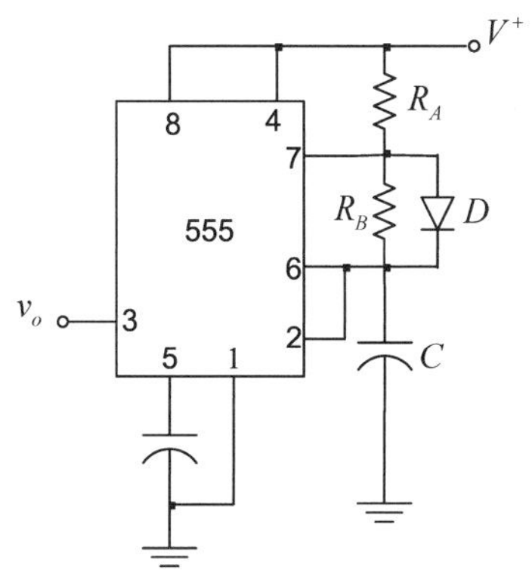

Figure 6.33 A pulse oscillator with D = 50% when $R_A = R_B$.

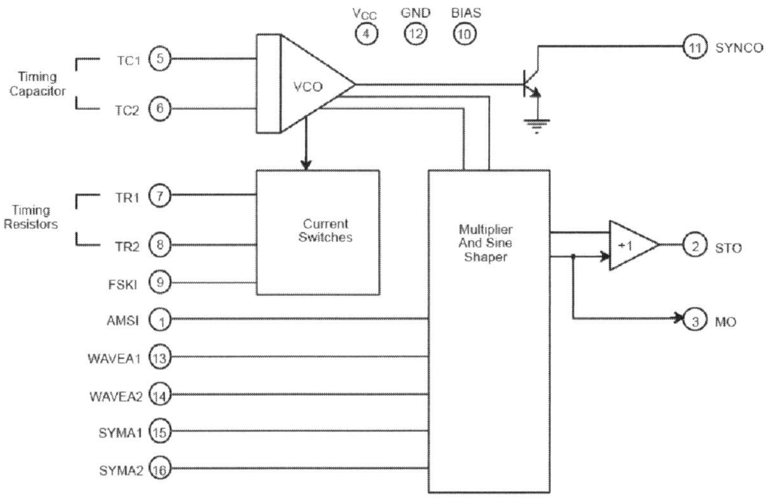

(a)

PIN DESCRIPTION

Pin #	Symbol	Type	Description
1	AMSI	I	**Amplitude Modulating Signal Input.**
2	STO	O	**Sine or Triangle Wave Output.**
3	MO	O	**Multiplier Output.**
4	V_{CC}		**Positive Power Supply.**
5	TC1	I	**Timing Capacitor Input.**
6	TC2	I	**Timing Capacitor Input.**
7	TR1	O	**Timing Resistor 1 Output.**
8	TR2	O	**Timing Resistor 2 Output.**
9	FSKI	I	**Frequency Shift Keying Input.**
10	BIAS	O	**Internal Voltage Reference.**
11	SYNCO	O	**Sync Output.** This output is a open collector and needs a pull up resistor to V_{CC}.
12	GND		**Ground pin.**
13	WAVEA1	I	**Wave Form Adjust Input 1.**
14	WAVEA2	I	**Wave Form Adjust Input 2.**
15	SYMA1	I	**Wave Symetry Adjust 1.**
16	SYMA2	I	**Wave Symetry Adjust 2.**

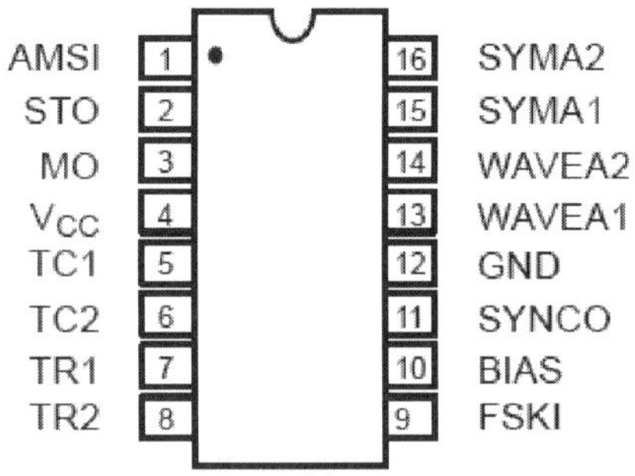

(b)

Figure 6.34 (a) Block diagram of the XR-2206 and (b) the pin configuration. (*From:* © Exar Corporation. Reproduced with the express authorization of Exar Corporation.)

multiplier with a sine wave shaper, a unity gain amplifier and a set of current switches. The output frequency of the VCO is proportional to its input current. The input current is controlled by the resistors connected between the timing pins 7 and 8 and ground. The FSKI input (pin 9) is the control pin for FSK signal modulation. The logic signal at pin 9 selects the timing resistor used. If pin 9 is open or high (≥2V) the resistor at pin 7 is used, and if pin 9 is low (≤1V) the resistor at pin 8 is used. The two timing resistors can produce two different output frequencies.

A circuit for generating sine, triangular, and square waves is shown in Figure 6.35. The frequency of oscillation depends of the external timing capacitor C connected across pins 5 and 6, and the timing resistor connected to pin 7, where $R = R_a + R_b$. It is given by

$$f_o = \frac{1}{RC} \tag{6.41}$$

The frequency of oscillation can be varied from about 0.01 Hz to 1 MHz.

The XR-2206 can be operated with a split power supply by replacing the ground connection with V^-. Also, the harmonic content of the sinusoidal output can be reduced by using R_Y to adjust the shape and R_X to adjust the symmetry. The resistor R_X is used when minimum distortion is desired.

The dc level at the output pin 2 is approximately the same as the dc bias at pin 3. In Figure 6.35, pin 3 is biased at $V_{CC}/2$ to provide a dc level at pin 2 of $V_{CC}/2$.

Example 6.7

Design the circuit in Figure 6.35 to produce a sinusoidal, triangle, and square waves in the 100-Hz to 100-kHz frequency range.

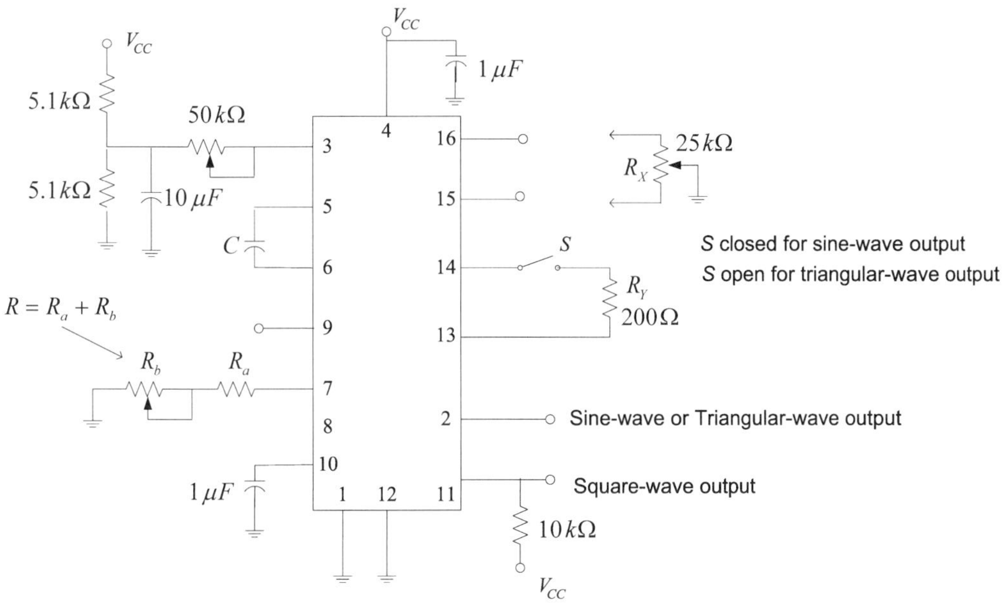

Figure 6.35 Circuit for sine, triangular, and square waves generation using the XR-2206.

Solution

Letting $C = 10$ nF, it follows from (6.41) that the minimum and maximum values of R are:

$$R_{\min} = \frac{1}{f_{o(\max)}C} = \frac{1}{(100 \times 10^3)(10 \times 10^{-9})} = 1\ \text{k}\Omega$$

and

$$R_{\max} = \frac{1}{f_{o(\min)}C} = \frac{1}{(100)(10 \times 10^{-9})} = 1\ \text{M}\Omega$$

This range of resistances can be implemented in Figure 6.35 with $R_a = 1$ kΩ and a variable potentiometer for R_b of 1 MΩ.

A circuit for pulse and sawtooth waveforms is shown in Figure 6.36. Observe that the FSK control pin 9 is connected to the square-wave output. Hence, the frequency shifts between two separate values as the square wave changes from low to high and from high to low. The pulse width can be adjusted from 1% to 99% by the timing resistors R_1 and R_2. In fact, the frequency of oscillation is given by

$$f_o = \frac{2}{(R_1 + R_2)C}$$

and the duty cycle by

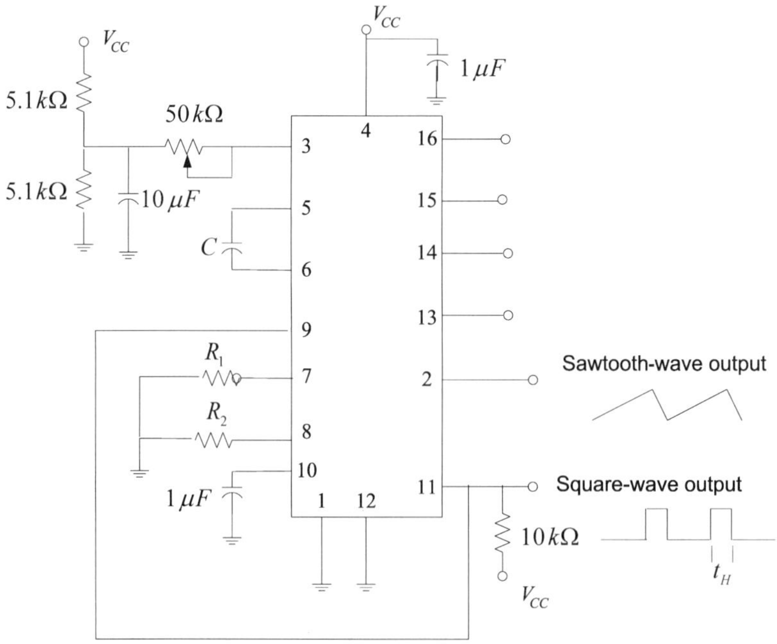

Figure 6.36 Circuit for pulse and sawtooth waves generation using the XR-2206.

$$D = \frac{t_H}{T} = \frac{R_1}{R_1 + R_2}$$

The block diagram of the ICL8038 is shown in Figure 6.37(a), and the pin configuration in Figure 6.37(b). This IC consists of two comparators, a flip-flop, buffer amplifiers, a sine converter, and two current sources. The current source I_2 is shown as being twice the value of I_1. An external timing capacitor C is connected to pin 10. This capacitor is charged and discharged by the current sources. The source I_2 switches on and off by the flip-flop, and I_1 is always on.

The capacitor charges through I_1 when I_2 is off, and its voltage rises linearly as a function of time. When the capacitor voltage reaches $2V^+/3$ the flip-flop closes the switch and I_2 connects to the capacitor. Hence, the capacitor discharges with a net current of $I_2 - I_1$, and its voltage decreases linearly with time. When the

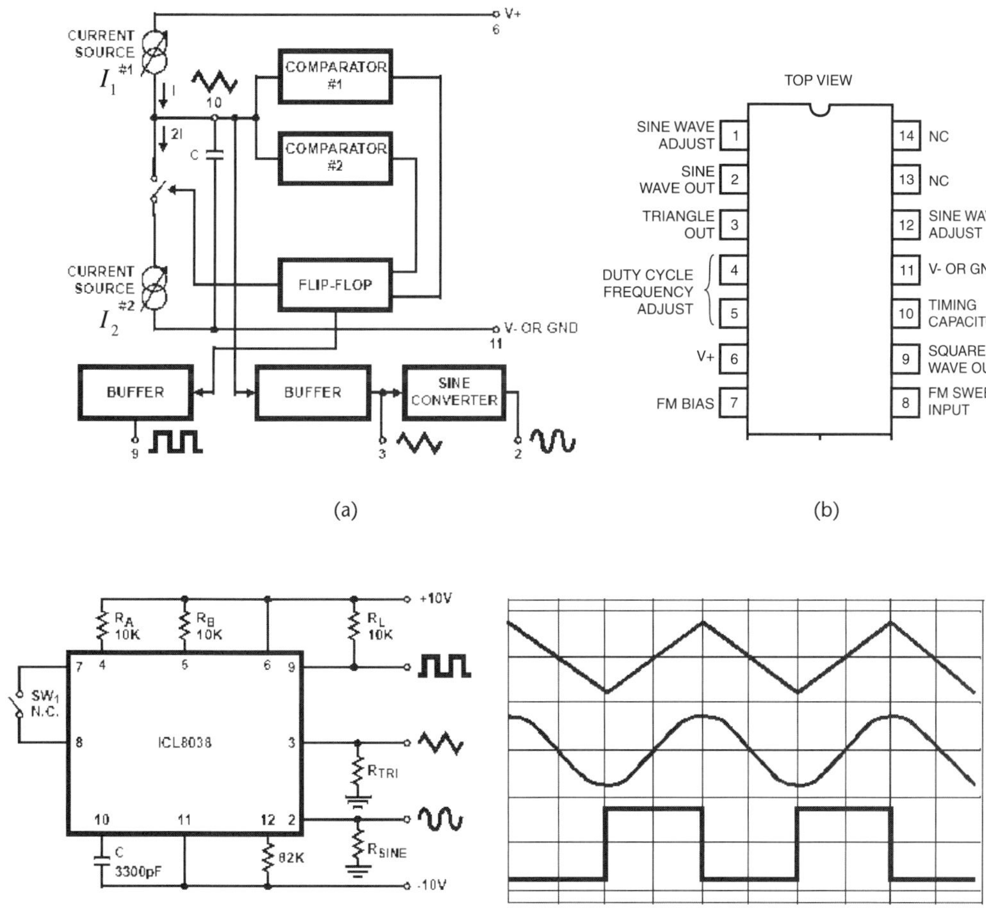

Figure 6.37 (a) Block diagram of the ICL8038, (b) the pin configuration, and (c) circuit for generating sine, triangular and square waves using the ICL8038. (*From:* © 2001 Intersil Americas Inc. Reproduced with permission.)

capacitor voltage reaches $V^+/3$, the flip-flop opens the switch, charging the capacitor through I_1 and the cycle is repeated.

A circuit diagram for the ICL8038 is shown in Figure 6.37(c), as well as the output waveforms for a duty cycle of 50%. The resistor R_A controls the current I_1; therefore, it controls the rising part of the output signal. The resistor R_B together with R_A controls the falling portion of the output signal. In Figure 6.37(c) the rising time of the triangular wave is given by

$$t_1 = \frac{C\left(\frac{V^+}{3}\right)}{0.22\frac{V^+}{R_A}} = \frac{R_A C}{0.66} \tag{6.42}$$

and the falling portion by

$$t_2 = \frac{C\left(\frac{V^+}{3}\right)}{2(0.22)\frac{V^+}{R_B} - 0.22\frac{V^+}{R_A}} = \frac{R_A R_B C}{0.66(2R_A - R_B)} \tag{6.43}$$

Equations (6.42) and (6.43) show that if $R_A = R_B$ then $I_2 = 2I_1$ and $t_1 = t_2$, or the duty cycle is 50%. The frequency of oscillation in this case is given by

$$f_o = \frac{1}{t_1 + t_2} = \frac{0.33}{R_A C} \tag{6.44}$$

If $R_A \neq R_B$ it follows that $I_2 \neq 2I_1$ and the frequency oscillation is given by

$$f_o = \frac{1}{\frac{R_A C}{0.66}\left(1 + \frac{R_B}{2R_A - R_B}\right)}$$

Best performances are obtained if the charging currents are between 10 μA to 1 mA. The device can be operated with dual supply voltages between ±5V to ±15V. Also, a single supply voltage of 10V to 30V can be used. Pins 7 and 8 are used to obtain an FM signal.

Example 6.8

Determine the frequency of oscillation in the circuit shown in Figure 6.37c.

Solution

Since $R_A = R_B = 10\ \text{k}\Omega$, it follows from (6.42) that

$$f_o = \frac{0.33}{(10 \times 10^3)(3{,}300 \times 10^{-12})} = 10\ \text{kHz}$$

6.7 UJTs and PUTs

The operation of the unijunction transistor (UJT) and the programmable unijunction transistor (PUT) is quite similar. Because of the programmable feature of the PUTs, they have replaced the UJTs. However, for a better understanding of the PUTs, the construction and operation of the UJT is first discussed.

The UJT construction, symbols, and characteristics are shown in Figure 6.38. The UJT has three terminals, called the emitter, base 1, and base 2 terminals. The name of the device denotes that there is only one pn junction. That is, the junction between the highly doped p-type emitter and the lightly doped bar of n-type material. The arrow in the symbol points in the direction of the current flow. The pn junction is closer to the base-2 contact than to the base-1 contact. In Figure 6.38(c), the $v_E - i_E$ characteristics of the UJT are shown for a constant value of the interbase voltage v_{BB}, denoted by $v_{BB} = V_{BB}$.

The operation of the UJT can be explained using the equivalent circuit shown in Figure 6.39. The bar of n-type silicon has a high resistivity, and its total resistance is

$$r_{BB} = r_{B1} + r_{B2}$$

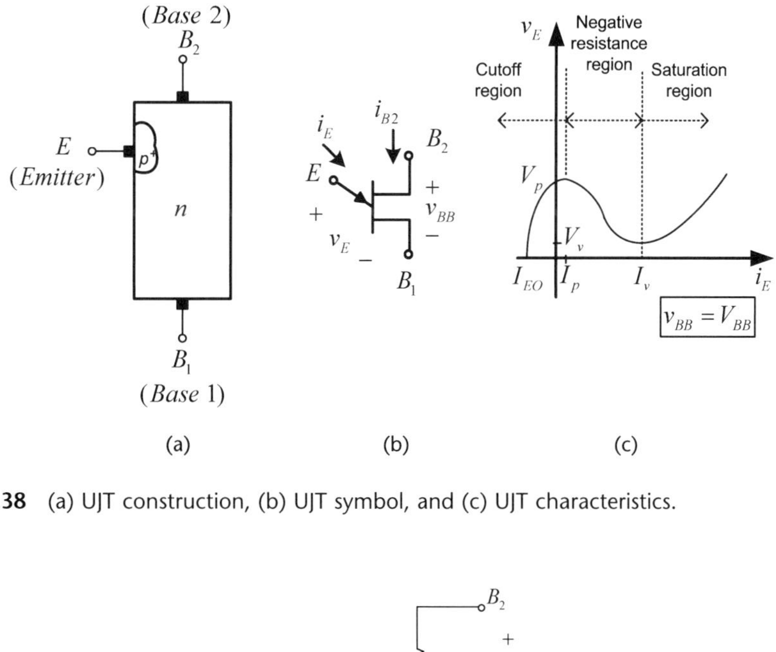

Figure 6.38 (a) UJT construction, (b) UJT symbol, and (c) UJT characteristics.

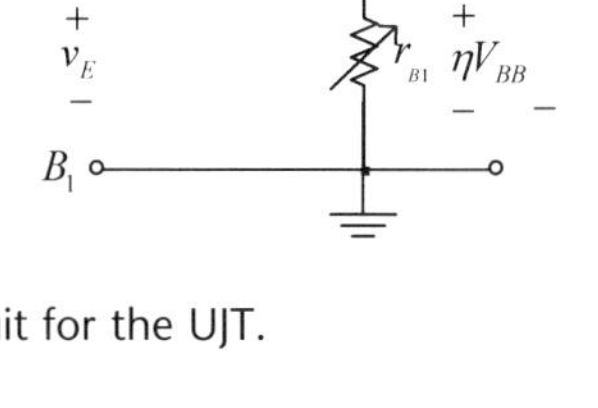

Figure 6.39 An equivalent circuit for the UJT.

The emitter contact determines the voltage division between r_{B1} and r_{B2}. The resistor r_{B1} is shown as a variable resistor to indicate its variable nature. The voltage across r_{B1} is given by ηV_{BB}, where

$$\eta = \frac{r_{B1}}{r_{B1} + r_{B2}}$$

The parameter η is called the intrinsic standoff ratio. Typical values of η are between 0.5 to 0.9, and values of r_{BB} range from 5 to 10 kΩ. If v_E is less than ηV_{BB}, the pn junction is reversed biased, and the diode D that represents the junction is open. Then, only a small reverse bias current I_{E0} flows in the emitter [see Figure 6.38(c)]. The pn junction remains reverse biased (i.e., the diode is open) until $v_E \approx 0.7 + \eta V_{BB}$. At this value of v_E the diode becomes forward biased. When the diode starts to conduct, the device enters a negative-resistance region in which the voltage v_E decreases as i_E increases. The explanation of this effect is complex. A simple description of this effect is to observe that when the UJT begins to conduct, the holes injected from the emitter into the region between the emitter and base 1 behave like minority carriers in this region. Thus, the conductivity in this region decreases because it depends on the total number of carriers. This decrease in conductivity produces a decrease in r_{B1} as the current i_E increases. A variable resistor r_{B1} is used in Figure 6.39 to show the variable nature of the conductivity between the emitter and base 1. The value of v_E at which the negative-resistance region begins is called the peak-point voltage V_p. At this point, v_E is given by

$$v_E = V_p = 0.7 + \eta V_{BB} \tag{6.45}$$

The peak-point current I_p is the emitter current when $v_E = V_p$.

The negative-resistance region extends until the valley point is reached, where the voltage and current are defined by V_v and I_v, respectively. Beyond the valley point the UJT behaves like a forward-biased diode and further increases in v_E produces an increase in i_E. In this region, called the saturation region, r_{B1} is fairly constant.

Figure 6.40 shows the effect of V_{BB} on the characteristics of the UJT. The voltage V_{BB} affects the value of V_p since, according to (6.45), if V_{BB} is increased the voltage v_E must also be increased in order for the UJT to enter the negative-resistance region.

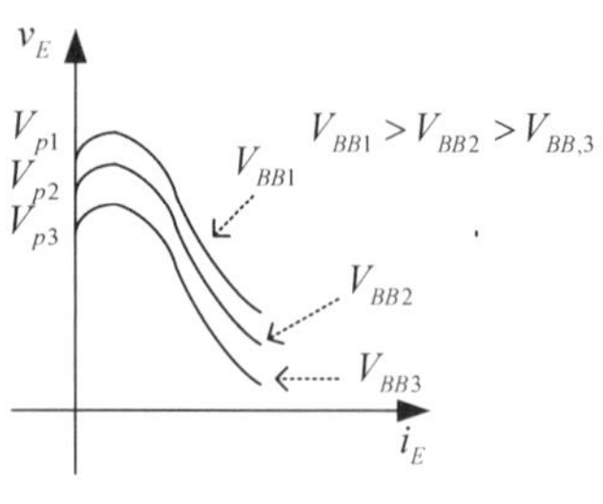

Figure 6.40 Effect of V_{BB} on the UJT characteristics.

The UJT can be used in a variety of applications, such as a voltage-controlled switch, in pulse generator circuits, in timing circuits, etc. The UJT used in an oscillator circuit is shown in Figure 6.41. This oscillator circuit is known as a UJT relaxation oscillator. The operation of this circuit is as follows. When the switch S is closed, the UJT is off and the voltage V_1 charges the capacitor C_x with a charging time constant of $\tau_c = R_x C_x$. Thus, v_x is given by

$$v_x = V_1\left(1 - e^{-t/R_x C_x}\right)$$

When v_x reaches the value V_p, the UJT turns on and the capacitor discharges very fast through R_1. The discharging time constant τ_d is

$$\tau_d = (R_L + r_{B1})C_x$$

The capacitor discharges until the voltage v_x decreases to a value that reduces the emitter current below I_v (i.e., $v_x \approx V_v$). At this point, the UJT opens and the capacitor begins to charge again.

When the UJT is off, since $R_L \ll r_B$, then $V_{BB} \approx V_1$, and the peak-point voltage is given by (6.45) with $V_{BB} \approx V_1$.

The oscillator waveforms are shown in Figure 6.41. The resulting pulses in v_{R_L} can be used for timing purposes.

At the time that the UJT turns on, the voltage across R_x is $V_1 - V_p$. The resistor R_x must be able to supply an emitter current greater or equal to I_p to guarantee that the UJT will turn on. Hence, the maximum value of R_x is bounded by

$$R_{x,\max} < \frac{V_1 - V_p}{I_p} \tag{6.46}$$

Also, if R_x is very small, the emitter current will always exceed I_V and the UJT will not turn off. Thus, the minimum value of R_x is

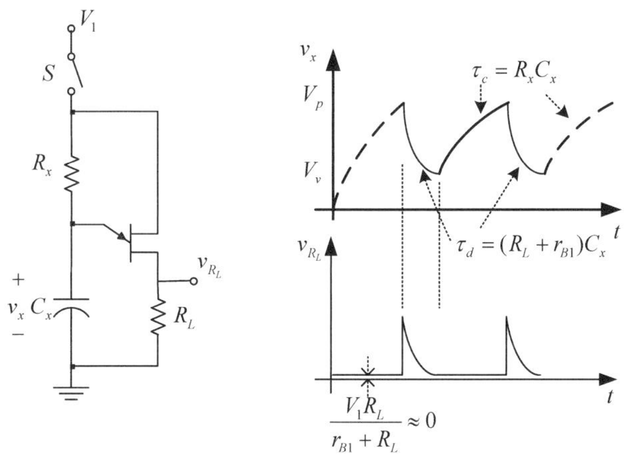

Figure 6.41 A UJT relaxation oscillator.

$$R_{x,\min} > \frac{V_1 - V_v}{I_v} \tag{6.47}$$

The UJT operates between I_p and I_V with the proper selection of R_x.

From Figure 6.41 it is seen that the frequency of oscillation of the UJT relaxation oscillator is given by

$$f_o = \frac{1}{\tau_d + \tau_c} \approx \frac{1}{(R_L + R_x)C_x} \approx \frac{1}{R_x C_x} \tag{6.48}$$

A more exact relation for the frequency of oscillation is

$$f_o = \frac{1}{R_x C_x \ln\left(\dfrac{V_1 - V_v}{V_1 - V_p}\right)} \tag{6.49}$$

Example 6.9

Design the UJT relaxation oscillator shown in Figure 6.41 to oscillate at 5 kHz. The supply voltage is $V_1 = 15$V and the UJT has $\eta = 0.65$, $r_{BB} = 5\ \text{k}\Omega$, $I_p = 2\ \mu$A, $I_V = 2$ mA, and $V_V = 2.5$V.

Solution

Selecting R_L to be 27Ω, it follows that $V_{BB} \approx V_1$, and from (6.45) the value of V_p is

$$V_p = 0.7 + \eta V_{BB} = 0.7 + 0.65(15) = 10.45\text{V}$$

The value of R_x must satisfy (6.46) and (6.47). From (6.46),

$$R_{x,\max} < \frac{V_1 - V_p}{I_p} = \frac{15 - 10.45}{2 \times 10^{-6}} = 2.27\ \text{M}\Omega$$

and from (6.47),

$$R_{x,\min} > \frac{V_1 - V_v}{I_v} = \frac{15 - 2.5}{2 \times 10^{-3}} = 6.25\ \text{k}\Omega$$

Hence, it follows that $6.25\ \text{k}\Omega < R_x < 2.27\ \text{M}\Omega$. The range of values for R_x is fairly large since I_V is much larger I_p.

From (6.48), letting $C_x = 0.01\ \mu$F, the approximate value of R_x is

$$R_x \approx \frac{1}{f_o C_x} = \frac{1}{5 \times 10^3 (0.01 \times 10^{-6})} = 20\ \text{k}\Omega$$

which is between the allowable values for R_x.

Using the exact relation in (6.49), the value of R_x

$$f_o = \frac{1}{R_x C_x \ln\left(\frac{V_1 - V_v}{V_1 - V_p}\right)} = \frac{1}{5 \times 10^3 (0.01 \times 10^{-6}) \ln\left(\frac{15 - 2.5}{15 - 10.45}\right)} = 19.8\ \text{k}\Omega$$

In practice, we can use a variable resistor for R_x to set the frequency exactly at 5 kHz.

Programmable Unijunction Transistors

The PUT is a four-layer device (i.e., a thyristor) whose characteristics are similar to those of the UJT. The construction, symbol, and characteristics of PUTs are shown in Figure 6.42.

Observe the location of the gate connection in Figure 6.42(a). Under normal operation, the gate-to-cathode voltage is positive. If the anode is approximately 0.7V above the gate voltage, the anode-to-gate pn junction is forward biased and the PUT switches on. In fact, the name "programmable" follows because the gate-to-cathode bias determines the value of v_{AK} that switches on the device. Once the device turns on, the gate loses its control. The PUT can be turned off by lowering i_A below its value at the valley.

A comparison of similarly rated UJTs and PUTs shows that PUTs have a higher interbase resistance than UJTs, and the peak and valley currents of PUTs are lower than UJTs.

A typical PUT relaxation oscillator is shown in Figure 6.43. The gate voltage is given by

$$V_G = \eta V_1$$

where

$$\eta = \frac{R_1}{R_1 + R_2}$$

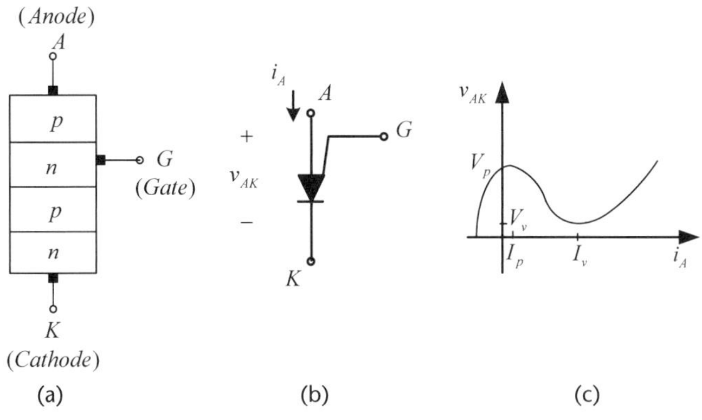

Figure 6.42 (a) PUT construction, (b) PUT symbol, and (c) PUT characteristics.

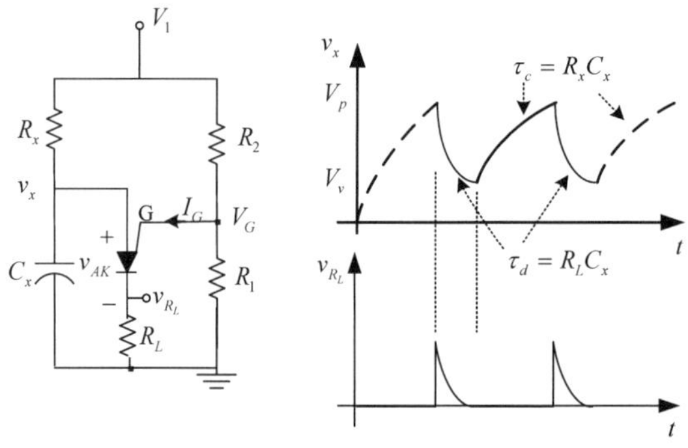

Figure 6.43 A PUT relaxation oscillator.

The peak-point voltage of v_{AK}, at which the PUT switches on, is given by

$$V_p = 0.7 + V_G = 0.7 + \eta V_1$$

Although the previous equations are similar in form to those of the UJT, it is observed that in the PUT the resistors R_1 and R_2 which determine η are external resistors; while in the UJT the resistors r_{B1} and r_{B2} which determine η are internal to the device. Hence, in the PUT the voltage V_p is programmable with R_1 and R_2.

The operation of a PUT relaxation oscillator is similar to that of a UJT relaxation oscillator. Hence, the frequency of oscillation is given by (6.49). Also, the resistor R_x is limited by $R_{x,\min} < R_x < R_{x,\max}$, where $R_{x,\max}$ and $R_{x,\min}$ are given by (6.46) and (6.47), respectively.

Manufacturers usually list typical values of I_p and I_v for specific test conditions. For example, for the 2N6027 PUT the manufacturer lists for a Thevenin's voltage at the gate of 10V, where

$$V_{TH} = \frac{V_1 R_1}{R_1 + R_2}$$

and a Thevenin's resistance of 10Ω, where

$$R_{TH} = \frac{R_1 R_2}{R_1 + R_2}$$

a typical value of I_p of 4 μA and of I_v of 150 μA. Also, the measured valley voltage is approximately $V_v = 0.8$V.

Example 6.10

Design the PUT relaxation oscillator in Figure 6.43 to oscillate at 100 Hz using the 2N6027 PUT.

Solution

Letting $V_1 = 20\text{V}$ and $R_1 = R_2 = 20\ \text{k}\Omega$ produces a Thevenin's voltage between the gate and ground of 10V and $R_{TH} = 10\ \text{k}\Omega$. Then, $V_p = 10.7\text{V}$, and with I_p of $4\ \mu\text{A}$ and I_v of $150\ \mu\text{A}$ we obtain from (6.46) and (6.47)

$$R_{x,\max} < \frac{V_1 - V_p}{I_p} = \frac{20 - 10.7}{4 \times 10^{-6}} = 2.33\ \text{M}\Omega$$

and

$$R_{x,\min} > \frac{V_1 - V_v}{I_v} = \frac{20 - 0.8}{150 \times 10^{-3}} = 128\ \text{k}\Omega$$

Hence, R_x is bounded by $128\ \text{k}\Omega < R_x < 2.33\ \text{M}\Omega$.

From (6.49), letting $C = 0.1\ \mu\text{F}$, the required value of R_x is

$$R_x \approx \frac{1}{f_o C_x \ln\left(\dfrac{V_1 - V_v}{V_1 - V_p}\right)} = \frac{1}{100(0.1 \times 10^{-6}) \ln\left(\dfrac{20 - 0.8}{20 - 10.7}\right)} = 144.3\ \text{k}\Omega$$

which is within the required bounds.

The value of R_L can be selected between 20 and 80Ω.

A more linear sawtooth oscillator using a PUT is shown in Figure 6.44(a). The PUT provides the voltage-controlled feature of the oscillator. Resistors R_a and R_b

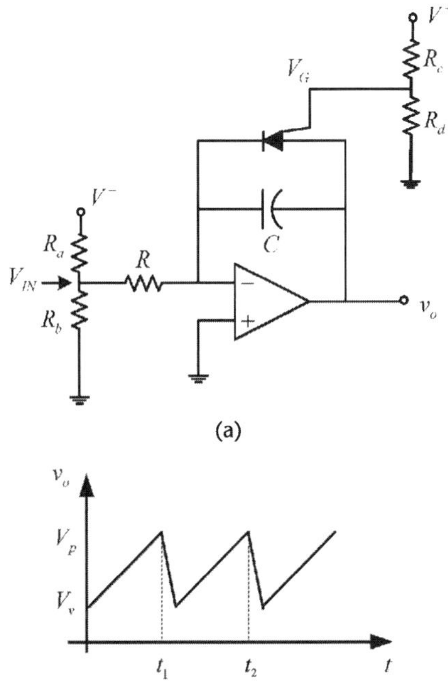

Figure 6.44 (a) A voltage controlled sawtooth oscillator using a PUT and (b) the output waveform.

set the value of V_{IN}. The negative input voltage V_{IN} gets integrated and produces an increasing ramp signal at the output, as shown in Figure 6.44(b). When the output voltage exceeds the gate voltage by approximately 0.7V, the PUT turns on, and the capacitor discharges through the PUT [see Figure 6.44(b)]. The discharge continues until the current in the PUT falls below its valley current value. At this time the PUT turns off and the capacitor charges again, repeating the cycle.

During the charging of the capacitor, the output voltage is given by

$$v_o(t) = \frac{-1}{RC}\int_0^t v_{IN}\,dt = \frac{-V_{IN}}{RC}t + V_v$$

Hence, the time $t = t_1$ in Figure 6.44(b) is given by

$$t_1 = \frac{RC(V_p - V_v)}{-V_{IN}} \tag{6.50}$$

If the discharge time is neglected, the period and frequency can be approximated by

$$f = \frac{1}{T} \approx \frac{-V_{IN}}{RC(V_p - V_v)}$$

Example 6.11

Design the sawtooth oscillator in Figure 6.44(a) to oscillate at 600 Hz with an amplitude of 5V.

Solution

Let the op amp supply voltages be $V^+ = 12$V and $V^- = -12$V, and use a PUT with $V_v = 1$V. For an amplitude of 5V and with $V_v = 1$V, it follows that $V_p = 6$V [see Figure 6.44(b)]. Hence, $V_G = 5.3$V, which can be obtained with $R_c = 6.32$ kΩ and $R_d = 5$ kΩ.

Equation (6.50) gives the value of RC that produces the desired frequency of oscillation for a give V_{IN}. For example, with $R_a = 50$ kΩ and $R_b = 10$ kΩ the value of V_{IN} is –2V. Then, from (6.50) it follows that

$$RC = \frac{2}{600(6-1)} = 667 \times 10^{-6}$$

which can be satisfied with $R = 100$ kΩ and $C = 6.67$ nF.

APPENDIX A

Conditions for a Stable Oscillation*

In this appendix the conditions for a stable oscillation are presented. Consider the one-port network shown in Figure A.1. Over a narrowband range of frequencies around the frequency of oscillation we assume that the input resistance $R_{IN}(A, \omega)$ (where $R_{IN}(A, \omega) < 0$) and the input reactance $X_{IN}(A, \omega)$ of the active device are only a function of the amplitude of $i(t)$. That is,

$$Z_{IN}(A, \omega) \approx Z_{IN}(A) = R_{IN}(A) + jX_{IN}(A)$$

where $R_{IN}(A) < 0$.

The current $i(t)$ can be written as

$$i(t) = A(t)\cos[\omega t + \theta(t)] = \text{Re}\left[A(t)e^{j[\omega t + \theta(t)]}\right] \tag{A.1}$$

where the time-dependent amplitude $A(t)$ and phase $\theta(t)$ are slowly varying functions of time. At steady state, $A(t)$ and $\theta(t)$ are constants and $i(t)$ becomes a sinusoidal signal with constant amplitude and phase.

Using (A.1) the voltage $v(t)$ is given by

$$\begin{aligned} v(t) &= \text{Re}\left[A(t)e^{j[\omega t + \theta(t)]} Z_{IN}(A)\right] \qquad \text{(A.2)} \\ &= A(t)R_{IN}(A)\cos[\omega t + \theta(t)] - A(t)X_{IN}(A)\sin[\omega t + \theta(t)] \end{aligned}$$

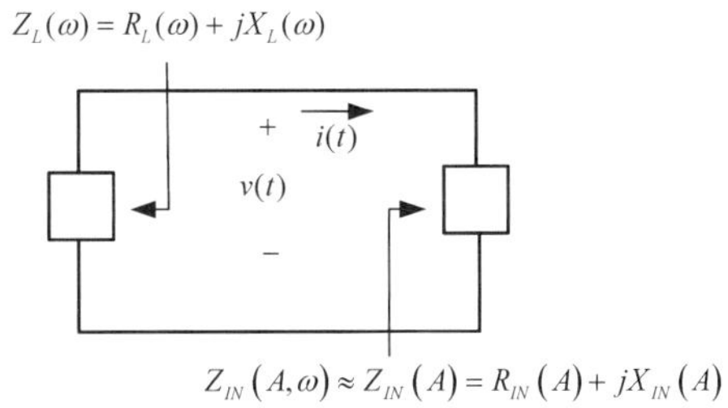

Figure A.1 One-port model of a negative-resistance oscillator.

* This appendix has been reproduced from Gonzalez, G., *Microwave Transistor Amplifiers—Analysis and Design*, Second Edition, Upper Saddle River, NJ: Prentice-Hall, 1997. Reprinted by permission of Pearson Education, Inc., NJ.

The voltage $v(t)$ can also be expressed in terms of $Z_L(\omega)$. This is a little more difficult to determine since $i(t)$ is not an exact periodic function, and a question arises as to the value of the frequency used to evaluate $Z_L(\omega)$. This frequency should be the instantaneous frequency of $i(t)$, denoted by ω_i. Since $A(t)$ and $\theta(t)$ are slowly varying functions of time, we can approximately express $v(t)$ in terms of $Z_L(\omega_i)$ as

$$v(t) \approx -\operatorname{Re}\left[A(t)e^{j[\omega t+\theta(t)]}Z_L(\omega_i)\right] \tag{A.3}$$

where ω_i is the instantaneous frequency of $i(t)$.

To find the instantaneous frequency of $i(t)$, we calculate

$$\begin{aligned}\frac{di(t)}{dt} &= \frac{d}{dt}\left\{\operatorname{Re}\left[A(t)e^{j[\omega t+\theta(t)]}\right]\right\}\\ &= \operatorname{Re}\left\{\left[j\left(\omega+\frac{d\theta(t)}{dt}\right)+\frac{1}{A(t)}\frac{dA(t)}{dt}\right]A(t)e^{j[\omega t+\theta(t)]}\right\}\\ &= \operatorname{Re}\left\{j\left[\omega+\frac{d\theta(t)}{dt}-j\frac{1}{A(t)}\frac{dA(t)}{dt}\right]A(t)e^{j[\omega t+\theta(t)]}\right\}\end{aligned} \tag{A.4}$$

From (A.4) it follows that the instantaneous frequency ω_i is complex and given by

$$\omega_i = \omega + \frac{d\theta(t)}{dt} - j\frac{1}{A(t)}\frac{dA(t)}{dt}$$

The value of $Z_L(\omega_i)$ is given by

$$\begin{aligned}Z_L(\omega_i) &= Z_L\left[\omega+\frac{d\theta(t)}{dt}-j\frac{1}{A(t)}\frac{dA(t)}{dt}\right] \qquad (A.5)\\ &= R_L\left[\omega+\frac{d\theta(t)}{dt}-j\frac{1}{A(t)}\frac{dA(t)}{dt}\right]+jX_L\left[\omega+\frac{d\theta(t)}{dt}-j\frac{1}{A(t)}\frac{dA(t)}{dt}\right]\end{aligned}$$

Since $A(t)$ and $\theta(t)$ are slowly varying functions of time, we have

$$\omega \gg \frac{d\theta(t)}{dt}$$

and

$$\omega \gg \frac{1}{A(t)}\frac{dA(t)}{dt}$$

Hence, (A.5) can be approximated using the first two terms of its Taylor's expansion about ω, namely,

$$
\begin{aligned}
Z_L(\omega_i) &\approx Z_L(\omega) + \frac{dZ_L(\omega)}{d\omega}\left[\frac{d\theta(t)}{dt} - j\frac{1}{A(t)}\frac{dA(t)}{dt}\right] \\
&= R_L(\omega) + \frac{dR_L(\omega)}{d\omega}\left[\frac{d\theta(t)}{dt} - j\frac{1}{A(t)}\frac{dA(t)}{dt}\right] \\
&\quad + jX_L(\omega) + j\frac{dX_L(\omega)}{d\omega}\left[\frac{d\theta(t)}{dt} - j\frac{1}{A(t)}\frac{dA(t)}{dt}\right] \\
&= R_L(\omega) + \frac{dR_L(\omega)}{d\omega}\frac{d\theta(t)}{dt} + \frac{dX_L(\omega)}{d\omega}\frac{1}{A(t)}\frac{dA(t)}{dt} \\
&\quad + j\left[X_L(\omega) + \frac{dX_L(\omega)}{d\omega}\frac{d\theta(t)}{dt} - \frac{dR_L(\omega)}{d\omega}\frac{1}{A(t)}\frac{dA(t)}{dt}\right]
\end{aligned}
\tag{A.6}
$$

Substituting (A.6) into (A.3) gives

$$
\begin{aligned}
v(t) =& -\left[R_L(\omega) + \frac{dR_L(\omega)}{d\omega}\frac{d\theta(t)}{dt} + \frac{dX_L(\omega)}{d\omega}\frac{1}{A(t)}\frac{dA(t)}{dt}\right]A(t)\cos[\omega t + \theta(t)] \\
&+\left[X_L(\omega) + \frac{dX_L(\omega)}{d\omega}\frac{d\theta(t)}{dt} - \frac{dR_L(\omega)}{d\omega}\frac{1}{A(t)}\frac{dA(t)}{dt}\right]A(t)\sin[\omega t + \theta(t)]
\end{aligned}
\tag{A.7}
$$

Equating (A.2) to (A.7) produces the relation

$$
R_L(\omega) + \frac{dR_L(\omega)}{d\omega}\frac{d\theta(t)}{dt} + \frac{dX_L(\omega)}{d\omega}\frac{1}{A(t)}\frac{dA(t)}{dt} = -R_{IN}(A) \tag{A.8}
$$

and

$$
X_L(\omega) + \frac{dX_L(\omega)}{d\omega}\frac{d\theta(t)}{dt} - \frac{dR_L(\omega)}{d\omega}\frac{1}{A(t)}\frac{dA(t)}{dt} = -X_{IN}(A) \tag{A.9}
$$

These coupled equations can be solved for $dA(t)/dt$ and $d\theta(t)/dt$. Multiplying (A.8) by $dX_L(\omega)/d\omega$ and (A.9) by $-dR_L(\omega)/d\omega$, and adding the resulting equations gives

$$
[R_L(\omega) + R_{IN}(A)]\frac{dX_L(\omega)}{d\omega} - [X_L(\omega) + X_{IN}(A)]\frac{dR_L(\omega)}{d\omega} + \left|\frac{dZ_L(\omega)}{d\omega}\right|^2 \frac{1}{A(t)}\frac{dA(t)}{dt} = 0 \tag{A.10}
$$

Similarly, multiplying (A.8) by $dR_L(\omega)/d\omega$ and (A.9) by $dX_L(\omega)/d\omega$, and adding the resulting equations gives

$$
[R_L(\omega) + R_{IN}(A)]\frac{dR_L(\omega)}{d\omega} + [X_L(\omega) + X_{IN}(A)]\frac{dX_L(\omega)}{d\omega} + \left|\frac{dZ_L(\omega)}{d\omega}\right|^2 \frac{d\theta(t)}{dt} = 0 \tag{A.11}
$$

At steady state the oscillator voltage $v(t)$ is such that $dA(t)/dt = 0$ and $d\theta(t)/dt = 0$. Hence, the steady-state solutions of (A.10) and (A.11), at which $A = A_o$ and $\omega = \omega_o$, are obtained when

$$[R_L(\omega_o) + R_{IN}(A_o)] \left.\frac{dX_L(\omega)}{d\omega}\right|_{\omega=\omega_o} + [X_L(\omega_o) + X_{IN}(A_o)] \left.\frac{dR_L(\omega)}{d\omega}\right|_{\omega=\omega_o} = 0$$

and

$$[R_L(\omega_o) + R_{IN}(A_o)] \left.\frac{dR_L(\omega)}{d\omega}\right|_{\omega=\omega_o} + [X_L(\omega_o) + X_{IN}(A_o)] \left.\frac{dX_L(\omega)}{d\omega}\right|_{\omega=\omega_o} = 0$$

These equations show that at steady state the oscillator must satisfy the following conditions:

$$R_L(\omega_o) + R_{IN}(A_o) = 0 \tag{A.12}$$

and

$$X_L(\omega_o) + X_{IN}(A_o) = 0 \tag{A.13}$$

or simply when

$$Z_L(\omega_o) + Z_{IN}(A_o) = 0$$

or equivalently when

$$\Gamma_{IN}(A_o)\Gamma_L(\omega_o) = 1$$

The oscillations might not reach the steady-state conditions described by (A.12) and (A.13) when a small variation in the amplitude produces an oscillation with increasing amplitude. Let A_o be the steady-state value of A. A small variation ΔA in the amplitude is given by $A = A_o + \Delta A$. Then, a Taylor expansion of $R_{IN}(A)$ and $X_{IN}(A)$ about A_o is

$$R_{IN}(A) = R_{IN}(A_o + \Delta A) \approx R_{IN}(A_o) + \left.\frac{\partial R_{IN}(A)}{\partial A}\right|_{A=A_o} \Delta A \tag{A.14}$$

and

$$X_{IN}(A) = X_{IN}(A_o + \Delta A) \approx X_{IN}(A_o) + \left.\frac{\partial X_{IN}(A)}{\partial A}\right|_{A=A_o} \Delta A \tag{A.15}$$

Substituting (A.12) and (A.13) into (A.14) and (A.15) gives

$$R_{IN}(A) = R_{IN}(A_o + \Delta A) \approx -R_L(\omega_o) + \left.\frac{\partial R_{IN}(A)}{\partial A}\right|_{A=A_o} \Delta A \tag{A.16}$$

and

$$X_{IN}(A) = X_{IN}(A_o + \Delta A) \approx -X_L(\omega_o) + \left.\frac{\partial X_{IN}(A)}{\partial A}\right|_{A=A_o} \Delta A \tag{A.17}$$

Then, substituting (A.16) and (A.17) into (A.10) gives

$$\Delta A \left[\left.\frac{\partial R_{IN}(A)}{\partial A}\right|_{A=A_o} \left.\frac{dX_L(\omega)}{d\omega}\right|_{\omega=\omega_o} - \left.\frac{\partial X_{IN}(A)}{\partial A}\right|_{A=A_o} \left.\frac{dR_L(\omega)}{d\omega}\right|_{\omega=\omega_o} \right]$$

$$+ \left.\left|\frac{dZ_L(\omega)}{d\omega}\right|^2\right|_{\omega=\omega_o} \frac{1}{A_o} \frac{d(\Delta A)}{dt} = 0$$

which can be conveniently written as

$$\frac{d(\Delta A)}{dt} + \beta(\Delta A) = 0 \tag{A.18}$$

where

$$\beta = \frac{A_o}{\left.\left|\frac{dZ_L(\omega)}{d\omega}\right|^2\right|_{\omega=\omega_o}} \left[\left.\frac{\partial R_{IN}(A)}{\partial A}\right|_{A=A_o} \left.\frac{dX_L(\omega)}{d\omega}\right|_{\omega=\omega_o} - \left.\frac{\partial X_{IN}(A)}{\partial A}\right|_{A=A_o} \left.\frac{dR_L(\omega)}{d\omega}\right|_{\omega=\omega_o} \right]$$

The solution of (A.18) is of the form $e^{-\beta t}$. This solution shows that the disturbance ΔA decays with time when $\beta > 0$. Since A_o and

$$\left.\left|\frac{dZ_L(\omega)}{d\omega}\right|^2\right|_{\omega=\omega_o}$$

are both positive quantities, it follows that a stable oscillation requires that

$$\left.\frac{\partial R_{IN}(A)}{\partial A}\right|_{A=A_o} \left.\frac{dX_L(\omega)}{d\omega}\right|_{\omega=\omega_o} - \left.\frac{\partial X_{IN}(A)}{\partial A}\right|_{A=A_o} \left.\frac{dR_L(\omega)}{d\omega}\right|_{\omega=\omega_o} > 0 \tag{A.19}$$

Equation (A.19) can be expressed in the form

$$\left[\frac{\partial Z_{IN}}{\partial A}\right] \times \left[\frac{\partial Z_L}{\partial \omega}\right] = \left[\frac{\partial R_{IN}}{\partial A}\right]\left[\frac{\partial X_L}{\partial \omega}\right]\hat{i}_z - \left[\frac{\partial X_{IN}}{\partial A}\right]\left[\frac{\partial R_L}{\partial \omega}\right]\hat{i}_z > 0$$

or

$$\left|\frac{\partial Z_L}{\partial \omega}\right|\left|\frac{\partial Z_{IN}}{\partial A}\right| \sin \gamma < 0 \tag{A.20}$$

and therefore, it follows that

$$0 \leq \gamma \leq 180°$$

In terms of reflection coefficients, (A.20) can be expressed in the form

$$\left|\frac{\partial \Gamma_L}{\partial \omega}\right|\left|\frac{\partial \Gamma_{IN}^{-1}}{\partial \mathrm{A}}\right| \sin \gamma > 0 \tag{A.21}$$

Equation (A.21) has a simple interpretation, for a stable oscillation a plot of $\Gamma_L(\omega)$ and $\Gamma_{IN}^{-1}(A)$ in the Smith chart should intersect at one point where the angle going counterclockwise from $\Gamma_L(\omega)$ and $\Gamma_{IN}^{-1}(A)$ is between 0° and 180°.

APPENDIX B

Analysis of the Series Feedback Circuit*

In this appendix a method that can be used to design the series-feedback network is discussed. The method is based on the fact that the three-port S parameters of the transistor can be used to calculate the two-port S parameters when a series-feedback impedance is connected to port 3. The three- and two-port representations of a transistor are shown in Figure B.1.

Letting $[\hat{S}]$ be the three-port scattering matrix and $[S]$ be the two-port scattering matrix, it follows that the relations between the $\hat{S}$ parameters and the S parameters are [1]

$$\begin{aligned}
&\hat{S}_{11} = S_{11} + \frac{\sigma_{11}\sigma_{12}}{4-\sigma} && \hat{S}_{12} = S_{12} + \frac{\sigma_{11}\sigma_{21}}{4-\sigma} && \hat{S}_{13} = \frac{2\sigma_{11}}{4-\sigma} \\
&\hat{S}_{21} = S_{21} + \frac{\sigma_{22}\sigma_{12}}{4-\sigma} && \hat{S}_{22} = S_{22} + \frac{\sigma_{22}\sigma_{21}}{4-\sigma} && \hat{S}_{23} = \frac{2\sigma_{22}}{4-\sigma} \\
&\hat{S}_{31} = \frac{2\sigma_{12}}{4-\sigma} && \hat{S}_{32} = \frac{2\sigma_{21}}{4-\sigma} && \hat{S}_{33} = \frac{\sigma}{4-\sigma}
\end{aligned} \tag{B.1}$$

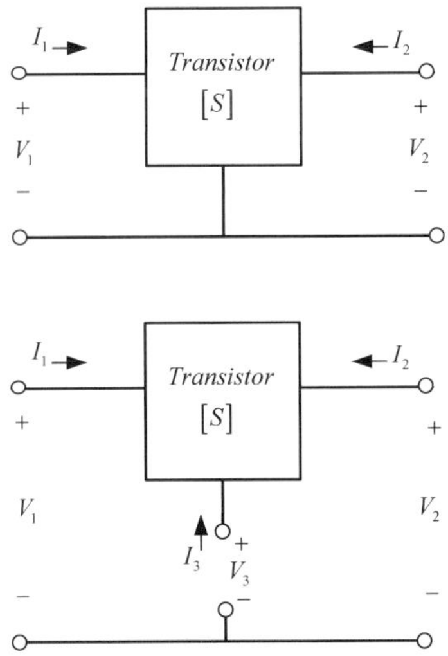

Figure B.1 Two-port and three-port representations of a transistor. (*From:* [2] © 1999 IEEE. Used with permission.)

* This appendix has been adapted from Gonzalez and Sosa.

where

$$\sigma_{11} = 1 - S_{11} - S_{12}$$
$$\sigma_{12} = 1 - S_{11} - S_{21}$$
$$\sigma_{21} = 1 - S_{22} - S_{12}$$
$$\sigma_{22} = 1 - S_{22} - S_{21}$$
$$\sigma = S_{11} + S_{12} + S_{21} + S_{22}$$

and

$$S_{11} = \hat{S}_{11} - \frac{\hat{S}_{13}\hat{S}_{31}}{1 + \hat{S}_{33}} \qquad S_{12} = \hat{S}_{12} - \frac{\hat{S}_{13}\hat{S}_{32}}{1 + \hat{S}_{33}} \tag{B.2}$$
$$S_{21} = \hat{S}_{21} - \frac{\hat{S}_{23}\hat{S}_{31}}{1 + \hat{S}_{33}} \qquad S_{22} = \hat{S}_{22} - \frac{\hat{S}_{23}\hat{S}_{32}}{1 + \hat{S}_{33}}$$

If the series-feedback impedance is used, as shown in Figure B.2, it follows that $a_3 = \Gamma_f b_3$ and the resulting S parameters are those in (B.2) with $1 + \hat{S}_{33}$ replaced by $\hat{S}_{33} - (1/\Gamma_f)$. Hence, S_{11} and S_{22} in (B.2) can be expressed in the form

$$S_{11} = \frac{\Delta_1 \Gamma_f - \hat{S}_{11}}{\hat{S}_{33}\Gamma_f - 1} \tag{B.3}$$

and

$$S_{22} = \frac{\Delta_2 \Gamma_f - \hat{S}_{22}}{\hat{S}_{33}\Gamma_f - 1} \tag{B.4}$$

where

$$\Delta_1 = \hat{S}_{11}\hat{S}_{33} - \hat{S}_{13}\hat{S}_{31}$$
$$\Delta_2 = \hat{S}_{22}\hat{S}_{33} - \hat{S}_{23}\hat{S}_{32}$$

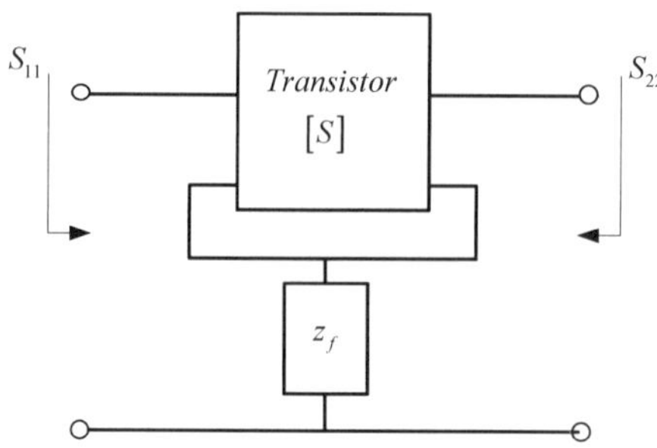

Figure B.2 Two-port network with series feedback. (*From:* © 1999 IEEE. Used with permission.)

Equations (B.3) and (B.4) are bilinear transformations and describe the mapping of the Γ_f plane onto the S_{11} and S_{22} planes, respectively. The mapping of $|\Gamma_f| = 1$ onto the S_{11} and S_{22} planes have centers and radii given by

$$C_1' = \frac{\hat{S}_{11} - \Delta_1 \hat{S}_{33}^*}{1 - |\hat{S}_{33}|^2} \tag{B.5}$$

$$r_1' = \frac{|\hat{S}_{13}\hat{S}_{31}|}{|1 - |\hat{S}_{33}|^2|}$$

and

$$C_2' = \frac{\hat{S}_{22} - \Delta_2 \hat{S}_{33}^*}{1 - |\hat{S}_{33}|^2} \tag{B.6}$$

$$r_2' = \frac{|\hat{S}_{23}\hat{S}_{32}|}{|1 - |\hat{S}_{33}|^2|}$$

for the S_{11} and S_{22} planes, respectively.

The maximum value of $|S_{11}|$ occurs at a point where

$$S_{11\,(\text{max})} = \left(|C_1'| + r_1'\right) \underline{|C_1'} \tag{B.7}$$

From (B.3) it follows that $x_{1\,(\text{max})}$ is given by

$$z_f = jx_{1\,(\text{max})} = \frac{\Delta_1 + \hat{S}_{11} - S_{11\,(\text{max})}(1 + \hat{S}_{33})}{\Delta_1 - \hat{S}_{11} + S_{11\,(\text{max})}(1 - \hat{S}_{33})} \tag{B.8}$$

To properly orient the mapping, the $z_f = \infty$ point maps onto the point

$$S_{11} = \frac{\Delta_1 - \hat{S}_{11}}{\hat{S}_{33} - 1} \tag{B.9}$$

and $z_f = 0$ maps onto the point

$$S_{11} = \frac{\Delta_1 + \hat{S}_{11}}{\hat{S}_{33} + 1} \tag{B.10}$$

Similarly, we obtain

$$S_{22\,(\text{max})} = \left(|C_2'| + r_2'\right) \underline{|C_2'} \tag{B.11}$$

and

$$z_f = jx_{2\,(\text{max})} = \frac{\Delta_2 + \hat{S}_{22} - S_{22\,(\text{max})}(1 + \hat{S}_{33})}{\Delta_2 - \hat{S}_{22} + S_{22\,(\text{max})}(1 - \hat{S}_{33})} \tag{B.12}$$

Moreover, the $z_f = \infty$ point maps onto

$$S_{22} = \frac{\Delta_2 - \hat{S}_{22}}{\hat{S}_{33} - 1} \tag{B.13}$$

and $z_f = 0$ point maps onto

$$S_{22} = \frac{\Delta_2 + \hat{S}_{22}}{\hat{S}_{33} + 1} \tag{B.14}$$

The design procedure is as follows:

1. Convert the two-port S parameters of the transistor to its corresponding three-port S parameters using (B.1).
2. Use (B.5) and (B.6) to calculate C_1', r_1', C_2', and r_2'.
3. Use (B.7), (B.8), (B.11), and (B.12) to calculate $S_{11(\max)}$, $x_{1(\max)}$, $S_{22(\max)}$, and $x_{2(\max)}$.
4. Plot the mapping of the Γ_f plane onto the S_{11} and S_{22} planes.
5. Select the appropriate z_f value.

Example B.1

The two-port S parameters of a transistor are

$$S_{11} = 0.386 \underline{|142^\circ}$$

$$S_{12} = 0.147 \underline{|81.3^\circ}$$

$$S_{21} = 1.380 \underline{|-45.4^\circ}$$

$$S_{22} = 1.110 \underline{|-28.5^\circ}$$

Using (B.1), the three-port $\hat{S}$ parameters are

$$\hat{S} = \begin{bmatrix} 0.413 \underline{|78.5^\circ} & 0.330 \underline{|61.67^\circ} & 1.031 \underline{|-42.42^\circ} \\ 0.919 \underline{|-61.71^\circ} & 0.904 \underline{|-37.98^\circ} & 1.374 \underline{|96.2^\circ} \\ 0.629 \underline{|39.99^\circ} & 0.296 \underline{|63.87^\circ} & 0.774 \underline{|-59.97^\circ} \end{bmatrix}$$

Then, from (B.5) to (B.12), we obtain

$$C_1' = 1.65 \underline{|62.7^\circ} \qquad C_2' = 1.07 \underline{|-84^\circ}$$

$$r_1' = 1.62 \qquad r_2' = 1.02$$

$$S_{11(\max)} = 3.27 \underline{|62.7^\circ} \qquad S_{22(\max)} = 2.08 \underline{|-84^\circ}$$

$$x_{1(\max)} = 1.71 \qquad x_{2(\max)} = 1.49$$

The mappings of the feedback impedance onto the S_{11} and S_{22} planes are shown in Figure B.3. In the S_{11} plane it is observed that using $z_f = j1.710$ produces $S_{11} = S_{11(\max)} = 3.27\,\underline{|62.7^\circ}$ and the associated value of S_{22} is $S_{22} = 1.88\,\underline{|-108.7^\circ}$. To properly orient the mapping in Figure B.3(a), we use (B.9) and (B.10) to obtain the values of S_{11} at points P_2 and P_3, which corresponds to the mapping of the points $z_f = \infty$ and $z_f = 0$, respectively. The mappings for $z_f = 1 + jx$ and $z_f = r \pm j1$ are also shown. These circles help in the visualization of the complete mapping. Similar details are provided for the mapping onto the S_{22} plane in Figure B.3(b) where (B.13) and (B.14) were used to obtain values of S_{22} at points P_2' and P_3'.

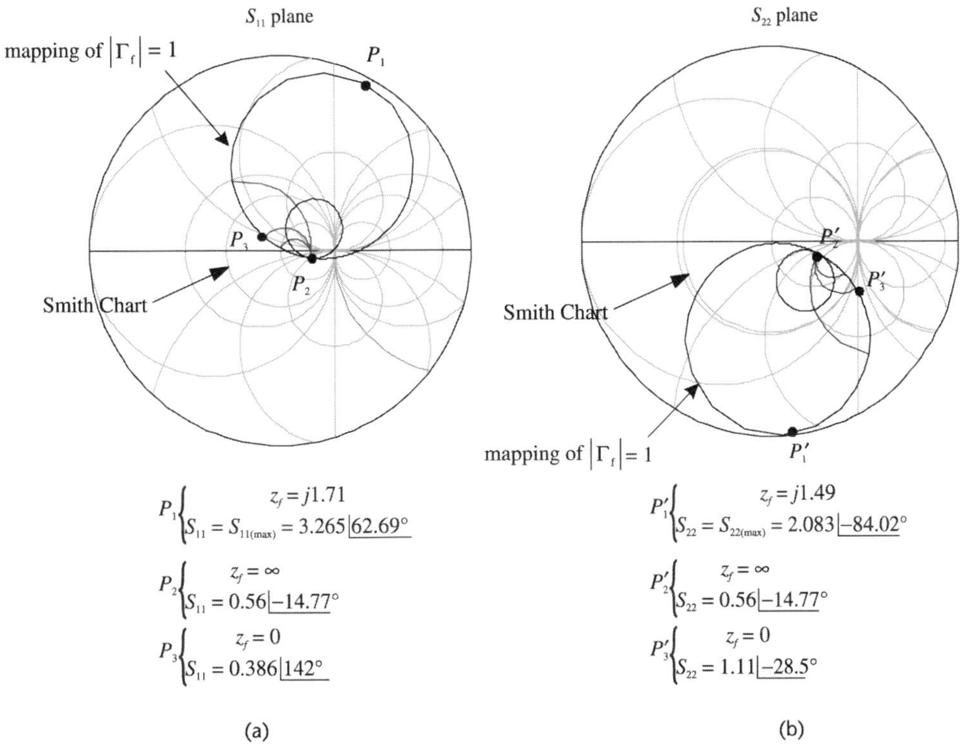

Figure B.3 (a) Mapping of the Γ_f plane onto the S_{11} plane and (b) mapping of the Γ_f plane onto the S_{22} plane. (*From:* [2] © 1999 IEEE. Used with permission.)

References

[1] Collins, R. E., *Foundations for Microwave Engineers*, Second Edition, New York: McGraw-Hill, 1992.

[2] Gonzalez, G., and O. J. Sosa, "On the Design of a Series-Feedback Network in a Transistor Negative-Resistance Oscillator," *IEEE Trans. on MTT*, Vol. 47, January 1999, pp. 42–47.

Selected Bibliography

Abrie, P., *Design of RF and MW Amplifiers and Oscillators*, Norwood, MA: Artech House, 1999.

Everard, J. K. A., *Fundamentals of RF Circuit Design with Low Noise Oscillators*, New York: Wiley, 2000.

Gilmore, R., and L. Besser, *Practical RF Circuit Design for Modern Wireless Systems, Volume II: Active Circuits*, Norwood, MA: Artech House, 2003.

Matthys, R., *Crystal Oscillator Circuits*, Melbourne, FL: Krieger Publishing Co., 1991.

Parzes, B., *Design of Crystal and Other Harmonic Oscillators*, New York: Wiley, 1982.

Rhea, R., *Oscillator Design and Computer Simulation*, New York: McGraw-Hill/Noble, 2000.

Rhode, U., A. Poddar, and G. Bock, *The Design of Modern Microwave Oscillators for Wireless Applications: Theory and Optimization*, New York: Wiley, 2005.

Odyniec, M., and M. Odyniec, (Eds.), *RF and MW Oscillator Design*, Norwood, MA: Artech House, 2002.

About the Author

Guillermo Gonzalez is a professor in the Department of Electrical and Computer Engineering at the University of Miami. He holds a Ph.D. in electrical engineering from the University of Arizona and an MSEE and a BSEE from the University of Miami. He worked in industry as a senior engineer at Bell Aerosystems Co., Tucson, before starting his teaching career. He has 36 years of teaching experience and has received, four times, the Alexander-Orr Jr. award for excellence in teaching at the University of Miami. He was elected to the following honor societies: Tau Beta Pi, Eta Kappa Nu, Phi Kappa Phi, and Sigma Xi. Early in his career he made important contributions to the wave-hop theory of LF/VLF propagation of waves around the Earth. He is the author of the well known text *Microwave Transistor Amplifiers: Analysis and Design,* Second Edition. Many of his former students are successful engineers in various electronics companies.

Index

Recent Titles in the Artech House Microwave Library

Active Filters for Integrated-Circuit Applications, Fred H. Irons

Advanced Production Testing of RF, SoC, and SiP Devices, Joe Kelly and Michael Engelhardt

Advanced Techniques in RF Power Amplifier Design, Steve C. Cripps

Automated Smith Chart, Version 4.0: Software and User's Manual, Leonard M. Schwab

Behavioral Modeling of Nonlinear RF and Microwave Devices, Thomas R. Turlington

Broadband Microwave Amplifiers, Bal S. Virdee, Avtar S. Virdee, and Ben Y. Banyamin

Classic Works in RF Engineering: Combiners, Couplers, Transformers, and Magnetic Materials, John L. B. Walker, Daniel P. Myer, Frederick H. Raab, and Chris Trask, editors

Computer-Aided Analysis of Nonlinear Microwave Circuits, Paulo J. C. Rodrigues

Design of FET Frequency Multipliers and Harmonic Oscillators, Edmar Camargo

Design of Linear RF Outphasing Power Amplifiers, Xuejun Zhang, Lawrence E. Larson, and Peter M. Asbeck

Design of RF and Microwave Amplifiers and Oscillators, Pieter L. D. Abrie

Distortion in RF Power Amplifiers, Joel Vuolevi and Timo Rahkonen

EMPLAN: Electromagnetic Analysis of Printed Structures in Planarly Layered Media, Software and User's Manual, Noyan Kinayman and M. I. Aksun

FAST: Fast Amplifier Synthesis Tool—Software and User's Guide, Dale D. Henkes

Feedforward Linear Power Amplifiers, Nick Pothecary

Foundations of Oscillator Circuit Design, Guillermo Gonzalez

Generalized Filter Design by Computer Optimization, Djuradj Budimir

High-Linearity RF Amplifier Design, Peter B. Kenington

High-Speed Circuit Board Signal Integrity, Stephen C. Thierauf

Integrated Circuit Design for High-Speed Frequency Synthesis, John Rogers, Calvin Plett, and Foster Dai

Intermodulation Distortion in Microwave and Wireless Circuits, José Carlos Pedro and Nuno Borges Carvalho

Lumped Elements for RF and Microwave Circuits, Inder Bahl

Microwave Circuit Modeling Using Electromagnetic Field Simulation, Daniel G. Swanson, Jr. and Wolfgang J. R. Hoefer

Microwave Component Mechanics, Harri Eskelinen and Pekka Eskelinen

Microwave Engineers' Handbook, Two Volumes, Theodore Saad, editor

Microwave Filters, Impedance-Matching Networks, and Coupling Structures, George L. Matthaei, Leo Young, and E.M.T. Jones

Microwave Materials and Fabrication Techniques, Second Edition, Thomas S. Laverghetta

Microwave Mixers, Second Edition, Stephen A. Maas

Microwave Radio Transmission Design Guide, Trevor Manning

Microwaves and Wireless Simplified, Thomas S. Laverghetta

Modern Microwave Circuits, Noyan Kinayman and M. I. Aksun

Neural Networks for RF and Microwave Design, Q. J. Zhang and K. C. Gupta

Nonlinear Microwave and RF Circuits, Second Edition, Stephen A. Maas

QMATCH: Lumped-Element Impedance Matching, Software and User's Guide, Pieter L. D. Abrie

Practical Analog and Digital Filter Design, Les Thede

Practical MMIC Design, Steve Marsh

Practical RF Circuit Design for Modern Wireless Systems, Volume I: Passive Circuits and Systems, Les Besser and Rowan Gilmore

Practical RF Circuit Design for Modern Wireless Systems, Volume II: Active Circuits and Systems, Rowan Gilmore and Les Besser

Production Testing of RF and System-on-a-Chip Devices for Wireless Communications, Keith B. Schaub and Joe Kelly

Radio Frequency Integrated Circuit Design, John Rogers and Calvin Plett

RF Design Guide: Systems, Circuits, and Equations, Peter Vizmuller

RF Measurements of Die and Packages, Scott A. Wartenberg

The RF and Microwave Circuit Design Handbook, Stephen A. Maas

RF and Microwave Coupled-Line Circuits, Rajesh Mongia, Inder Bahl, and Prakash Bhartia

RF and Microwave Oscillator Design, Michal Odyniec, editor

RF Power Amplifiers for Wireless Communications, Steve C. Cripps

RF Systems, Components, and Circuits Handbook, Second Edition, Ferril A. Losee

Stability Analysis of Nonlinear Microwave Circuits, Almudena Suárez and Raymond Quéré

TRAVIS 2.0: Transmission Line Visualization Software and User's Guide, Version 2.0, Robert G. Kaires and Barton T. Hickman

Understanding Microwave Heating Cavities, Tse V. Chow Ting Chan and Howard C. Reader

For further information on these and other Artech House titles, including previously considered out-of-print books now available through our In-Print-Forever® (IPF®) program, contact:

Artech House
685 Canton Street
Norwood, MA 02062
Phone: 781-769-9750
Fax: 781-769-6334
e-mail: artech@artechhouse.com

Artech House
46 Gillingham Street
London SW1V 1AH UK
Phone: +44 (0)20 7596-8750
Fax: +44 (0)20 7630 0166
e-mail: artech-uk@artechhouse.com

Find us on the World Wide Web at: www.artechhouse.com